MECHANICAL ALLOYING AND MILLING

C. Suryanarayana
Department of Mechanical, Materials and Aerospace Engineering
University of Central Florida
Orlando, FL 32816–2450, USA

CRC Press is an imprint of the
Taylor & Francis Group, an **informa** business

CRC Press
Taylor & Francis Group
6000 Broken Sound Parkway NW, Suite 300
Boca Raton, FL 33487-2742

First issued in paperback 2019

ISBN-13: 978-0-8247-4103-7 (hbk)
ISBN-13: 978-0-367-39386-1 (pbk)

Library of Congress Cataloging-in-Publication Data
A catalog record for this book is available from the Library of Congress.

Visit the Taylor & Francis Web site at
http://www.taylorandfrancis.com

and the CRC Press Web site at
http://www.crcpress.com

Foreword

Chemists sometimes talk of 'brute force methods' and of 'soft chemistry', with a clear preference for the latter. At first sight, hurling small samples of metallic powder violently between hard grinding balls would seem the apotheosis of brute-force methods, and yet that process (called 'mechanical alloying' or 'mechanical milling', according to the nature of the powder) has attracted ever-growing numbers of metallurgical researchers in the past two decades. Several kinds of metastable structures, crystalline or amorphous, can be prepared in a wide range of compositions, and the literature has burgeoned.

Professor Suryanarayana, known to all his many friends and admirers as Surya for short, has been involved with this type of research since the beginning, and has an unequalled familiarity with the entire literature. In this book, he has exploited this expertise to present the most thorough treatment of the field to see the light of day. I commend it cordially to the metallurgical profession.

Robert W. Cahn, FRS
(University of Cambridge)

Preface

Mechanical alloying and milling continues to be a fascinating field of investigation for nearly four decades. Originally developed in the late 1960's for the production of oxide dispersion strengthened (ODS) nickel-base superalloys, the technique of mechanical alloying has now branched out to the synthesis of a variety of equilibrium and non-equilibrium alloy phases, including solid solutions, intermetallics, quasicrystals, amorphous alloys, and bulk metallic glasses. Nanostructured materials, hydrogen storage materials, and even exotic materials are being synthesized using this simple, inexpensive, and scaleable technique. It is a continually growing field with over 500 research papers being published annually. It is also an accepted industrial processing method with varied applications for these materials.

Several dedicated conferences have been held on this topic and their Proceedings have been published. A few monographs have also been published on this topic. But, in majority of these, the treatment has been rather cursory, with focus on selected topics and not all the topics have been covered. The purpose of this present book is to serve as a resource for both students and researchers who require a timely and comprehensive treatment of the subject.

The present book surveys the vast field of Mechanical Alloying (MA) (and the related techniques) from a scientific and technological point of view. The organization of the book progresses logically from the historical perspective of the technique, through a description of the process, different metastable effects produced, mechanisms of phase formation, to applications of mechanically alloyed products. The book comprises 19 chapters. Chapter 1 presents an overview of the different non-equilibrium processing techniques presently available to synthesize advanced materials and sets the tone for the rest of the book. A brief historical introduction of the development of the technique of MA is presented in Chapter 2 followed by introduction of the different variants of MA in Chapter 3. A detailed description of the mills and the effect of process variables on phase constitution are discussed in Chapters 4 and 5, respectively. Chapter 6 describes the mechanism of alloying in different types of powder combinations. A detailed description of the techniques to measure particle size, lattice strain, phase constitution, and microstructure of the as-milled powder and consolidated products is presented in Chapter 7. The temperature rise during milling is discussed in

Chapter 8. The next four Chapters deal with the metastable constitutional effects achieved in the milled powders of different alloy systems. These include Solid Solutions (Chapter 9), Intermetallics (including Quasicrystals) (Chapter 10), Disordering of Intermetallics (Chapter 11), and Amorphous Alloys (Chapter 12). Formation of nanostructures and nanocomposites, their properties, and potential applications are described in Chapter 13. Synthesis of alloy phases, purification of metals by exchange reactions, preparation of refractory compounds, etc. by mechanochemical methods is discussed in Chapter 14. The ubiquitous problem of powder contamination and ways of avoiding/minimizing it are presented in Chapter 15. Chapter 16 briefly explains the different modeling attempts that have been undertaken to-date and their current limitations. The existing and potential applications of the mechanically alloyed products are described in Chapter 17. Safety hazards associated with powder handling are outlined in Chapter 18 and the last Chapter makes some concluding remarks on the present status of mechanical alloying and milling.

This book is primarily intended for use by graduate students and research personnel involved with this technique of powder processing. Industry personnel connected with the production and characterization of mechanically alloyed products, and wishing to exploit these materials for potential applications will also find this book very useful in understanding the basic scientific features. Scientists beginning their research work in this area can also find this book helpful from the clear introduction to the several different topics (the mills, process variables, methods of characterization of mechanically alloyed powders, types of phases that could be produced, etc.) and also the detailed explanations of the various aspects associated with this technique. Persons involved in the non-equilibrium processing of materials in general will also benefit from a comparison of this technique with other techniques. Because of these attractive features, this book could also be of use for a graduate level course in processing of materials.

I have strived to make the discussion of topics as comprehensive and self-contained as possible. For this reason, I have included unusually extensive (and most comprehensive at the time of writing) listing of the results of different metastable effects obtained in various alloy systems. Extensive lists of references have also been provided at the end of each Chapter. These are important features of this book. Because of this, the book could be an excellent source of references for the literature on MA.

The treatment of the subject matter has been simple and has been presented in an easy to understand manner. However, the scientific accuracy has not been compromised. The most recent literature has also been cited, including several papers published in 2003! The ISMANAM conference proceedings have been very good sources of the most recent literature, in addition to the archival journals in which research papers have been published.

The field of Mechanical Alloying has become quite diversified with active contributions from materials scientists, chemists, physicists, and engineers. Therefore, in surveying such a diverse field, it is possible that some errors have crept in. I would be most pleased if the discerning readers, who spot any mistakes, bring them to my notice at csuryana@mail.ucf.edu.

C. Suryanarayana
Orlando, October 2003

Acknowledgements

In writing any book it is unlikely that the author has worked entirely in isolation and without assistance from colleagues and friends. I am certainly not an exception and it is with great pleasure that I acknowledge those people that have contributed, in various ways, to the successful completion of this book.

In my long journey in the field of non-equilibrium processing, I had the good fortune to interact with a great number of colleagues, from all of whom I have learned immensely and benefited greatly in understanding the complexities of metastable phases. In alphabetical order, they are T.R. Anantharaman, R.W. Cahn, A. Inoue, E. Ivanov, C.C. Koch, E.J. Lavernia, T. Masumoto, J.J. Moore, P. Ramachandrarao, S. Ranganathan, and R. Sundaresan. I am particularly thankful to Professors T.R. Anantharaman, R.W. Cahn, T. Masumoto, and J.J. Moore for their guidance and advice at different stages in my professional career. I am grateful to them for sharing with me the excitement of working in the general field of non-equilibrium processing of materials. Particular mention may be made of E. Ivanov, C.C. Koch, E.J. Lavernia, and R. Sundaresan for collaborating with me at various stages in working in the field of mechanical alloying and sharing their expertise with me. I was also fortunate to work with many talented undergraduate and graduate students and I am thankful to them for their dedicated efforts in joining me in pushing the frontiers of mechanical alloying farther. I hope that they have learned from me at least a fraction of what I have learnt from them.

A large number of friends have helped me by supplying figures that have been included in different Chapters of the book. I am thankful to them. Additionally, many students have helped in the arduous task of drawing some of the figures. They are Chandrasen Rathod, Bhaskar Srivastava, Ms. Honey Dandwani, Soon-Jik Hong, and K.V. Krishna Murthy. I am particularly indebted to them for their help. The aesthetics of the figures are mostly due to their efforts.

Parts of the book were written while the author was a Guest Scientist at the GKSS Research Center in Geesthacht, Germany, during the summers of 2002 and 2003. I am deeply obliged to Thomas Klassen and Rüdiger Bormann for providing kind hospitality and for the several useful discussions during the post-lunch walks in the woods. Thomas Klassen has also read some Chapters and made constructive comments to improve the readability and clarity of the book. I am thankful to him for this.

I would like to thank David Nicholson and Ranganathan Kumar, successive Chairmen of the Department of Mechanical, Materials, and Aerospace Engineering at the University of Central Florida for providing a conducive environment to complete the draft of the Book.

I wish to thank the staff of the publisher, Marcel Dekker, Inc., for their high level of co-operation and interest in successfully producing a high quality and aesthetically pleasing book. I am particularly grateful to B.J. Clark for his patience in waiting for the delivery of the manuscript.

Last, but by no means least, I owe a huge debt of gratitude to my wife, Meena, who encouraged me and supported me with love, understanding, and patience throughout this endeavor.

Finally, I dedicate this book to my spiritual mother, Karunamayi, Sri Sri Sri Vijayeswari Devi, for Her invisible and ever-inspiring encouragement and blessings.

C. Suryanarayana
Orlando, FL; October 2003

Table of Contents

1

Introduction

1.1 MOTIVATION

The search for new and advanced materials has been the major preoccupation of metallurgists, ceramicists, and materials scientists for the past several centuries. Scientific investigations during the past few decades have been continuously directed to improving the properties and performance of materials. Significant improvements in mechanical, chemical, and physical properties have been achieved by alloying and through chemical modifications and by subjecting the materials to conventional thermal, mechanical, and thermomechanical processing methods. Several exotic materials, such as metallic glasses, quasi-crystals, nanocrystalline materials, and high-temperature superconductors, have been synthesized. An important offshoot of these materials syntheses is the development of advanced characterization techniques to observe the microstructures, determine the crystal structures, and analyze for the composition of phases of ever-decreasing dimensions and with higher and higher resolutions. However, the rapid progress of technology has been constantly putting forward ever-increasing demands for materials that have higher strength or improved stiffness, and those that could be used at much higher temperatures and in more aggressive environments than is possible with the traditional and commercially available materials. This has led to the design and development of advanced materials that are "stronger, stiffer, hotter, and lighter" than the existing materials. Synthesis and development of such materials has been facilitated by exploring the interrelationship among processing, structure, properties, and performance of materials—the underpinning theme of materials science and engineering. The high-technology industries have certainly provided an added stimulus to accelerate these efforts.

1.2 ADVANCED MATERIALS

Advanced materials have been defined as those where first consideration is given to the systematic synthesis and control of crystal structure and microstructure of materials in order to provide a precisely tailored set of properties for demanding applications [1]. The attraction of advanced materials is that they could be synthesized with improved properties and performance. Thus, as indicated in Figure 1.1, the materials behavior trend band of "basic" materials could be raised to higher levels of strength, stiffness, and high-temperature capability, with concomitant improvement of "forgiveness" and reduction in cost. This is made possible by "tailoring" or "engineering" the properties of advanced materials through innovative chemistries, processes, and microstructures [2].

It is well recognized that the structure and constitution of advanced materials can be better controlled by processing them under nonequilibrium (or far from equilibrium) conditions [3]. This realization has led to the development of several nonequilibrium processing techniques during the second half of the twentieth century. Among these, special mention may be made of rapid solidification processing [3–6], mechanical alloying [7,8], plasma processing [3,9], vapor deposition [3,10], and spray forming [3,11]. Significant research effort is being spent on each and every one of these technologies as evidenced by the increasing number of publications every year and the number of conferences devoted to these topics.

1.3 THERMODYNAMIC STABILITY

Let us briefly examine the factors determining whether a phase will be stable or metastable under the given conditions of temperature, pressure, and composition. The stability of a system at constant temperature and pressure is determined by its Gibbs free energy, G, defined as:

$$G = H - TS \tag{1.1}$$

where H is enthalpy, T absolute temperature, and S entropy. Thermodynamically, a system will be in stable equilibrium, i.e., it will not transform to any other phase(s)

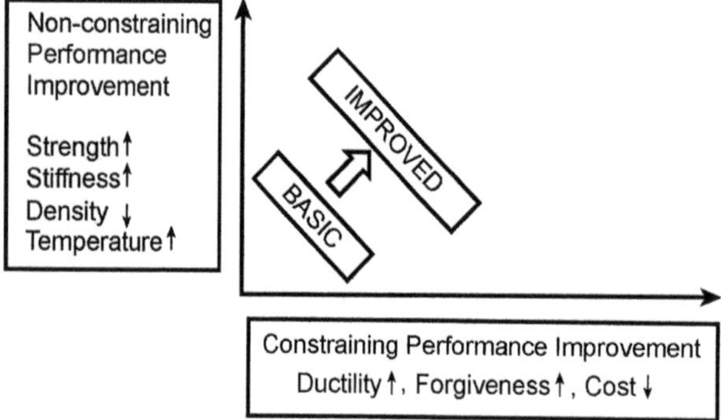

Figure 1.1 Trend band exhibited by basic materials and the enhanced trend band characteristic of advanced materials. Note that advanced materials exhibit improved and more desirable properties than basic materials.

under the given conditions of temperature and pressure, if it has the lowest possible value of the Gibbs free energy. The above equation predicts that a system at any temperature can be most stable either by increasing the entropy or decreasing the enthalpy or both. Consequently, solids are the most stable phases at low temperatures since they have the strongest atomic bonding and therefore the lowest H. On the other hand, the $-TS$ term dominates at high temperatures; therefore, phases with more freedom of atomic movement, i.e., liquids and gases, become most stable. Thus, during solid-state transformations, a close packed structure is more stable at low temperatures, while a less close packed structure is most stable at higher temperatures. For example, titanium with the hcp structure is the stable phase at low temperatures whereas the bcc structure is the stable modification at temperatures higher than 882°C. A phase can transform into another phase if ΔG, the change in free energy, is negative, i.e., if the product phase has a lower free energy than the parent phase.

Let us consider the phase stability in a pure metal. At a constant pressure, the free energy, G, decreases with increasing temperature because of the larger contribution of the entropy term S at high temperatures. But the rate at which G decreases with temperature is different for the solid and liquid phases due to the difference in the specific heat and entropy. Thus, as indicated in Figure 1.2, the solid phase will be more stable up to T_m, the melting point of the system (since the solid phase has a lower Gibbs free energy than the liquid phase) and above this temperature the liquid phase will be more stable. At still higher temperatures (not shown in Figure 1.2), the Gibbs free energy of the vapor phase (at atmospheric pressure) may be lower than that of the liquid or solid phase, and consequently the vapor phase will be the most stable.

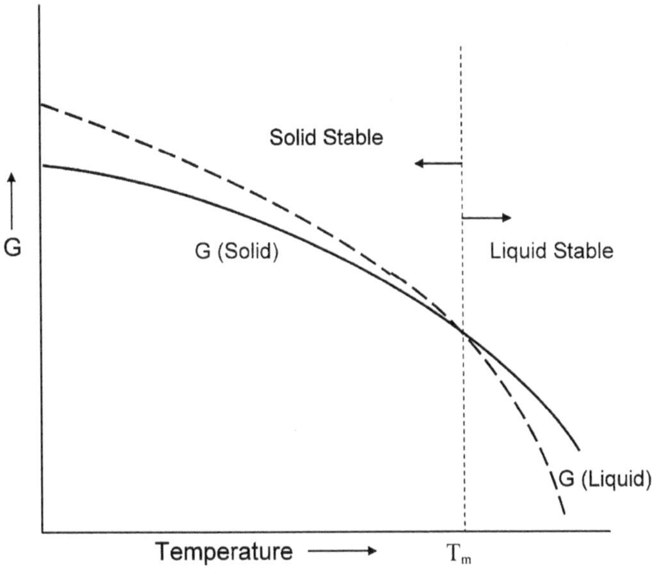

Figure 1.2 Variation of free energy of solid and liquid phases as a function of temperature. Note that the free energies of the two competing phases are equal at the melting point. Under equilibrium conditions, the solid phase is more stable than the liquid phase below the melting temperature, T_m, and the liquid phase is more stable above T_m.

If we consider a binary alloy system A-B with limited solid solubility of B in A (α phase) and of A in B (β phase) at room temperature, e.g., as in a eutectic system (Fig. 1.3a), then variation of free energy, at room temperature, of the α and β phases with composition will be as shown in Figure 1.3b. It may be noted that the free energy of the α phase will continue to decrease with increasing B content until the composition C_α is reached, which represents the minimum in the G_α vs. composition curve. Similarly, the free energy of the β phase will continue to decrease with increasing A content until the composition C_β is reached, which represents the minimum in the G_β vs. composition curve. In an ideal situation, when the minima in the G_α and G_β curves are equal to each other, the compositions C_α and C_β represent the phase boundaries in the system. In other words, any alloy with a B content less than C_α will exist as the α phase, and any alloy with a B content greater than C_β will exist as the β phase. At intermediate B contents, the alloy will exist as a mixture of the α and β phases. On the other hand, when the minima in the G_α and G_β curves are not equal to each other, the phase stability is determined by drawing a common tangent to the free energy curves of the α and β phases. In such a case, the α phase will be stable up to the composition

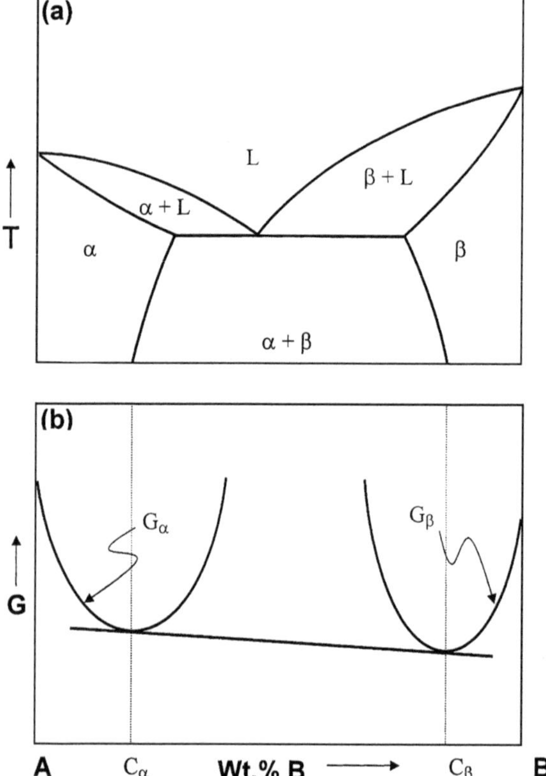

Figure 1.3 (a) A typical eutectic phase diagram showing limited solid solubility of B in A (α phase) and of A in B (β phase). (b) Variation of free energy of the α and β phases as a function of composition. At compositions below C_α the α phase exists homogeneously and above C_β the β phase exists homogeneously. In between these two compositions, a mixture of the α and β phases is the most stable constitution.

representing the point of intersection of the common tangent and the G_α curve and the β phase will be stable beyond the composition (B content) representing the point of intersection of the common tangent and the G_β curve. A mixture of the α and β phases will be stable in between these two composition extremes. A full description of the thermodynamic stability of alloy phases may be found in standard textbooks on thermodynamics (see, for example, Ref. 12).

A phase is considered nonequilibrium or metastable if it has a higher Gibbs free energy than in the equilibrium state for the given composition. If the Gibbs free energy of this phase is lower than that of other competing phases (or mixtures thereof), then it can exist in a metastable equilibrium. Consequently, nonequilibrium phases can be synthesized and retained at room temperature and pressure when the free energy of the stable phases is raised to a higher level than under equilibrium conditions but is maintained at a value below those of other competing phases. It may also be noted in this context that kinetics plays an important role. If the kinetics during synthesis is not fast enough to allow the formation of equilibrium phase(s), i.e., the equilibrium phase(s) are suppressed by suitable processing conditions, then metastable phases could form.

1.4 BASIS OF NONEQUILIBRIUM PROCESSING

The central underlying theme to synthesize materials in a nonequilibrium state is to "energize and quench," as proposed by Turnbull [13]. Processes such as solid-state quenching, rapid solidification from the melt, irradiation, and condensation from vapor were considered by Turnbull to evaluate the departure from equilibrium. However, there are several other methods of nonequilibrium processing that do not involve quenching. These include, among others, static undercooling of liquid droplets, electrodeposition of alloys, mechanical alloying, and application of high pressures. Therefore, instead of calculating the quench rate, it may be desirable to evaluate the maximal departure from equilibrium in each processing method. As depicted in Figure 1.4, the process of energization involves bringing the equilibrium crystalline material, with a free energy G_0, into a highly nonequilibrium (metastable) state, with a free energy G_2. This could be achieved by some external dynamic forcing, e.g., through increase of temperature T (melting or evaporation), irradiation, application of pressure P, or storing of mechanical energy E by plastic deformation [3,8]. Such energized materials were referred to as "driven materials" by Martin and Bellon [14]. The energization also usually involves a possible change of state from solid to liquid (melting) or gas (evaporation). For example, during rapid solidification processing the starting solid material is melted and during vapor deposition the material is vaporized. The energized material is then "quenched" into a configurationally frozen state by methods such as rapid solidification processing or mechanical alloying, such that the resulting phase is in a highly metastable condition, having a free energy G_1. This phase could then be used as a precursor to obtain the desired chemical constitution (other less metastable phases) and/or microstructure (e.g., nanocrystalline material) by subsequent heat treatment/processing. It has been shown that materials processed in this way possess improved physical and mechanical characteristics in comparison to conventional ingot (solidification) processed materials. These metastable phases can also be subsequently transformed to the equilibrium crystalline phase(s) by long-term annealing.

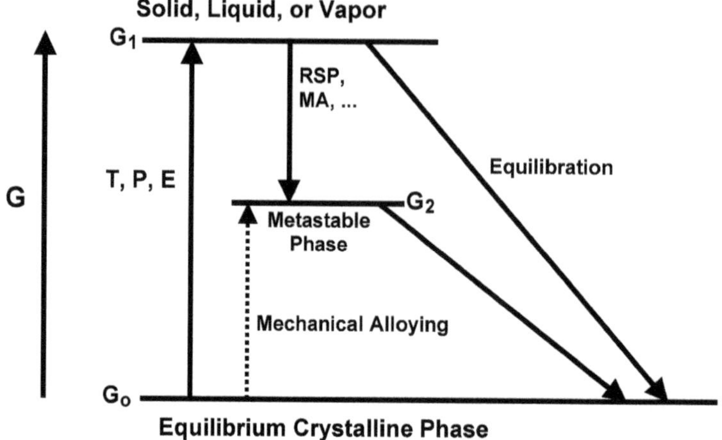

Figure 1.4 The basic concept of "energize and quench" to synthesize nonequilibrium materials.

The ability of the different processing techniques to synthesize nonequilibrium structures may be conveniently evaluated by measuring or estimating the achieved departure from equilibrium, i.e., the maximal energy that can be stored in excess of that of the equilibrium/stable structure. This has been done by different groups for different nonequilibrium processing techniques [13,15–18]. While the excess energy is expressed in kJ/mol in Refs. 15–17, Turnbull [13] expressed this as an "effective quenching rate." The way the departure is calculated is different in these different approaches (the reader is encouraged to consult the original papers for details of calculations) and therefore the results do not correspond exactly in all the cases. Table 1.1 summarizes the departures calculated for the different nonequilibrium processing techniques. It is clear from the table that vapor deposition and ion implantation techniques have very large departures from equilibrium (or effective quench rates). It is also clear that mechanical alloying is a technique that allows the material to be processed much farther from equilibrium than, for example, rapid solidification processing, which has been shown to have a tremendous potential in

Table 1.1 Departure from Equilibrium Achieved in Different Nonequilibrium Processing Techniques

Technique	Effective quench rate (K/s) Ref. 13	Maximal departure from equilibrium (kJ/mol)	
		Ref. 15	Refs. 16, 17
Solid state quench	10^3	—	16
Rapid solidification processing	10^5–10^8	2–3	24
Mechanical alloying	—	30	30
Mechanical cold work	—	—	1
Irradiation/ion implantation	10^{12}	—	30
Condensation from vapor	10^{12}	—	160

developing industrially useful nonequilibrium materials, e.g., amorphous transformer core laminations [3–5].

1.5 SOME NONEQUILIBRIUM PROCESSING METHODS

As mentioned earlier, a number of nonequilibrium processing methods have been developed during the past few decades to synthesize novel and advanced materials. We will now briefly describe some of the different nonequilibrium processing techniques that have become popular in recent years. The reader is advised to consult the references listed for each technique to get a full and better appreciation of the potential and limitations of these techniques.

1.5.1 Rapid Solidification Processing

Rapid solidification processing (RSP) is the method developed by Pol Duwez at CalTech in 1960 [19]. In this method, a molten metal or alloy is cooled very rapidly at rates of about 10^6 K/s, but at least about 10^4 K/s. This has been traditionally done by (1) allowing molten droplets to solidify either in the form of splats (on good thermally conducting substrate, e.g., as in "gun" quenching) or by impinging a cold stream of air or an inert gas against the molten droplets (as, e.g., in atomization), or (2) stabilizing a flowing melt stream so that it freezes as a continuous filament, ribbon, or sheet in contact with a moving chill surface (melt spinning and its variants), or (3) surface melting technologies involving rapid melting at a surface followed by rapid cooling sustained by rapid heat extraction into the unmelted block (laser surface treatments). A number of techniques based on these three categories have been developed over the years, and these have been summarized in some reviews [20,21].

The technique RSP has revolutionized many traditional concepts of metallurgy and materials science. For example, it has been shown that metallic materials can be made to exist in either a glassy (noncrystalline) state [4,5] or in a quasi-crystalline state (in which the traditionally forbidden crystal symmetries could be observed) [6], in addition to their normal crystalline state. Furthermore, it is possible to produce a variety of other metastable phases, such as supersaturated solid solutions and nonequilibrium intermediate phases. Rapidly solidified materials have been finding a multitude of applications, including a range of soft (for transformer core laminations) and hard magnetic materials, wear-resistant light alloys, materials with enhanced catalytic performance and for fuel cell applications, powder metallurgy tool steels and superalloys, and new alloys for medical implants and dental amalgams [22].

1.5.2 Mechanical Alloying

Mechanical alloying, the subject matter of this book, is a powder processing technique that was developed in the mid-1960s by John Benjamin [7] to produce nickel-based oxide dispersion strengthened (ODS) superalloys for gas turbine applications [8]. Subsequently, it was realized that mechanical alloying can also be used to synthesize a variety of both equilibrium and nonequilibrium materials at room temperature and starting from blended elemental powders. This technique has attracted the attention of a large number of researchers during the past 15–20 years or so. The processing involves repeated cold welding, fracturing, and rewelding of powder particles in a high-energy ball mill resulting in the formation of alloy phases. This technique is also capable of synthesizing a variety of equilibrium and nonequilibrium alloy phases starting from prealloyed powders. In fact, all the nonequilibrium effects achieved by RSP of metallic melts have also been observed in mechanically alloyed powders. Consequently, interest in this technique has been constantly growing [23]. Mechanical alloying is presently one of the most popular nonequilibrium processing techniques.

1.5.3 Plasma Processing

The high temperature, high enthalpy, and fast quench rates associated with plasma processing offer unlimited potential in synthesizing novel and advanced materials with improved properties. The quench rates associated with this technique are typically in the range of 10^5–10^7 K/s. Plasma processes are usually one-step processes with the capability of handling large throughputs in small reactor volumes in relatively short processing times. This technique has been used to synthesize a variety of alloys, intermetallics, and refractory compounds [24].

1.5.4 Vapor Deposition

Vapor deposition methods have been used for many decades to produce nonequilibrium phases, including amorphous phases. The rate at which the vapor transforms into a solid is very high and can reach values of about 10^{12} K/s. Under these conditions, many different types of nonequilibrium phases have been produced. Variations of this technique, including physical vapor deposition (PVD) and chemical vapor deposition (CVD), are being regularly exploited to study the formation and characteristics of both stable and metastable phases, in addition to producing coatings for decorative purposes and also to enhance wear and oxidation resistance [25,26].

1.5.5 Spray Forming

In spray forming, highly energetic gas jets impinge on a stream of molten metallic material and disintegrate the melt into small, irregular ligaments. These almost immediately transform into spherical droplets, which are then cooled by the atomization gas, which facilitates momentum transfer and causes the droplets to cool down and solidify during their flight toward the substrate. Deposition of one layer of metal over the other results in the formation of a layered "splat" structure, similar to that obtained in the "gun" technique of rapid solidification of metallic melts. In this respect, the types of microstructure and nature of phases produced by spray forming are somewhat similar to those obtained at the lower end of the cooling rate of RSP [11].

There are several other nonequilibrium processing techniques such as laser processing [3,27], ion mixing [3,28], combustion synthesis [3,29], and application of high pressures [30]. But due to limitations of space, we will not discuss these methods of producing nonequilibrium effects.

1.6 OUTLINE OF THE BOOK

The outline of the book is as follows. In Chapter 2, we briefly discuss the historical background that has led to the development of the technique. This is followed by the nomenclature of the different mechanical alloying methods explored so far (Chapter 3). Chapter 4 describes the different types of equipment available for mechanical alloying.

The mechanical alloying process and the effect of process variables in achieving different types of materials is described in Chapter 5. The mechanism of mechanical alloying is discussed in Chapter 6, and Chapter 7 briefly describes the different methods of characterizing the mechanically alloyed powders. The temperature rise observed during milling of powders is discussed in Chapter 8. The synthesis of stable and metastable phases (supersaturated solid solutions and intermediate phases) is discussed in Chapters 9 and 10, respectively. Disordering of ordered intermetallics is discussed in Chapter 11, whereas the synthesis of amorphous alloys by solid-state amorphization techniques is described in Chapter 12. Formation of nanostructured materials is considered in Chapter 13, and reduction of oxides, chlorides, and the like to pure metals and synthesis of nanocomposites by mechanochemical synthesis is discussed in Chapter 14. The ubiquitous problem of powder contamination is discussed in Chapter 15. Recent developments in understanding the process of mechanical alloying through modeling and milling maps is described in Chapter 16. Applications of mechanically alloyed products are described in Chapter 17, and the problem of safety hazards in handling fine powders, such as those produced by mechanical alloying, are discussed in Chapter 18. The last chapter presents concluding remarks and future prospects for this area.

REFERENCES

1. Bloor, D., Brook, R. J., Flemings, M. C., Mahajan, S., eds. (1994). *The Encyclopedia of Advanced Materials*. Oxford, UK: Pergamon.
2. Froes, F. H., deBarbadillo, J. J., Suryanarayana, C. (1990). In: Froes, F. H., deBarbadillo, J. J., eds. *Structural Applications of Mechanical Alloying*. Materials Park, OH: ASM International, pp. 1–14.
3. Suryanarayana, C. ed. (1999). *Nonequilibrium Processing of Materials*. Oxford, UK: Pergamon.
4. Liebermann, H. H. ed. (1993) *Rapidly Solidified Alloys: Processes, Structures, Properties, Applications*. New York: Marcel Dekker.
5. Anantharaman, T. R., Suryanarayana, C. (1987). *Rapidly Solidified Metals: A Technological Overview*. Aedermannsdorf, Switzerland: Trans Tech.
6. Suryanarayana, C., Jones, H. (1988). *Int. J. Rapid. Solidif.* 3:253–293.
7. Benjamin, J. S. (1970). *Metall. Trans.* 1:2943–2951.
8. Suryanarayana, C. (2001). *Prog. Mater. Sci.* 46:1–184.
9. Upadhya, K. ed. (1993) *Plasma Synthesis and Processing of Materials*. Warrendale, PA: TMS.
10. Bickerdike, R. L., Clark, D., Easterbrook, J. N., Hughes, G., Mair, W. N., Partridge, P. G., Ranson, H. C. (1984–85). *Int. J. Rapid. Solidif.* 1:305–325.
11. Lavernia, E. J., Wu, Y. (1996). *Spray Atomization and Deposition*. Chichester, UK: Wiley.
12. Gaskell, D. R. (1995). *Introduction to the Thermodynamics of Materials*. 3rd ed. Washington, DC: Taylor & Francis.
13. Turnbull, D. (1981). *Metall. Trans.* 12A:695–708.
14. Martin, G., Bellon, P. (1997). *Solid State Phys.* 50:189–331.
15. Shingu, P. H. In: Henein, H., Oki, T., eds. *Processing Materials for Properties*. Warrendale, PA: TMS, pp. 1275–1280.
16. Froes, F. H., Suryanarayana, C., Russell, K., Ward-Close, C. M. In: Singh, J., Copley, S. M., eds. *Novel Techniques in Synthesis and Processing of Advanced Materials*. Warrendale, PA: TMS, pp. 1–21.

17. Froes, F. H., Suryanarayana, C., Russell, K., Li, C-G. (1995). *Mater. Sci. Eng.* A 192/193:612–623.
18. Klassen, T., Oehring, M., Bormann, R. (1997). *Acta Mater.* 45:3935–3948.
19. Duwez, P. (1967). *Trans. ASM Q.* 60:607–633.
20. Suryanarayana, C. (1991). In: Cahn, R. W., ed. *Processing of Metals and Alloys. Materials Science and Technology: A Comprehensive Treatment.* Vol. 15. Weinheim, Germany: VCH, pp. 57–110.
21. Jones, H. (2001). *Mater. Sci. Eng.* A 304/306:11–19.
22. Suryanarayana, C. (2002). In: Buschow, K. H. J., et al. eds., *Encyclopedia of Materials: Science and Technology—Updates.* Oxford, UK: Pergamon.
23. Suryanarayana, C. (1995). *Bibliography on Mechanical Alloying and Milling.* Cambridge, UK: Cambridge International Science Publishing.
24. Ananthapadmanabhan, P. V., Ramani, N. (1999). In: Suryanarayana, C., ed. *Nonequilibrium Processing of Materials.* Oxford, UK: Pergamon, pp. 121–150.
25. Colligon, J. S. (1999). In: Suryanarayana, C., ed., *Nonequilibrium Processing of Materials.* Oxford, UK: Pergamon, pp. 225–253.
26. Teyssandier, F., Dollet, A. (1999). In: Suryanarayana, C., ed. *Nonequilibrium Processing of Materials.* Oxford, UK: Pergamon, pp. 257–285.
27. Singh, J. (1994). *J. Mater. Sci.* 29:5232–5258.
28. Liu, B. X. (1997). *Phys. Stat. Sol. (a)* 161:3–33.
29. Moore, J. J., Feng, H. J. (1995). *Prog. Mater. Sci.* 39:243–316.
30. Sharma, S. M., Sikka, S. K. (1996). *Prog. Mater. Sci.* 40:1–77.

2

Historical Perspective

2.1 INTRODUCTION

Mechanical alloying (MA) is a powder processing technique that allows production of homogeneous materials starting from blended elemental powder mixtures. John S. Benjamin and his colleagues at the Paul D. Merica Research Laboratory of the International Nickel Company (INCO) developed the process around 1966. The technique was the result of a long search to produce a nickel-based superalloy for gas turbine applications that combined the high-temperature strength of oxide dispersion and the intermediate-temperature strength of γ' precipitate. The required corrosion and oxidation resistance were also included in the alloy by suitable alloying additions. Benjamin [1–3] has summarized the historical background of the process and the background work that led to the development of the present process.

2.2 HISTORICAL BACKGROUND

Both precipitation hardening and oxide dispersion strengthening in alloy systems were known by the 1960s. Precipitation hardening in nickel was first reported by Chevenard in 1929 [4]. Rapid development of the γ' age-hardened alloys took place in the 1940s as an enabling materials technology for the aircraft gas turbine [5]. Multicomponent alloys, containing 10 or more elements, and simultaneously containing more than 50% of the γ' strengthening phase, were in common use by 1966.

It was also known that the strength of metals at high temperatures could be increased by the deliberate addition of a fine dispersion of insoluble refractory oxides [6,7]. The origin could be traced back to the early work of Coolidge on thoria-dispersed tungsten [8]. The principle was also applied to aluminum by Irmann in 1949

[9] and to nickel by Alexander et al. in 1961 [10]. Although many methods could be used to produce such dispersions in simple metal systems, these techniques were not applicable to the production of more highly alloyed materials such as those required for gas turbine engines. For example, conventional powder metallurgy techniques, either did not produce an adequate dispersion or did not permit the use of reactive alloying elements such as chromium and aluminum, which confer the needed property characteristics including corrosion resistance and intermediate-temperature strength.

In the early 1960s, INCO developed a process for manufacturing graphitic aluminum alloys by injecting nickel-coated graphite particles into a molten aluminum bath by argon sparging. A modification of the same technique was tried to inoculate nickel-based alloys with a dispersion of nickel-coated, fine refractory oxide particles. The purpose of nickel coating was to render the normally unwetted oxide particles wettable by a nickel-chromium alloy. Early experiments used metal-coated zirconium oxide, which did not yield the desired result. A thorough analysis revealed that the experiment failed because the vendor had supplied powder that was zirconia-coated nickel rather than nickel-coated zirconia! Since the reaction of aluminum with nickel is strongly exothermic, the heat generated cleansed the surface of the graphite and lowered the surface energy. On this basis it was assumed that coating of the refractory oxide with aluminum would be ideal to produce the exothermic reaction. This also did not prove successful. When some other attempts also failed to yield the desired result, out of desperation, researchers turned their attention to the ball milling process.

2.3 DEVELOPMENT OF HIGH-ENERGY BALL MILLING

Ball milling has been applied to the coating of tungsten carbide with cobalt for a long time [11]. Thus, it was known that ball milling could be used to coat hard phases such as tungsten carbide with a soft phase such as cobalt or nickel. It was also known that metal powder particles could be fractured when subjected to heavy plastic deformation. However, if ductile metal powder particles are used, cold welding among them would prevent fragmentation of the particles. Hence, at some stage cold welding could be as rapid as fracturing. This cold welding could be avoided or minimized by employing special chemicals that act as surfactants, such as stearic acid. Consequently, cold welding could be minimized, allowing fracturing to take place more easily and effectively to produce finer particles. Another problem with very fine-powder particles, especially those containing reactive elements such as chromium or aluminum, was that they are pyrophoric at worst or pick up large amounts of oxygen at best. The reactivity of the element also had to be considered; aluminum in a dilute nickel-aluminum alloy is orders of magnitude less reactive than pure aluminum. Taking all these factors into consideration, Benjamin decided to produce composite powder particles by:

Using a high-energy ball mill to favor plastic deformation required for cold welding and reduce the process times

Using a mixture of elemental and master alloy powders (the latter to reduce the activity of the element, since it is known that the activity in an alloy or a compound could be much less than in a pure metal)

Eliminating the use of surface-active agents that would produce fine pyrophoric powder as well as contaminate the powder, and

Relying on the constant interplay between welding and fracturing to yield a powder with a refined internal structure, but having an overall particle size, which was relatively coarse and therefore stable. Refinement of microstructure is a common feature of the powders produced by MA

This method of making the composite powders reproduced the properties of thoria-dispersed nickel synthesized by a completely different process. Encouraged by this success, Benjamin and coworkers conducted experiments to produce a nickel-chromium-aluminum-titanium alloy containing thoria as the dispersoid. This was also successfully produced, first in a small high-speed shaker mill and later in a 1-gallon stirred ball mill. This heralded the birth of MA as a method to produce oxide dispersion strengthened (ODS) alloys on an industrial scale.

Figure 2.1 presents a chronology of the development of mechanical alloying technology. In passing, it may be mentioned that this process, as developed by Benjamin and coworkers, was referred to as "milling/mixing," but Ewan C. MacQueen, a patent attorney for INCO, coined the term *mechanical alloying* to describe the process in the first patent application, and this term has remained in the literature.

For successful production of ODS γ'-hardened superalloys, synthesis of the powder alloy with a fine oxide dispersion is only the first step. Commercial use of the alloy requires that this powder be consolidated to full density, proper microstructure be developed, and the desired properties be obtained. These are not trivial problems for these multicomponent alloys with fine microstructural features. Thus, the additional problems to be solved include the following:

Choice of the method to consolidate the powders—cold pressing and sintering, or hot isostatic pressing, or vacuum hot pressing, or hot extrusion,

Whether precompaction of the powders is required

Proper choice of the can material during extrusion to match the plastic deformation characteristics of the alloy

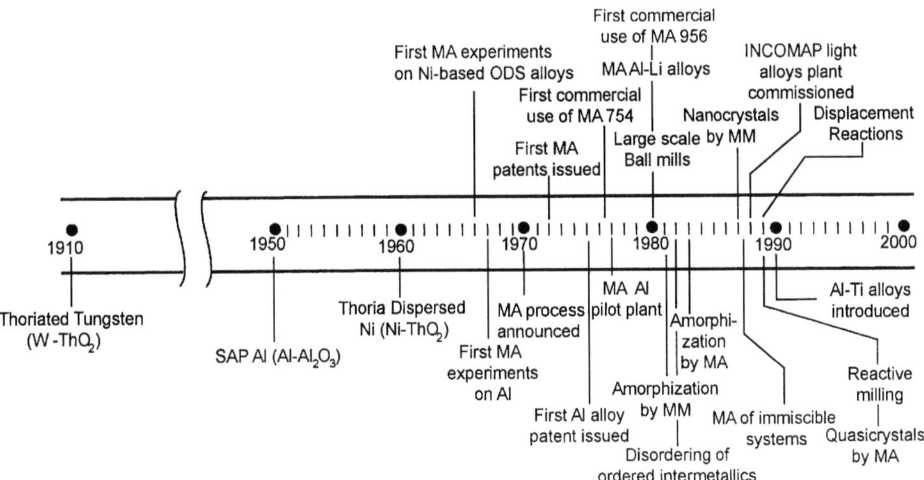

Figure 2.1 Chronology of development of mechanical alloying.

Whether evacuation of the powder is required before consolidation and, if so, whether the powder should be degassed prior to canning, evacuated within the can, or heated during evacuation

Proper selection of hot extrusion conditions—temperature, pressure, extrusion ratio, and so forth

Whether any postextrusion working was required to develop optimal high-temperature properties

2.4 POTENTIAL OF MECHANICAL ALLOYING

Mechanical alloying is normally a dry, high-energy ball milling technique that has been employed in the production of a variety of commercially useful and scientifically interesting materials. The formation of an amorphous phase by mechanical grinding of a Y-Co intermetallic compound in 1981 [12] and in the Ni-Nb system by ball milling of blended elemental powder mixtures in 1983 [13] brought about the recognition of MA as a potential nonequilibrium processing technique. Beginning from the mid-1980s, several investigations have been carried out to synthesize a variety of stable and metastable phases, including supersaturated solid solutions, crystalline and quasicrystalline intermediate phases, and amorphous alloys [14–18]. In addition, it has been recognized that powder mixtures can be mechanically activated to induce chemical reactions, i.e., mechanochemical reactions, at room temperature or at least at much lower temperatures than normally required to produce pure metals, nanocomposites, and a variety of commercially useful materials [19,20]. Efforts have also been under way since the early 1990s to understand the process fundamentals of MA through modeling studies [21]. Because of all these special attributes, this simple but effective processing technique has been applied to metals, ceramics, polymers, and composite materials. The attributes of mechanical alloying are listed in Table 2.1, and some important milestones in the development of the field are presented in Table 2.2.

2.5 POTENTIAL RESOURCES OF MECHANICAL ALLOYING LITERATURE

The results of application of MA to different materials have been published in archival journals and also in different conference proceedings. A number of stand-alone conferences have been organized on this topic [22–31]. MA has now become an integral

Table 2.1 Attributes of Mechanical Alloying

1. Production of fine dispersion of second-phase (usually oxide) particles
2. Extension of solid solubility limits
3. Refinement of grain sizes down to nanometer range
4. Synthesis of novel crystalline and quasi-crystalline phases
5. Development of amorphous (glassy) phases
6. Disordering of ordered intermetallics
7. Possibility of alloying of difficult to alloy elements/metals
8. Inducement of chemical (displacement) reactions at low temperatures
9. Scalable process

Table 2.2 Important Milestones in the Development of Mechanical Alloying

1966	Development of ODS nickel-based alloys
1981	Amorphization of intermetallics
1982	Disordering of ordered compounds
1983	Amorphization of blended elemental powder mixtures
1987/88	Synthesis of nanocrystalline phases
1989	Occurrence of displacement reactions
1989	Synthesis of quasi-crystalline phases

part of the triennial international conferences on rapidly quenched (RQ) metals since RQ VI held in Montreal, Canada in 1987 [32–36]. These conferences are now redesignated as rapidly quenched and metastable materials (RQMM) since RQ-10 in Bangalore, India, and their proceedings are contained in the international journal *Materials Science and Engineering A* published by Elsevier. In addition, proceedings of the International Symposia on Metastable, Mechanically Alloyed and Nanocrystalline Materials (ISMANAM) contain many papers on mechanical alloying, and these are regularly published in *Materials Science Forum* (and also as special volume(s) of the *Journal of Metastable and Nanocrystalline Materials*) by Trans Tech Publications [37–45]. A few books have been recently published on MA and also on mechanochemical processing [17,19,46–51]. The literature on MA up to 1994 has been collected together in an annotated bibliography published in 1995 [15]. The journal *International Journal of Mechanochemistry and Mechanical Alloying* was started in 1994 but was short lived. Several reviews have also appeared during the past 10 years

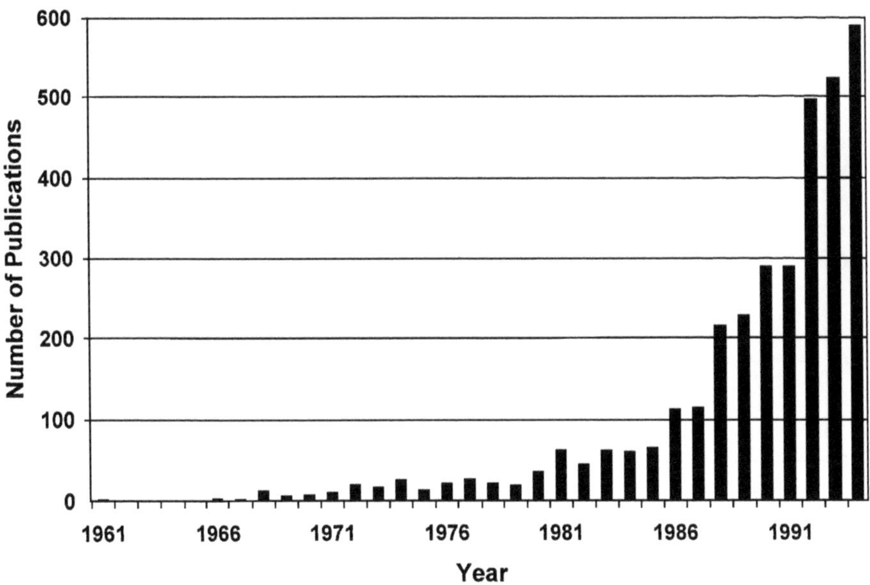

Figure 2.2 Growth of annual number of publications in the field of mechanical alloying.

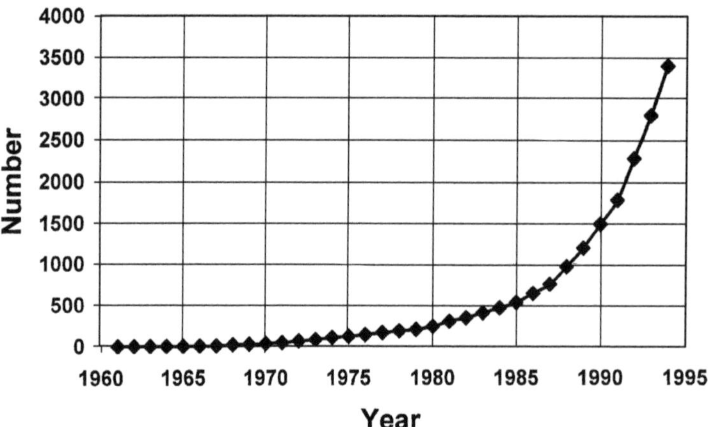

Figure 2.3 Growth of cumulative number of publications in the field of mechanical alloying.

with emphasis on a particular topic [14,16,18,52–63]. However, the present book is an attempt to present all aspects of MA in a comprehensive and critical manner in one place, as well as to present the potential and limitations of this technique as a nonequilibrium processing tool and its current and future applications.

Figure 2.2 presents the growth of annual publications in the field of MA, and Figure 2.3 shows the cumulative list of publications in the field of MA. It may be noted that, as in every field of endeavor, the activity in this area was at a very low level for the

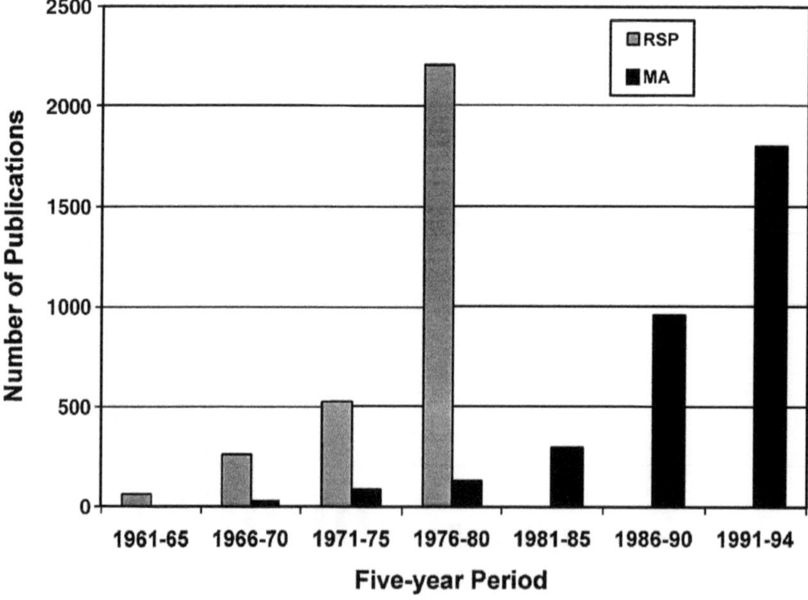

Figure 2.4 Comparison of the growth of publications in the fields of mechanical alloying and rapid solidification processing, two popular nonequilibrium processing techniques.

first few years and picked up a rapid pace from 1988 onward. Whereas only seven papers were published during the calendar year in 1970, the number increased to 61 in 1981, 214 in 1988, and 522 in 1993. The number of papers published is increasing every year, and this trend is continuing today. It is estimated that more than 7000 papers have been published in this area until now and that the annual rate of publications nowadays exceeds 500 papers. It is also of interest to note that the growth of publications in the field of RSP, another nonequilibrium processing technique, introduced in 1960, shows a similar trend. A comparison between the growth of publications in these two areas (Fig. 2.4) suggests that the take-off point in the rapid growth of publications in the field of MA is offset by about 15 years from that of RSP. Accordingly, the sudden spurt in the rate of publications started in the 1970s for RSP and around 1988 for MA.

Majority of the publications in the area of MA are generally found in the proceedings of the specific conferences devoted to this topic, especially the ISMANAM series. A good number of research and scientific/technical papers may also be found in several archival journals. The two journals—*Materials Science and Engineering A* and *Journal of Alloys and Compounds*—appear to be publishing a majority of the papers in the field of MA. Some of the other journals where papers on MA are frequently published are *Metallurgical and Materials Transactions A*, *Journal of Materials Research*, *Journal of Materials Science*, *Journal of Materials Science Letters*, *Journal of Materials Synthesis and Processing* (discontinued in 2003), *Scripta Materialia*, and *Acta Materialia*.

REFERENCES

1. Benjamin, J. S. (1976). *Sci. Am.* 234(5):40–48.
2. Benjamin, J. S. (1990). *Metal Powder Rep.* 45:122–127.
3. Benjamin, J. S. (1992). In: Capus, J. M., German, R. M., eds. *Advances in Powder Metallurgy and Particulate Materials*. Vol. 7 (Novel Powder Processing). Princeton, NJ: Metal Powder Industries Federation, pp. 155–168.
4. Chevenard, P. (1929). *Compte. Rendu.* 189:846.
5. Betteridge, W. (1959). In: *The Nimonic Alloys*. London, UK: Edward Arnold, p. 5.
6. Ansell, G. S. (1968). In: *Oxide Dispersion Strengthening*. New York: Gordon and Breach, p. 61.
7. Wilcox, B. A., Clauer, A. H. (1972). In: Sims, C. T., Hagel, W. C., eds. *The Superalloys*. New York: Wiley, p. 197.
8. Coolidge, W. D. (1910). *Proc. Am. Inst. Elect. Eng. 961.*
9. Irmann, R. (1949). *Techn. Rundschau.* 41(36):19.
10. Alexander, G. B., Iler, R. K., West, S. F. US Patent 2,972,529, February 21, 1961.
11. Hoyt, S. L. (1930). *Trans. AIME* 89:9–30.
12. Eermakov, A. E., Yurchikov, E. E., Barinov, V. A. (1981). *Phys. Met. Metallogr.* 52(6): 50–58.
13. Koch, C. C., Cavin, O. B., McKamey, C. G., Scarbrough, J. O. (1983). *Appl. Phys. Lett.* 43:1017–1019.
14. Koch, C. C. (1991). In: Cahn, R. W., ed. *Processing of Metals and Alloys, Materials Science and Technology—A Comprehensive Treatment*. Vol. 15. Weinheim, Germany: VCH, pp. 193–245.
15. Suryanarayana, C. (1995). *Bibliography on Mechanical Alloying and Milling*. Cambridge, UK: Cambridge International Science Publishing.

16. Suryanarayana, C. (1996). *Metals Mater.* 2:195–209.
17. Lu, L., Lai, M. O. (1998). *Mechanical Alloying*. Boston, MA: Kluwer.
18. Murty, B. S., Ranganathan, S. (1998). *Int. Mater. Rev.* 43:101–141.
19. Heinicke, G. (1984). *Tribochemistry*. Berlin: Akademie Verlag.
20. McCormick, P. G. (1995). *Mater. Trans. Jpn. Inst. Metals* 36:161–169.
21. Maurice, D. R., Courtney, T. H. (1990). *Metall. Trans.* A21:289–303.
22. Benjamin, J. S., ed. (1981). *Frontiers of High-Temperature Materials*. New York: INCO Alloys International.
23. Benjamin, J. S., Benn, R. C., eds. (1983). *Frontiers of High-Temperature Materials II*. New York: INCO Alloys International.
24. Arzt, E., Schultz, L., eds. (1989). *New Materials by Mechanical Alloying Techniques*. Oberursel, Germany: DGM Informationgesellschaft.
25. Clauer, A. H., deBarbadillo, J. J., eds. (1990). *Solid State Powder Processing*. Warrendale, PA: TMS.
26. Froes, F. H., deBarbadillo, J. J., eds. (1990). *Structural Applications of Mechanical Alloying*. Materials Park, OH: ASM International.
27. Yavari, A. R., Desré, P. J., eds. (1990). Proceedings of the Conference on Multilayer Amorphisation by Solid-State Reaction and Mechanical Alloying. *J Physique Colloq C4 Suppl* 14, 51:C4-1–C4-310.
28. Shingu, P. H., ed. (1992). *Mechanical Alloying. Mater. Sci. For.* Aedermannsdorf, Switzerland: Trans. Tech. Publications, vol. 88–90.
29. deBarbadillo, J. J., Froes, F. H., Schwarz, R., eds. (1993). *Mechanical Alloying for Structural Applications*. Materials Park, OH: ASM International.
30. Baláz, P., Plesingerova, B., Sepelk, V., Stevulova, N., eds. (1994). Proceedings of the First International Conference on Mechanochemistry. Cambridge, UK: Cambridge International Science Publishing.
31. Soni, P. R., Rajan, T. V., eds. (2002). *Trends in Mechanical Alloying*. Ensield, NH: Science Publishers.
32. Cochrane, R. W., Sröm-Olsen, J. O., eds. (1988). Proceedings Sixth of the International Conference on Rapidly Quenched Metals. *Mater. Sci. Eng.* vol. 97–99.
33. Frederiksson, H., Savage, S. J., eds. (1991). Proceedings of the Seventh International Conference on Rapidly Quenched Metals. *Mater. Sci. Eng.* vol. A133–134.
34. Masumoto, T., Hashimoto, K., eds. (1994). Proceedings of the Eighth International Conference on Rapidly Quenched Metals. *Mater. Sci. Eng.* vol. A179–182.
35. Duhaj, P., Mrafko, P., Švec, P., eds. (1997). Proceedings of the Ninth International Conference on Rapidly Quenched Metals. *Mater. Sci. Eng.* vol. A226–228.
36. Chattopadhyay, K., Ranganathan, S., eds. (2001). Proceedings of the Tenth International Conference on Rapidly Quenched and Metastable Materials. *Mater. Sci. Eng.* vol. A304–306.
37. Yavari, A. R., ed. (1995). Proceedings of the International Symposium on Metastable, Mechanically Alloyed and Nanocrystalline Materials (ISMANAM – 94). *Mater. Sci. For.* Zurich: Trans Tech Publications, vol. 179–181.
38. Schulz, R., ed. (1996). Proceedings of the International Symposium on Metastable, Mechanically Alloyed and Nanocrystalline Materials (ISMANAM – 95). Mater. Sci. For. Zurich: Trans Tech Publications, vol. 225–227.
39. Fiorani, D., Magini, M., eds. Proceedings of the International Symposium on Metastable, Mechanically Alloyed and Nanocrystalline Materials (ISMANAM – 96). *Mater. Sci. For.* Zurich: Trans Tech Publications, vol. 235–238.
40. Baró, M. D., Suriñach, S., eds. (1998). Proceedings of the International Symposium on Metastable, Mechanically Alloyed and Nanocrystalline Materials (ISMANAM – 97). *Mater. Sci. For.* Zurich: Trans Tech Publications, vol. 269–272.
41. Calka, A., Wexler, D., eds. (1999). Proceedings of the International Symposium on

Metastable, Mechanically Alloyed and Nanocrystalline Materials (ISMANAM – 98). *Mater. Sci. For.* Zurich: Trans Tech Publications, vol. 312–314 (and also in *J. Metastable Nanocryst. Mater.* 2–6, 1999).
42. Eckert, J., Schlörb, H., Schultz, L., eds. (2000). Proceedings of the International Symposium on Metastable, Mechanically Alloyed and Nanocrystalline Materials (ISMANAM – 99). *Mater. Sci. For.* Zurich: Trans Tech Publications, vol. 343–346. (and also in *J. Metastable Nanocryst. Mater.* 8, 2000).
43. Schumacher, P., Warren, P., Cantor, B., eds. (2001). Proceedings of the International Symposium on Metastable, Mechanically Alloyed and Nanocrystalline Materials (ISMANAM – 2000). *Mater. Sci. For.* Zurich: Trans Tech Publications, vol. 360–362. (and also in *J. Metastable Nanocryst. Mater.* 10, 2001).
44. Ma, E., Atzmon, M., Koch, C. C., eds. (2002). Proceedings of the International Symposium on Metastable, Mechanically Alloyed and Nanocrystalline Materials (ISMANAM – 01). *Mater. Sci. Forum.* Zurich: Trans Tech Publications, vol. 386–388 (and also in *J. Metastable Nanocryst. Mater.* 13, 2002).
45. Ahn, J. -H., Hahn, Y. -D., eds. (2003). Proceedings of the International Symposium on Metastable, Mechanically Alloyed and Nanocrystalline Materials (ISMANAM–02). *J. Metastable Nanocryst. Mater*, vol. 15–16.
46. Avvakumov, E. G. (1986). *Mechanical Methods of the Activation of Chemical Processes.* Novosibirsk, Russia: Nauka (in Russian).
47. Avvakumov, E. G., Senna, M., Kosova, N. V. (1998). *Soft Mechanical Synthesis: A Basis for New Chemical Technologies.* Boston, MA: Kluwer.
48. Soni, P. R. (2000). *Mechanical Alloying: Fundamentals and Applications.* Cambridge, UK: Cambridge International Science Publishing.
49. Besterci, M. (1998). *Dispersion-Strengthened Aluminum Prepared by Mechanical Alloying.* Cambridge, UK: Cambridge International Science Publishing.
50. Gutman, E. M. (1998). *Mechanochemistry of Materials.* Cambridge, UK: Cambridge International Science Publishing. (See also the Special Issues of *J. Mater. Synth. Proc.* 2000; 8(Issues 3–6)).
51. Sherif El-Eskanadarany, M. (2001). *Mechanical Alloying for Fabrication of Advanced Engineering Materials.* New York: William Andrew.
52. Gilman, P. S., Benjamin, J. S. (1983). *Annu. Rev. Mater. Sci.* 13:279–300.
53. Singer, R. F., Gessinger, G. H. (1984). In: Gessinger, G. H., ed. *Powder Metallurgy of Superalloys.* London: Butterworths, pp. 213–292.
54. Sundaresan, R., Froes, F. H. (1987). *J. Metals* 39(8):22–27.
55. Weeber, A. W., Bakker, H. (1988). *Physica.* B153:93–135.
56. Koch, C. C. (1989). *Annu. Rev. Mater. Sci.* 19:121–143.
57. Schaffer, G. B., McCormick, P. G. (1992). *Mater. For.* 16:91–97.
58. Bakker, H., Zhou, G. F., Yang, H. (1995). *Prog. Mater. Sci.* 39:159–241.
59. Shingu, P. H., ed. (1995). *Special Issue on Mechanical Alloying. Mater. Trans. Jpn. Inst. Metals*, 36:83–388.
60. Schwarz, R. B., ed. (1996). *Viewpoint set on mechanical alloying. Scripta Mater.* 34:1–73.
61. Ivanov, E., Suryanarayana, C. (2000). *J. Mater. Synth. Proc.* 8:235–244.
62. Suryanarayana, C. (2001). *Prog. Mater. Sci.* 46:1–184.
63. Takacs, L. (2002). *Prog. Mater. Sci.* 47:355–414.

3

Nomenclature

3.1 INTRODUCTION

The synthesis and development of novel alloy phases by mechanical means [through mechanical alloying (MA) methods] has been in existence for more than four decades. During this period, several new versions of the original process of mechanical alloying developed by Benjamin [1,2] have come up. Consequently, the different processes have been designated differently, and it would be useful to have a clear knowledge and understanding of the different terms and acronyms used in the literature. This will help in clearly identifying the actual process used in synthesizing the material.

3.2 MECHANICAL ALLOYING

Mechanical alloying is the generic term for processing of metal powders in high-energy ball mills. However, depending on the state of the starting powder mix and the processing steps involved, different terms have been used in the powder metallurgy literature. Two different terms are most commonly used to describe the processing of powder particles in high-energy ball mills. Mechanical alloying describes the process when mixtures of powders (of different metals or alloys/compounds) are milled together. Thus, if powders of pure metals A and B are milled together to produce a solid solution (either equilibrium or supersaturated), intermetallic, or amorphous phase, the process is referred to as MA. Material transfer is involved in this process to obtain a homogeneous alloy.

3.3 MECHANICAL MILLING/DISORDERING

When powders of uniform (often stoichiometric) composition, such as pure metals, intermetallics, or prealloyed powders, are milled in a high-energy ball mill, and material transfer is *not* required for homogenization, the process has been termed mechanical milling (MM). It may be noted that when a mixture of two intermetallics is processed, and then alloying occurs, this will be referred to as MA because material transfer is involved. However, if a pure metal or an intermetallic is processed only to reduce particle (or grain) size and increase the surface area, then this will be referred to as MM because material transfer is not involved. The destruction of long-range order in intermetallics to produce a disordered intermetallic (solid solution) or an amorphous phase has been referred to as mechanical disordering (MD) [3]. The advantage of MM/MD over MA is that since the powders are already alloyed and only a reduction in particle size and/or other transformations have to be induced mechanically, the time required for processing is short. For example, MM requires half the time required for MA to achieve the same effect [4]. In fact, the time required to achieve the same constitution could be reduced by a factor of 10 when prealloyed $Al_{75}Ti_{25}$ powder was used rather than the blended elemental Al-Ti powder mix [5]. Whereas a solid solution formed in about 5 h when one started with prealloyed powders, it took more than 50 h for the same phase to be formed on MA of blended elemental powders [6]. The milling conditions were identical in both cases. An additional advantage of MM over MA of powders is that MM reduces oxidation of the constituent powders, related to the shortened time of processing [4]. Some investigators have referred to MM as mechanical grinding (MG) [7]. Since "grinding" is normally thought of as an abrasive machining process that involves mainly shear forces and chip formation, the term "milling" is preferred to include the more complex triaxial, perhaps partly hydrostatic, stress states that can occur during ball milling of powders [8]. It should be realized that MA is a generic term, and some investigators use this term to include both MA and MM/MD/MG. However, for the sake of clarity, it is better to distinguish among these terms by using MA or MM depending on whether material transfer is involved or not during the processing.

Some other terms are also used in the literature on MA. These include reaction (or reactive ball) milling, cryomilling, rod milling, mechanically activated annealing (M2A), double mechanical alloying (dMA), and mechanically activated self-propagating high-temperature synthesis (MASHS).

3.4 REACTION MILLING

Reaction milling (RM), pioneered by Jangg et al. [9], is the MA process accompanied by a solid-state reaction. In this process the powder is milled without the aid of any process control agent (see Chapter 5 for its function during milling) to produce fine dispersions of oxides and carbides in aluminum [10]. The dispersion of carbides is achieved by adding lamp black or graphite during milling of aluminum. Adjusting the oxygen content via close control of the milling atmosphere (oxygen, argon, nitrogen, air, etc.) results in production of the oxides. Thus, the final product of milling contains a dispersion of Al_4C_3 and Al_2O_3 in an aluminum matrix, and these alloys are given the trade name DISPAL. Milling of metal powders in the presence of reactive solids/ liquids/gases (enabling a chemical reaction to take place) is now regularly done to synthesize metal oxides, nitrides, and carbides [11,12]. Thus, milling of titanium in a nitrogen atmosphere has resulted in production of titanium nitride [13,14]. Several other compounds have also been produced in a similar way. Milling of tungsten with carbon (graphite) has resulted in production of tungsten carbide [15]. Milling of metal powders with boron has resulted in production of borides, e.g., TiB_2 [16]. These oxides, carbides, borides, or nitrides could then be incorporated into the alloy matrix providing additional strength and high-temperature stability.

3.5 CRYOMILLING

Another variation of milling that is being increasingly used nowadays is cryomilling [17] whereby the milling operation is carried out at cryogenic (very low) temperatures and/or milling of materials is done in a cryogenic medium such as liquid nitrogen. In this process, the vessel is cooled by a continuous flow of liquid nitrogen through the "water cooling" jacket of the mill. In addition, or alternatively, liquid nitrogen could be introduced into the milling chamber itself throughout the run. After completion of the milling run, the powder is removed from the mill in the form of a slurry and transferred to a glove box containing dry argon. When the liquid nitrogen is allowed to evaporate, a residue of the milled powder is left behind.

The original intention of cryomilling was to alloy Al_2O_3 into the aluminum matrix. Mechanical property measurements of hot-pressed and swaged materials indicated that the strength of the composite was independent of the Al_2O_3 content. However, the strength was related to a fine dispersion of aluminum oxynitride particles. These particles were found to form in situ during the milling process by coadsorption of nitrogen and oxygen onto clean aluminum surfaces. During subsequent annealing, the mechanically trapped nitrogen and oxygen interacted with aluminum to form Al(O,N) particles, which are typically 2–10 nm in diameter. These particles have been reported to be very effective in inhibiting grain growth in the aluminum matrix, with the result that cryomilled alloys have a typical grain size of 50–300 nm, and the grains continue to be stable even on prolonged annealing at high temperatures.

It was noted that the powder quality was poor and the yield low when cryomilling was conducted in a standard Szegvari-type attritor. In addition, formation of dead zones in the tank, excessive powder loss due to liquid nitrogen evaporation and flow control, excessive seal wear, jamming of the stir arms, and freezing of the apparatus were some of the problems encountered. Aikin and Juhas [18] modified the attritor to minimize the above problems and reduce the oxygen pickup. These modifications improved the properties of the cryomilled product, including the homogeneity of the powder. They showed that by a proper choice of the process parameters it is possible to make materials with the desired AlN content in the powder.

3.6 ROD MILLING

Rod milling is a technique that was developed in Japan [19] essentially to reduce powder contamination during the processing stage. In this process, the grinding medium is in the form of rods rather than spherical balls. In a conventional ball mill, impact forces scratch the surfaces of the milling media and the debris from the milling media contaminates the powder being milled. On the other hand, if shear forces predominate, they are more effective in kneading the powder mixtures and the resulting powder is much less contaminated. To achieve this minimized contamination, the balls were replaced by long rods in the rod mill because long rods rotating in a cylindrical vial predominantly exert shear forces on the material. Yet another advantage reported for the rod-milled powders is that the level of contamination from the milling medium is much less than that obtained for the powder milled in a conventional ball mill with spherical balls. For example, it was reported that the iron contamination for the $Al_{30}Ta_{70}$ powder mechanically alloyed for 400 h was 16 at% in the ball mill, compared to only about 5 at% in the rod mill [19].

It has also been reported [3] that the degree of agglomeration of the powder particles is less in rod milling, resulting in smaller average diameters. Furthermore, subsequent disintegration into fine powders during later stages of milling proceeds at a much faster rate leading to the formation of fine powders with a narrow size distribution.

3.7 MECHANICALLY ACTIVATED ANNEALING

Mechanically activated annealing (M2A) is a process that combines short MA duration with a low-temperature isothermal annealing. The combination of these two steps has been found to be effective in producing different refractory materials such as silicides based on niobium, molybdenum, etc. [20,21]. For example, MA of molybdenum and silicon powders for 1–2 h in a planetary ball mill followed by a 2- to 24-h annealing at 800 °C produced the $MoSi_2$ phase [21]. A consequence of this method is that optimization of the M2A process could lead to a situation in which isothermal annealing can be carried out inside the milling container to avoid air contamination of the end product.

3.8 DOUBLE MECHANICAL ALLOYING

Double mechanical alloying (dMA) involves two stages of milling. In the first stage, the constituent elemental powder sizes are refined and uniformly distributed as an intimate mixture. This mixture is then subjected to a heat treatment at high temperatures during which intermetallic phases are formed. The size of the intermetallics formed ranges from less than 1 μm to a few micrometers. During the second stage, the heat-treated powder is milled again to refine the powder size of the intermetallic phases and reduce the grain size of the matrix. After degassing, the powders are consolidated to a bulk shape [22,23]. Figure 3.1 shows the flow sheet of the dMA process and the type of microstructures developed in an Al-5 wt% Fe-4 wt% Mn powder mixture [22]. This appears to be a useful process to produce fine intermetallics in alloy systems that cannot be directly produced by milling.

3.9 MECHANICALLY ACTIVATED SELF-PROPAGATING HIGH-TEMPERATURE SYNTHESIS

Yet another term recently coined is mechanically activated self-propagating high-temperature synthesis (MASHS), which is based on a combination of MA and self-propagating high-temperature synthesis (SHS) processes. SHS is a well-known method to produce advanced materials as a practical alternative to the conventional methods. SHS offers advantages with respect to process economics and process simplicity. In the typical SHS reaction, the mixed reactant powders are pressed into

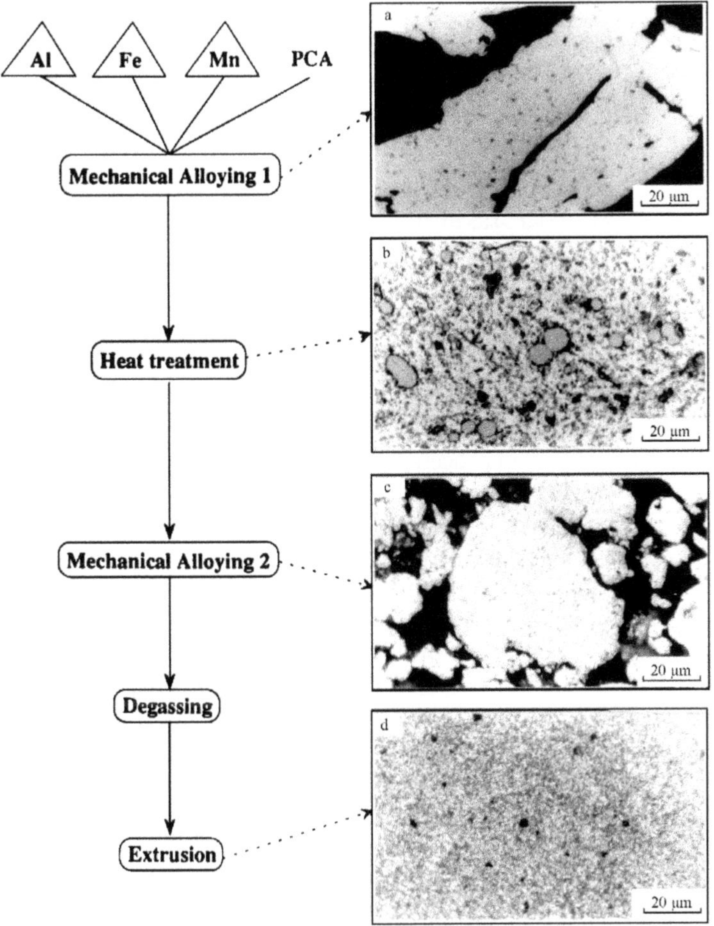

Figure 3.1 Process flow sheet and the microstructures developed during double mechanical alloying (dMA) of an Al-5 wt%Fe-4 wt% Mn powder mixture.

a pellet of certain green density and subsequently ignited to the ignition temperature. Due to the highly exothermic nature of the reaction, the heat generated allows the reaction to become self-sustained [24]. In the MASHS process the powder mixture is mechanically alloyed to produce a nanocrystalline structure and then the SHS reaction is initiated by pressing the powder into a pellet and igniting it in a furnace. The previous MA step decreases the ignition temperature by as much as 100°C. Several iron aluminide and niobium aluminide intermetallics were synthesized in this way [25,26]. It is possible that an SHS reaction can sometimes take place during the milling operation, and this aspect is discussed in Chapter 14.

3.10 OXIDATION–ATTRITION MILLING–REDUCTION

Oxidation–attrition milling–reduction (OMR) is another term that has been used recently in the literature [27]. In this method, the commercially available powders (of micrometer size) are oxidized at appropriate temperatures to produce brittle oxide powders. These powders are subjected to attrition milling to obtain nanoscale oxide powders, which are subsequently reduced to metallic powders with nanometer grain sizes. This method should be particularly useful to produce nanocrystalline powders of ductile metals and materials. It is necessary to ensure that the oxides of such materials are brittle and that they can also be easily reduced to pure metals.

3.11 MECHANOCHEMICAL PROCESSING

Mechanochemical processing (MCP) or mechanochemical synthesis is the term applied to powder processing in which chemical reactions and phase transformations take place during milling due to the application of mechanical energy [11,28–30]. An important feature of the process is that plastic deformation and chemical processes occur almost *simultaneously*. This process has a long history with the first publication dating back to 1892 [31]. The mechanochemical reactions could result in the synthesis of novel materials, reduction/oxidation processes, exchange reactions, decomposition of compounds, and phase transformations in both organic and inorganic solids. The materials produced in this way have already found applications in areas such as hydrogen storage materials, gas absorbers, fertilizers, and catalysts, and this technique has become a large effort in the general field of mechanical alloying and milling. In fact, a series of conferences are being regularly held on this aspect, and the fourth in this series was held on September 7–11, 2003 in Braunschweig, Germany.

A recent development in this broad area of MCP of materials is the synthesis of inorganic compounds based on the use of solid acids, bases, hydrated compounds, crystalline hydrates, and base and acid salts as the raw materials [32]. This process has been termed soft mechanochemical synthesis [32,33]. In contrast, the process will be referred to as hard mechanochemical synthesis if anhydrous oxides are used as the starting materials. The soft mechanochemical synthesis technique has been used to synthesize a variety of nanocrystalline oxides such as $ZrTiO_4$, Al_2TiO_5, and $ZrSiO_4$ [34]. The size of the particles obtained by this method is in the range of a few nanometers and is close to that obtained by the sol-gel method. For example, the particle size of $ZrTiO_4$ obtained by soft mechanochemical synthesis is 12 nm, while that obtained by the sol-gel process is 14 nm [34].

3.12 OTHER METHODS

Several new methods have been developed in recent years to produce materials in a nonequilibrium condition. The majority of these methods produce effects very similar to those obtained in mechanically alloyed powders. The major difference is that while MA produces the material in the powder form, most of these new methods deal with bulk materials. Yet another difference is that these new methods focus on producing nanocrystalline and amorphous materials in view of their novelty and potential applications. A common theme among all of these methods is that heavy plastic deformation is involved except in the case of multilayer amorphization. Even so, this method is included here because it is so similar to MA in producing amorphous alloys and other nonequilibrium phases. Rapid solidification processing (RSP) is another method for producing nonequilibrium alloys, and comparisons of the results obtained by RSP and MA are given in subsequent chapters. Some of these new methods (other than RSP) will be briefly described below.

3.12.1 Repeated Cold Rolling

Repeated cold rolling produces effects very similar to those of MA. As will be shown later, MA results in repeated rolling and folding over of the individual ductile metal powder particles. Thus, if we consider two metals A and B, and if a stack of these two metals (bimetal layer) is assumed to have a thickness of d_o, then rolling of this stack in one pass reduces the thickness to $d_o \times (1/a)$, where a is the reciprocal of the reduction in thickness in one pass. Subsequent rolling passes reduce the thickness further. After n passes, the thickness of the bimetal layer is reduced to $d_o \times (1/a)^n$. This is shown schematically in Figure 3.2 [35]. The actual process involves stuffing a mixture of the elemental powders into a stainless steel tube of 20 mm outside diameter and 18 mm inside diameter. The tube containing the powder is then pressed to a thickness of 8 mm by a hydraulic press and then cold rolled to a thickness of 0.5 mm. The thickness of the sample after removal of the stainless sheath is about 0.2 mm. The rolled sample is packed again into the stainless steel tube of the same size, pressed, and rolled in a similar way. The process is repeated up to 30 times. Figure 3.3 shows a schematic of the process. It was shown that supersaturated solid solutions could be obtained in Ag-70 at% Cu samples both by MA and by repeated cold rolling methods, suggesting that both methods yield similar types of metastable phases [36].

Since plenty of disorder (chaos) occurs as a result of the repetition of a simple deterministic process, the MA process can also be described in terms of the philosophy governing the occurrence of chaos in the area-preserving mapping. The occurrence of refined microstructures during MA can then be easily explained using this approach. A slight deviation from the exact linear mapping was found to produce apparently more realistic MA structures [37].

Instead of stuffing the powders into a stainless steel tube, one could also take thin foils of the constituent metals, stack them alternately, and press them repeatedly. Alternatively, one could also take powder mixtures, press them repeatedly, break them into smaller pieces, and repeat the process. Metastable structures similar to those obtained by conventional MA have been achieved in these cases also [38]. Figure 3.4 shows schematic views of the repeated pressing of layered thin foils and of powders.

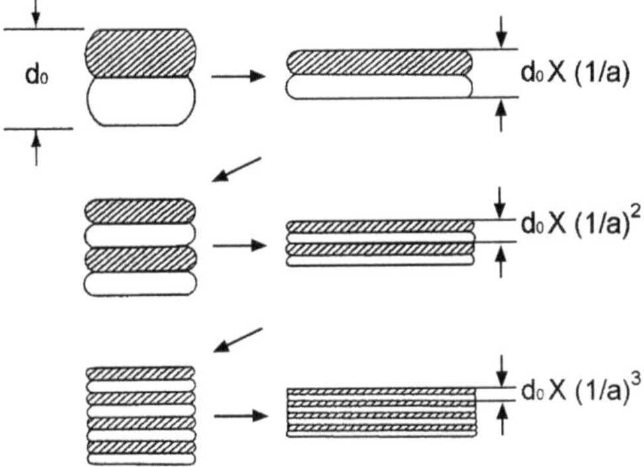

Figure 3.2 Thickness reduction by repeated rolling and folding operations. When the reduction in thickness during one rolling pass is $1/a$ and the number of passes is n, then the total reduction in thickness is $d_o \times (1/a)^n$, where d_o is the original thickness.

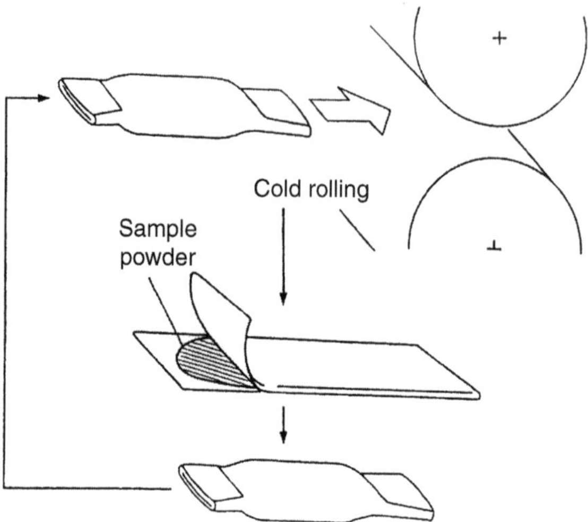

Figure 3.3 Schematic diagram showing the experimental procedure for repeated cold rolling.

3.12.2 Multilayer Amorphization

Amorphous phases could be synthesized by solid-state reaction between pure polycrystalline metal thin films [39]. This phenomenon is again somewhat similar to the repeated cold rolling process described above (where solid solution formation was reported) and also to that of MA, except that the departure from equilibrium is quite extensive. Two conditions must be satisfied for the formation of the amorphous phase in this method. One is that one of the species must be a fast diffuser in the other, and the second is that the alloys must exhibit a large negative heat of mixing providing the

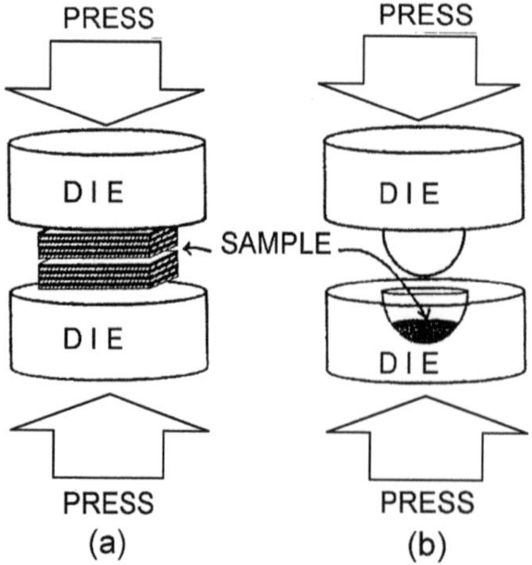

Figure 3.4 Schematic views of the repeated pressing procedure. (a) Repeated pressing of layered thin foils and (b) repeated pressing of powders.

required driving force for the formation of the amorphous phase. The process consists of putting alternating layers of the constituent metals in a stack (with close contact with each other) and annealing them at a relatively low temperature. The above authors had demonstrated that $Au_{1-x}La_x$ alloys in the composition range of $0.3 \leq x \leq 0.5$ could be amorphized by solid-state interdiffusion of pure crystalline Au and La thin films by annealing them at temperatures of 50–80°C. Similar results have been reported later in several other alloy systems [40].

3.12.3 Severe Plastic Deformation

It has been well known that heavy deformations, e.g., by cold rolling or drawing or multiple forging, result in significant refinement of microstructures at low temperatures. However, the microstructures formed are usually substructures of a cellular type having low-angle grain boundaries. However, the grain boundaries in nanocrystalline materials are of the high-angle type, and significant improvement in the properties of materials can be achieved only in these cases. The majority of methods available for the synthesis of nanostructures produce them in the form of powder, often requiring consolidation of the powder to handle and characterize the properties.

Severe plastic deformation (SPD) has been developed to produce ultrafine-grained materials in the submicrometer range, with a typical minimal grain size of about 200–300 nm. In these methods, very large deformations (with true strains ≥10) are provided at relatively low temperatures, mostly by two processes: torsion straining under high pressure and equal channel angular pressing. Figure 3.5 shows a schematic of these two types of processes. Many other variations of these two basic methods have been subsequently developed, but equal channel angular pressing (ECAP) has attracted the most attention.

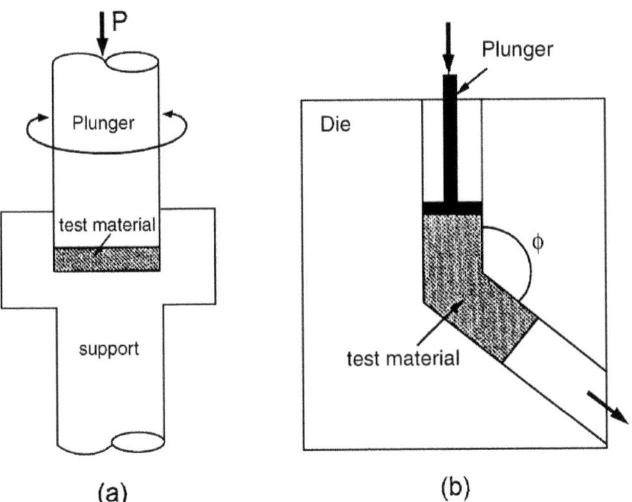

Figure 3.5 Principles of severe plastic deformation by (a) torsion under high pressure and (b) equal channel angular pressing.

The method of SPD has some advantages over other methods in synthesizing ultrafine-grained materials. First, ultrafine-grained materials with high-angle grain boundaries could be synthesized. Second, the samples are dense, and there is no porosity in them since they have not been produced by consolidation of powders. Third, the grain size is fairly uniform throughout the structure. Last, the technique may be directly applied to commercial cast metals. However, the limitations of this technique are that the minimal grain size is not very small and that amorphous phases or other metastable phases have not been synthesized. Nevertheless, these methods are very useful for producing bulk ultrafine-grained materials.

The principle of ECAP was initially developed by Segal and coworkers [41] in the former Soviet Union. The process is based on the observation that metal working by simple shear provides close to an ideal method for the formation of microstructure and texture [42]. Figure 3.6a schematically illustrates the principle of ECAP. A solid die contains a channel that is bent into an L-shaped configuration. A sample that just fits into the channel is pressed through the die using a plunger. Thus the sample undergoes shear deformation but comes out without a change in the cross-sectional dimensions, suggesting that repetitive pressings can be conducted to obtain very high total strains. The strain introduced into the sample is primarily dependent on the angle subtended by the two parts of the channel.

Four different processing routes have been commonly used during ECAP. These are illustrated in Figure 3.6b. In route A, the sample is pressed repetitively without any rotation. In route B, the sample is rotated by 90° between each pass. Depending on the direction in which the rotations are made, one can visualize two alternatives for this. If the rotation is made in alternate directions through 0°-90°-0°-90°, this processing route is referred to as B_A. On the other hand, if the rotation is made in the same direction through 0°-90°-180°-270°, then this is known as B_C process. In route C, the sample is rotated by 180° between each pass.

The reader is advised to refer to a recent review [43] for the details about processing, properties, and applications of materials obtained by SPD. The keynote reviews presented at the Second International Conference on "Nanomaterials by Severe Plastic Deformation: Fundamentals—Processing—Applications" have been published in the May 2003 issue of *Advanced Engineering Materials* [44]. It is estimated that about 200 research papers have been published annually in the field of SPD during the past couple of years.

3.12.4 Accumulative Roll Bonding

The methods of SPD described above have two drawbacks. First, production machines with large load capacities and expensive dies is required. Second, the productivity is relatively low. To overcome these two limitations, Saito et al. [45,46] developed the accumulative roll bonding process. In this process, shown schematically in Figure 3.7, stacking of materials and conventional roll bonding are repeated. First, a strip is neatly placed on the top of another strip. The surfaces of the strips are treated, if necessary, to enhance the bond strength. The two layers of the material are joined together by rolling as in a conventional roll bonding process. Then the length of the rolled material is sectioned into two halves. The sectioned strips are again surface treated, stacked to the initial dimensions, and roll bonded. The whole process is repeated several times.

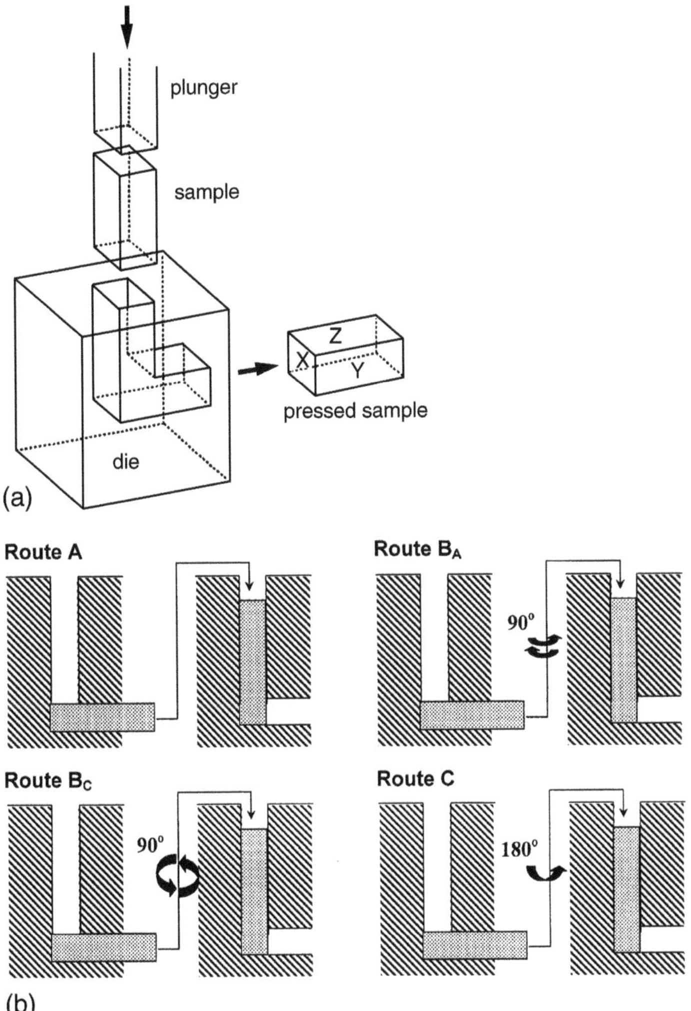

Figure 3.6 (a) Principle of equal channel angular pressing (ECAP) method. (b) The four different processing routes for ECAP.

The accumulative roll bonding process is conducted at elevated temperatures, but below the recrystallization temperature, since recrystallization cancels out the accumulated strain. Very low temperatures would result in insufficient ductility and bond strength. It has been shown that if the homologous temperature of the roll bonding is less than 0.5, sound joining can be achieved for reductions greater than 50%. Very-high-strength materials were obtained by this process in 1100 aluminum, 5083 Al-Mg alloys, and interstitial-free steels. The accumulative roll-bonded materials also showed large elongations of up to 220% and large strain–rate sensitivity (m) values of more than 0.3 even at 200°C for the 5083 Al-Mg alloy, suggesting that low-temperature superplasticity could be achieved due to the formation of ultrafine grains. Tsuji et al. [47] recently presented a brief review of the developments in this process.

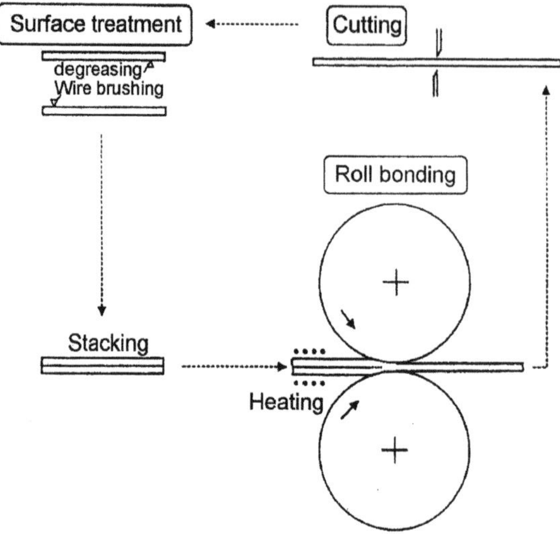

Figure 3.7 Schematic representation of the accumulative roll bonding (ARB) process.

3.13 ACRONYMS USED IN THE LITERATURE

The technique of MA is growing rapidly (see Figure 2.2 for the rate of growth of publications in this area). Consequently, new approaches to obtaining the desired microstructure, properties, and applications for the mechanically alloyed materials are in development. Several new terms have been coined to explain these processes, and acronyms have become common. The following list explains the expansions of some of the most commonly used acronyms:

ARB	accumulative roll bonding
BM	ball milling
BPR	ball-to-powder weight ratio
CR	charge ratio
dMA	double mechanical alloying
ECAE	equal channel angular extrusion
ECAP	equal channel angular pressing
MA	mechanical alloying
MASHS	mechanically activated self-propagating high-temperature synthesis
MCP	mechanochemical processing
MD	mechanical disordering
MG	mechanical grinding
MIGFR	mechanically induced glass formation reaction
MISSD	mechanically induced solid-state devitrification
MM	mechanical milling
M2A	mechanically activated annealing
ODS	oxide dispersion strengthened (materials)
OMR	oxidation-attrition milling-reduction
PCA	process control agent
RM	reaction milling

SPD	severe plastic deformation
SSAR	solid-state amorphization reaction
TEGFR	thermally enhanced glass formation reaction

REFERENCES

1. Benjamin, J. S. (1976). *Sci. Am.* 234(5):40–48.
2. Benjamin, J. S. (1970). *Metall. Trans.* 1:2943–2951.
3. Sherif El-Eskandarany, M., Aoki, K., Suzuki, K. (1991). *J. Alloys Compounds* 177:229–244.
4. Weeber, A. W., Bakker, H., deBoer, F. R. (1986). *Europhys. Lett.* 2:445–448.
5. Klassen, T., Oehring, M., Bormann, R. (1997). *Acta Mater.* 45:3935–3948.
6. Klassen, T., Oehring, M., Bormann, R. (1994). *J. Mater. Res.* 9:47–52.
7. Ermakov, A. E., Yurchikov, E. E., Barinov, V. A. (1981). *Phys. Met. Metallogr.* 52(6):50–58.
8. Koch, C. C. (1991). In: Cahn, R. W., ed. *Processing of Metals and Alloys, Materials Science and Technology—A Comprehensive Treatment.* Vol. 15. Weinheim, Germany: VCH, pp. 193–245.
9. Jangg, G., Kuttner, F., Korb, G. (1975). *Aluminium* 51:641–645.
10. Jangg, G. (1989). In: Arzt, E., Schultz, L., eds. *New Materials by Mechanical Alloying Techniques.* Oberursel, Germany: DGM Informationgesellschaft, pp. 39–52.
11. Avvakumov, E. G. (1986). *Mechanical Methods of Activation of Chemical Processes.* Novosibirsk, Russia: Nauka.
12. Calka, A., Nikolov, J. J., Williams, J. S. (1996). *Mater. Sci. For.* 225–227:527–532.
13. Calka, A. (1991). *Appl. Phys. Lett.* 59:1568–1569.
14. Suryanarayana, C. (1995). *Intermetallics* 3:153–160.
15. Radlinski, A. P., Calka, A. (1991). *Mater. Sci. Eng.* A134:1376–1379.
16. Calka, A., Radlinski, A. P. (1990). *J. Less Common Metals* 161:L23–L26.
17. Luton, M. J., Jayanth, C. S., Disko, M. M., Matras, M. M., Vallone, J. (1989). In: McCandlsih, L. E., Polk, D. E., Siegel, R. W., Kear, B. H., eds. *Multicomponent Ultrafine Microstructures.* Vol. 132. Pittsburgh, PA: Mater Res Soc, pp. 79–86.
18. Aikin, B. J. M., Juhas, J. J. (1997). In: Froes, F. H., Hebeisen, J. C., eds., *Advanced Particulate Materials and Processes–1997.* Princeton, NJ: Metal Powder Industries Federation, pp. 287–294.
19. El-Eskandarany, M. S., Aoki, K., Suzuki, K. (1990). *J. Less Common Metals* 167:113–118.
20. Gaffet, E., Malhouroux, N., Abdellaoui, M. (1993). *J. Alloys Compounds* 194:339–360.
21. Gaffet, E., Malhouroux-Gaffet, N. (1994). *J. Alloys Compounds* 205:27–34.
22. Froyen, L., Delaey, L., Niu, X. -P., LeBrun, P., Peytour, C. (1995). *JOM* 47(3):16–19.
23. LeBrun, P., Froyen, L., Delaey, L. (1992). *Mater. Sci. Eng.* A157:79–88.
24. Moore, J. J., Feng, H. J. (1995). *Prog. Mater. Sci.* 39:243–316.
25. Charlot, F., Gaffet, E., Zeghmati, B., Bernanrd, F., Niepce, J. C. (1999). *Mater. Sci. Eng.* A262:279–288.
26. Gauthier, V., Josse, C., Bernard, F., Gaffet, E., Larpin, J. P. (1999). *Mater. Sci. Eng.* A265:117–128.
27. Gutmanas, E. Y., Trudler, A., Gotman, I. (2002). *Mater. Sci. For.* 386–388:329–334.
28. McCormick, P. G. (1995). *Mater. Trans. Jpn. Inst. Metals* 36:161–169.
29. Gutman, E. M. (1998). *Mechanochemistry of Materials.* Cambridge, UK: Cambridge International Science Publishing.
30. Takacs, L. (2002). *Prog. Mater. Sci.* 47:355–414.
31. Carey-Lea, M. (1892). *Phil. Mag.* 34:465.

32. Avvakumov, E. G., Senna, M., Kosova, N. V. (2001). *Soft Mechanochemical Synthesis: A Basis for New Chemical Technologies*. Boston, MA: Kluwer.
33. Senna, M. (1993). *Solid State Ionics* 63–65, 3.
34. Avvakumov, E. G., Karakchiev, L. G., Gusev, A. A., Vinokurova, O. B. (2002). *Mater. Sci. For.* 386–388:245–250.
35. Shingu, P. H., Ishihara, K. N., Kuyama, J. (1991). *Proc 34th Japan Congress on Materials Research. Soc Mater Sci, Japan, Kyoto,* pp. 19–28.
36. Shingu, P. H., Ishihara, K. N., Uenishi, K., Kuyama, J., Huang, B., Nasu, S. (1990). In: Clauer, A. H., deBarbadillo, J. J., eds. *Solid State Powder Processing*. Warrendale, PA: TMS, pp. 21–34.
37. Shingu, P. H., Ishihara, K. N., Otsuki, A. (1995). *Mater. Sci. For.* 179–181:5–10.
38. Shingu, P. H., Ishihara, K. N., Kondo, A. (1993). In: deBarbadillo, J. J., et al., eds. *Mechanical Alloying for Structural Applications*. Materials Park, OH: ASM International, pp. 41–44.
39. Schwarz, R. B., Johnson, W. L. (1983). *Phys. Rev. Lett.* 51:415–418.
40. Yavari, A. R., Desré, P. eds. (1990) Multilayer Amorphisation by Solid-State Reaction and Mechanical Alloying. Les Ulis Cedex, France: Les editions de Physique. (Colloque de Physique, 51, Colloque C4, Supplement # 14, 1990).
41. Segal, V. M., Reznikov, V. I., Drobyshevskiy, A. E., Kopylov, V. I. (1981). *Russian Metallurgy (Metally)* 1:99–105.
42. Segal, V. M. (1995). *Mater. Sci. Eng.* A197:157–164.
43. Valiev, R. Z., Islamgaliev, R. K., Alexandrov, I. V. (2000). *Prog. Mater. Sci.* 45:103–189.
44. Zehetbauer, M. J., ed. (2003). *Adv. Eng. Mater.* 5:277–378.
45. Saito, Y., Tsuji, N., Utsunomiya, H., Sakai, T., Hong, R. G. (1998). *Scripta Mater.* 47: 1221–1227, 1998.
46. Saito, Y., Utsunomiya, H., Tsuji, N., Sakai, T. (1999). *Acta Mater.* 47:579–583.
47. Tsuji, N., Saito, Y., Lee, S. -H., Minamino, Y. (2003). *Adv. Eng. Mater.* 5:338–344.

4

Equipment for Mechanical Alloying

4.1 INTRODUCTION

The process of mechanical alloying (MA) starts with mixing of the powders in the desired proportion and loading of the powder mix into the mill along with the grinding medium (generally steel balls). Sometimes a process control agent (PCA) is added to prevent or minimize excessive cold welding of powder particles among themselves and/or to the milling container and the grinding medium. This mix (with or without the PCA) is then milled for the required length of time until a steady state is reached. At this stage alloying occurs and the composition of every powder particle is the same as the proportion of the elements in the starting powder mix. However, alloying is not required to occur during mechanical milling (MM) (since one starts with prealloyed powders or materials with uniform composition throughout), but only particle/grain refinement and/or some phase transformations should take place. The milled powder is then consolidated into a bulk shape and subsequently heat treated to obtain the desired microstructure and properties. Figure 4.1 shows a schematic of the different steps involved in preparing a component starting from the constituent powders by the process of MA. Thus, the important components of the MA process are the raw materials, the mill, and the process variables. We next discuss the different parameters involved in the proper selection of raw materials and describe the different types of mills available for conducting MA/MM investigations. A brief description of the selection process of the grinding media is also presented. The role of process variables during MA in achieving the desired constitution in the final product is discussed separately in Chapter 5.

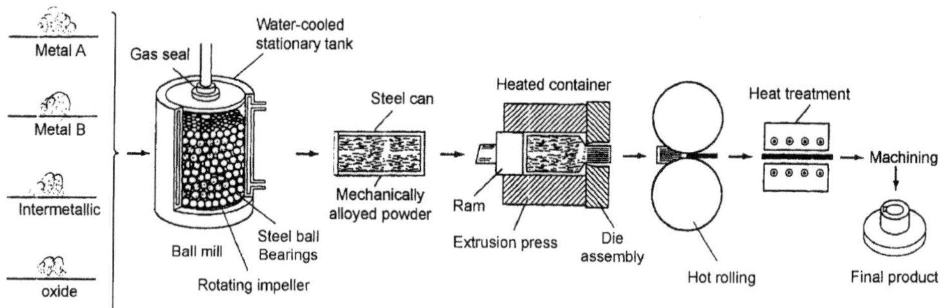

Figure 4.1 Schematic of the different steps involved in producing a product from powders by the mechanical alloying route.

4.2 RAW MATERIALS

The raw materials used for MA are the widely available commercially pure powders that have particle sizes in the range of 1–200 μm. However, the powder particle size is not very critical, except that it should be smaller than the grinding ball size. This is because the powder particle size decreases exponentially with milling time and reaches a small value of a few micrometers only after a short period (typically a few minutes) of milling.

The raw powders fall into the broad categories of pure metals, master alloys, prealloyed powders, and refractory compounds. The oxygen content of the commercially pure metal powders ranges from 0.05 to 2 wt% Therefore, if one is interested in studying phase transformations in the milled powders, it is necessary to choose reasonably high-purity powders for the investigations. This is important because most commonly the nature and amount of impurities in the system decides the type of the final phase formed, chemical constitution of the alloy, and the kinetics of transformations. Raw material powders of different purity and sizes can be obtained from standard suppliers like Goodfellow Metals, Alfa Aesar, and so forth.

Dispersion-strengthened materials usually contain additions of oxides, carbides, and nitrides. Oxides are the most common and these alloys are known as oxide dispersion strengthened (ODS) materials. In the early days of MA (the 1970s), the powder charge for MA consisted of at least 15 vol% of a ductile, compressibly deformable metal powder to act as a host or a binder. However, in recent years, mixtures of fully brittle materials have been milled successfully resulting in alloy formation [1]. Thus, the earlier requirement of having a ductile metal powder during milling is no longer necessary. Consequently, ductile-ductile, ductile-brittle, and brittle-brittle powder mixtures have been milled to produce novel alloys. Recently, mixtures of solid powder particles and liquids have also been milled [2,3]. In these cases, the liquid phase participates in alloying with the powder particles. For example, copper (solid) and mercury (liquid) have been milled together at room temperature to produce Cu-Hg solid solutions [2]. This process should be distinguished from cryomilling [4] (see Sec. 3.5 in Chapter 3), where a cryogenic medium like liquid nitrogen is used to reduce the temperature of milling.

Occasionally, metal powder mixtures are milled with a liquid medium (here the liquid only facilitates milling but does not take part in alloying with the powder), and this is referred to as wet grinding [5–7]; if no liquid is involved the process is called dry grinding. Cryomilling also is wet grinding except that the liquid used is maintained at cryogenic temperatures and usually interacts with the powder being milled. It has been reported that wet grinding is a more suitable method than dry grinding to obtain more finely ground products because the solvent molecules are adsorbed on the newly formed surfaces of the particles and lower their surface energy. The less agglomerated condition of the powder particles in the wet condition is also a useful factor. It has been reported that the rate of amorphization is faster during wet grinding than during dry grinding [8]. However, a disadvantage of wet grinding is increased contamination of the milled powder. Thus, most of the MA/MM operations have been carried out dry. In addition, dry grinding is more efficient than wet grinding in some cases, e.g., during the decomposition of $Cu(OH)_2$ to Cu under mechanical activation [9].

We next discuss the different types of mills available for powder processing.

4.3 TYPES OF MILLS

Different types of high-energy milling equipment are used to produce mechanically alloyed/milled powders. They differ in their design, capacity, efficiency of milling, and additional arrangements for cooling, heating, and so forth. A brief description of the different mills available for MA may be found in Refs. 10 and 11. While several high-energy mills are commercially available and can be readily purchased for standard milling operations, researchers have also designed special mills for specific purposes. The following sections describe some of the more common mills currently in use for MA/MM, which are also readily available in the market.

4.3.1 Spex Shaker Mills

Shaker mills, such as SPEX mills (Figure 4.2), which mill about 10–20 g of the powder at a time, are most commonly used for laboratory investigations and for alloy screening purposes. These mills are manufactured by SPEX CertPrep (203 Norcross Avenue, Metuchen, NJ 08840, Tel: 1-800-522-7739 or 732-549-7144; www.spexcsp.com). The common version of the mill has one vial, containing the powder sample and grinding balls, secured in the clamp and swung energetically back and forth several thousand times a minute. The back-and-forth shaking motion is combined with lateral movements of the ends of the vial, so that the vial appears to be describing a figure of 8 or infinity symbol as it moves. With each swing of the vial the balls impact against the sample and the end of the vial, both milling and mixing the sample. Because of the amplitude (about 5 cm) and speed (about 1200 rpm) of the clamp motion, the ball velocities are high (on the order of 5 m/s) and consequently the force of the ball's impact is unusually great. Therefore, these mills can be considered as high-energy variety.

The most recent design of the SPEX mills has provision for simultaneously milling the powder in two vials to increase the throughput. This machine incorporates forced cooling to permit extended milling times. A variety of vial materials is available for the SPEX mills; these include hardened steel, alumina, tungsten carbide, zirconia, stainless steel, silicon nitride, agate, plastic, and methacrylate. A typical example of a tungsten carbide vial, gasket, and grinding balls for the SPEX mill is shown in Figure 4.3. SPEX mills are the most commonly used, and most of the research on the fundamental aspects of MA has been carried out using some version of these mills.

Figure 4.2 SPEX 8000 mixer/mill in the assembled condition. (Courtesy of SPEX CertiPrep, Metuchen, NJ.)

SPEX mills have certain disadvantages. First, the balls may roll around the end of the vial rather than hitting it; this decreases the intensity of milling. Second, the powder may collect in the eyes of the “8” and remain unprocessed [12]. Last, the round-ended vial is rather heavy; the flat-ended vial is 30% lighter, and impacts at the ends of the vial dominate the milling action. However, some powder may collect at the edges and remain unprocessed.

Figure 4.3 Tungsten carbide vial set consisting of the vial, lid, gasket, and balls. (Courtesy of SPEX CertiPrep, Metuchen, NJ.)

4.3.2 Planetary Ball Mills

Another popular mill for conducting MA experiments is the planetary ball mill (referred to as Pulverisette) in which a few hundred grams of the powder can be milled at the same time (Fig. 4.4). These are manufactured by Fritsch GmbH (Industriestraβe 8, D-55743 Idar-Oberstein, Germany; +49-6784-70 146;www.FRITSCH.de) and marketed by Gilson Co. in the United States and Canada (P.O. Box 200, Lewis Center, OH 43085-0677, USA, Tel: 1-800-444-1508 or 740-548-7298;www.globalgilson.com). The planetary ball mill owes its name to the planet-like movement of its vials. These are arranged on a rotating support disk, and a special drive mechanism causes them to rotate around their own axes. The centrifugal force produced by the vials rotating around their own axes and that produced by the rotating support disk both act on the vial contents, consisting of the material to be ground and the grinding balls. Since the vials and the supporting disk rotate in opposite directions, the centrifugal forces alternately act in like and opposite directions. This causes the grinding balls to run down the inside wall of the vial—the friction effect, followed by the material being ground and the grinding balls lifting off and traveling freely through the inner chamber of the vial and colliding with the opposing inside wall—the impact effect (Fig.4.5). The grinding balls impacting with each other intensify the impact effect considerably.

The grinding balls in the planetary mills acquire much higher impact energy than is possible with simple pure gravity or centrifugal mills. The impact energy acquired depends on the speed of the planetary mill and can reach about 20 times the earth's acceleration. As the speed is reduced, the grinding balls lose the impact energy, and when the energy is sufficiently low there is no grinding involved; only mixing occurs in the sample.

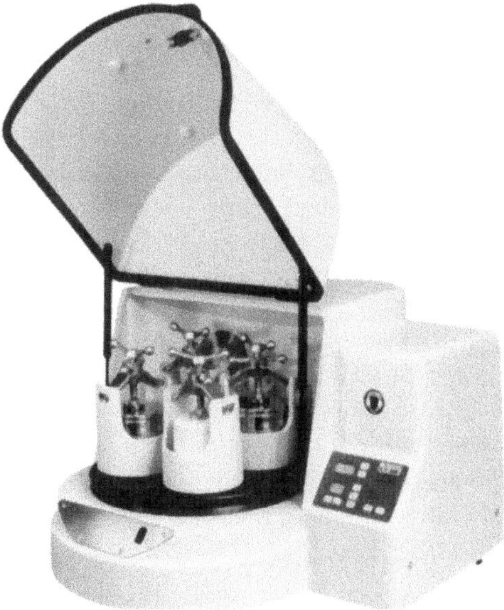

Figure 4.4 Fritsch Pulverisette P-5 four-station planetary ball mill. (Courtesy of Gilson Company, Worthington, OH.)

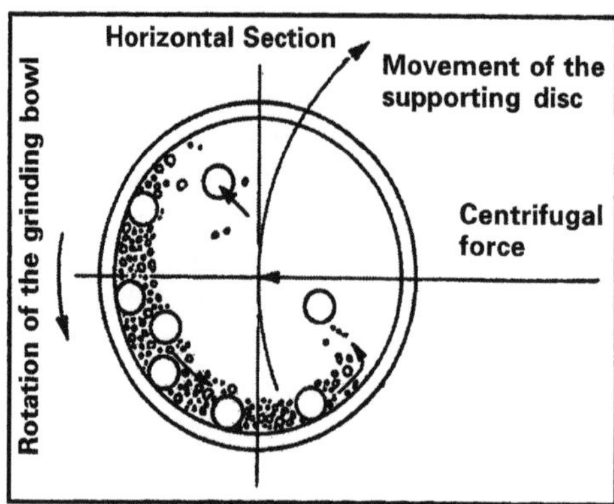

Figure 4.5 Schematic depicting the ball motion inside the planetary ball mill. (Courtesy of Gilson Company, Worthington, OH.)

Even though the disk and the vial rotation speeds could not be independently controlled in the early versions, it is possible to do so in the modern versions of the Fritsch planetary ball mills. In a single mill one can have either two (Pulverisette 5 or 7) or four (Pulverisette 5) milling stations. Recently, a single-station mill was also developed (Pulverisette 6). Three different sizes of containers, with capacities of 80, 250, and 500 ml, are available. Grinding vials and balls are available in eight different materials—agate, silicon nitride, sintered corundum, zirconia, chrome steel, Cr-Ni steel, tungsten carbide, and plastic polyamide. Even though the linear velocity of the balls in this type of mill is higher than that in the SPEX mills, the frequency of impacts is much less than in the SPEX mills. Hence, in comparison to SPEX mills, Fritsch Pulverisette can be considered as lower energy mills.

Some high-energy planetary ball mills have been developed by Russian scientists, and these have been designated as AGO mills, such as AGO-2U [13,14] and AGO-2M [15]. The high energy of these mills is derived from the very high rotation speeds that are achievable. For example, Salimon et al. [14] used their planetary ball mill at a rotation speed of 1235 rpm corresponding to the mill energy intensity of 50 W/g. It has been reported that some of these mills can be used at rotation speeds greater than 2000 rpm [13].

A recent development in the design of the Fritsch mills has been the incorporation of a gas pressure and temperature measuring system (GTM) for in situ data acquisition during milling [16]. Generally, the occurrence of phase changes in the milled powder is interpreted or inferred by analyzing the powder constitution *after* milling has been stopped. Sometimes a small quantity of the powder is removed from the charge in the mill and analyzed to obtain information on the progress of alloying and/or phase transformations. This method could lead to some errors because the state of the powder during milling could be different from what it is after the milling has been stopped. To overcome this difficulty, Fritsch GmbH developed the GTM system to enable the operator to obtain data *during* milling.

The basic idea of this measuring system is the quick and continuous determination of temperature and pressure during the milling process. The temperature measured corresponds to the total temperature rise in the system due to the combination of grinding, impact, and phase transformation processes. Since the heat capacity of the container and the grinding medium is much higher than the mass of the powder, it is necessary to have a sensitive temperature measurement in order to derive meaningful information. Accordingly, a continuous and sensitive measurement of gas pressure inside the milling container is carried out to measure very quickly and detect small temperature changes. The measured gas pressure includes not only information about the temperature increase due to friction, impact forces, and phase transformations, but also the interaction of gases with the fresh surfaces formed during the milling operation (adsorption and desorption of gases). The continual and highly sensitive measurement of the gas pressure within the milling container facilitates detection of abrupt and minute changes in the reactions occurring inside the vial. The pressure could be measured in the range of 0–700 kPa, with a resolution of 0.175 kPa, which translates to a temperature resolution of 0.025 K.

4.3.3 Attritor Mills

A conventional ball mill consists of a rotating horizontal drum half-filled with small steel balls. As the drum rotates the balls drop on the metal powder that is being ground; the rate of grinding increases with the speed of rotation. At high speeds, however, the centrifugal force acting on the steel balls exceeds the force of gravity, and the balls are pinned to the wall of the drum. At this point the grinding action stops. An attritor (a ball mill capable of generating higher energies) consists of a vertical drum containing a series of impellers. A powerful motor rotates the impellers, which in turn agitate the steel balls in the drum. Set progressively at right angles to each other, the impellers energize the ball charge. The dry particles are subjected to various forces such as impact, rotation, tumbling, and shear. This causes powder size reduction because of collisions between balls, between balls and container wall, and between balls, agitator shaft, and impellers. Therefore, micrometer-range fine powders can be easily produced. Some size reduction, though small in proportion, appears to take place by interparticle collisions and by ball sliding. In addition, combination of these forces creates a more spherical particle than other impact-type milling equipment.

Attritors, also known as stirred ball mills, are the mills in which large quantities of powder (from a few pounds to about 100 lb) can be milled at a time (Fig. 4.6). Commercial attritors are available from Union Process (1925 Akron-Peninsula Road, Akron, OH 44313, Tel: 330-929-3333;www.unionprocess.com; and Zoz GmbH, D-57482 Wenden, Germany, Tel: +49-2762-97560; office in Newark, DE, Tel: 302-369-6761; www.zoz.de). The velocity of the grinding medium in the attritors is much lower (about 0.5 m/s) than in the planetary or SPEX mills, and consequently the energy of milling in the attritors is low. Attritors of different sizes and capacities are available. The grinding tanks or containers are available either in stainless steel or stainless steel coated inside with alumina, silicon carbide, silicon nitride, zirconia, rubber, and polyurethane. A variety of grinding media also is available—glass, flint stones, steatite ceramic, mullite, silicon carbide, silicon nitride, sialon, alumina, zirconium silicate, zirconia, stainless steel, carbon steel, chrome steel, and tungsten carbide.

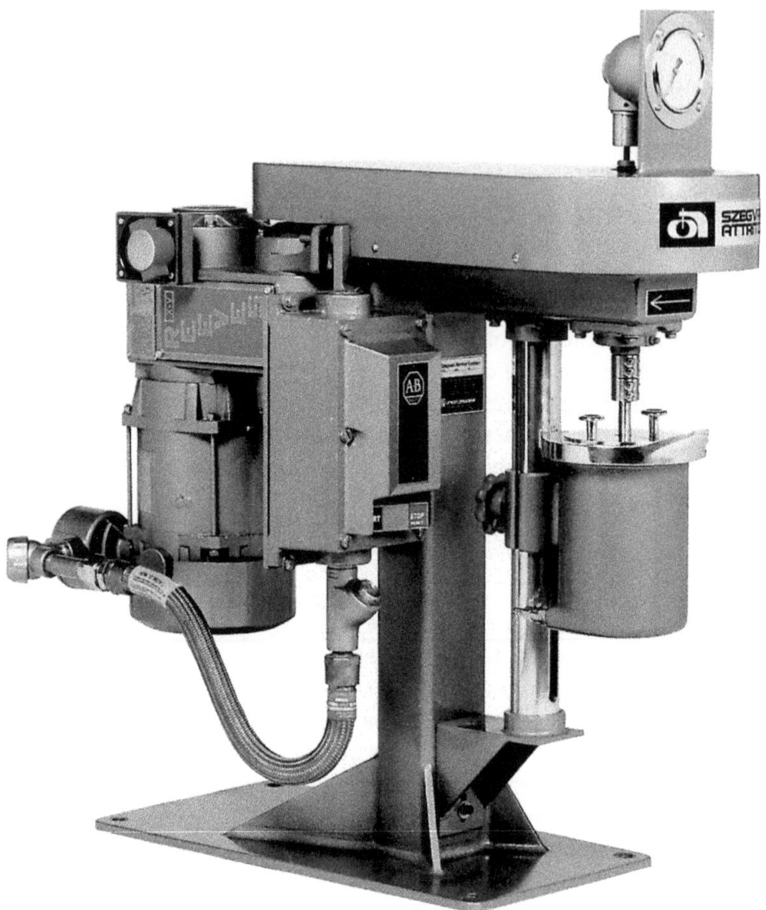

Figure 4.6 Model 01-HD attritor. (Courtesy of Union Process, Akron, OH.)

The operation of an attritor is simple. All the elemental powders and oxides to be milled are placed in a stationary tank with the grinding media. No premixing is needed. The maximal feed material size can be up to 10 mm, provided the material is relatively brittle. Otherwise, only material smaller than 2 mm can be processed. This mixture is then agitated by a shaft with arms, rotating at a high speed of about 250 rpm (Fig. 4.7). This causes the media to exert both shearing and impact forces on the material. The laboratory attritor works up to 10 times faster than conventional ball mills.

While the attritors are generally batch type, they can be made to work in a continuous mode. Since the attritors are open at the top, the material is charged at the top by a preset rate feeder, ground, and removed from the lower side of the grinding tank. By constantly taking out the fine product and reloading the oversize particles, one can achieve very efficient narrow size distribution of particles. This process is best suited for the continuous production of large quantities of material. However, a well-premixed slurry is needed for this type of process and consequently cannot be used for dry grinding purposes.

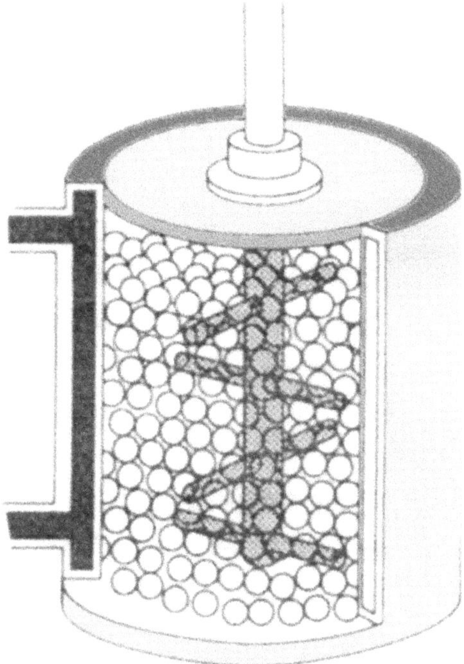

Figure 4.7 Arrangement of rotating arms on a shaft in the attrition ball mill. (Courtesy of Union Process, Akron, OH.)

Conventional ball mills use large media, normally $\frac{1}{2}$in. or larger, and run at a low speed (10–50 rpm). Attritors, on the other hand, use grinding medium of about 1/8 to 3/8 in. size and run typically at moderate speeds of about 60 rpm in the largest production size units, but at about 250 rpm for the laboratory size units. The most important concept in the attritor is that the power input is used directly for agitating the media to achieve grinding and is not used for rotating or vibrating a large, heavy tank in addition to the media. Figure 4.8 shows a comparison of the effectiveness of the various grinding devices in ultrafine grinding of chalcopyrite concentrate. It may be noted that the attritor is much more effective than a conventional ball mill or a vibratory ball mill. For the specific energy input of around 100 kWh/ton, the median particle size achieved through the attritor is nearly half smaller than that obtained from conventional ball mills, and is about one-third smaller than that obtained from vibratory mills. Furthermore, for specific energy input exceeding 200 kWh/ton, attritors continue to grind into the submicrometer range, while other grinding machines can no longer effectively produce any smaller particles.

While the Union Process attritors are vertical in configuration, the Zoz attritors are horizontal (Fig. 4.9). The main features of the Zoz attritor are as follows:

1. It supplies three times higher relative velocity of the grinding media than conventional devices.
2. It can be operated and also charged and discharged under vacuum or inert atmosphere.

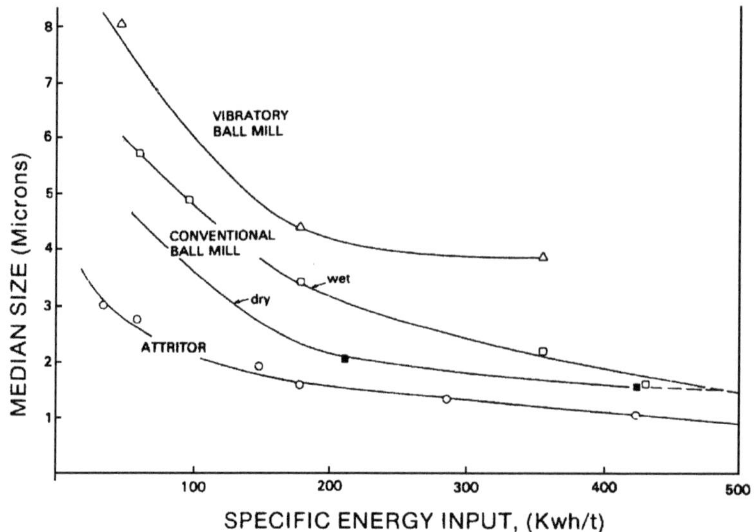

Figure 4.8 Comparison of the effectiveness of various grinding devices for ultrafine grinding of chalcopyrite concentrate.

3. It allows processing without dead zones and negative gravity influence.
4. It can be operated at high temperatures due to separated cooling/heating systems.
5. It is usually software operated, and allows parameter measurement, control, and history.
6. It is scalable from 0.5 to 900 liters processing chamber capacity.
7. It can be operated in batch, semicontinuous, and auto-batch mode.
8. It is available with tools of stainless steel/satellite, WC-Co, and Si_3N_4.

Figure 4.10 shows the principle of powder flow during processing in a Zoz horizontal attritor.

The main problem of the horizontal attritors is that very fine powder prepared in the attritor is not forced by gravity to places away from the rotary seals. This technical problem was solved by means of an adjustable preseal unit, an extension chamber, and finally a redesign of the rotary seal itself. Thus, advantages of the horizontal attritor are as follows:

No dead zones due to gravity
Possibility of extremely high-energy impact
Charging and discharging under controlled atmosphere

4.3.4 Commercial Mills

Commercial mills for MA are much larger in size than the mills described above and can process several hundred pounds of powder at a time. MA for commercial production is carried out in ball mills of up to about 3000 lb (1250 kg) capacity (Fig. 4.11).

(a)

(b)

Figure 4.9 (a) Horizontal rotary ball mill (Simoloyer CM100-S2), production unit for nanocrystalline metal hydrides. (b) Horizontal rotary ball mill (Simoloyer CM01-2l) with air lock for loading, operation, and unloading under vacuum and/or inert gas. (Courtesy of Zoz GmbH, Wenden, Germany.)

The milling time decreases with an increase in the energy of the mill. It has been reported that 20 min of milling in a SPEX mill is equivalent to 20 h of milling in a low-energy mill of the Invicta BX 920/2 type [17]. As a rule of thumb, it can be estimated that a process that takes only a few minutes in the SPEX mill may take hours in an attritor and a few days in a commercial mill, even though the details can differ depending on the efficiency of the different mills. Figure 4.12 shows the times required

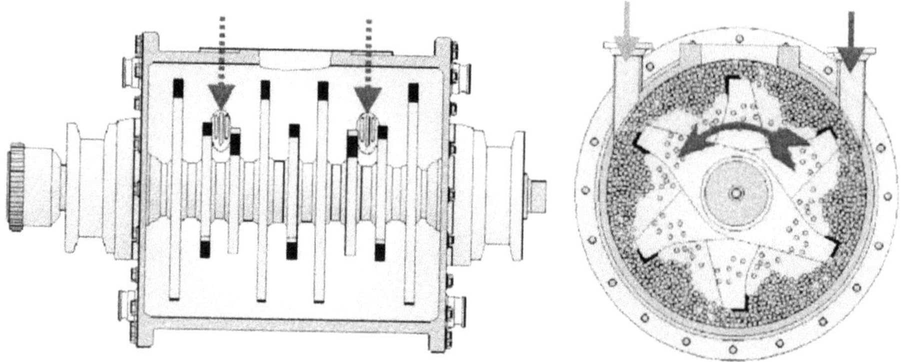

Figure 4.10 Detail of the rotor in operation showing the principle of powder flow during processing in a Zoz horizontal attritor.

Figure 4.11 Commercial production size ball mills used for mechanical alloying. (Courtesy of Inco Alloys International.)

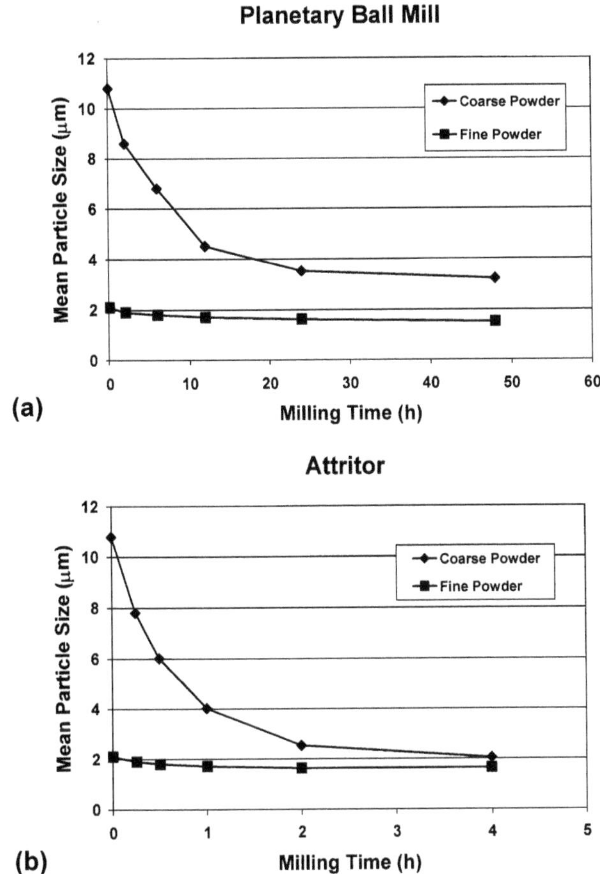

Figure 4.12 Times required to reach similar particle sizes during milling of TiB_2 powder in (a) planetary ball mill and (b) attritor.

to reach a specific particle size during milling in a planetary ball mill and an attritor. It may be noted that the times are an order of magnitude shorter in the attritor [18]. Similarly, by comparing the powder yield it becomes clear that the yield is much higher and that alloying is complete sooner in the Zoz Simoloyer than in the Fritsch P5 planetary ball mill (Fig. 4.13).

4.3.5 New Designs

Several new designs of mills have been developed in recent years for specialized purposes. These include the rod mills, vibrating frame mills, and the equipment available from Dymatron (Cincinnati, OH), Nisshin Giken (Tokyo, Japan), Australian Scientific Instruments (Canberra, Australia), and M.B.N. srl (Rome, Italy).

The rod mills are very similar to the ball mills except that they use steel rods instead of balls as grinding medium [19,20]. It has been claimed that the powder

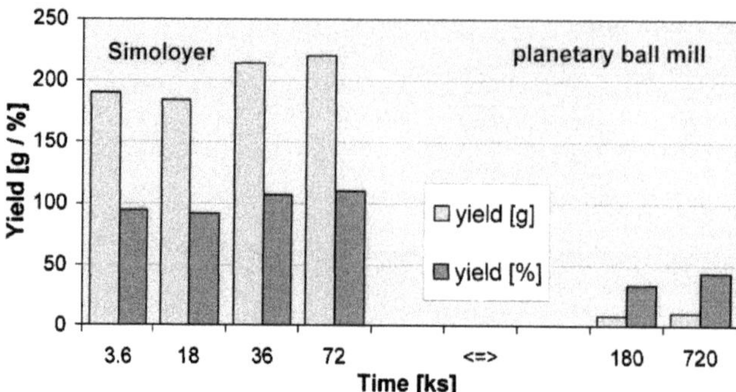

Figure 4.13 Powder yield as a function of milling time for the Zoz Simoloyer and Fritsch P5 planetary ball mill.

contamination during rod milling is much lower than during ball milling. This has been explained to be due to the increased proportion of shear forces over the impact forces during rod milling.

The Megapact and Megamill-5 units supplied by Dymatron are high-energy vibration mills. The Megapact mill (Fig. 4.14a) could be obtained with either two large (1400 cm^3) or two small (280 cm^3) stainless steel chambers. The cradles support up to four short chambers, two on each cradle. The milling chambers permit milling under controlled conditions and/or atmospheres. The Megamill-5 (Fig. 4.14b) is equipped with two steel or stainless steel chambers, each with an internal volume of 5 liters. The chambers are secured to each cradle of the mill by a roller chain and clamp, thereby facilitating quick loading and unloading of the chambers. The leak-proof cover seals provided with the milling chambers permit operation in vacuum, inert, or controlled atmospheres under ambient, heated, or cryogenic conditions. Reactive gas or fluid atmospheres could also be used, but provision must be made to release high internal chamber pressures.

The Super Misuni NEV-MA-8 from Nisshin Giken (Fig. 4.15) has been a popular milling unit with several Japanese researchers. This machine has some special features such as variety of selections for milling energy (up to a speed of 1600 rpm), variety of selections for atmospheres (vacuum or any flowing gas), ability to control the temperature of milling from very low temperatures of −190°C (obtained by spraying liquid nitrogen) up to a high temperature of 300°C (achieved by electrical heating), and in situ observation of the grinding medium during milling.

The Uni-Ball mill from Australian Scientific Instruments (Canberra, Australia) is another type of mill wherein it is possible to control the nature and magnitude of impact of the balls by controlling the field strength with the help of adjustable magnets. The incorporation of permanent magnets generates a magnetic field, which acts on the ferromagnetic balls used for milling. By adjusting the distance between the balls and the magnet, the impact energy can be varied. Thus, one can choose the desired impact energy depending on the requirement and the materials to be milled. By proper adjustment of the magnetic field, the effective mass of the grinding ball can be increased by a factor of 80. Furthermore, changing the positions of the magnets can alter the type of energy transferred to the powder and one can obtain impact or shear modes or a combination of both.

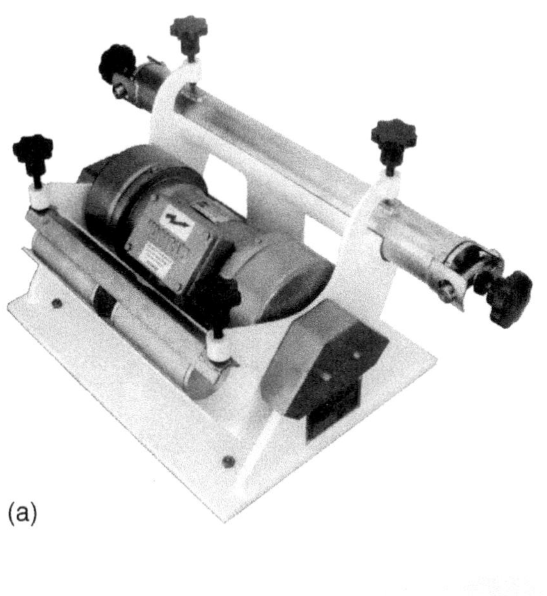

(a)

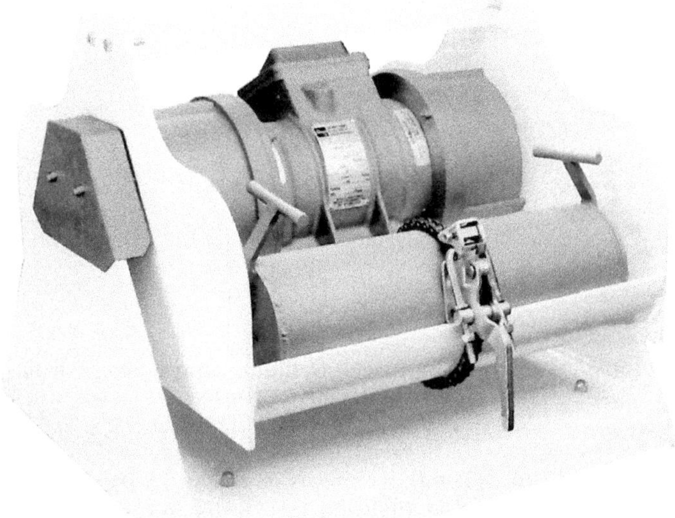

(b)

Figure 4.14 Mechanical alloying equipment manufactured by Dymatron. (a) Megapact mill and (b) Megamill-5. (Courtesy of Dymatron, Cincinnati, OH.)

The Uni-Ball mill has the following special features:

Multiple mill chambers allowing large amount of powder to be processed at a time.

Precision control of milling energy by properly adjusting the position, quantity, and proximity of the magnets around each mill chamber. This allows the user to select the impact energy, shear force and the proportion of these.

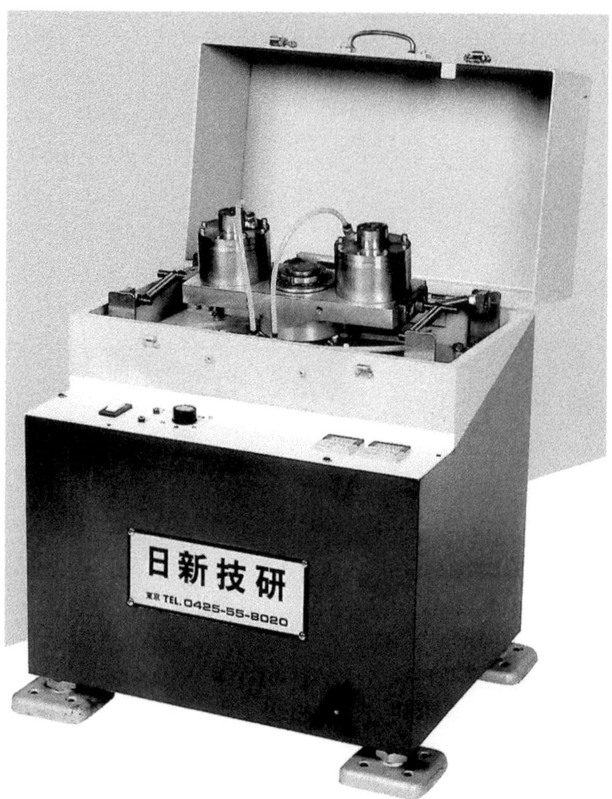

Figure 4.15 Super Misuni NEV mill.

Variable milling speed.
Fully controlled atmosphere (argon, nitrogen, helium, ammonia, hydrogen, etc.).
Fully controlled temperature (from low to high temperatures of about 200 °C).

Figure 4.16a shows the Uni-Ball mill with the provision of a spatially adjustable high-strength magnet (on the left side) and Figure 4.16b shows the mill chamber components. Calka and Wexler [21] recently modified the Uni-Ball mill 5 to achieve milling modes involving impact or shearing of hardened curved surfaces on particles under a controlled electric discharge within a controlled atmosphere. The effects of low-frequency and high-voltage electrical impulses during milling were shown to be very efficient in synthesizing intermetallics, promoting reduction reactions, and forming nitride phases of several different metals not possible by conventional milling under regular impact mode.

The HE (high-energy) mill from M.B.N. srl (Fig. 4.17) has high impact velocities of up to 4 m/s, high impact frequency of the balls with the vial walls of 20 Hz, and large mill capacities of up to 10-kg ball charge. The HE mill can operate with one large vial or with two smaller vials in which independent (materials/atmosphere/ball-to-powder weight ratios) milling can be performed. The vial motion has a fixed amplitude of 30 mm, but its frequency can be continuously varied and the work cycle can be

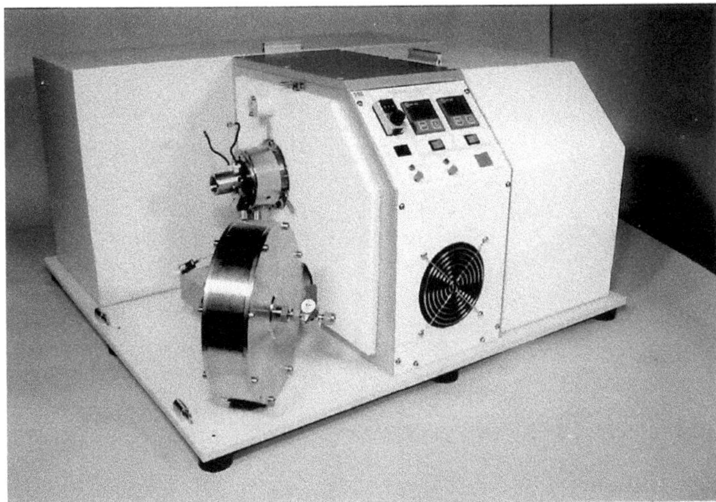

(a)

(b)

Figure 4.16 (a) Uni-Ball mill with the provision of a spatially adjustable high-strength magnet. (b) Chamber components. (Courtesy of Australian Scientific Instruments, Fyshwick, Australia.)

programmed with on/off periods having a resolution of 1 min and total milling time of up to 4 days. The vial could be filled with different types of gases and the internal pressure could be measured with a transducer.

The operating details and special features of the commercially available mills can be found in the respective brochures of the manufacturers.

Some special equipment is also designed for specific laboratory applications. Bakker and colleagues [22] used a mill in which a single hardened steel ball with a diameter of 6 cm is kept in motion by a vibrating frame on which the vial is mounted. A similar, but instrumented, variety of the mill has been used by Pochet et al. [23] (Fig. 4.18). The vial is made of stainless steel with a hardened steel bottom, the central part of which consists of a tungsten carbide disk of 2 cm diameter. This mill has a low efficiency since the ball is moving from one side to another (and the gap between the

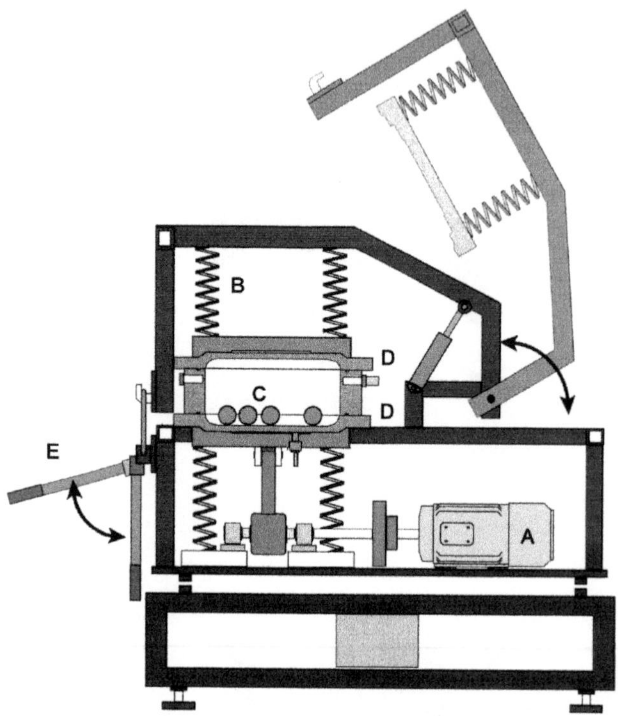

Figure 4.17 Schematic of the HE (high energy) mill manufactured by MBN srl., Rome, Italy. A, Motor; B, springs; C, grinding medium; D, vial lids; and E, lever.

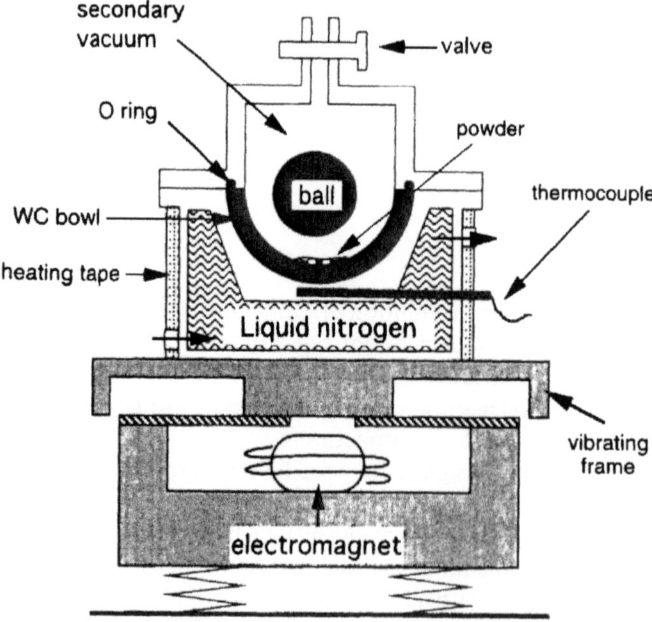

Figure 4.18 Instrumented vertical vibrating frame.

vial walls and the ball is very small), has a long repetition time, and the specimen gets milled only during the collisions. On the other hand, the shaker mills, e.g., SPEX mills, have a much higher efficiency. However, even in the SPEX mills, several non-head-on collisions occur, and these reduce the maximal possible efficiency. Furthermore, because of friction, the surface of the grinding medium gets eroded and this debris contaminates the sample.

To overcome some of the above-mentioned problems, Szymański et al. [24] designed a friction-free mechanical grinder (Fig. 4.19). This grinder consists of seven steel rods (10 mm in diameter and 100 mm in height) arranged in a hexagonal motif with one rod in the center. These rods stand in a vertical position between a pair of parallel horizontal steel anvils separated by a distance of 116 mm. The device moves up and down with an amplitude of 3.5 mm and a frequency of 25 Hz. Powders milled in this device have a low contamination level.

Taking into consideration the fact that a high-energy ball mill should have high impact velocities and high impact frequencies of the grinding media, one can design new mills for specific applications. For example, Basset et al. [25] and Kimura et al. [26]

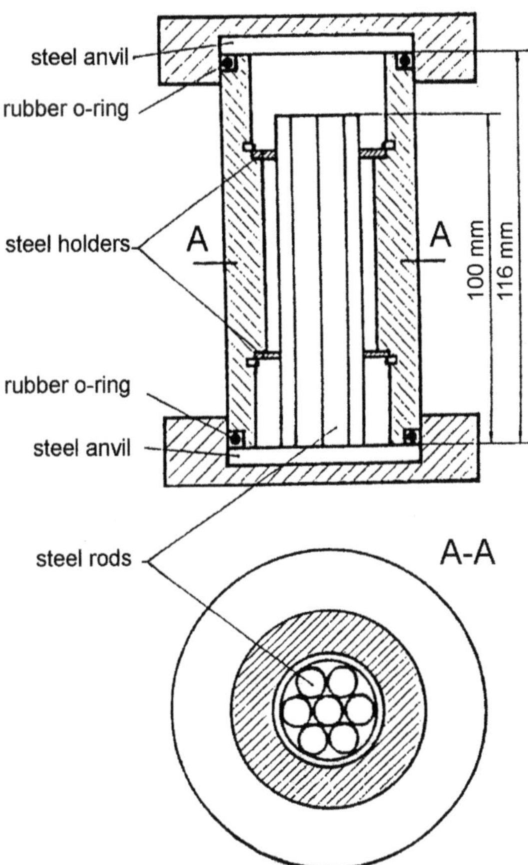

Figure 4.19 Schematic of the friction-free grinding mill.

designed special high-energy ball mills to produce nanocrystalline materials and amorphous alloys, respectively. One could also design mills of higher capacity to produce large quantities of the mechanically alloyed powder at a time or to increase the productivity by increasing the MA speed (or reducing the MA times) [27].

Aizawa et al. [28,29] reported development of a high-speed MA route to synthesize large amounts of powder. The salient features of this process are as follows:

Cyclic and continuous application of mechanical power and energy with the prescribed amount per cycle or period.

Combination of hydrostatic pressing mode with high shear deformation mode into the specified pass schedule per cycle. Two pressing modes and one severe shear deformation mode are included in one cycle.

Ability to change the above loading sequences.

Conducting the operation for a number of cycles results in true alloying. This has been demonstrated for Cu-Ag, Cu-Co, Ni-Al, Mo-Si, and several other systems [28,29]. It has been claimed that the processing times could be significantly reduced using this equipment.

One of the difficulties encountered during milling of powders, mostly in the attritors and sometimes in other mills, is the development of dead zones. These are the areas where the powder gets accumulated without getting milled. This is because the attritor arms do not reach those places (or the grinding medium does not exert any force on the powder); this can be attributed mainly to a decrease in the radial velocity of the grinding medium just near the edges of the attritor arms [30]. Similarly, a dead zone exists at the bottom of the attritor also due to the inaccessibility of the powder at the bottom of the attritor chamber to the rotating arms. It is desirable to avoid the occurrence of dead zones and facilitate powder circulation to shorten the milling times and also to achieve homogeneity in the powder. The following three methods are effective in avoiding dead zones in an attritor [30]:

1. If the grinding balls are of the same size, they form a close-packed array, resulting in the predominance of rolling of one ball over the others. This leads to reduced impact events and so effectively reduces deformation and missing of the powder particles. This can be overcome by using more than one size of the grinding medium. In fact, it has been shown through cinematographic studies that a ball array does not form when a mix of the different grinding ball sizes is used and that milling takes place more efficiently [31].
2. The bottom of the attritor chamber could be contoured in such a way so as to steer the balls and the powder up into the more active mill region.

Table 4.1 Typical Capacities of Different Types of Mills

Type of mill	Media size (in.)	Rotation speed (rpm)	Sample weight
Mixer	1/8 to 1/2	1200	Up to 2 × 20 g
Planetary	3/16 or larger	100–2500	Up to 4 × 250 g
Attritor	1/8 to 3/8	60–300	0.5 to 100 kg
Uni-Ball	1/4 or larger	Variable	Up to 4 × 2000 g

Device	Simoloyer®	Planetary Ball Mill	Attritor®	Drum (ball) mill
Maximum diameter [m]	0.9	0.2	1	3
Maximum total volume [l]	400	8	1000	20000
Maximum relative velocity [m/s]	14	5	4.5 - 5.1	< 5
Graphic (cross section)	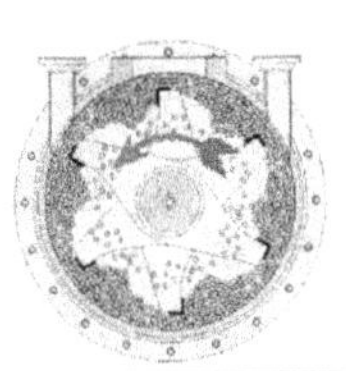	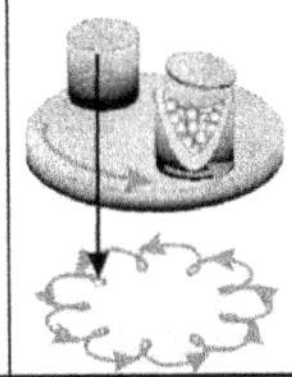	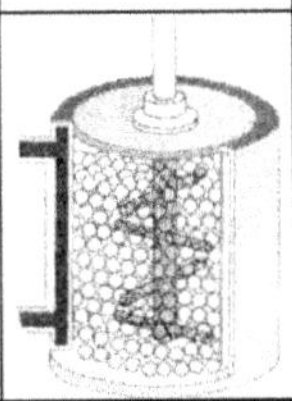	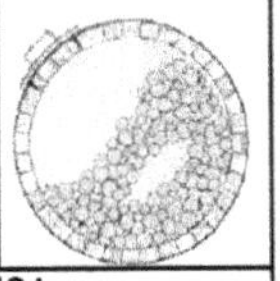
Simoloyer® is a brand of Zoz GmbH, Germany; Attritor® is a brand of Union Process, USA;				

Figure 4.20 Comparison of the different devices used for ball milling of powders.

3. Another alternative is to place an additional arm at the bottom of the attritor chamber. This arm could be machined into a wedge shape as opposed to the round shape of the other arms. This design would force the balls and the powder (like a scoop action) into the more active regions of the attritor.

Table 4.1 summarizes the capacities of the different mills for MA/MM that are commercially available [32]. Figure 4.20 compares the different devices (Simoloyer, planetary ball mill, attritor, and a conventional ball mill) regarding their sizes and capacities.

The main goal of the design of any MA/MM device is to produce homogeneous powder of the desired stoichiometry in an economical way and with the required purity (minimal or no contamination). It would also be useful to increase the productivity without increasing the milling time, and it is most desirable for the powder to be produced continuously.

4.4 SELECTION OF GRINDING MEDIUM

Proper selection of the nature, size, and size distribution of the grinding medium is an important step in achieving efficient milling of powder particles. The selection depends on several factors, some of which are interrelated.

Specific Gravity

In general, high-density media give better results. This is because the kinetic energy of the balls is higher and consequently higher amounts of energy are transferred to the milled powder. The media should be denser than the powder being ground. Also, highly viscous materials require media with a higher density to prevent floating.

Initial Feed Size

Since smaller media cannot break up large particles, the grinding medium should be larger than the powder particle size. It will also be shown in Chapter 16 that it is desirable to use a mixture of different sizes of the balls rather than one size. A mixture of different sizes of the grinding balls helps to improve the efficiency of the attritor.

Final Particle Size

The grinding medium should be smaller when very fine particles are desired; the smaller the grinding medium, the smaller is the final particle size.

Hardness

The harder the media the lesser is the powder contamination, and consequently the medium lasts longer. However, if the medium is brittle, then edges of the medium may be chipped off and get incorporated into the milled powder and contaminate it. However, if the media is magnetic in nature and the powder is not, then the material resulting from the wear of the grinding medium may be separated by magnetic means.

pH Value

Some strong acid or basic slurries may react with certain metallic media.

Discoloration

Certain media result in color development and are, therefore, not suitable in the production of some materials such as white coatings.

Table 4.2 Properties of the Different Grinding Media Available for Mechanical Alloying/Mechanical Milling

Material	Density (g/cm^3)	Abrasion resistance	Material to be ground
Agate (99.9% SiO_2)	2.65	Good	Soft to medium hard samples
PTFE with steel core	3.0	Adequate	Soft, brittle samples and for homogenization
Silicon nitride (90% Si_3N_4)	3.1	Extremely good	Abrasive sample, iron-free grinding
Sintered corundum (99.7% Al_2O_3)	>3.8	Fairly good	Medium-hard as well as fibrous samples
Sintered corundum-2 (85–90% Al_2O_3)	>3.8	Fairly good	Medium-hard as well as fibrous samples
Stainless steel (12.5 to 14.5% Cr + 1% Ni)	7.8	Fairly good	Medium hard as well as brittle samples
Tempered steel (1.0 to 1.65% Cr)	7.9	Good	Hard, brittle samples
Tungsten carbide (93.2% WC + 6% Co)	14.7	Very good	Hard, abrasive samples
Zirconium oxide (94.8% ZrO_2)	5.7	Very good	Fibrous as well as abrasive samples

Contamination

This could be a serious consideration since material resulting from the wear of the media affects the product and may contaminate it. Therefore, it should be removed by a magnetic separator, chemicals, or sintering.

Cost

Media that are two to three times more expensive may last considerably longer. These may be well worth the extra cost over the long run.

Table 4.2 presents the different types of grinding media available for MA/MM purposes and their properties.

REFERENCES

1. Davis, R. M., McDermott, B., Koch, C. C. (1988). *Metall. Trans.* 19A:2867–2874.
2. Ivanov, E. (1992). *Mater. Sci. For.* 88–90:475–480.
3. Yamazaki, T., Terayama, K., Shimazaki, T., Sugimoto, K. (1997). *J. Mater. Sci. Lett.* 16:1357–1359.
4. Luton, M. J., Jayanth, C. S., Disko, M. M., Matras, M. M., Vallone, J. (1989). In: McCandlsih, L. E., Polk, D. E. Siegel, R. W., Kear, B. H., eds. *Multicomponent Ultrafine Microstructures*. Pittsburgh, PA: Mater. Res. Soc. 132:79–86.
5. Okada, K., Kikuchi, S., Ban, T., Otsuka, N. (1992). *J. Mater. Sci. Lett.* 11:862–864.
6. Nicoara, G., Fratiloiu, D., Nogues, M., Dormann, J. L., Vasiliu, F. (1997). *Mater. Sci. For.* 235–238:145–150.
7. Bellosi, A., Montverde, F., Botti, S., Martelli, S. (1997). *Mater. Sci. For.* 235–238:255–260.
8. Dolgin, B. P., Vanek, M. A., McGory, T., Ham, D. J. (1986). *J. Non-Cryst. Solids* 87:281–289.
9. Blaskov, V., Radev, D. D., Klissurski, D., Yordanov, N. D. (1994). *J. Alloys Compounds* 206:267–270.
10. Suryanarayana, C. (1998). In: *Powder Metal Technologies and Applications*. ASM Handbook, Vol. 7. Materials Park, OH: ASM International, pp. 80–90.
11. Suryanarayana, C. (2001). *Prog. Mater. Sci.* 46:1–184.
12. Takacs, L. (1996). In: Suryanarayana, C., et al., eds. *Processing and Properties of Nanocrystalline Materials*. Warrendale, PA: TMS, pp. 453–464.
13. Uchrin, J., Uchrin, R., Avvakumov, E. G. (1995). *Mater. Sci. For.* 179–181:425–430.
14. Salimon, A. I., Korsunsky, A. M., Kaloshkin, S. D., Tcherdyntsev, V. V., Shelekhov, E. V., Sviridova, T. A. (2001). *Mater. Sci. For.* 360–362:137–142.
15. Kwon, Y.-S., Gerasimov, K. B., Lomovsky, O. I., Pavlov, S. V. (2003). *J. Alloys Compounds* 353:194–199.
16. Scholl, R., Wegerle, R., Mutter, W. (2000). *Mater. Sci. For.* 343-346:964–972.
17. Yamada, K., Koch, C. C. (1993). *J. Mater. Res.* 8:1317–1326.
18. Thümmler, F., Oberacker, R. (1993). In: *Introduction to Powder Metallurgy*. London, UK: Institute of Materials, p. 12.
19. Shoji, K., Austin, L. G. (1974). *Powder Technol.* 10:29–35.
20. Sherif El-Eskandarany, M., Aoki, K., Suzuki, K. (1990). *J. Less Common Metals* 167:113–118.
21. Calka, A., Wexler, D. (2002). *Mater. Sci. For.* 386–388:125–134.
22. Bakker, H., Zhou, G. F., Yang, H. (1995). *Prog. Mater. Sci.* 39:159–241.
23. Pochet, P., Tominez, E., Chaffron, L., Martin, G. (1995). *Phys. Rev.* B52:4006–4016.

24. Szymański, K., Zaleski, P., Rećko, K., Waliszewski, J. (1997). *Mater. Sci. For.* 235–238:223–228.
25. Basset, D., Matteazzi, P., Miani, F. (1993). *Mater. Sci. Eng.* A168:149–152.
26. Kimura, H., Kimura, M., Takada, F. (1988). *J. Less Common Metals* 140:113–118.
27. Kobayashi, O., Aizawa, T., Kihara, J. (1996). *Mater. Trans. Jpn. Inst. Metals* 37:1497–1504.
28. Aizawa, T., Tatsuzawa, K., Kihara, J. (1993). *Metal Powder Rep.* 48:30–33.
29. Aizawa, T., Kihara, J., Benson, D. (1995). *Mater. Trans. Jpn. Inst. Metals* 36:138–149.
30. Rydin, R. W., Maurice, D., Courtney, T. H. (1993). *Metall. Trans.* 24A:175–185.
31. Cook, T. M., Courtney, T. H. (1995). *Metall. Mater. Trans.* 26A:2389–2397.
32. Kerr, I. (1993). *Metal Powder Rep.* 48:36–38.

5

Process Variables in Milling

5.1 INTRODUCTION

Mechanical alloying (MA) is a complex process involving optimization of a number of process variables to achieve the desired product phase, microstructure, and/or properties. However, we will not consider the nature and composition of the powder mix as a variable here. These will decide the nature of the phase formed (solid solution, intermetallic, or amorphous phase, etc.) in the milled powder, and discussion about these is presented in subsequent chapters. For a given composition of the powder, some of the important variables that have an important effect on the final constitution of the milled powder are as follows:

Type of mill
Milling container
Milling energy/speed
Milling time
Type, size, and size distribution of grinding medium
Ball-to-powder weight ratio
Extent of vial filling
Milling atmosphere
Process control agent
Temperature of milling

These process variables are not completely independent. For example, the optimal milling time depends on the type of mill, size of the grinding medium, temperature of milling, ball-to-powder ratio (BPR), etc. Further, more energy could be imparted to the milled powder by increasing the BPR or by milling the powder for a longer time. Similarly, it could be shown that some of the other parameters are also interdependent. Even so, we will discuss in the following paragraphs the effect of these variables on the final product obtained after MA assuming mostly that the other variables have no significant effect on the specific variable being discussed.

5.2 TYPE OF MILL

As discussed earlier (Chapter 4), a number of different types of mills are available for MA/MM. These mills differ in their capacity, speed of operation, and capacity to control the operation by varying the temperature of milling and the extent of minimizing contamination of the milled powders. Depending on the type of powder, the quantity of powder to be milled, and the final constitution required, a suitable mill can be chosen. Most commonly, however, the SPEX shaker mills are used for exploratory and alloy screening purposes. The Fritsch Pulverisette planetary ball mills or attritors are used to produce larger quantities of the milled powder. Specially designed mills are used for specific applications.

It has been recently shown [1] that the degree of contamination, amount of the amorphous phase formed, crystallization temperature, and the activation energy for crystallization of the amorphous phase depend on the type of mill used. For example, the investigators conducted MA experiments on a Mo-47 at% Ni powder mix in a SPEX 8000D mill and the Zoz Simoloyer mill. They noted that the oxygen level in the powder milled in the Simoloyer was much less than in the SPEX mill. This resulted in changes in the nature of the amorphous phase formed. For example, for powders processed in the SPEX mill, the crystallization temperature decreased with increasing milling time and the activation energy for crystallization was lower than for the powder milled in the Zoz Simoloyer mill. Both of these effects were explained on the basis of increased contamination levels of oxygen at longer milling times.

5.3 MILLING CONTAINER

The material used for the milling container (grinding vessel, vial, jar, and bowl are some of the other terms used) is important because due to impact of the grinding medium on the inner walls of the container some material will be dislodged and be incorporated into the powder. This can contaminate the powder and/or alter the chemistry of the milled powder. If the material of the grinding vessel is different from that of the powder being milled, then the powder may be contaminated with the material of the grinding vessel. On the other hand, if the materials of the container and the powder being milled are the same, then the chemistry of the final powder may be altered. For example, during milling of Cu-In-Ga-Se powder mixtures in a copper container, it was noted that the copper content in the milled powder was higher than the copper content in the initial powder mix. This was due to the incorporation of copper into the powder from the container and the grinding balls [2]. Even though this is not "contamination," the chemistry of the milled powder is different from the desired composition. This problem could be solved by taking a lower copper content in the initial powder mix, noting that the copper uptake in the milled powder depends on the processing conditions.

Hardened steel, tool steel, hardened chromium steel, tempered steel, stainless steel, WC-Co, WC-lined steel [3], and bearing steel are the most common types of materials used for the grinding vessels. Some specific materials are used for specialized purposes; these include copper [2], titanium [4], sintered corundum, yttria-stabilized zirconia (YSZ) [5], partially stabilized zirconia + yttria [6,7], sapphire [8,9], agate [10–12], hard porcelain, Si_3N_4 [13], and Cu-Be [14–18].

The shape of the container also seems to be important, especially the internal design of the container. Both flat-ended and round-ended SPEX mill containers have been used. Alloying was found to occur at significantly higher rates in the flat-ended vial than in the round-ended container [19]. The time required to reach a constant intensity and shape of the (111) peak in the X-ray diffraction (XRD) pattern of the Si-Ge mixture was 9 h for the flat-ended vial and 15 h for the round-ended vial. However, a problem with the round-ended vial is that the balls may roll around the end of the vial rather than hitting it, thereby decreasing the milling intensity. The balls may even roll around the inside surface of the vial repeatedly tracing out a bent-8 shape. Powder may collect in the eyes of the "8" and remain unprocessed [20]. Proper design of the milling container is important in avoiding the formation of dead zones, areas where the powder does not get milled because the grinding medium cannot reach there.

5.4 MILLING ENERGY/SPEED

It is easy to realize that the faster the mill rotates the higher will be the energy input into the powder. This is because the kinetic energy of the grinding medium ($E = 1/2\ mv^2$, where m is the mass and v is the relative velocity of the grinding medium) is imparted to the powder being milled. Therefore, the kinetic energy supplied to the powder is higher at higher relative velocities of the grinding medium. However, depending on the design of the mill there are certain limitations to the maximal speed that could be employed. For example, in a conventional ball mill increasing the speed of rotation will increase the speed with which the balls move. Above a critical speed, the balls are pinned to the inner walls of the vial and do not fall down to exert any impact force. Therefore, the maximal speed should be just below this critical value so that the balls fall down from the maximal height to produce the maximal collision energy.

Another limitation to the maximal speed is that at high speeds (or intensity of milling) the temperature of the vial may reach a high value. This may be advantageous in some cases where diffusion is required to promote homogenization and/or alloying in the powders. However, in some cases this increase in temperature may be a disadvantage because the increased temperature accelerates the transformation processes and results in the decomposition of supersaturated solid solutions or other metastable phases formed during milling [21]. In addition, the high temperatures generated may also contaminate the powders. It has been reported that during nanocrystal formation, the average crystal size increased and the internal strain decreased at higher milling intensities due to the enhanced thermal effects and consequent dynamic recrystallization [22]. The maximal temperature reached is different in different types of mills and the values vary widely. (This aspect is discussed in detail in Chapter 8.)

Yet another disadvantage of increased milling speed is the excessive wear of the milling tools (grinding medium and container), which could also lead to increased powder contamination. The powder yield could also be lower if the powder gets stuck to the inner walls of the milling container due to increase in cold welding caused by the higher degree of plastic deformation.

It is important to realize that it may not be possible to change the milling speed in all types of mills. For example, while SPEX mills usually run at a constant speed, the speed of rotation of the rotating disk and the containers could be changed in the Fritsch Pulverisette. Similarly, by changing the position and intensity of the magnets, the milling intensity as well as the nature of the forces (shear, impact, or a combination of these) could be altered in the Uni-Ball mill.

As was mentioned in Chapter 4, the disk and the vials can rotate either in the same direction (normal direction) or in the opposite direction (counter direction). Depending on the direction in which they move, the impact energy acquired by the balls is going to vary. Mio et al. [23] investigated the effect of rotation speed of the disk and the ratio of the speeds of the vial and the disk on the specific impact energy of the balls and the efficiency of milling talc in a planetary ball mill. They noted that the specific impact energy increased with increasing speed of the disk (in the investigated range of 300–600 rpm). But the highest value obtained was about 15 kJ/s.kg when the rotating direction was normal and as high as 80 kJ/s.kg when the disk was rotating in the counter direction. Because of this, they concluded that rotation of the vial in the counter direction to the revolution of the disk was more effective in fine grinding of materials as well as its mechanochemical activation. As an example, they showed that talc amorphized in 2 min when milling was done in the counter direction, and it took more than 8 min when milling was done in the normal direction. It was also reported that the specific impact energy of the balls increase rapidly with the ratio of the rotation (of the vial) and revolution (of the disk) speeds when milled in the counter direction, whereas for the normal direction there was only a gradual (and slow) increase. At a critical speed ratio the specific energy of the balls reached the highest value; therefore, it is desirable to mill the powder at that ratio. This ratio is dependent on the radius of the revolving disk but is about 2 in majority of the cases.

Some investigators prefer to use the term milling intensity rather than milling speed or milling energy to describe the phase changes that take place on alloying. Chen et al. [24] defined the intensity of milling as a function of BPR ($=M_b/M_p$), and velocity and frequency of the balls as:

$$I = \frac{M_b V_{max} f}{M_p} \tag{5.1}$$

where M_b is the mass of the ball, V_{max} is the maximum velocity of the ball, f is the impact frequency, and M_p is mass of the powder in the vial. Thus, the milling intensity increases rapidly with the impact frequency, velocity, and mass of the balls. Using this as a parameter, Pochet et al. [25] established the dependence of the induced order–disorder transformation on milling intensity and milling temperature. Gonzalez et al. [26] modified the SPEX mill to vary the frequency of vibration in the range of 1000–3500 Hz. They noted that at high impact frequencies (3000 rpm), a solid solution formed in an Fe-15 wt% Al powder mix after 1 h of milling, whereas at the lower frequencies (1800 rpm), alloy formation was not complete even after 5 h of milling. Thus, for the same mill and balls, alloy formation could be achieved in shorter milling periods at higher frequencies.

Calka et al. [27] reported that when vanadium and carbon powders were milled together at different energy levels (by adjusting the positions of the magnets in the Uni-Ball mill), the final constitution of the powder was different. For example, at very low milling energy (or speed), the powder consisted of nanometer-sized grains of vanadium and amorphous carbon, which on annealing formed either V_2C or a mixture of V + VC. At an intermediate energy level, the as-milled powder contained a nanostructure, which on annealing transformed to VC. At the highest energy level, the carbide phase, VC formed directly on milling. Similarly, a fully amorphous phase formed in a Ni-Zr powder mixture at high milling energy whereas a mixture of crystalline and amorphous phases formed at low and intermediate milling energies [28].

El-Eskandarany et al. [29] studied the effect of milling speed in a Fritsch Pulverisette 5 mill. They reported that the time required for the formation of an amorphous phase in the Co-Ti system decreased with an increase in the rotation speed. For example, while it took 200 h for the formation of the amorphous phase at a speed of 65 rpm, it took only 100 h at 125 rpm and 24 h at 200 rpm. They had observed a cyclic crystalline-amorphous-crystalline transformation on continued milling.

The effect of milling speed during milling of an Fe-7 wt% C-6 wt% Mn powder blend in a planetary ball mill was investigated at three different speeds of 80, 100, and 120 rpm by Rochman et al. [30]. They noted that the solid solubility of carbon in iron increased with increasing speed of operation of the mill for the same milling time. For example, they reported that on 100 h of milling, the solid solubility of C in α-Fe was 0.2 at% at 80 rpm, 0.5 at% at 100 rpm, and 1.3 at% at 120 rpm. (The highest solid solubility of C in α-Fe was mentioned as 1.3 wt% (*not* at%) in an earlier reference [31].)

Eckert et al. [32] reported that the nature of the phase formed in the mechanically alloyed Al-Cu-Mn powder blend was different depending on the intensity of milling in the Fritsch Pulverisette 5. They reported that an amorphous phase formed when the blended elemental powder mix was milled at an intensity of 5, and an equilibrium intermetallic phase formed at an intensity of 9. A quasi-crystalline phase formed at intermediate intensities. These results were explained on the basis of the increased temperature experienced by the powder at higher milling intensities. Details of these investigations on the effect of process variables will also be discussed in the individual sections on the phases formed during milling.

Riffel and Schilz [33] investigated the alloying behavior of Mg and Si powders in a planetary ball mill and examined the optimal conditions for synthesizing the Mg_2Si intermetallic. They noted that an improvement in the mill setting could be obtained by changing the ratio of the angular velocity of the planetary wheel, ω, to the system wheel angular velocity, Ω. A ratio ω/Ω of at least 3 was found to be necessary to offset the effects of slip.

Bab et al. [34] investigated the absorption of nitrogen in Hf during MA in a Retsch MM2 horizontal oscillatory ball mill and noted that the amount of nitrogen absorbed at a frequency of 21 Hz was several times smaller than for higher frequencies, suggesting that nitrogen diffusion into the metal lattice was limited by the production rate of fresh surfaces and lattice defects during milling. Delogu et al. [35] noted that variation in milling intensity only affected the rate at which the reaction took place and not the basic features of the transformation process. A similar result was also reported for a ternary Cu-Fe-Co powder blend by Galdeano et al. [36], who noted that there was no significant effect of milling intensity on the as-milled nanostructure.

Gerasimov et al. [37] conducted MA of Zr-Co alloys in a planetary ball mill at accelerations of 300, 600, and 900 m/s^2 and also using steel balls of 5, 7, and 10 mm diameter. They observed that increased kinetic energy of the balls due to an increase in their diameter and/or rotating speed of the vials led to a rise in the degree of crystallinity of the milled samples. Accordingly, the blended elemental powder mix of Zr-33 at% Co became completely amorphous only under the "softest" milling conditions—a centrifugal acceleration of 300 m/s^2 and a ball diameter of 5 mm. Under all other milling conditions, amorphization was either incomplete or a mixture of equilibrium crystalline phases was obtained [37].

Mechanical alloying and milling (MA/MM) are important processing techniques to synthesize metastable phases, including amorphous phases, supersaturated solid solutions, and others. However, it has been shown repeatedly that only "soft" milling conditions (i.e., low milling energy, lower BPR values, smaller size, and lighter grinding media) produce metastable phases in general, and less metastable (including equilibrium) phases are produced when "hard" milling conditions are employed [2,37–41]. For example, a metastable high-pressure cubic phase formed on milling a Cu-In-Ga-Se powder mixture at 150 rpm in a Fritsch Pulverisette 5, while the equilibrium tetragonal phase had formed on milling the powder mix at 300 rpm [2].

5.5 MILLING TIME

The time of milling is one of the most important parameters in milling of powders. Normally the time is chosen to achieve a steady state between fracturing and cold welding of powder particles to facilitate alloying. The times required vary depending on the type of mill used, mill settings, intensity of milling, BPR, and temperature of milling. The necessary times have to be decided for each combination of the above parameters and for the particular powder system under consideration. However, it should be realized that the level of contamination increases and some undesirable phases form if the powder is milled for much longer than required, especially in reactive metals like titanium and zirconium [42]. Therefore, it is desirable that the powder be milled just for the required duration and no longer. As a general rule, it may be appreciated that the times taken to achieve the steady-state conditions are short for high-energy mills and longer for low-energy mills. Again, the times are shorter for high BPR values and longer for low BPR values.

5.6 GRINDING MEDIUM

In this section we discuss details about the efficiency of alloying based on the nature of the grinding medium as well as its size and size distribution. Hardened steel, tool steel, hardened chromium steel, tempered steel, stainless steel, WC-Co, and bearing steel are the most common types of materials used for the grinding medium. (See Table 12.1 for details of the grinding media used in different investigations during amorphous phase formation.) The density of the grinding medium should be high enough that the balls create enough impact force on the powder to affect alloying. However, as in the case of the grinding vessel, some special materials are used for the grinding medium; these include copper [2], titanium [4], niobium [43], zirconia (ZrO_2) [44,45], agate [11,12], YSZ [5], partially stabilized zirconia + yttria [6,7], sapphire [9], silicon nitride (Si_3N_4) [13], and Cu-Be [18]. It is desirable, when possible, to have the grinding vessel and the grinding medium made of the same material as the powder being milled to avoid cross contamination of the milled powder (see also Chapter 15).

The material of the grinding medium is an important variable. The higher the density of the grinding medium, the more kinetic energy it acquires during milling and this can be transferred to the powder. Thus, WC balls are frequently used, instead of steel, to generate higher impact energy. Gonzalez et al. [46] compared the milling efficiencies of steel and WC grinding media during milling of Fe-50 at% Co powders and noted that a solid solution of Co in Fe formed after 5 h of milling in both cases. However, they noted differences in the microstructural features of the final powders. They reported that when steel grinding medium was used, a kneading process was observed; small powder particles of ε-Fe (a high-pressure form of iron) were embedded into larger agglomerates, i.e., a heterogeneous structure was obtained. On the other hand, when WC grinding medium was used, stronger attrition effect was noted; the powder particles were more fragmented and isolated. They also observed that small oxide particles were observed in the powders milled with WC tools, which showed a superparamagnetic effect not observed with steel balls. This is an expected result. But, the same authors also noted in another investigation [47] that milling of Fe-27 at% Al powders with a BPR of 8:1 resulted in alloying after 3 h with steel balls, whereas it required 5 h with the WC balls. Generally, it is expected that the heavier grinding medium (WC in this case) imparts more kinetic energy to the powder and so alloying should have occurred faster. This is a surprising result, but the authors ascribe it to possible differences in the input energies of the two different grinding media.

The size of the grinding medium also has an influence on the milling efficiency. Generally speaking, a larger size (and higher density) of the grinding medium is useful since the larger weight of the balls transfers more impact energy to the powder particles. It has also been reported that the final constitution of the powder is dependent on the size of the grinding medium used for milling. For example, when balls of 15 mm diameter were used to mill the blended elemental Ti-Al powder mixture, a solid solution of aluminum in titanium was formed. On the other hand, use of 20- and 25-mm-diameter balls resulted in a mixture of only the titanium and aluminum phases (and no alloying occurred), even after a long milling duration [48]. In another set of investigations [49,50] it was reported that an amorphous phase could be produced faster in Ti-Al alloys by using steel balls of 3/16 in. diameter than by using balls of 3/4 in. diameter. In fact, in some cases an amorphous phase was not produced and only the stable crystalline compound formed when milling was done with large steel balls [49]. In yet another investigation it was reported that an amorphous phase formed only when the Ti-Al powder mixture was milled using either 5- or 8-mm-diameter balls; an amorphous phase did not form when 12-mm-diameter balls were used for milling [39,50]. A similar situation was also reported in the Pd-Si system where it was reported that a smaller ball size favored amorphous phase formation [38]. It was suggested that the smaller balls produced intense frictional action, which promoted the amorphous phase formation.

It was also reported [51] that use of smaller balls (4.8 mm diameter) resulted in the formation of larger amounts of Cr-based solid solution than with larger balls (8.8 mm diameter). It was suggested that the larger balls produced a higher temperature, which resulted in decomposition of the metastable solid solution. They also reported that the grain size of the solid solution was finer when smaller balls were used. Faster kinetics also is expected when larger balls are used because the impact energy is higher [26].

In fact, as mentioned earlier, it appears that soft milling conditions (smaller ball sizes, lower energies, and lower BPR) favor formation of metastable phases (amorphous or metastable crystalline phases), while the hard milling conditions allow the formation of less metastable or even the equilibrium phases [2,37–39,52].

Even though most of the investigators generally use only one size of the grinding medium, there have been instances when different sized balls have been used in the same experiment [53]. It has been predicted that the highest collision energy can be obtained if balls with different diameters are used [54], most likely due to the interference between balls of different diameters. In the initial stages of milling the powder being milled gets coated onto the surface of the grinding medium and also gets cold welded. This is advantageous because it prevents excessive wear of the grinding medium and prevents contamination of the powder due to the wear of the grinding medium. However, the thickness of this layer must be kept to a minimum to avoid formation of a heterogeneous final product [55]. The disadvantage of this powder coating is that it is difficult to detach this and so the powder yield is low. It has also been reported that a combination of large and small balls during milling minimizes the amount of cold welding and the amount of powder coated onto the surface of the balls [56]. Although no specific explanation has been given for the improved yield under these conditions, it is possible that the different sized balls produce shearing forces that may help in detaching the powder from the surface of the balls.

A different effect is also reported on the use of a mixture of different sizes of the grinding balls [33]. By investigating the synthesis of Mg_2Si from an elemental mixture of Mg and Si under different experimental conditions, it was noted that the best results were obtained on milling with 10-mm-diameter hardened steel balls. Use of larger diameter balls led to a significant increase in iron contamination, whereas milling with smaller diameter balls did not produce enough kinetic energy to achieve alloying. More specifically, they observed that a mixture of different sizes of the balls did not improve the milling process, noting that if there were large differences in the sizes of the balls, the smaller balls could even be destroyed by the larger ones.

Use of grinding balls of the same size in either a round- or a flat-bottomed vial has been shown to produce tracks. Consequently, the balls roll along a well-defined trajectory instead of hitting the end surfaces randomly. Therefore it is necessary to use several balls, generally a combination of smaller and larger balls, to "randomize" the motion of the balls [20].

5.7 BALL-TO-POWDER WEIGHT RATIO

The ratio of the weight of the balls to the powder (BPR), sometimes referred to as charge ratio (CR), is an important variable in the milling process. This has been varied by different investigators from a value as low as 1:1 [57] to one as high as 1000:1 [58]. Very high values of BPR are uncommon, and values such as 1000:1 or 220:1 [59] are used only in special cases to achieve certain desired features or to achieve the effects faster. Even though a ratio of 10:1 is most commonly used while milling the powder in a small capacity, high-energy mill such as a SPEX mill, BPR values ranging from 4:1 to 30:1 have been generally used. However, when milling is conducted in a large-capacity, low-energy mill, like an attritor, a higher BPR of up to 50:1 or occasionally even 100:1 is used to achieve the desired alloying in a reasonable time. Increased BPR values can be obtained either by increasing the number (weight) of balls or decreasing the powder weight. Higher BPR can also be obtained (for the same number of balls) either by increasing the diameter of the balls or by using higher density materials such as WC rather than steel.

The BPR has a significant effect on the time required to achieve a particular phase in the powder being milled. The higher the BPR, the shorter is the time required, and this has been repeatedly shown to be true in a number of instances. For example, formation of an amorphous phase was achieved in a Ti-33 at% Al powder mixture milled in a SPEX mill in 7 h at a BPR of 10:1, in 2 h at a BPR of 50:1, and in 1 h at a BPR of 100:1 [60]. Similarly, it has been shown [61] that milling of an $Al_{50}Si_{30}Fe_{15}Ni_5$ powder mix for 40 h at a BPR of 15:1 produced similar results to milling for 95 h at a BPR of 10:1. It was also reported [47] that no alloying occurred in an Fe-Al powder mixture at a BPR of 1:1 when either steel or WC grinding medium was used; alloying occurred at a BPR of 8:1 in 3 h for steel and 5 h for WC medium. At a high BPR, because of an increase in the weight proportion of the balls, the mean free path of the grinding balls decreases and the number of collisions per unit time increases. Consequently, more energy is transferred to the powder particles resulting in faster alloying. Several other investigators also have reported similar results [46]. It is also possible that due to the higher energy more heat is generated and this could raise the powder temperature resulting in faster kinetics of alloying. Alternatively, the higher temperature could change the constitution of the powder. In an extreme situation, the amorphous phase formed may even crystallize if the temperature rise is substantial, or the supersaturated solid solution formed could decompose to precipitate either transition or equilibrium phases or even achieve complete equilibration.

It will be shown later that the crystallite size during milling decreases with milling time. It has been frequently reported that the BPR has a significant influence on the rate of decrease of the crystallite size. Figure 5.1a shows that the crystallite size decreases with milling time and that the rate of decrease is higher for higher BPR values [62]. The shape of the grinding medium does not seem to have any effect on the rate of decrease of crystallite size. It may be noted from Figure 5.1b that whether balls or rods are used, the rate of crystallite size reduction simply increases with increasing BPR.

It may be appreciated that use of higher BPR results in a faster refinement of the crystallite size, but the amount of powder synthesized is low. On the other hand, use of a lower BPR increases the amount of powder produced, but the time required to reach the same crystallite size is longer. Thus, optimization of BPR and milling time should be done to achieve a particular crystallite size in a reasonable time.

A detailed study of the effect of BPR on the powder hardness and mean particle size was conducted by Niu [63]. It was observed that the microhardness of the powder increased with increasing BPR due to an enhanced and accelerated plastic deformation of the powder particles. However, the rate of increase in the microhardness was lower at higher BPR values. The mean particle size, on the other hand, was found to decrease initially when the BPR was increased from 2:1 to 5:1, and then it started to increase to its maximal value when the BPR was increased to 11:1; a steady decrease was observed with further increase in the BPR. The powder morphology was also different at these different BPR values—flaky at 3:1, equiaxed at 10:1 or higher, and flaky plus equiaxed at intermediate values.

As mentioned earlier, soft conditions (e.g., low BPR values, low speeds of rotation, etc.) during MA produce metastable phases, whereas hard conditions

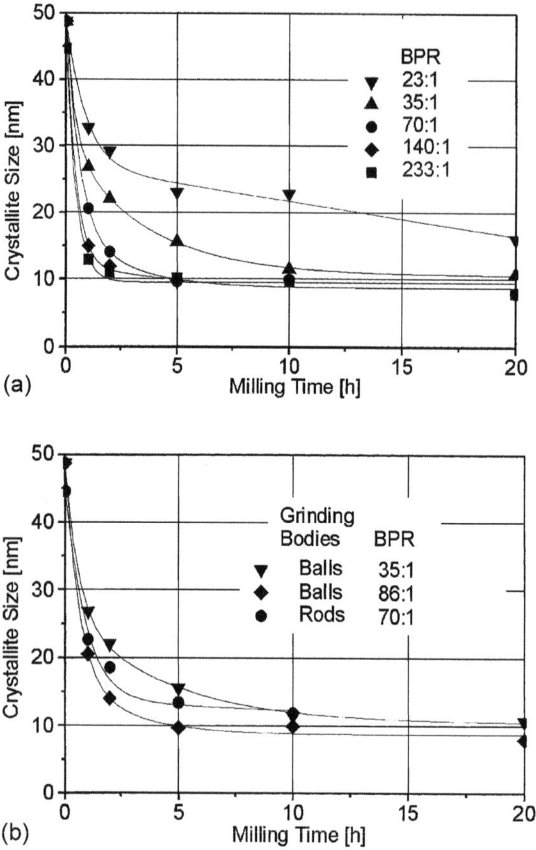

Figure 5.1 (a) Variation of crystallite size with milling time and ball-to-powder weight ratio (BPR). Note that the crystallite size decreases with increasing milling time and that the rate of decrease is higher at higher BPR values [62]. (b) The shape of the grinding medium does not appear to affect the rate of decrease of crystallite size. Note that the rate of decrease follows the same trend whether balls or rods are used [62].

produce the equilibrium or near-equilibrium phases. In a similar manner, it was also reported that the BPR value has a significant effect on the occurrence of the SHS (self-propagating high-temperature synthesis) reactions. Lower BPR values and shorter times induced the SHS reaction in an Al-Zr powder blend, whereas higher BPR did not induce the SHS reaction [64].

Proper selection of the BPR is important. In general, the higher the BPR, the faster is the MA process. However, this faster processing also can introduce high contamination levels into the powder, and this should be avoided or at least minimized.

5.8 EXTENT OF FILLING THE VIAL

Since alloying among the powder particles occurs mainly due to impact forces exerted on them, it is necessary that there be enough space for the balls and the powder particles to move around freely in the milling container. Therefore, the extent of filling the vial with the powder and the balls is important. If the quantity of the balls and the powder is very small, then the production rate is very low. On the other hand, if the quantity is large, then there is not enough space for the balls to move around and so the energy of the impact is less. Consequently, alloying may not occur, and even if it occurs, it may take a very long time. Thus, care must be taken not to overfill the vial; generally about 50% or a little more of the vial space is left empty.

5.9 MILLING ATMOSPHERE

The MA process is normally conducted under vacuum or in an inert atmosphere to prevent/minimize oxidation and/or contamination of the milled powder. The major effect of the milling atmosphere is on the nature and extent of contamination of the powder. Therefore, powders are milled generally in containers that have been evacuated or filled with an inert gas such as argon or helium. (Nitrogen has been found to react with metal powders to produce nitride phases and consequently cannot be used to prevent contamination during milling, unless one is interested in producing nitrides.)

Normally, loading and unloading of the powders into the vial is carried out inside evacuated or atmosphere-controlled glove boxes. These glove boxes are usually repeatedly evacuated and refilled with an inert gas such as argon. Some investigators have even conducted the milling operation in mills that have been placed inside evacuated glove boxes [65]. This would prevent the atmosphere from contaminating the powder in the container during milling. An alternative method is to load the powder into the container in air but to evacuate the container before starting to mill the powder [66]; this is much less convenient than the other methods and may not always be possible.

High-purity argon is the most common ambient used to prevent oxidation and/or contamination of the powder. It has also been noted that oxidation can be generally prevented or minimized in the presence of nitrogen ambient. But this does not appear to be true when reactive powders such as titanium or its alloys are milled. It has been reported that Ti-48Al-2W (at%) powders milled in an oxygen atmosphere picked up 1.5 wt% oxygen after 20 h of milling. The same powder milled for the same length of time, but in a nitrogen atmosphere, picked up 4.7 wt% oxygen [67]—a difficult to explain observation.

Different atmospheres have been used during milling for specific purposes. Nitrogen or ammonia atmosphere has been used to produce nitrides [68,69]. Hydrogen atmosphere was used to produce hydrides [70]. The presence of air in the vial has been shown to produce oxides and nitrides in the powder, especially if the powders are reactive in nature. Thus, care must be taken to use an inert atmosphere during milling.

In addition to contaminating the powder, the milling atmosphere influences the kinetics of alloying, transformation behavior, and the nature of the product phase. It was reported that the crystallization temperature of the $Ni_{60}Nb_{40}$ amorphous alloy prepared in air was lower than that prepared in a helium atmosphere, a direct consequence of the increased oxygen content [71]. A similar observation was made in the Mo-Ni system [1]. The transformation paths were found to be different when the $Ni_{60}Ti_{40}$ alloy powder mix was processed in different types of atmospheres—argon, air, or nitrogen [72–75]. It has been reported that an amorphous phase could be obtained by directly milling the Ni-40 at% Ti powder mix in an argon atmosphere.

Nitrogen was shown to have only a slight effect. But oxygen, even though present in small quantities, had a significant effect and no amorphous phase was formed; instead Ni-Ti crystalline intermetallic phases and oxide phases were formed.

The type of atmosphere also seems to affect the nature of the final phase. For example, it was shown that when Cr-Fe powder mixtures were milled in different types of atmosphere, the constitution of the final powder was different [76]. When the powder was milled in an argon atmosphere, no amorphous phase formed and Cr peaks continued to be present in the XRD patterns. On the other hand, when the powder was milled in either air containing argon or a nitrogen atmosphere, the powder became completely amorphous. Similarly, oxygen was shown to enhance the kinetics of amorphization in the Ni-Nb system [77]. It was also reported [78] that alloying between Ni and Nb did not take place when milling was done in water, alcohol, or acetone. The Nb metal formed a bcc phase, identified as an interstitial phase of oxygen in a niobium matrix. On the other hand, when argon was used, the authors noted formation of a supersaturated solid solution of Nb in Ni up to 15 at% Nb.

5.10 PROCESS CONTROL AGENTS

The powder particles get cold welded to each other, if they are ductile and sometimes excessively, due to the heavy plastic deformation they undergo during milling. However, true alloying among powder particles can occur only when a balance is maintained between cold welding and fracturing of powder particles. A process control agent (PCA) (also referred to as lubricant or surfactant) is added to the powder mixture during milling to reduce the effect of excessive cold welding. PCAs can be solid, liquid, or gaseous. They are mostly, but not necessarily, organic compounds, which act as surface-active agents. They adsorb on the surface of the powder particles and minimize cold welding among powder particles, thereby inhibiting agglomeration. The surface-active agents adsorbed on particle surfaces interfere with cold welding and lower the surface tension of the solid material. The energy required for the physical process of size reduction, E, is given by

$$E = \gamma . \Delta S \tag{5.2}$$

where γ is the specific surface energy and ΔS is the increase in surface area. Thus, a reduction in the surface energy results in the use of shorter milling times to obtain a particular particle size. Alternatively, finer powder particles will be generated for the given milling duration.

5.10.1 Nature of PCAs

A wide range of PCAs has been used in practice at a level of about 1–5 wt% of the total powder charge. The most important PCAs include stearic acid, hexane, methanol, and ethanol. A partial listing of the PCAs used in different investigations and their quantities is presented in Table 5.1. In addition, other exotic PCAs such as sodium-1,2-bis(dodecylcarbonyl)ethane-1-sulfonate, lithium-1,2-bisdodecyloxycarbonylsulfasuccinate, didodecyldimethylammonium acetate (DDAA), didodecyldimethyl ammonium bromide (DDAB), trichlorotrifluoroethane, and others such as polyethylene glycol, dodecane, ethyl acetate, oxalic acid, boric acid, borax, alumina, and aluminum nitrate have also been used. Most of these compounds have low melting

Table 5.1 Process Control Agents and the Quantities Used in Different Selected Investigations

Process control agent	Chemical formula	Quantity	Selected ref.
Alcohol	—	—	79
Benzene	C_6H_6	—	80, 81
C wax	$H_{35}C_{17}CONHC_2H_4NHCOC_{17}H_{35}$	1.5 wt%	82
Cyclohexane	—	—	83
Didodecyl dimethyl ammonium acetate	$C_{28}H_{59}NO_2$	—	84
Dihexadecyl dimethyl ammonium acetate	$C_{36}H_{75}NO_2$	—	84
Dodecane	$CH_3(CH_2)_{10}CH_3$	—	85
Ethanol	C_2H_5OH	4 wt%	86
Ethyl acetate	$CH_3CO_2C_2H_5$	—	87, 88
Ethylenebisdistearamide Nopcowax–22 DSP	$C_2H_2\text{-}2(C_{18}H_{36}ON)$	2 wt%	89
Graphite	C	0.5 wt%	87, 90
Heptane	$CH_3(CH_2)_5CH_3$	0.5 wt%	91, 92
Hexane	$CH_3(CH_2)_4CH_3$	—	80, 84, 93–95
		5 wt%	96, 97
Lithium-1,2 bisdodecyloxy-carbonyl sulfasuccinate	—	—	84
Methanol	CH_3OH	—	98–103
		1 wt%	104
		3 wt%	105
		4 wt%	106
Octane	$CH_3(CH_2)_6CH_3$	1 wt%	107
Paraffin	—	—	108
Petrol		3 drops	111
Polyethylene glycol	$HO(CH_2CH_2)_nOH$	0.5–2 wt%	109, 110
Silicon grease	—	—	108
Sodium chloride	NaCl	2 wt%	112
Sodium-1,2 bis(dodecyl carbonyl) ethane-1-sulfonate	—	—	84
Stearic acid	$CH_3(CH_2)_{16}COOH$	1 wt%	113–116
		4 wt%	117
		0.5–3 wt%	110
Tetrahydrofuran	—	—	80
Toluene	$C_6H_5CH_3$	5 ml	118–122
Vacuum grease	—	—	123

and boiling points (Table 5.2) [124], and consequently they either melt or evaporate during milling owing to the temperature increase. Furthermore, they decompose during milling, interact with the powder and form interstitial compounds, and these get incorporated in the form of inclusions and/or dispersoids into the powder particles during milling. Thus, hydrocarbons containing hydrogen and carbon, and carbohydrates containing hydrogen, carbon, and oxygen are likely to introduce carbon and/or oxygen into the powder particles, resulting in the formation of carbides and oxides,

Table 5.2 Melting and Boiling Points of the Different Process Control Agents Used in Mechanical Alloying

PCA	Melting point (°C)	Boiling point (°C)
Dodecane	−12	216.2
Ethyl acetate	−84	76.5–77.5
Ethyl alcohol	−130	78
Ethylenebisdistearamide	141	259
Heptane	−91	98
Hexane	−95	68–69
Methyl alcohol	−98	64.6
Polyethylene glycol	59	205
Stearic acid	67–69	183–184

which are uniformly dispersed in the matrix. These are not necessarily harmful to the alloy system since they can contribute to dispersion strengthening of the material resulting in increased strength and higher hardness [125]. The hydrogen subsequently escapes as a gas or is absorbed into the metal lattice on heating or sintering. Even though hydrogen gas primarily serves as a surfactant and does not usually participate in the alloying process [126], some reports indicate that hydrogen acts as a catalyst for amorphous phase formation in titanium-rich alloys [127,128]. In some other cases, hydrogen assists in embrittling the powders and consequently finer particles could be produced. It has also been reported that PCAs affect the final phase formation, changing the solid solubility levels [129], modifying the glass-forming range [102, 129,130], and altering the contamination levels.

The presence of air in the milling container or milling of powders at very low temperatures (cryomilling) also has been shown to minimize welding, most probably due to the increased brittleness of the powder particles at such low temperatures [131,132]. Metal powders (with an fcc structure) milled in a hydrogen atmosphere become brittle and do not stick to themselves or the container, probably due to the formation of a hydride phase [133].

5.10.2 Quantity of PCA

The quantity of PCA used and the type of powder milled determine the final size, shape, and purity of the powder particles. Zhang et al. [110] used two different PCAs—stearic acid and polyethylene glycol—in the range 0.5–4.0 wt% of the powder weight to investigate their effect on the size of the powder particles of aluminum and magnesium mechanically milled in a planetary ball mill. It was noted that the amount of the powder recovered increased with increasing amount of PCA and that longer milling times led to an increase in the amount of the powder recovered. This was attributed to the decreased proportion of cold welding due to the use of the PCA and consequent increase in the fracturing of powder particles. The authors have also noted that the efficiency of reducing the particle size was better with stearic acid than with polyethylene glycol. Stated differently, the amount of polyethylene glycol required to achieve a particular particle size is higher than the amount of stearic acid.

The quantity of PCA used is another important variable. It was reported [110] that the particle size increased for small quantities of PCA. However, with the use of larger quantities of PCA the particle size decreased. It was observed that use of a larger quantity of PCA normally reduces the particle size by two to three orders of magnitude. For example, Zhang et al. [110] reported that milling of aluminum for 5 h produced a particle size of about 500 μm when 1 wt% stearic acid was used as a PCA. But when 3 wt% of stearic acid was used, the particle size was only about 10 μm. The variation of the Al particle size as a function of the amount of polyethylene glycol is shown in Figure 5.2 [110]. Similar results were reported for other PCAs as well. It was also shown that an increase of the PCA content led to an exponential decrease of the powder size for a given milling duration. For example, the powder particle size reached a value of 1000 μm without the use of a PCA; with 2.3 wt% of PCA, the mean powder particle size reduced to 18 μm [134]. Alternatively, the powder particle size decreased exponentially with milling time, and the larger the quantity of the PCA used, the smaller the powder particle size that resulted [117].

Results similar to the above may not be observed when brittle materials are milled; no large particles were observed even if a small quantity of the PCA was used. In fact, use of a PCA is not required for milling of brittle materials. Niu [63] noted that a homogeneous distribution of particle size could be easily achieved when the PCA is in the liquid state (e.g., ethyl acetate) than when it is in the solid state (e.g., stearic acid). A detailed discussion on the effect of PCAs during milling of metal powders is available in Ref. 48.

Lee and Kwun [81] conducted a detailed investigation on the effects of the nature and quantity of PCA on the constitution of mechanically alloyed Ti-48 at% Al powders. They observed that an amorphous phase formed after milling for 300 h without a PCA and a metastable fcc phase formed after milling for 500 h. But when 0.3 wt% methanol was used as a PCA, a metastable disordered Ti_3Al phase formed after 300 h of milling and an amorphous phase after 1000 h. On the other hand, when 3 ml of benzene was used as a PCA, a metastable fcc phase formed after milling for 1000 h. From these observations Lee and Kwun [81] concluded that formation of the

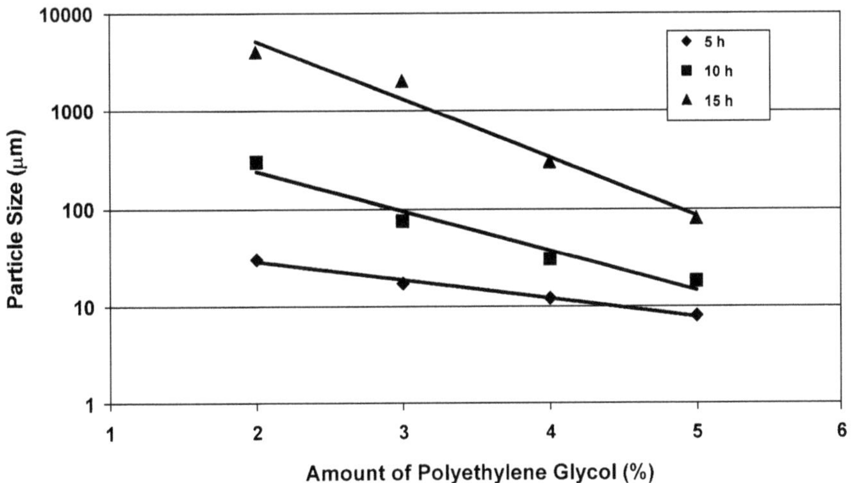

Figure 5.2 Mean particle size of Al as a function of the amount of polyethylene glycol and milling time [110].

metastable fcc phase was caused by atomic penetration of carbon and/or hydrogen atoms into the interstitial sites of the lattice. They also noted that the activation energy for crystallization of the amorphous phase increased (from 281 kJ/mol without a PCA to 411 kJ/mol when 3 wt% methanol was used) as the number of impurity atoms, especially oxygen, in the PCA increased.

5.10.3 Constitution of the Milled Powder

The use of a PCA can also alter the constitution of the final milled powder. It was reported that milling of gas-atomized $Fe_{82}Nb_6B_{12}$ powder for 220 h, using cyclohexane as a PCA, resulted in the formation of an amorphous phase [83]. However, the authors noted that the carbon content in the milled amorphous powder was as high as 11 wt%. Consequently, when milling was done without cyclohexane (to avoid carbon contamination), only the iron solid solution was obtained, with no detectable amorphous phase in the material.

The results reported on the effect of the amount of PCA used on the kinetics of alloying are not consistent. Some investigators reported that alloying takes place more quickly whereas others reported that the kinetics is slowed down. When only a small quantity of the PCA is used, the surface area of the powder particles covered by it is limited. Therefore, excessive cold welding still takes place and alloying does not occur. On the other hand, when sufficient PCA is available to cover the whole surface area of the powder particles, then excessive cold welding does not take place, a balance is achieved between cold welding and fracturing, and alloying occurs rapidly.

It is also possible in some cases that when a small amount of PCA is used it cannot cover the whole surface area of the powder particles, so that alloying takes place more quickly due to direct contact between the reactant particles. That is, the kinetics of alloying is faster. A large amount of PCA may cover the surfaces of all the particles preventing direct contact between the components and possibly preventing rapid alloying. To prove this point it was shown that the fraction of the Al_3Mg_2 phase formed during milling of Al-Mg powders in a planetary ball mill was 0.3 when 1 wt% PCA was used, 0.28 with 2 wt%, PCA, and only 0.07 with 4 wt% PCA [117]. Thus, the quantity of the PCA is an important variable.

The powder particles produced are large when very small quantities of PCA are used. It was reported that when only 0.5 wt% PCA (stearic acid) is used, the particle size of Al was reported to be as large as 3–6 mm after 10 h of MA [117]. A minimal concentration of PCA is required during alloying. Below this value the particle size produced is large, and above this the particle size is small.

Shaw et al. [135] have conducted a detailed study of the effects of stearic acid and methanol on the MA of nanostructured Al-3Fe-2Ti-2Cr (at%) alloy powders. They investigated the effect of the nature and amount of the PCA on powder particle sizes, grain sizes, atomic level strain, lattice parameter, formation of solid solutions, and microstructural evolution. They have reported that 2 wt% of stearic acid could completely prevent excessive cold welding even at the early stage of milling and that the more the PCA in the powder the smaller was the crystallite size. For example, Figure 5.3 shows the variation of the crystallite size of fcc-Al as a function of milling time and the amount of stearic acid. It may be noted that the crystallite size decreases with increasing milling time at all concentrations of the PCA, but the actual crystallite size increases as the amount of PCA increases. Between stearic acid and methanol as

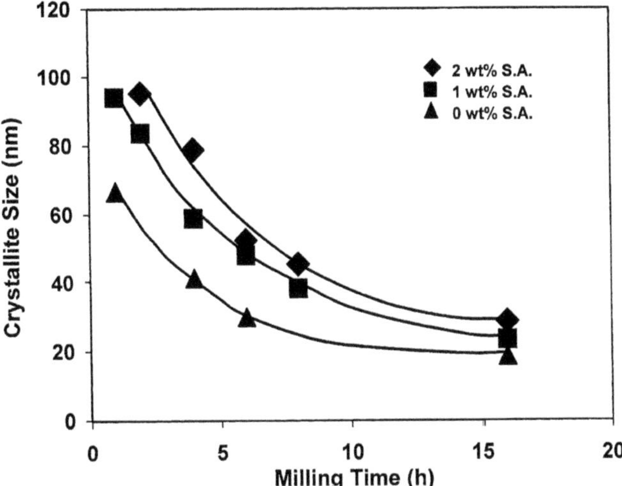

Figure 5.3 Calculated crystallite size of fcc-Al in the Al-3Fe-2Ti-2Cr (at%) powder mix with and without stearic acid as a function of milling time [135].

PCAs, Shaw et al. [135] reported that stearic acid is more effective in reducing the crystallite size. This was explained on the basis that a smaller quantity of methanol was more effective than stearic acid in preventing excessive cold welding of the powder particles. This difference was related to the structure of the two molecules. For the same weight of the PCA, methanol provides more molecular layers than stearic acid due to its smaller molecular size, and so methanol is more effective in preventing excessive cold welding among the powder particles. One of the important conclusions the authors have arrived at was that use of a PCA can prevent excessive cold welding, but at the expense of reduction in grain size, kinetics of formation of solid solutions, and rate of microstructural refinement, all of which are desired in MA processing.

On the basis of experimental observations of processing of Al and Mg powders with stearic acid and polyethylene glycol as PCAs, Zhang et al. [110] tried to predict the minimum amount of PCA required using the back-propagation neural network architectures. The results showed a fairly good agreement between the back-propagation-network predictions and the experimental data.

5.10.4 Choice of PCA

The choice of a PCA for milling depends on the nature of the powder being milled and the purity of the final product desired. The nature and amount of PCA used during milling would determine the final powder particle size and powder yield. In fact, one way of determining the effectiveness of the PCA is to determine the powder yield after MA. If the powder yield is high, the PCA is effective. If the powder yield is not high, then either the amount of PCA used is not sufficient or, more likely, it is not the right PCA. It has been reported [48] that after milling for 15 h, only 50% of the powder was recovered if 2 wt% of polyethylene glycol was used, whereas almost 100% of the powder was recovered if stearic acid was used.

It should be realized that there is no universal PCA. The amount of PCA is dependent on the (1) cold welding characteristics of the powder particles, (2) chemical and thermal stability of the PCA, and (3) amount of the powder and grinding medium used. The powder particle size tends to increase if the weight proportion of the PCA to the powder is below a critical value, whereas above this value the particle size tends to decrease. One has to decide on a PCA by looking at the possible interactions between the metal and the components in the PCA. A critical discussion of the role of PCAs in the milling of Al-Cu powder mixtures is presented in Ref. 136.

5.11 TEMPERATURE OF MILLING

The temperature of milling is another important variable in deciding the constitution of the milled powder. Since diffusion processes are involved in the formation of alloy phases, irrespective of whether the final product phase is a solid solution, intermetallic, nanostructure, or an amorphous phase, it is expected that the temperature of milling will have a significant effect in any alloy system. For example, it was noted that the XRD pattern of Si milled in anhydrous ammonia for 48 h at 100 °C was similar to that milled for 168 h at room temperature [137].

There have been only a few investigations reported where the temperature of milling has been intentionally varied. This was done by dripping either liquid nitrogen or a nitrogen-alcohol mixture on the milling container to lower the temperature or electrically heating the milling vial with a tape to increase the temperature of milling. These investigations were undertaken either to ascertain the effect of milling temperature on the variation in solid solubility levels, or to determine whether an amorphous phase or a nanocrystalline structure forms at different temperatures. During the formation of nanocrystals, it was reported that the root mean square (rms) strain in the material was lower and the grain size larger for materials milled at higher temperatures [138]. The extent of solid solubility was reported to decrease at higher milling temperatures. For example, during planetary ball milling of a Cu-37 at% Ag powder mixture, it was noted that a mixture of an amorphous and crystalline (supersaturated solid solution) phases was obtained on milling at room temperature; instead, only a Cu-8 at% Ag solid solution was obtained on milling the powder at 200 °C [139]. Similar results were also reported by others in Cu-Ag [140], Zr-Al [141], and Ni-Ag [142] alloy systems and explained on the basis of increased diffusivity and equilibration effects at higher temperatures of milling.

Results different from the above were reported by Rochman et al. [31]. They reported that during milling of Fe-6 wt% C powder mixture at 253, 293, and 323 K, the solid solubility of C in α-Fe increased with increasing temperature, reaching a maximum value of 1.3 wt% at 323 K. They also noted that the time required to start forming a solid solution at 253 K was longer than that at 293 K, but the time to reach the nonequilibrium solid solubility was shorter than at 293 K. Thus, the kinetics appears to be faster at 253 K than at 293 K.

There have been conflicting reports on the formation of an amorphous phase as a function of the milling temperature. As explained later, amorphization during MA involves formation of microdiffusion couples of the constituent powders followed by a solid-state amorphization reaction. Thus, higher milling temperatures should enhance amorphization kinetics. This has been observed in the Ni-Ti [143] and Ni-Zr [144] systems. During the milling of a Ni-50 at% Zr powder mixture in a vibrating mill,

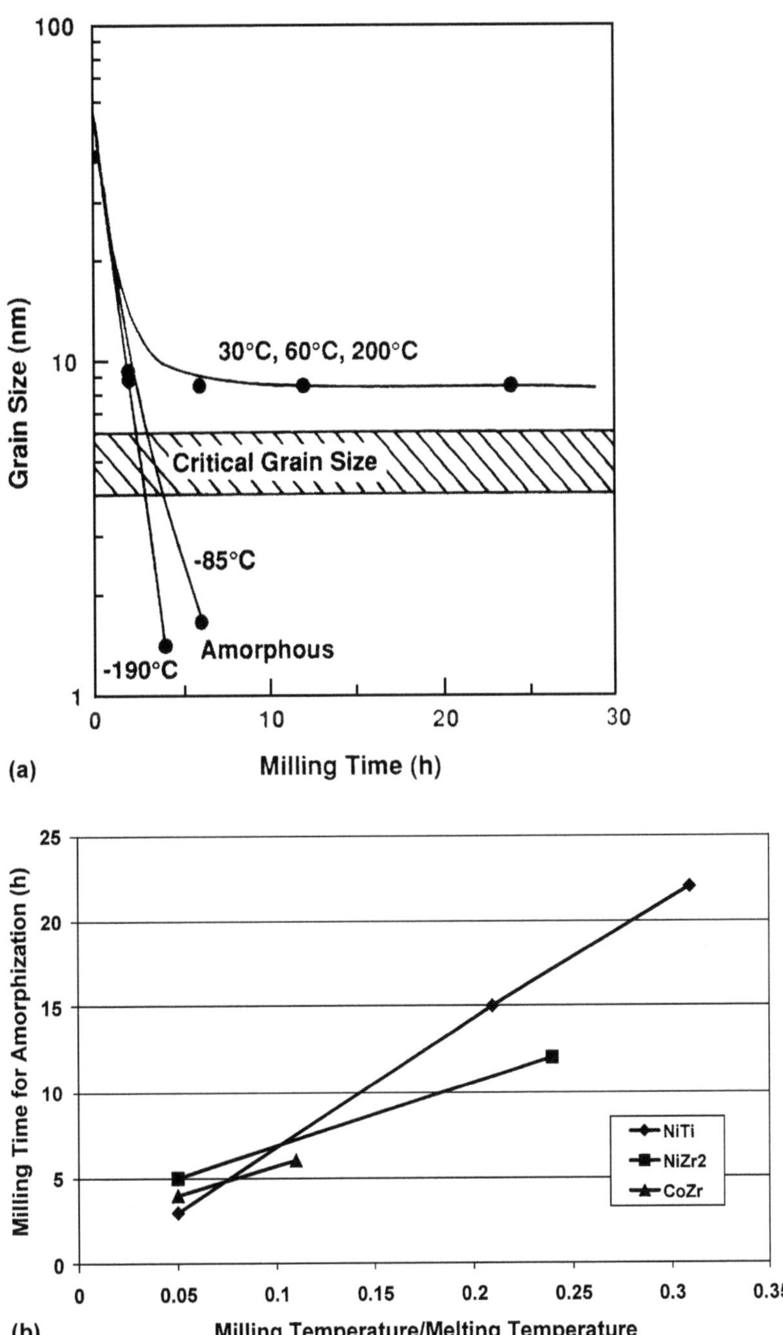

Figure 5.4 (a) Grain size vs. milling time for CoZr milled at different temperatures in a SPEX mill. The shaded area represents the grain size below which the alloy becomes amorphous [148]. (b) Milling time for amorphization vs. normalized milling temperature for CoZr, NiTi, and $NiZr_2$ intermetallics. A direct relationship exists between these two parameters.

amorphous phase formation was not observed when the powder was milled at liquid nitrogen temperature for 15 h. On the other hand, milling for the same period of time produced a fully amorphous phase at 200°C; a partially amorphous phase was produced on milling at room temperature.

Formation of an amorphous phase by MM occurs by a different mechanism than during MA. The increase in free energy of the crystalline phase by the introduction of defects such as antisite chemical disorder or increased grain boundary area through formation of a nanocrystalline structure is considered responsible for amorphization during MM. Thus, lower milling temperatures are expected to favor amorphization. However, both increased and decreased kinetics have been reported. Shorter milling times for amorphization were reported for NiTi at 170°C than at 60°C [143]. Enhanced amorphization kinetics was also observed for $NiZr_2$ intermetallic [16,144]. In contrast to these observations, others have reported reduced amorphization kinetics for the $Ni_{10}Zr_7$ and $Ni_{11}Zr_9$ intermetallics [145,146]. Koch et al. [147,148] also reported reduced amorphization kinetics with temperature during milling of the NiTi intermetallic. They reported that whereas it took 2 h for amorphization at liquid nitrogen temperature, it required 18 h of milling at 220°C. Similar results were reported for the CoZr and $NiZr_2$ intermetallics. These results were rationalized on the basis that amorphization occurred in this system due to increased grain boundary energy through formation of a nanocrystalline structure and that the nanocrystalline structure formed more rapidly at lower milling temperatures. Figure 5.4a shows the variation of grain size with milling time for CoZr milled for different times in the SPEX mill [148]. It can be easily seen that reduction in grain size occurs much more rapidly at liquid nitrogen temperature than at higher temperatures. The band representing the critical grain size, below which amorphization occurs, is also included. Figure 5.4b shows a plot of the milling time for amorphization against the normalized milling temperature ($T_{milling}/T_{melting}$). It may be noted that the milling time for amorphization increases with increased normalized milling temperature for the NiTi, $NiZr_2$, and CoZr intermetallics [149].

REFERENCES

1. Martinez-Sánchez, R., Estrada-Guel, I., Jaramillo-Vigueras, D., De la Torre, S. D., Gaona-Tiburcio, C., Guerrero-Paz, J. (2002). *Mater. Sci. For.* 386–388:135–140.
2. Suryanarayana, C., Ivanov, E., Noufi, R., Contreras, M. A., Moore, J. J. (1999). *J. Mater. Res.* 14:377–383.
3. Di, L. M., Bakker, H. (1991). *J. Phys. C: Condens. Matter.* 3:3427–3432.
4. Chu, B.-L., Chen, C.-C., Perng, T. -P. (1992). *Metall. Trans.* 23A:2105–2110.
5. Tokumitsu, K. (1997). *Mater. Sci. For.* 235–238:127–132.
6. Yen, B. K., Aizawa, T., Kihara, J. (1996). *Mater. Sci. Eng.* A220:8–14.
7. Yen, B. K., Aizawa, T., Kihara, J. (1997). *Mater. Sci. For.* 235–238:157–162.
8. Sherif El-Eskandarany, M. (1994). *J. Alloys Compounds* 203:117–126.
9. Sherif El-Eskandarany, M., Sumiyama, K., Aoki, K., Masumoto, T., Suzuki, K. (1994). *J. Mater. Res.* 9:2891–2898.
10. Tonejc, A., Duzevic, D., Tonejc, A. M. (1991). *Mater. Sci. Eng.* A134:1372–1375.
11. Ohtani, T., Maruyama, K., Ohshima, K. (1997). *Mater. Res. Bull.* 32:343–350.
12. Sherif El-Eskandarany, M. (1996). *Metall. Mater. Trans.* 27A:2374–2382.
13. Abe, O., Suzuki, Y. (1996). *Mater. Sci. For.* 225–227:563–568.

14. Fukunaga, T., Nakamura, K., Suzuki, K., Mizutani, U. (1990). *J. Non-Cryst. Solids* 117–118:700–703.
15. Fukunaga, T., Mori, M., Inou, K., Mizutani, U. (1991). *Mater. Sci. Eng.* A134:863–866.
16. Lee, C.H., Fukunaga, T., Mizutani, U. (1991). *Mater. Sci. Eng.* A134:1334–1337.
17. Lee, C. H., Mori, M., Fukunaga, T., Sakurai, K., Mizutani, U. (1992). *Mater. Sci. For.* 88–90:399–406.
18. Sakurai, K., Lee, C. H., Kuroda, N., Fukunaga, T., Mizutani, U. (1994). *J. Appl. Phys.* 75:7752–7755.
19. Harringa, J. L., Cook, B. A., Beaudry, B. J. (1992). *J. Mater. Sci.* 27:801–804.
20. Takacs, L. (1996). In: Suryanarayana, C., et al., eds. *Processing and Properties of Nanocrystalline Materials.* Warrendale, PA: TMS, pp. 453–464.
21. Kaloshkin, S. D., Tomlin, I. A., Andrianov, G. A., Baldokhin, U. V., Shelekhov, E. V. (1997). *Mater. Sci. For.* 235–238:565–570.
22. Kuhrt, C., Schropf, H., Schultz, L., Arzt, E. (1993). In: deBarbadillo, J. J., et al., eds. *Mechanical Alloying for Structural Applications.* Materials Park, OH: ASM. International, pp. 269–273.
23. Mio, H., Kano, J., Saito, F., Kaneko, K. (2002). *Mater. Sci. Eng.* A332:75–80.
24. Chen, Y., Bibole, M., Le Hazif, R., Martin, G. (1993). *Phys. Rev.* B48:14–21.
25. Pochet, P., Tominez, E., Chaffron, L., Martin, G. (1995). *Phys. Rev.* B52:4006–4016.
26. Gonzalez, G., D'Angelo, L., Ochoa, J., Lara, B., Rodriguez, E. (2002). *Mater. Sci. For.* 386–388:159–164.
27. Calka, A., Nikolov, J. I., Ninham, B. W. (1993). In: deBarbadillo, J. J., et al., eds. *Mechanical Alloying for Structural Applications.* Materials Park, OH: ASM International, pp. 189–195.
28. Calka, A., Radlinski, A. P. (1991). *Mater. Sci. Eng.* A134:1350–1353.
29. Sherif El-Eskandarany, M., Aoki, K., Sumiyama, K., Suzuki, K. (2002). *Acta Mater.* 50:1113–1123.
30. Rochman, N. T., Kuramoto, S., Fujimoto, R., Sueyoshi, H. (2003). *J. Mater. Proc. Technol.* 138:41–46.
31. Rochman, N. T., Kawamoto, K., Sueyoshi, H., Nakamura, Y., Nishida, T. (1999). *J. Mater. Proc. Technol.* 89–90:367–372.
32. Eckert, J., Schultz, L., Urban, K. (1990). *Z. Metallkde.* 81:862–868.
33. Riffel, M., Schilz, J. (1998). *J. Mater. Sci.* 33:3427–3431.
34. Bab, M. A., Mendoza-Zélis, L., Damonte, L. C. (2001). *Acta Mater.* 49:4205–4213.
35. Delogu, F., Schiffini, L., Cocco, G. (2001). *Mater. Sci. For.* 360–362:337–342.
36. Galdeano, S., Chaffron, L., Mathon, M. -H., Vincent, E., De Novion, C. -H. (2001). *Mater. Sci. For.* 360–362:367–372.
37. Gerasimov, K. B., Gusev, A. A., Ivanov, E. Y., Boldyrev, V. V. (1991). *J. Mater. Sci.* 26:2495–2500.
38. Padella, F., Paradiso, E., Burgio, N., Magini, M., Martelli, S., Guo, W., Iasonna, A. (1991). *J. Less Common Metals* 175:79–90.
39. Guo, W., Iasonna, A., Magini, M., Martelli, S., Padella, F. (1994). *J. Mater. Sci.* 29:2436–2444.
40. Ahn, J. H., Zhu, M. (1998). *Mater. Sci. For.* 269–272:201–206.
41. Kaloshkin, S. D. (2000). *Mater. Sci. For.* 342–346:591–596.
42. Suryanarayana, C. (1995). *Intermetallics* 3:153–160.
43. Larson, J. M., Luhman, T. S., Merrick, H. F. (1977). In: Meyerhoff, R. W., ed. *Manufacture of Superconducting Materials.* Materials Park, OH: ASM International, pp. 155–163.
44. Biswas, A., Dey, G. K., Haq, A. J., Bose, D. K., Banerjee, S. (1996). *J. Mater. Res.* 11: 599–607.

45. Katamura, J., Yamamoto, T., Qin, X., Sakuma, T. (1996). *J. Mater. Sci. Lett.* 15:36–37.
46. Gonzalez, G., Sagarzazu, A., Villalba, R., Ochoa, J., D'Onofrio, L. (2001). *Mater. Sci. For.* 360–362:355–360.
47. Gonzalez, G., D'Angelo, L., Ochoa, J., D'Onofrio, L. (2001). *Mater. Sci. For.* 360–362:349–354.
48. Lu, L., Lai, M.O. (1998). *Mechanical Alloying*. Boston, MA: Kluwer.
49. Watanabe, R., Hashimoto, H., Park, Y. -H. (1991). In: Pease, L. F., Sansoucy, R. J. III, eds. *Advances in Powder Metallurgy 1991*. Vol. 6. Princeton, NJ: Metal Powder Industries Federation, pp. 119–130.
50. Park, Y.-H., Hashimoto, H., Watanabe, R. (1992). *Mater. Sci. For.* 88–90:59–66.
51. Tcherdyntsev, V. V., Kaloshkin, S. D., Tomilin, I. A., Shelekhov, E. V., Serdyukov, V. N. (2001). *Mater. Sci. For.* 360–362:361–366.
52. Liu, L., Casadio, S., Magini, M., Nannetti, C. A., Qin, Y., Zheng, K. (1997). *Mater. Sci. For.* 235–238:163–168.
53. Atzmon, M. (1990). *Phys. Rev. Lett.* 64:487–490.
54. Gavrilov, D., Vinogradov, O., Shaw, W. J. D. (1995). Poursartip, A.Street, K., eds., *Proc. Int. Conf. on Composite Materials, ICCM-10*. Vol. 3. Cambridge, UK: Woodhead, p. 11.
55. Gilman, P. S., Benjamin, J. S. (1983). *Annu. Rev. Mater. Sci.* 13:279–300.
56. Takacs, L., Pardavi-Horvath, M. (1994). *J. Appl. Phys.* 75:5864–5866.
57. Chin, Z. H., Perng, T. P. (1997). *Mater. Sci. For.* 235–238:121–126.
58. Umemoto, M., Liu, Z. G., Masuyama, K., Tsuchiya, K. (1999). *Mater. Sci. For.* 312–314:93–102.
59. Kis-Varga, M., Beke, D. L. (1996). *Mater. Sci. For.* 225–227:465–470.
60. Suryanarayana, C., Chen, G. H., Froes, F. H. (1992). *Scripta Metall. Mater.* 26:1727–1732.
61. Sá Lisboa, R. D., Perdgão, M. N. R. V., Kiminami, C. S., Botta F, W. J. (2002). *Mater. Sci. For.* 386–388:59–64.
62. Eigen, N. (2003). GKSS Research Center, Geesthacht, Germany, private communication.
63. Niu, X. P. (1991). PhD Thesis, Katholieke Universiteit Leuven, Belgium.
64. Pallone, E. M. J. A., Hanai, D. E., Tomasi, R., Botta F, W. J. (1998). *Mater. Sci. For.* 269–272:289–294.
65. Klassen, T., Oehring, M., Bormann, R. (1997). *Acta Mater.* 45:3935–3948.
66. Zhou, G. F., Bakker, H. (1995). *Mater. Sci. For.* 179–181:79–84.
67. Goodwin, P. S., Mukhopadhyay, D. K., Suryanarayana, C., Froes, F. H., WardClose, C. M. (1996). In: Blenkinsop, P., et al., ed. *Titanium '95*. Vol. 3. London, UK: Institute of Materials, pp. 2626–2633.
68. Miki, M., Yamasaki, T., Ogino, Y. (1992). *Mater. Trans. Jpn. Inst. Metals* 33:839–844.
69. Calka, A., Williams, J. S. (1992). *Mater. Sci. For.* 88–90:787–794.
70. Chen, Y., Williams, J. S. (1996). *Mater. Sci. For.* 225–227:881–888.
71. Koch, C. C., Cavin, O. B., McKamey, C. G., Scarbrough, J. O. (1983). *Appl. Phys. Lett.* 43:1017–1019.
72. Wang, K. Y., Shen, T. D., Quan, M. X., Wang, J. T. (1992). *J. Mater. Sci. Lett.* 11:129–131.
73. Wang, K. Y., Shen, T. D., Quan, M. X., Wang, J. T. (1992). *Scripta Metall. Mater.* 26:933–937.
74. Wang, K. Y., Shen, T. D., Quan, M. X., Wang, J. T. (1992). *J. Mater. Sci. Lett.* 11: 1170–1172.
75. Wang, K. Y., Shen, T. D., Quan, M. X., Wang, J. T. (1992). *J. Non-Cryst. Solids* 150: 456–459.
76. Ogino, Y., Yamasaki, T., Maruyama, S., Sakai, R. (1990). *J. Non-Cryst. Solids* 117–118: 737–740.
77. Lee, P. Y., Koch, C. C. (1987). *J. Non-Cryst. Solids* 94:88–100.

78. Portnoy, V. K., Fadeeva, V. I., Zaviyalova, I. N. (1995). *J. Alloys Compounds* 224:159–161.
79. Schneider, M., Pischang, K., Worch, H., Fritsche, G., Klimanek, P. (2000). *Mater. Sci. For.* 343–346:873–879.
80. Imamura, H., Sakasai, N., Kajii, Y. (1996). *J. Alloys Compounds* 232:218–223.
81. Lee, W., Kwun, S. I. (1996). *J. Alloys Compounds* 240:193–199.
82. Rodriguez, J. A., Gallardo, J. M., Herrera, E. J. (1997). *J. Mater. Sci.* 32:3535–3539.
83. Caamaño, Z., Pérez, G., Zamora, L. E., Suriñach, S., Muñoz, J. S., Baró, M. D. (2001). *J. Non-Cryst. Solids* 287:15–19.
84. Millet, P., Calka, A., Ninham, B. W. (1994). *J. Mater. Sci. Lett.* 13:1428–1429.
85. Harris, A. M., Schaffer, G. B., Page, N. W. (1993). In: deBarbadillo, J. J., et al., ed. *Mechanical Alloying for Structural Applications*. Materials Park, OH: ASM International, pp. 15–19.
86. Huang, B., Ishihara, K. N., Shingu, P. H. (1997). *Mater. Sci. Eng.* A231:72–79.
87. Froyen, L., Delaey, L., Niu, X. P., LeBrun, P., Peytour, C. (1995). *JOM* 47(3):16–19.
88. Lu, L., Lai, M. O., Wu, C. F., Breach, C. (1999). *Mater. Sci. For.* 312–314:357–362.
89. Öveçoğlu, M. L., Nix, W. D. (1986). *Internat. J. Powder Metall.* 22:17–30.
90. Morris, D.G., Morris, M.A. (1990). *Mater. Sci. Eng.* A125:97–106.
91. Ameyama, K., Okada, O., Hirai, K., Nakabo, N. (1995). *Mater. Trans. Jpn Inst. Metals* 36:269–275.
92. Suzuki, T. S., Nagumo, M. (1995). *Mater. Sci. For.* 179–181:189–194.
93. Srinivasan, S., Desch, P. B., Schwarz, R. B. (1991). *Scripta Metall. Mater.* 25:2513–2516.
94. Tracy, M. J., Groza, J. R. (1992). *Nanostructured Mater.* 1:369–378.
95. Fan, G. J., Gao, W. N., Quan, M. X., Hu, Z. Q. (1995). *Mater. Lett.* 23:33–37.
96. Keskinen, J., Pogany, A., Rubin, J., Ruuskanen, P. (1995). *Mater. Sci. Eng.* A196: 205–211.
97. Révész, A., Lendvai, J., Ungár, T. (2000). *Mater. Sci. For.* 343–346:326–331.
98. Zdujic, M., Kobayashi, K. F., Shingu, P. H. (1990). *Z. Metallkde.* 81:380–385.
99. Cabañas-Moreno, J. G., Dorantes, H., López-Hirata, V. M., Calderón, H. A., Hallen-López, J. M. (1995). *Mater. Sci. For.* 179–181:243–248.
100. López-Hirata, V. M., Juárez-Martinez, U., Cabañas-Moreno, J. G. (1995). *Mater. Sci. For.* 179–181:261–266.
101. Saji, S., Neishi, Y., Araki, H., Minamino, Y., Yamane, T. (1995). *Metall. Mater. Trans.* 26A:1305–1307.
102. Enayati, M. H., Chang, I. T. H., Schumacher, P., Cantor, B. (1997). *Mater. Sci. For.* 235–238:85–90.
103. López-Hirata, V. M., Zhu, Y. H., Daucedo-Munoz, M. L., Hernández Santiago, F. (1998). *Z. Metallkde.* 89:230–232.
104. Arce Estrada, E. M., Díaz De la Torre, S., López Hirata, V. M., Cabañas-Moreno, J. G. (1996). *Mater. Sci. For.* 225–227:807–812.
105. Li, F., Ishihara, K. N., Shingu, P. H. (1991). *Metall. Trans.* 22A:2849–2854.
106. Kobayashi, K. F., Tachibana, N., Shingu, P. H. (1990). *J. Mater. Sci.* 25:801–804.
107. Fair, G. H., Wood, J. V. (1993). *Powder Metall.* 36:123–128.
108. Kobayashi, O., Aizawa, T., Kihara, J. (1996). *Mater. Trans. Jpn. Inst. Metals* 37:1497–1504.
109. Lu, L., Lai, M. O., Zhang, S. (1995). *Key Eng. Mater.* 104–107:111.
110. Zhang, Y. F., Lu, L., Yap, S. M. (1999). *J. Mater. Proc. Technol.* 89–90:260–265.
111. Salimon, A. I., Korsunsky, A. M., Kaloshkin, S. D., Tcherdyntsev, V. V., Shelekhov, E. V., Sviridova, T. A. (2001). *Mater. Sci. For.* 360–362:137–142.
112. Hida, M., Asai, K., Takemoto, Y., Sakakibara, A. (1997). *Mater. Sci. For.* 235–238: 187–192.
113. Wang, J. S. C., Donnelly, S. G., Godavarti, P., Koch, C. C. (1988). *Int. J. Powder Metall.* 24:315–325.

114. Suryanarayana, C., Froes, F. H. (1990). *J. Mater. Res.* 5:1880–1886.
115. Wolski, K., Le Caër, G., Delcroix, P., Fillit, R., Thévenot, F., Le Coze, J. (1996). *Mater. Sci. Eng.* A207:97–104.
116. Liang, G., Li, Z., Wang, E. (1996). *J. Mater. Sci.* 31:901–904.
117. Lu, L., Zhang, Y. F. (1999). *J. Alloys Compounds* 290:279–283.
118. McCormick, P. G. (1995). *Mater. Trans. Jpn. Inst. Metals* 36:161–169.
119. Pabi, S. K., Murty, B. S. (1996). *Mater. Sci. Eng.* A214:146–152.
120. Chitralekha, J., Raviprasad, K., Gopal, E. S. R., Chattopadhyay, K. (1995). *J. Mater. Res.* 10:1897–1904.
121. Murty, B. S., Joardar, J., Pabi, S. K. (1996). *Nanostructured Mater.* 7:691–697.
122. Chattopadhyay, P. P., Pabi, S. K., Manna, I. (2001). *Mater. Sci. Eng.* A304–306:424–428.
123. Hwang, S., Nishimura, C., McCormick, P. G. (2001). *Mater. Sci. Eng.* A318:22–33.
124. Lu, Lai, M. O. (1998). *Mechanical Alloying*. Boston, MA: Kluwer, p. 29.
125. Frazier, W. E., Koczak, M. J. (1987). *Scripta Metall.* 21:129–134.
126. Chen, G., Wang, K., Wang, J., Jiang, H., Quan, M. (1993). In: deBarbadillo, J. J., et al., eds. *Mechanical Alloying for Structural Applications.* Materials Park, OH: ASM. International, pp. 183–187.
127. Ivison, P. K., Cowlam, N., Soletta, I., Cocco, G., Enzo, S., Battezzati, L. (1991). *Mater. Sci. Eng.* A134:859–862.
128. Ivison, P. K., Soletta, I., Cowlam, N., Cocco, G., Enzo, S., Battezzati, L. (1992). *J. Phys. C: Condens. Matter.* 4:1635–1645.
129. Gaffet, E., Harmelin, M., Faudot, F. (1993). *J. Alloys Compounds* 194:23–30.
130. Ivison, P. K., Soletta, I., Cowlam, N., Cocco, G., Enzo, S., Battezzati, L. (1992). *J. Phys. C: Condens. Matter.* 4:5239–5248.
131. Hwang, S. J., Nash, P., Dollar, M., Dymek, S. (1992). *Mater. Sci. For.* 88–90:611–618.
132. Huang, B. L., Perez, R. J., Crawford, P. J., Nutt, S. R., Lavernia, E. J. (1996). *Nanostructured Mater.* 7:57–65.
133. Eckert, J., Holzer, J. C., Krill, C. E. III, Johnson, W. L. (1992). *Mater. Sci. For.* 88–90:505–512.
134. LeBrun, P., Froyen, L., Munar, B., Delaey, L. (1990). *Scand. J. Metall.* 19:19–22.
135. Shaw, L., Zawrah, M., Villegas, J., Luo, H., Miracle, D. (2003). *Metall. Mater. Trans.* 34A:159–170.
136. Weber, J. H. (1990). In: Clauer, A. H., deBarbadillo, J. J., eds., *Solid State Powder Processing.* Warrendale, PA: TMS, pp. 227–239.
137. Li, Z. L., Williams, J. S., Calka, A. (1998). *Mater. Sci. For.* 269–272:271–276.
138. Hong, L. B., Bansal, C., Fultz, B. (1994). *Nanostructured Mater.* 4:949–956.
139. Qin, Y., Chen, L., Shen, H. (1997). *J. Alloys Compounds* 256:230–233.
140. Klassen, T., Herr, U., Averback, R. S. (1997). *Acta Mater.* 45:2921–2930.
141. Fu, Z., Johnson, W. L. (1993). *Nanostructured Mater.* 3:175–180.
142. Mishurda, J. C. (1993). University of Idaho, Moscow, ID, Unpublished results.
143. Kimura, H., Kimura, M. (1990). In: Clauer, A. H., deBarbadillo, J. J., eds., *Solid State Powder Processing.* Warrendale, PA: TMS, pp. 365–377.
144. Lee, C. H., Mori, M., Fukunaga, T., Mizutani, U. (1990). *Jpn. J. Appl. Phys.* 29: 540–544.
145. Chen, Y., Le Hazif, R., Martin, G. (1992). *Mater. Sci. For.* 88–90:35–41.
146. Gaffet, E., Yousfi, L. (1992). *Mater. Sci. For.* 88–90:51–58.
147. Yamada, K., Koch, C. C. (1993). *J. Mater. Res.* 8:1317–1326.
148. Koch, C. C., Pathak, D., Yamada, K. (1993). In: deBarbadillo, J. J., et al., eds. *Mechanical Alloying for Structural Applications.* Materials Park, OH: ASM International, pp. 205–212.
149. Koch, C. C. (1995). *Mater. Trans. Jpn. Inst. Metals* 36:85–95.

6

Mechanism of Alloying

6.1 INTRODUCTION

A variety of alloys with different constitutions have been synthesized by mechanical alloying. These alloy phases include solid solutions (both equilibrium and metastable), intermediate phases (quasi-crystalline and equilibrium or metastable crystalline phases), and amorphous alloys. The types of materials investigated include metallic, ceramic, polymeric, and composites. Even though the number of phases reported to form in different alloy systems is unusually large [1], and property evaluations have been done in only some cases, and applications have been explored (see Chapters 9, 10, 12, and 17 for details), the number of investigations devoted to an understanding of the mechanism through which the alloy phases form is very limited. This chapter summarizes the information available in this area.

6.2 BALL-POWDER-BALL COLLISIONS

Mechanical alloying (MA) and mechanical milling (MM) involve loading the blended elemental or prealloyed powder particles along with the grinding medium in a vial and subjecting them to heavy deformation. During this process the powder particles are repeatedly flattened, cold welded, fractured, and rewelded. The processes of cold welding and fracturing, as well as their kinetics and predominance at any stage, depend mostly on the deformation characteristics of the starting powders.

The effects of a single collision on each type of constituent powder particle are shown in Figure 6.1. The initial impact of the grinding ball causes the ductile metal powders to flatten and work harden. The severe plastic deformation increases the

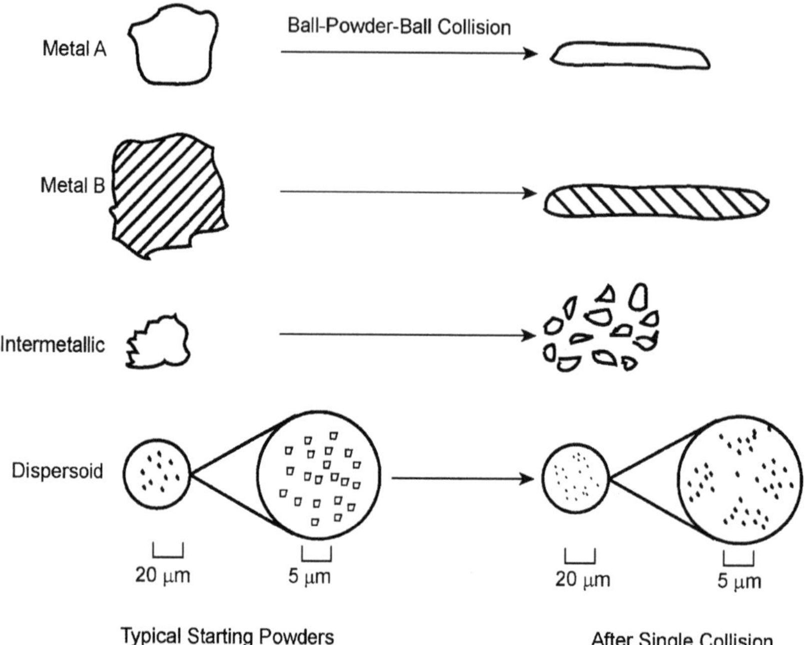

Figure 6.1 Deformation characteristics of representative constituents of starting powders in mechanical alloying. Note that the ductile metal powders (metals A and B) get flattened, whereas the brittle intermetallic and dispersoid particles get fragmented into smaller particles.

surface-to-volume ratio of the particles and ruptures the surface films of adsorbed contaminants. The brittle intermetallic powder particles get fractured and are refined in size. The oxide dispersoid particles are comminuted more severely.

6.3 DIFFERENT STAGES OF PROCESSING

Whenever two grinding balls collide, a small amount of powder is trapped in between them. Typically, around 1000 particles with an aggregate weight of about 0.2 mg are trapped during each collision (Fig. 6.2). During this process, the powder morphology can be modified in two different ways. If the starting powders are soft metal particles, the flattened layers overlap and form cold welds. This leads to formation of layered composite powder particles consisting of various combinations of the starting ingredients. The more brittle constituents tend to become occluded by the ductile constituents and trapped in the composite. The work-hardened elemental or composite powder particles may fracture at the same time. These competing events of cold welding (with plastic deformation and agglomeration) and fracturing (size reduction) continue repeatedly throughout the milling period. Eventually, a refined and homogenized microstructure is obtained, and the composition of the powder particles is the same as the proportion of the starting constituent powders.

Along with the cold welding event described above, some powder may also coat the grinding medium and/or the inner walls of the container. A thin layer of the

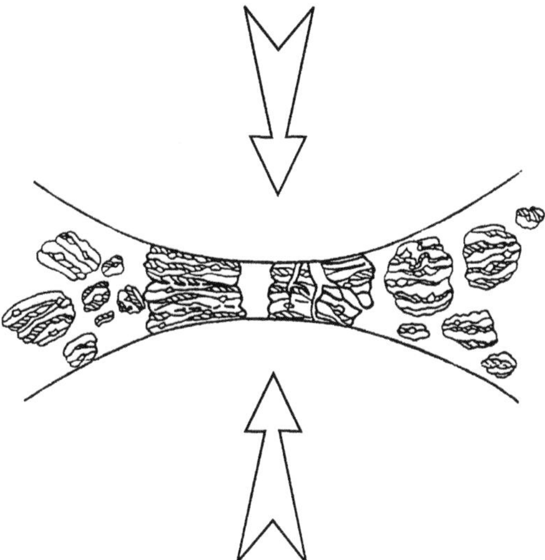

Figure 6.2 Ball–powder–ball collision of powder mixture during mechanical alloying.

coating is beneficial in preventing wear-and-tear of the grinding medium and also in preventing contamination of the milled powder with the debris. However, too thick a layer results in compositional inhomogeneity of the powder and should be avoided.

Let us now look at the different stages of processing an oxide dispersion strengthened material. The starting powders to fabricate this material will consist of two or more metal powders and an oxide dispersoid.

6.3.1 Early Stage of Processing

In the early stages of processing each of the soft metal powder particles is flattened. The different flattened particles form layered composites of the starting elements (Fig. 6.3a); this is referred to as lamellar structure in the MA literature. The size of these composite particles varies significantly from a few micrometers to a few hundred micrometers. The lamellar spacing is also large.

Since it is only the beginning of processing, not all of the individual powder particles get flattened and form composites. Consequently, some unwelded, original starting metal powder particles may also exist. The dispersoids are now closely spaced along the lamellar boundaries. Further, the chemical composition of the different composite particles varies significantly from particle to particle and within the particles.

6.3.2 Intermediate Stage of Processing

With continued milling, the cold welding and fracturing events continue to take place leading to microstructural refinement. At this stage, the particles consist of convoluted lamellae (Fig. 6.3b). Due to the increased amount of cold working, the

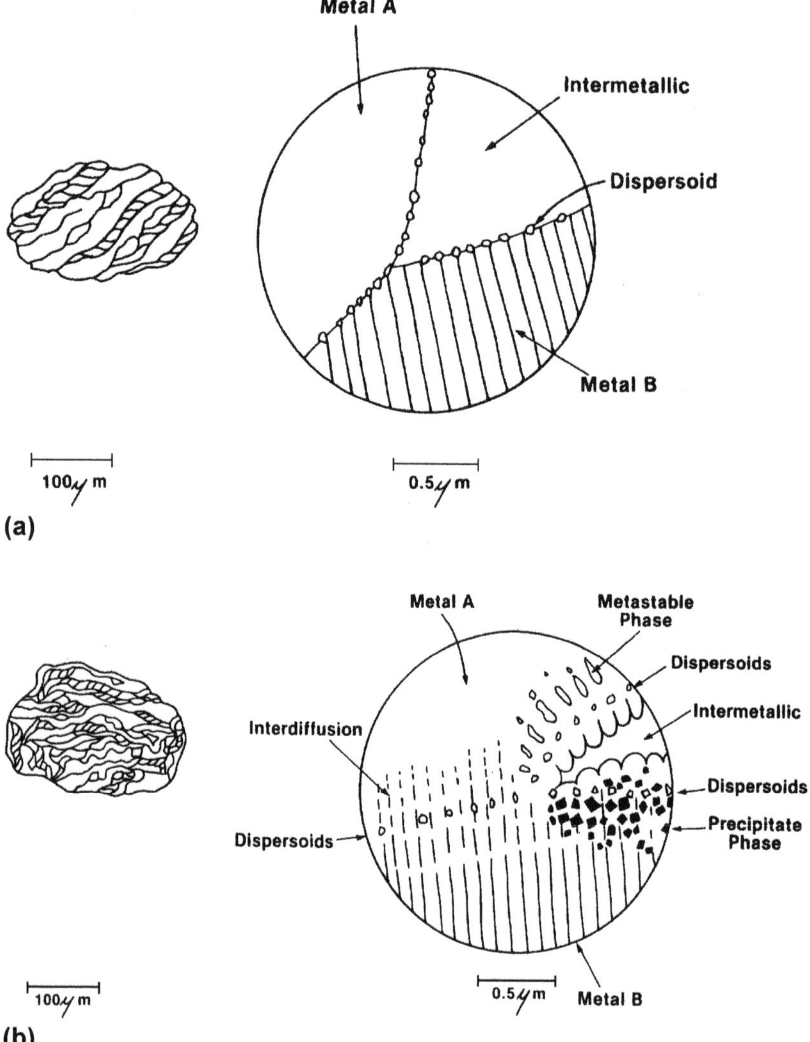

Figure 6.3 (a) Early stage of processing: The individual powder particles are layered composites of the starting elements. (b) Intermediate stage of processing: Individual powder particles consist of convoluted lamellae. A small quantity of new-phase formation may also occur due to some short-range interdiffusion of the constituents. (c) Final stage of processing: All the powder particles and each powder particle approach the composition of the starting powder blend, and the lamellar spacing approaches the dispersoid spacing [5].

number of crystal defects introduced—dislocations, vacancies, grain boundaries, etc.—increases with time and these provide short-circuit diffusion paths. The impact of the ball-ball, ball-powder, and ball-wall collisions also causes a rise in the powder temperature and this further facilitates diffusion. Alloy formation (either stable or metastable phases) occurs due to the combined effect of all these factors. Microstructural refinement continues to take place at this stage, and the oxide dispersion becomes more uniform.

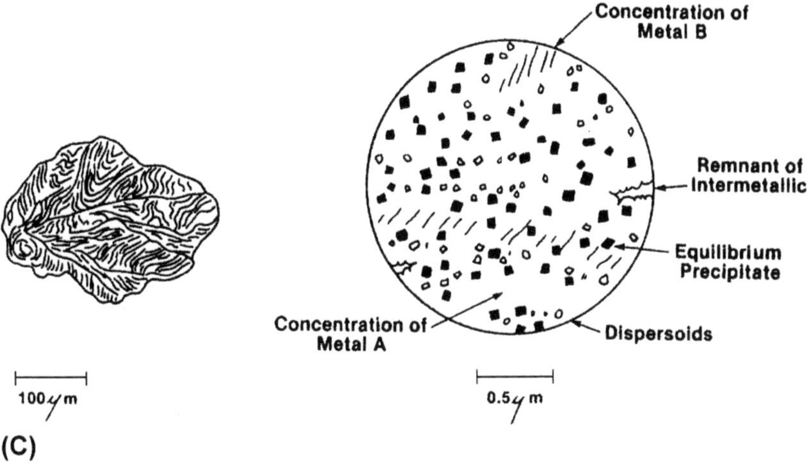

Figure 6.3 Continued.

6.3.3 Final Stage of Processing

At this stage, the lamellae become finer and more convoluted (Fig. 6.3c), and the composition of the individual powder particles approaches that of the overall composition of the starting blend. The lamellar spacing and dispersoid spacing is less than 1 μm, and the microhardness of the particles attains a saturation level, due to accumulation of strain energy.

At the completion of the process, the lamellae are no longer resolvable under the optical microscope. Further processing will not improve the distribution of the dispersoid, and the compositions of the individual powders are now equivalent to the starting powder blend. This is called the steady state.

6.4 EVOLUTION OF PARTICLE SIZE

As mentioned above, the powder particles are plastically deformed and get cold welded if they are soft. Furthermore, work hardening takes place due to the force of impact of the grinding medium and eventually the work-hardened powder particles fracture. The fresh and atomically clean surfaces created during fracture allow the particles to weld together in the cold condition, and this leads to an increase in particle size. Since in the early stages of milling the particles are soft (if we are using either ductile-ductile or ductile-brittle material combination), their tendency to weld together and form large particles is high. A broad range of particle sizes develops, with some as large as three times bigger than the starting particles (Fig. 6.4). The composite particles at this stage have a characteristic layered structure consisting of various combinations of the starting constituents. Even if one is using a single component or prealloyed powder (such as during MM), the particle size increases due to the cold welding of the smaller particles. With continued deformation, the particles get work hardened and fracture by a fatigue failure mechanism and/or by fragmentation of the fragile flakes. Fragments generated by this mechanism may continue to reduce in size

Figure 6.4 Broad distribution of particle sizes in an Al-30 at% Mg powder sample mechanically alloyed for 2 h.

in the absence of strong agglomerating forces. At this stage, the tendency to fracture predominates over cold welding. Due to the continued impact of grinding balls, the structure of the particles is steadily refined, but the particle size continues to be the same. Consequently, the interlayer spacing decreases and the number of layers in a particle increases.

However, it should be remembered that the efficiency of particle size reduction is very low (about 0.1% in a conventional ball mill). The efficiency may be somewhat higher in high-energy ball milling processes but is still less than 1%. The remaining energy is lost mostly in the form of heat, but a small amount is also utilized in the elastic and plastic deformation of the powder particles.

After milling for a certain length of time, steady-state equilibrium is attained when a balance is achieved between the rate of welding, which tends to increase the average particle size, and the rate of fracturing, which tends to decrease the average composite particle size. Smaller particles can withstand deformation without fracturing and tend to be welded into larger pieces, with an overall tendency to drive both very fine and very large particles toward an intermediate size [2]. The particle size distribution at this stage is narrow because particles larger than average are reduced in size at the same rate that fragments smaller than average grow through agglomeration of smaller particles (Fig. 6.5) [3].

From the foregoing it is clear that during MA, heavy deformation is introduced into the particles. This is manifested by the presence of a variety of crystal defects such as dislocations, vacancies, stacking faults, and increased number of grain boundaries. The presence of this defect structure enhances the diffusivity of solute elements into the matrix. Furthermore, the refined microstructural features decrease the diffusion distances. In addition, the slight rise in temperature during milling further aids the diffusion behavior; consequently, true alloying takes place among the constituent elements. While this alloying generally takes place nominally

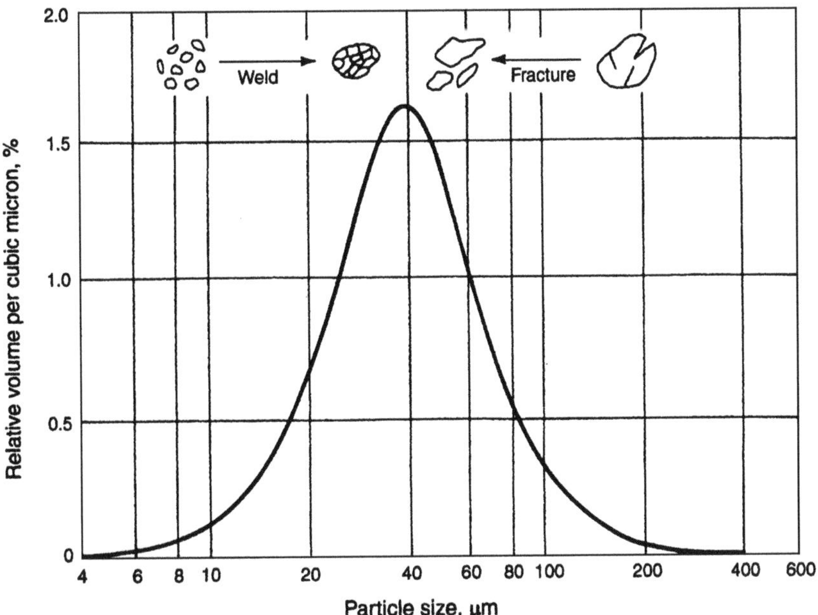

Figure 6.5 Narrow particle size distribution caused by tendency of small particles to weld together and large particles to fractue under steady-state conditions.

at room temperature, sometimes it may be necessary to anneal the mechanically alloyed powder at a slightly elevated temperature for alloying to take place. This is particularly true when formation of intermetallics is desired.

The specific times required to develop a given structure in any system would be a function of the initial particle size and mechanical characteristics of the ingredients as well as the specific equipment used for conducting the MA operation and the operating parameters of the equipment. However, in most cases, the rate of refinement of the internal structure (particle size, crystallite size, lamellar spacing, etc.) is roughly logarithmic with processing time and therefore the size of the starting particles is relatively unimportant. In a few minutes to an hour, the lamellar spacing usually becomes small and the crystallite (or grain) size is refined to nanometer (1 nm $= 10^{-9}$ m or 10 Å) dimensions (Fig. 6.6). The ease with which nanostructured materials can be synthesized is one reason why MA has been extensively employed to produce nanocrystalline materials (see Chapter 13).

As mentioned above, it is possible to conduct MA of three different combinations of metals and alloys: (1) ductile-ductile, (2) ductile-brittle, and (3) brittle-brittle. Therefore, it is convenient to discuss the mechanism of MA in terms of these categories.

6.5 DUCTILE-DUCTILE COMPONENTS

Ductile-ductile is the ideal combination of materials for MA. Benjamin [2] suggested that it was necessary to have at least 15% of a ductile component to achieve alloying. This was because true alloying results from the repeated action of cold welding and

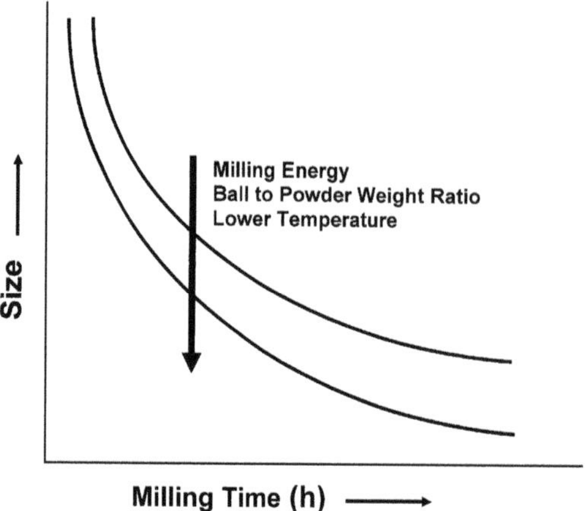

Figure 6.6 Refinement of particle/grain/crystallite size with milling time. The rate of refinement increases with increasing milling energy, ball-to-powder weight ratio, and lower temperatures.

fracturing of powder particles; cold welding cannot occur if the particles are not ductile. However, we will see later that this factor is really not critical since alloying occurs even in completely brittle materials.

Benjamin and Volin [4] first described the mechanism of alloying in a system involving two different ductile components. It was suggested that in the early stages of MA the ductile components get flattened to platelet/pancake shapes by a microforging process. A small quantity of the powder, usually one or two particle thicknesses, also gets welded onto the ball surfaces. This coating of the powder on the grinding medium is advantageous because it prevents excessive wear of the grinding medium. However, the thickness of the powder layer on the grinding medium must be kept to a minimum to avoid formation of a heterogeneous product [5]. In the next stage, these flattened particles get cold welded together and form a composite lamellar structure of the constituent metals. An increase in particle size is also observed at this stage. With increasing MA time, the composite powder particles get work hardened, the hardness and consequently the brittleness increases, and the particles get fragmented resulting in particles with more equiaxed dimensions. With further milling, the elemental lamellae of the welded layer and both the coarse and fine powders become convoluted rather than being linear (Fig. 6.7). This is due to the random welding together of equiaxed powder particles without any particular preference to the orientation with which they weld. Alloying begins to occur at this stage due to the combination of decreased diffusion distances (interlamellar spacing), increased lattice defect density, and any heating that may have occurred during the milling operation. The hardness and particle size tend to reach a saturation value at this stage, called the steady-state processing stage. With further milling, true alloying occurs at the atomic level resulting in the formation of solid solutions, intermetallics, or even amorphous phases. The layer spacing becomes so fine at this stage that it is no longer visible under an optical or scanning electron microscope.

An indication of the completion of the MA process and of the attainment of a homogeneous structure in the powder is the ease with which the powder could be

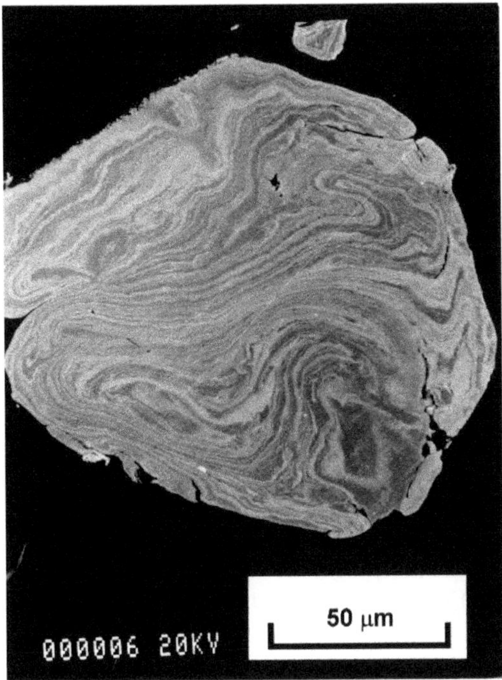

Figure 6.7 Scanning electron micrograph depicting the convoluted lamellar structure obtained during milling of ductile-ductile component system (Ag-Cu).

removed from the grinding medium. Benjamin [2] showed that it was possible to produce a true Ni-Cr alloy starting from elemental powders by demonstrating that the magnetic behavior of the mechanically alloyed powder was identical to that of a homogeneous Ni-Cr alloy produced by melting and working.

Even though structural refinement is a statistical process since a wide variety of structures exist, especially in the initial stages of MA, the rate of structural refinement was found to depend on the rate of mechanical energy input into the process and the work hardening rate of the material being processed [4].

6.6 DUCTILE-BRITTLE COMPONENTS

The traditional ODS alloys fall in this category because the brittle oxide particles are dispersed in a ductile matrix. The microstructural evolution in this type of system was also described by Benjamin and others [5,6]. In the initial stages of milling, the ductile metal powder particles get flattened by the ball-powder-ball collisions, whereas the brittle oxide or intermetallic particles get fragmented/comminuted (Fig. 6.1). These fragmented brittle particles tend to become occluded by the ductile constituents and trapped in the ductile particles. The brittle constituent is closely spaced along the interlamellar spacings (Fig. 6.8a). With further milling, the ductile powder particles get work hardened, and the lamellae get convoluted and refined (Fig. 6.8b) (as described in Sec. 6.4). The composition of the individual particles converges toward the overall composition of the starting powder blend. With

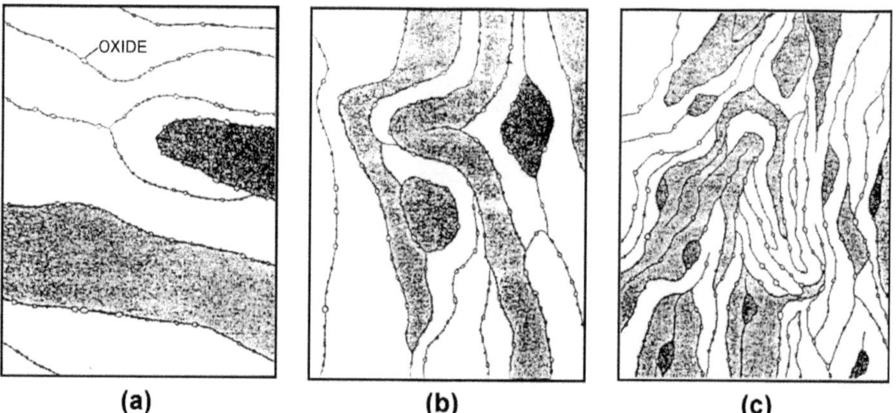

Figure 6.8 Schematics of microstructural evolution during milling of a ductile-brittle combination of powders. This is typical of an oxide dispersion–strengthened case. (a) The oxide particles become trapped in the welds of the composite particles. The weld spacing is large and the concentration of the oxide particles fairly high. (b) The weld spacing decreases and the oxide particle spacing has increased. (c) The oxide is evenly distributed along the welds which have a small spacing.

continued milling, the lamellae get further refined, the interlamellar spacing decreases, and the brittle particles get uniformly dispersed, if they are insoluble, in the ductile matrix, e.g., as in an ODS alloy (Fig. 6.8c). A typical transmission electron micrograph showing the dispersion of Er_2O_3 in a mechanically milled α_2-titanium aluminide matrix is shown in Figure 6.9. On the other hand, if the brittle phase is soluble, alloying occurs between the ductile and brittle components also and chemical homogeneity is achieved. Formation of an amorphous phase on milling a mixture of pure Zr (ductile) and $NiZr_2$ intermetallic (brittle) powder particles is a typical example of this type of system [7]. Whether or not alloying occurs in a ductile-brittle system also depends on the solid solubility of the brittle component in the ductile matrix. If a component has a negligible solid solubility then alloying is unlikely to occur, e.g., boron in iron. Thus, alloying of ductile-brittle components during MA requires not only fragmentation of brittle particles to facilitate short-range diffusion, but also reasonable solid solubility in the ductile matrix component.

6.7 BRITTLE-BRITTLE COMPONENTS

It was mentioned earlier that intimate mixing of the component powder particles occurs due to interplay between the cold welding and fracturing events. Therefore, from an intuitive stand point it would appear unlikely that alloying occurs in a system consisting of two or more brittle components and with no ductile component present. This is because the absence of a ductile component prevents any welding from occurring, and in its absence alloying is not expected to occur. However, alloying has occurred in brittle-brittle component systems such as Si-Ge and Mn-Bi [8,9]. Milling of mixtures of brittle intermetallics [10] also produced amorphous phases.

As mentioned above, the brittle components get fragmented during milling and their particle size gets reduced continuously. However, at very small particle sizes the

Figure 6.9 Transmission electron micrograph showing a uniform dispersion of Er_2O_3 particles in a mechanically alloyed α_2-titanium aluminide alloy matrix.

powder particles behave in a ductile fashion, and further reduction in size is not possible; this is termed the limit of comminution [11].

During milling of brittle-brittle component systems, it has been observed that the harder (more brittle) component gets fragmented and gets embedded in the softer (less brittle) component. Thus, the harder Si particles are embedded in the softer Ge matrix (Fig. 6.10). Once an intimate mixture of the two components is achieved, then

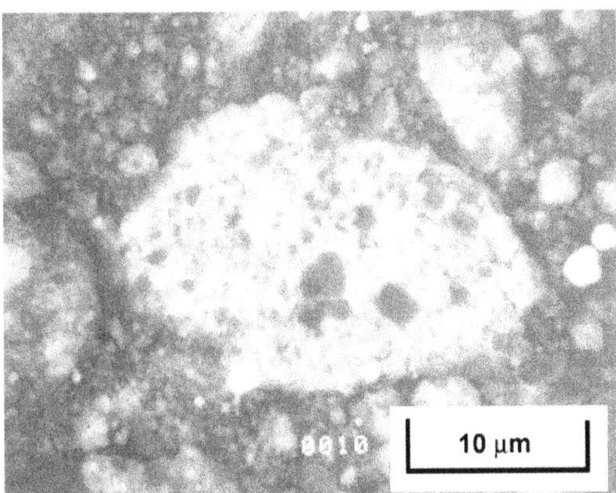

Figure 6.10 Scanning electron micrograph showing that the harder Si particles are incorporated in a softer Ge matrix after mechanical alloying of the Si-Ge powder mix for 12 h.

alloying occurs by diffusional processes. Thus, it was reported that alloying did not occur in the brittle-brittle systems (Si-Ge) at very low temperatures, e.g., liquid nitrogen temperatures, whereas alloying did occur at subambient temperatures in the ductile-ductile and ductile-brittle systems. This may be due to the longer diffusion distances required in the brittle-brittle granular vs. ductile-ductile lamellar geometry, and/or the enhanced diffusion paths provided by dislocations, grain boundaries, and other defects introduced via severe plastic deformation in the ductile-ductile systems.

Possible mechanisms that may contribute to material transfer during milling of brittle components may include plastic deformation, which is made possible by (1) local temperature rise, (2) microdeformation in defect-free volumes, (3) surface deformation, and/or (4) hydrostatic stress state in the powders during milling [9].

REFERENCES

1. Suryanarayana, C. (2001). *Prog. Mater. Sci.* 46:1–184.
2. Benjamin, J. S. (1976). *Sci. Am.* 234(5):40–48.
3. Lee, P. Y., Yang, J. L., Lin, H. M. (1998). *J. Mater. Sci.* 33:235–239.
4. Benjamin, J. S., Volin, T. E. (1974). *Metall. Trans.* 5:1929–1934.
5. Gilman, P. S., Benjamin, J. S. (1983). *Annu. Rev. Mater. Sci.* 13:279–300.
6. Benjamin, J. S. (1990). *Metal Powder Rep.* 45:122–127.
7. Lee, P. Y., Koch, C. C. (1988). *J. Mater. Sci.* 23:2837–2845.
8. Davis, R. M., Koch, C. C. (1987). *Scripta Metall.* 21:305–310.
9. Davis, R. M., McDermott, B., Koch, C. C. (1988). *Metall. Trans.* A19:2867–2874.
10. Lee, P. Y., Koch, C. C. (1987). *Appl. Phys. Lett.* 50:1578–1580.
11. Harris, C. C. (1967). *Trans. Soc. Min. Engrs.* 238:17.

7

Characterization of Powders

7.1 INTRODUCTION

The product of mechanical alloying (MA) or mechanical milling (MM) is in powder form. Successful application of these materials, either in the as-synthesized powder condition or after consolidation to bulk shape, requires that the powders and/or the products of consolidation be fully characterized. Consolidation of the mechanically alloyed/milled powders requires their exposure to high temperatures and/or pressures. This can be done by a number of techniques such as hot pressing, hot isostatic pressing, hot extrusion, plasma-activated sintering, or shock consolidation. Consolidation of the "nonequilibrium" powders synthesized by MA/MM techniques (containing supersaturated solid solutions; metastable intermediate phases; quasicrystalline, amorphous, or nanocrystalline phases) to full density is a nontrivial problem. Achievement of full density requires that the powders be exposed to high temperatures and/or high pressures for extended periods. This could result in the crystallization of amorphous phases, formation of equilibrium phases from the metastable phases produced, and coarsening of nanometer-sized grains. However, if one consolidates the powders at low temperatures and/or pressures to retain "nonequilibrium" features, sufficient bonding between particles may not exist; consequently, the material will be porous and not fully dense. Therefore, innovative methods must be adopted to achieve full densification while retaining the "nonequilibrium" effects in these mechanically alloyed/milled powders. Sufficient literature is available on this aspect, and the reader is referred to a recent review for additional details [1].

The powders obtained after MA/MM are characterized for their size and shape, surface area, phase constitution, and microstructural features. Measurement of the crystallite size and lattice strain in the mechanically alloyed powders is also very important since the phase constitution and transformation characteristics appear to be critically dependent on both of them, but more significantly on the crystallite size. The nature and density of the chemical defects in the powders also play a significant role. In addition, one could characterize the transformation behavior of the mechanically alloyed/milled powders on annealing or after consolidation to full density, or after other treatments. However, it should be realized that not all investigators characterize their powders/products for all of the above features. The extent of characterization appears to be limited depending on the final requirements and applications of the product. In principle, most of the general techniques used to characterize powders synthesized by conventional methods can also be used to characterize the mechanically alloyed/milled powders. These techniques have been described in detail in standard textbooks and/or reference books on powder metallurgy, so that only a brief description will be given here (see, for example, Refs. 2 and 3 for full details).

7.2 SIZE AND SHAPE

The size and shape of the powder particles may be determined accurately using direct methods of either scanning electron microscopy (SEM) for relatively coarse powders or transmission electron microscopy (TEM) for finer powders.

The powder particles synthesized by MA are expected to have a uniform size and narrow particle size distribution, once steady-state conditions of processing have been reached. Furthermore, since a number of different types of mill are available for milling powders, one may suspect that the particle size and size distribution will be significantly different in powders processed in different mills. However, it has been recently shown that by milling Al_2O_3 powder in three different types of mill (a vibrational mill containing only one large ball, a Fritsch planetary ball mill, and a horizontal Uni-Ball mill equipped with magnets to achieve strong impact, shearing, or chaotic ball movement) and controlling the milling conditions, it was possible to synthesize spherical powders. Furthermore, it was shown that the powder size distributions are similar and that the particle size tended to be bimodal, with the same size distribution, for the case of extended milling times [4]. It has also been possible to produce hollow spherical powders in some cases [4,5]. Figure 7.1 shows typical SEM images of mechanically alloyed Al-20 at% Mg powder milled for 24 h and pure Ni powder milled for 12 h in a SPEX 8000 mixer mill. It may be noted that the powder particles have a range of sizes and that most of the particles have an irregular morphology.

If the powders are spherical, their size is defined by the particle diameter. However, very rarely the mechanically alloyed powder particles have a perfect spherical shape. In the early stages of milling (and also in the late stages in some cases) the powder is flaky and has an irregular shape [6]. In such cases, the equivalent spherical diameter can be determined from the volume, surface area, or projected area, or settling rate measurements. For example, if the measured volume is V, then the equivalent spherical diameter D_v is given as:

$$D_v = (6V/\pi)^{1/3} \tag{7.1}$$

(a)

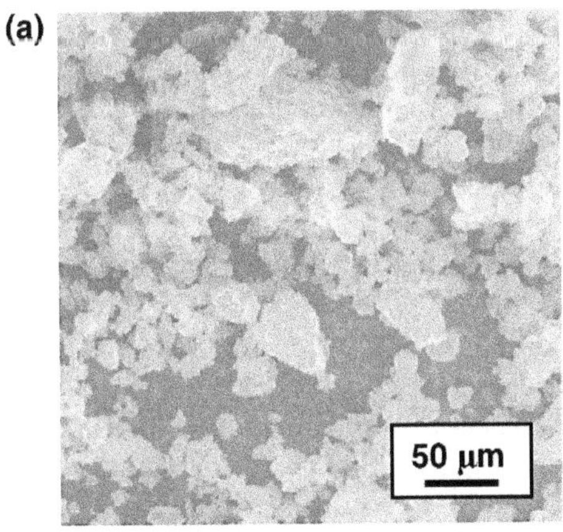

(b)

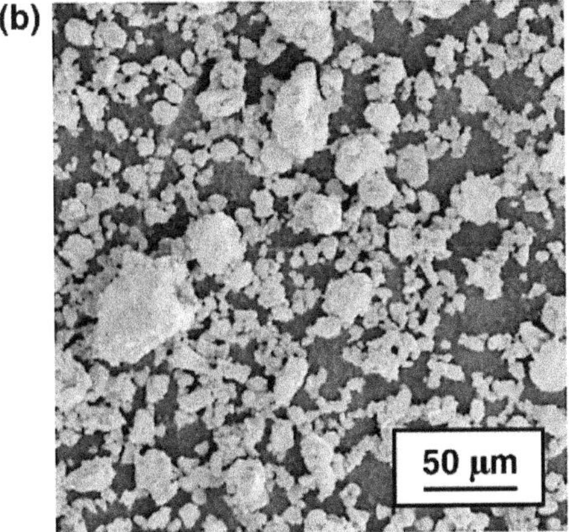

Figure 7.1 Scanning electron micrographs of (a) Al-20 at% Mg powder mechanically alloyed for 24 h and (b) pure Ni powder milled for 12 h in a SPEX mill. Note that the powders have an irregular morphology and different size ranges.

Alternatively, for a measured surface area S, the equivalent spherical surface diameter D_s is simply

$$D_s = (S/\pi)^{1/2} \tag{7.2}$$

If the particles are not spherical, the equivalent spherical projected diameter D_A is calculated by setting the projected area equal to the equivalent area of a circle A. Then

$$D_A = (4A/\pi)^{1/2} \tag{7.3}$$

The particle size distribution can be conveniently obtained by the technique of screening (or sieving). Screening is a mechanical method of separating the different size fractions of a metal powder. The screen surface allows particles smaller than the surface opening to pass through while the larger particles remain on the top of the screening surface. Thus, to determine the size and size distribution of the particles, the powder mix is screened through successively finer sieves of known screen opening size. The screen is often specified by its mesh size, which is the number of wires per inch. The opening size varies inversely with the mesh size; large mesh numbers imply small opening sizes and vice versa. For example, if the powders pass through a 325 mesh, then they are referred to as having a −325-mesh size; in this case, the particle size is less than 44 μm. Table 7.1 lists the typical mesh numbers and the related particle sizes.

It is shown in Figure 7.1 that the as-milled powder contains particles of different sizes. However, by sieving through different mesh sizes, it will be possible to obtain particles of specific size distributions. Figure 7.2a–c shows scanning electron micrographs of mechanically alloyed blended elemental powder mixtures of Al-20 at% Mg powder particles, separated as described above, showing particle size ranges of 88–105, 44–53, and 20–25 μm, respectively. Figure 7.3 shows a histogram of the distribution of weight fraction of the particles classified according to the particle size. It may be noted that the distribution is essentially bimodal with majority of the particles belonging to the very small (<20 μm) or very large (>105 μm) size range. Particles classified as above will be useful for applications requiring a specific size, e.g., to study the effect of particle size on the combustion behavior.

By measuring the particle size and calculating the weight fraction of each particle size range (percentage of particles having that specific size, also known as frequency) a histogram relating the weight fraction and particle size can be constructed. It is of interest to note that the particle size distribution of mechanically alloyed/milled powders is generally Gaussian (log normal), i.e., a bell-shaped curve is obtained when the frequency is plotted on a linear scale against the logarithm of the particle size.

Table 7.1 Relationship among Typical Mesh Numbers, Screen Sizes, and Related Particle Sizes

Mesh no.	U.S. standard sieve opening, mm (in.)	Particle size (μm)
4	4.750 (0.1870)	4750
10	2.000 (0.0787)	2000
18	1.000 (0.0394)	1000
35	0.500 (0.0197)	500
50	0.300 (0.0118)	300
60	0.250 (0.0098)	250
80	0.177 (0.0070)	177
100	0.149 (0.0059)	149
120	0.125 (0.0049)	125
140	0.105 (0.0041)	105
170	0.088 (0.0035)	88
200	0.074 (0.0029)	74
230	0.063 (0.0024)	63
270	0.053 (0.0021)	53
325	0.044 (0.0017)	44
400	0.037 (0.0015)	37
500	0.025 (0.00098)	25
600	0.020 (0.00079)	20

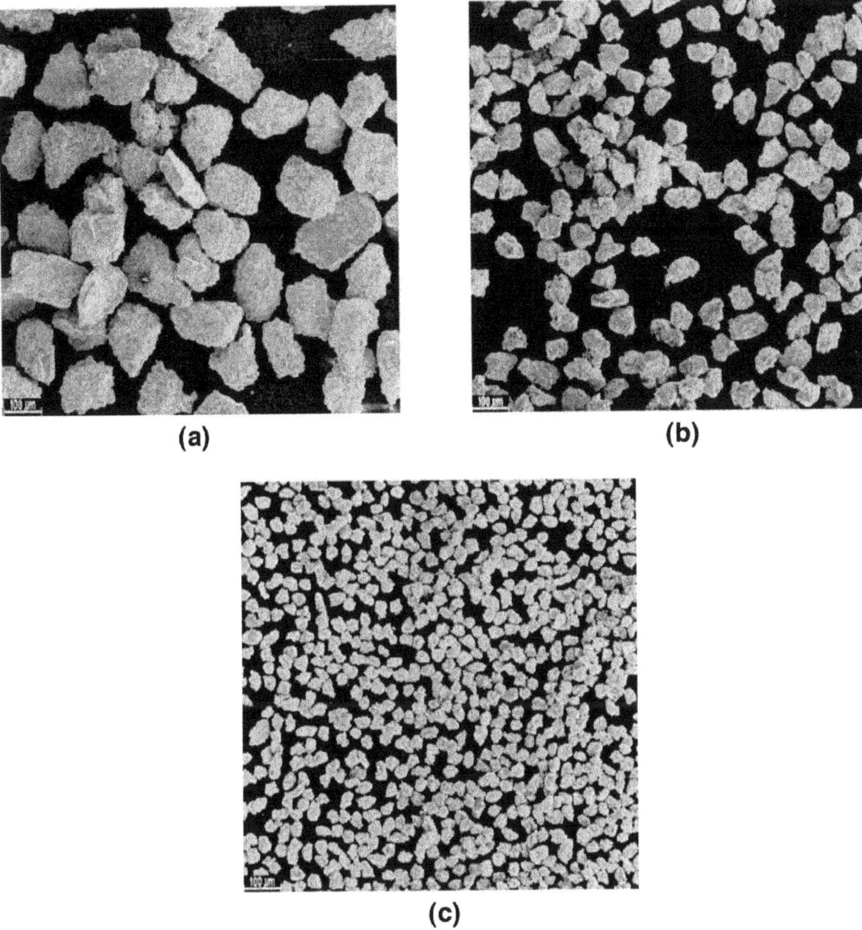

Figure 7.2 Scanning electron micrographs of an Al-20 at% Mg powder mechanically alloyed for 15 h in a SPEX mill and sieved to different sizes. Note the almost spherical shape of the particles present. (a) 88–105 μm, (b) 44–53 μm, and (c) 20–25 μm.

It is important to realize that the powder particles are usually agglomerated and therefore care must be exercised in determining the correct particle size. Thus, a commonly observed powder particle may consist of several individual smaller particles. Furthermore, an individual powder particle may contain a number of grains or crystallites defined as coherently diffracting domains. Microscopic examination normally provides information on the particle size (or even the grain size, if sufficient resolution is available), whereas diffraction techniques (e.g., X-ray) give measures of the crystallite size.

7.3 SURFACE AREA

Knowledge about the surface area of powder particles is useful in understanding the sintering behavior because reduction of surface area is the main driving force for the

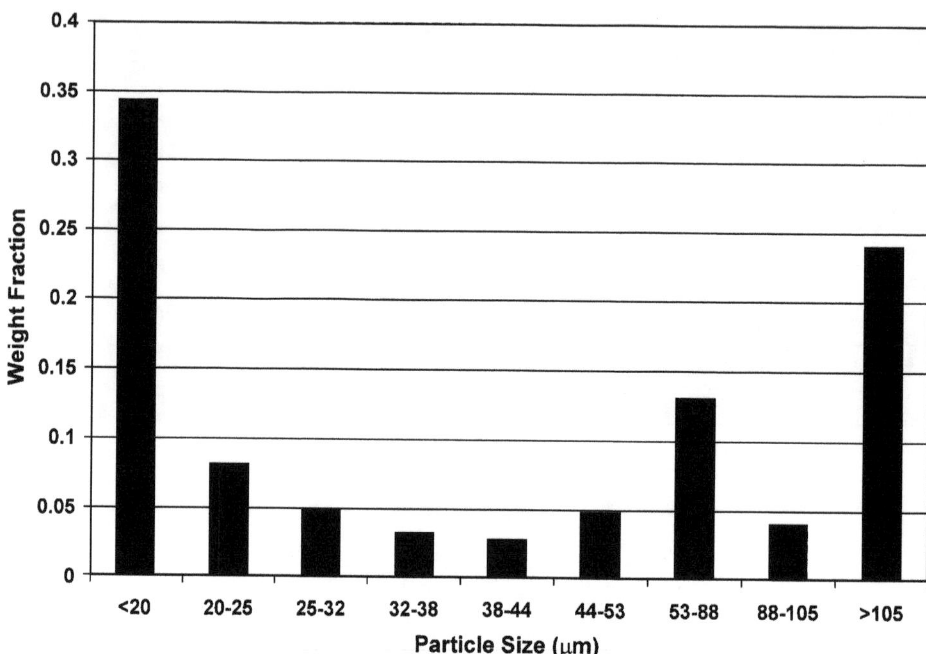

Figure 7.3 Histogram plot of the particle size distribution of the powder (after sieving the powder shown in Fig. 7.2) showing the weight percent of the powder particles as a function of particle size range.

sintering process. Information about the surface area of the powders is also important in evaluating the catalysis aspects. Measurement of surface area of powders relies on calculating the amount of gas absorbed on the sample surface at low temperatures and then estimating the monolayer capacity using the "known" size of the adsorbed molecules. Nitrogen, at liquid nitrogen temperature, is the most common gas used, whereas krypton, at liquid nitrogen temperature, is used for samples with an area less than 1 m^2/g [7].

The Brunauer-Emmett-Teller (BET) method [8] of measuring the specific surface area is based on the above principle. The specific surface area (m^2/g) determined by this method includes both the external and internal (pores) surface area. The surface area of closed pores cannot be determined because the adsorbing gas molecules have no physical path to the surface. Usually, the quantity of gas adsorbed (cm^3) per gram of adsorbent is measured at constant temperature as a function of pressure, p. The BET equation can then be written as

$$\frac{p}{v(p - p_o)} = \frac{1}{v_m c} + \left(\frac{c - 1}{v_m c}\right)\frac{p}{p_0} \tag{7.4}$$

where v is the volume of the gas adsorbed in cm^3 at standard temperature and pressure (STP) conditions per gram of solid at a pressure p, v_m is the monolayer volume, and p_o is the vapor pressure. The constant c is related to the heat of adsorption. By plotting $p/[v(p-p_o)]$ against p/p_o, one obtains a straight line, with the slope $(1/v_m)$ and the intercept $1/v_m c$, from which both c and v_m can be calculated. The

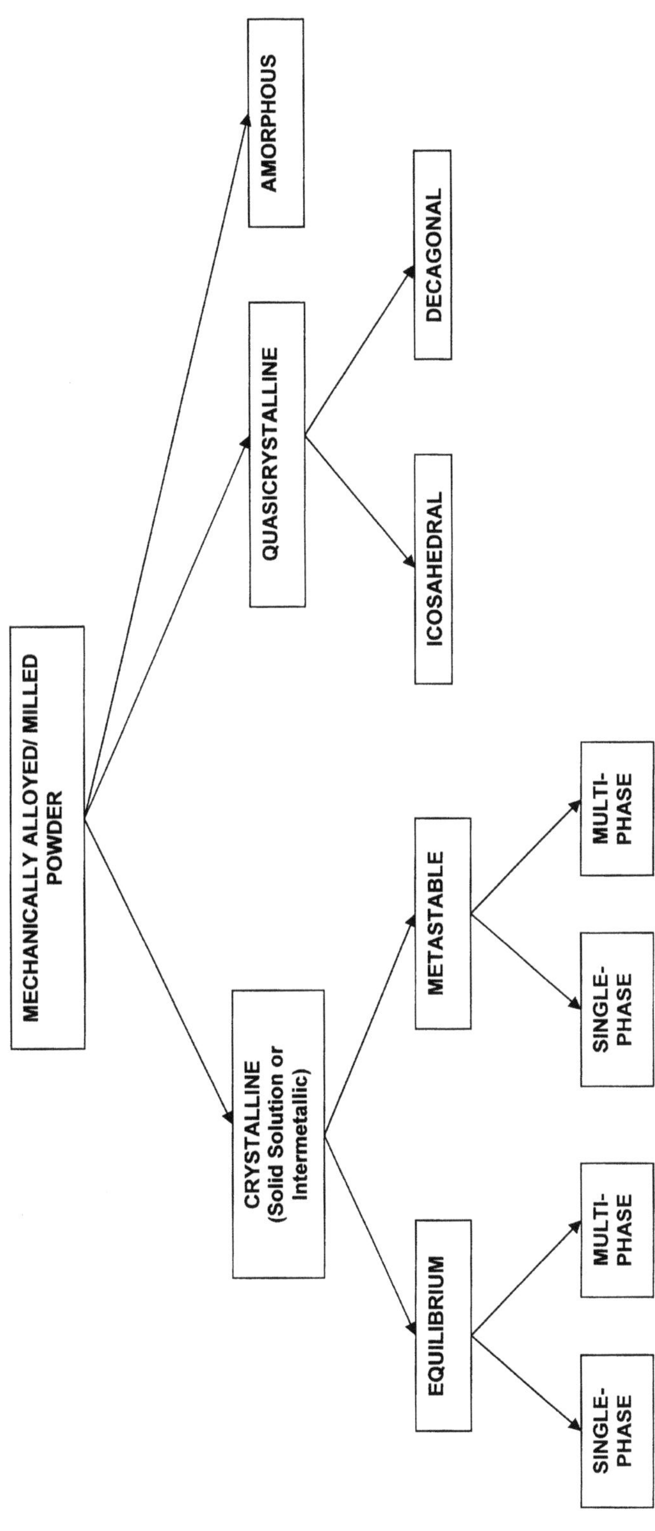

Figure 7.4 Scheme representing the possible constitution of mechanically alloyed/milled powders.

specific surface area Σ (m^2/g) is then obtained from the relation $\Sigma = 0.269\ \sigma v_m$, where σ is the molecular area of nitrogen (0.162 nm^2) and v_m has the units of cm^3/g. The numerical factor of 0.269 comes from Loschmidt's number (molecules per cm^3 at STP). It has been mentioned in the literature that when the pressures are measured near the boiling point of the adsorbed gas (77 K for nitrogen), the range of validity of the BET equation is in the relative pressure (p/p_o) range of 0.05 and 0.35 and that at least four points should be measured in this region. Details of calculations and the precautions to be taken to obtain accurate results may be found in Ref. 7.

7.4 PHASE CONSTITUTION

The powder synthesized by MA/MM techniques may be crystalline, quasi-crystalline, or amorphous in nature. Different possibilities exist in each of these broad categories (Fig. 7.4). If the powder is crystalline, then it may contain one or more phases. Furthermore, if the sample contains a quasi-crystalline phase, then it may be either an icosahedral or a decagonal phase. If the sample contains an amorphous phase, usually it is homogeneous, even though the sample may, in principle, contain more than one amorphous phase (e.g., during decomposition of an amorphous phase two amorphous phases may coexist, i.e., phase separation could occur). However, this has not been actually observed so far in mechanically alloyed/milled materials. Evaluation of the nature and amount of the phase(s) is important in correlating the structure, properties, and transformation behavior of the powders.

7.4.1 Crystalline Phases

Crystalline phases produce well-defined peaks in their diffraction patterns, noting that the width and the maximal intensity of the peaks depend on lattice strain, crystallite size, and other imperfections such as stacking faults in the powders. Figure 7.5 shows typical X-ray and electron diffraction patterns and an electron micrograph from a crystalline phase produced by the MA technique.

X-ray diffraction (XRD) is perhaps the most convenient and common technique used to determine the number and type of crystalline phases. The angular positions of the peaks in the XRD patterns can be used to determine the interplanar spacings using the Bragg equation:

$$\lambda = 2d_{\mathrm{hkl}} \sin \theta \tag{7.5}$$

where λ is the wavelength of the X-ray beam, d_{hkl} is the interplanar spacing for the plane with the Miller indices hkl, and θ is the diffraction angle. XRD patterns could be indexed and the phases present identified by comparing the observed angular positions (or the interplanar spacings) and intensities of the peaks with those of standard patterns available in the Powder Diffraction Files [9]. If all the peaks present and their intensities in the observed diffraction pattern could be matched

Figure 7.5 Typical X-ray and electron diffraction patterns and an electron micrograph from a crystalline phase obtained by MA. (a) Typical X-ray diffraction pattern from a crystalline material. (b) Electron micrograph. (inset) Ring diffraction pattern obtained from a fine-grained (nanocrystalline) material. (c) Single-crystal electron diffraction pattern.

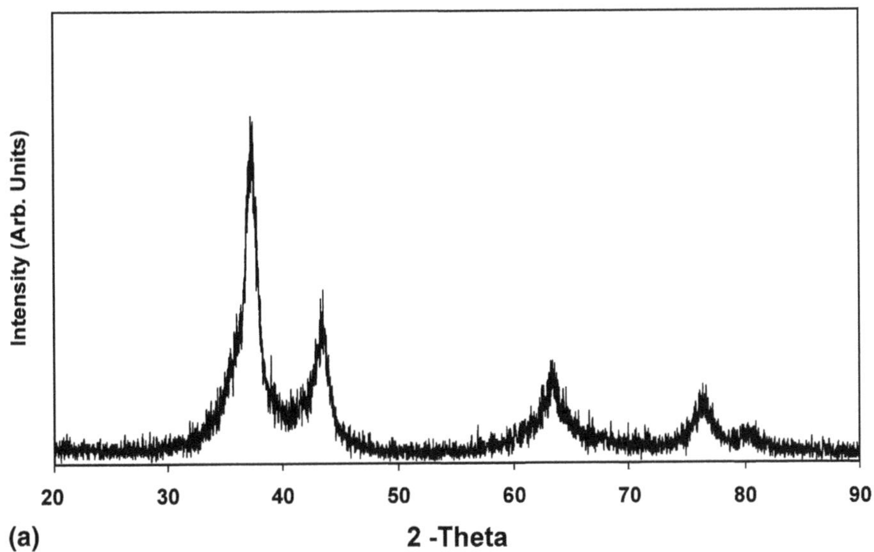
Intensity (Arb. Units)
20
30
40
50
60
70
80
90
(a)
2 -Theta

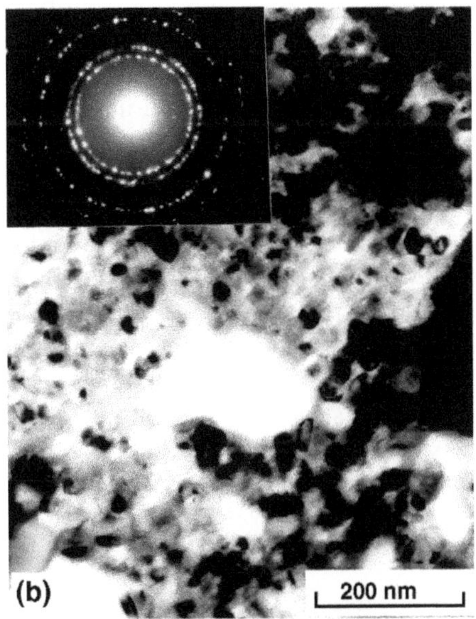
(b)
200 nm

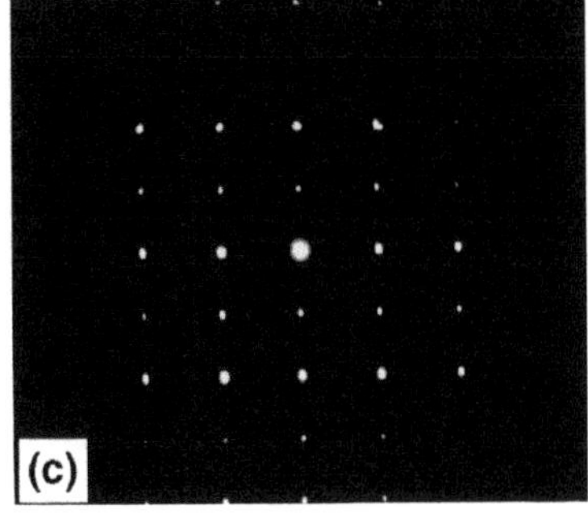
(c)

with those of the standard pattern in the Powder Diffraction Files, then the identification is accurate. It can then be assumed that a single phase was obtained in the mechanically alloyed/milled powder. If this is not the case, then it can be safely inferred that the mechanically alloyed/milled powder contains more than one phase. The XRD pattern could then be indexed, according to the standard procedures, assuming it contains more than one phase [10].

Since formation of metastable phases has been reported to be of common occurrence in mechanically alloyed/milled materials [11], it is possible that the XRD pattern corresponds to the presence of one or more metastable phase(s) in the milled powder. In such a case, there are again two possibilities—the phase (or a similar phase in a homologous alloy system) may have been identified earlier or it may be an entirely new and unidentified phase. If a similar phase has been identified earlier and its structural details have been listed in the Powder Diffraction Files [9] or in the literature, then the structural details (crystal structure and lattice parameters) of the phase obtained in the mechanically alloyed/milled powders could be determined by comparing its diffraction pattern with the available standard patterns, ensuring that the interplanar spacings and the intensities of *all* the diffraction peaks match perfectly.

In case a metastable phase with an unknown structure has formed, then its structure has to be determined by following standard procedures for identifying crystal structures from first principles. So far this has not been necessary for mechanically alloyed/milled powders. Full details of the procedures to identify crystal structures of both known and unknown phases could be found in standard X-ray diffraction textbooks (see, for example, Refs. 10 and 12). It has been frequently reported that some stable or metastable transition phases form during MA of blended elemental powders. If the stability of such phases is reasonably high, then conventional XRD methods, as mentioned above, could be used to identify their crystal structures. However, these methods will be ineffective if the phase under question has a low stability. To overcome this difficulty, time-resolved X-ray diffraction (TRXRD) techniques have been utilized to record XRD patterns with a temporal resolution of about 50 ms [13,14]. The TRXRD method, coupled with infrared thermography, provides information on the structural phase transformations and direct observations of the heat release. This combination has been found to be very effective in detecting the formation of transitional phases of short "lifetime," e.g., in Fe-Al alloys, and in investigating the mechanically activated self-propagating high-temperature synthesis (MASHS) reactions [13].

7.4.2 Quasi-crystalline Phases

It will be shown in Chapter 10 that MA can also synthesize quasi-crystalline phases from blended elemental powders. Even though both XRD and electron diffraction techniques have been employed earlier to identify and index these phases in rapidly solidified alloys, electron diffraction methods should be preferred. This is especially true for the mechanically alloyed materials. Since the mechanically alloyed powder contains very small particles and a significant amount of lattice strain, the X-ray peaks are broadened. The peak positions are very close to each other, especially at lower angles, in most of the quasi-crystalline materials. The combined effect of the close spacing and peak broadening makes an unambiguous identification of the quasi-crystalline phase difficult by XRD methods. Since electron diffraction patterns can be obtained from very small areas, and also because the aperiodic spacing of the diffraction spots is very easy to identify, electron diffraction is the preferred method to identify the formation of quasi-crystalline phases in mechanically alloyed materials. Quasi-crystalline phases may be either the icosahedral (with fivefold rotation symmetry) or the decagonal (with tenfold rotation symmetry) type (among other possibilities). So far only the icosahedral-type phases have been reported in mechanically alloyed powders.

7.4.3 Amorphous Phases

It will be shown in Chapter 12 that MA and MM techniques have produced amorphous phases in a number of alloy systems [11]. Amorphous phases give rise to diffuse halos in their diffraction patterns. It is desirable to confirm the formation of the amorphous phase using direct transmission electron microscopic techniques. Transmission electron micrographs from amorphous phases do not show any diffraction contrast and their electron diffraction patterns show broad and diffuse halos. Figure 7.6 shows typical X-ray and electron diffraction patterns from an amorphous phase in mechanically alloyed powders. Differentiation between a "truly" amorphous (i.e., without translational symmetry as in a liquid) and micro- or nanocrystalline structure (i.e., an assembly of randomly oriented fragments of a bulk crystalline phase) has not been easy on the basis of diffraction studies alone; considerable confusion exists in the literature. In the diffraction experiment of an "amorphous" structure, the intensity but not the phase of the scattered radiation is measured. Fourier inversion of these data yields only the radial distribution function of the structure, which cannot uniquely specify the atomic positions. To determine the structure, the experimentally determined radial distribution function must be compared with the radial distribution functions calculated from the structural models being considered [15,16].

The occurrence of an amorphous phase is generally inferred by observing the presence of broad and diffuse peaks in the XRD patterns. It should be noted, however, that XRD patterns present only an average picture. Thus, by observing the broad X-ray peaks alone, it is not possible to distinguish among materials that are (1) truly amorphous, (2) extremely fine grained, or (3) a material in which a very small volume fraction of crystals is embedded in an amorphous matrix. Hence, in recent years, it has been the practice to designate these identifications as "X-ray amorphous," indicating that the conclusions were reached only on the basis of observation of a broad peak (or a few diffuse peaks) in the XRD pattern. There have been several examples of reports of a phase reported as amorphous on the basis of XRD studies alone. But, based on supplementary investigations by neutron diffraction and/or TEM techniques, which have specific advantages and/or a higher resolution capability than the XRD technique, it could be unambiguously confirmed that the phase produced was not truly amorphous [17,18].

Neutron diffraction techniques have the advantage of detecting lighter atoms in the presence of heavy atoms, whereas the scattering intensity of the lighter atoms may be completely masked by that of the heavy atoms in techniques such as XRD or electron diffraction. In addition, the technique of neutron diffraction has the ability to distinguish between neighboring elements in the periodic table [19]. XRD techniques are unsuitable for this because neighboring elements have their atomic

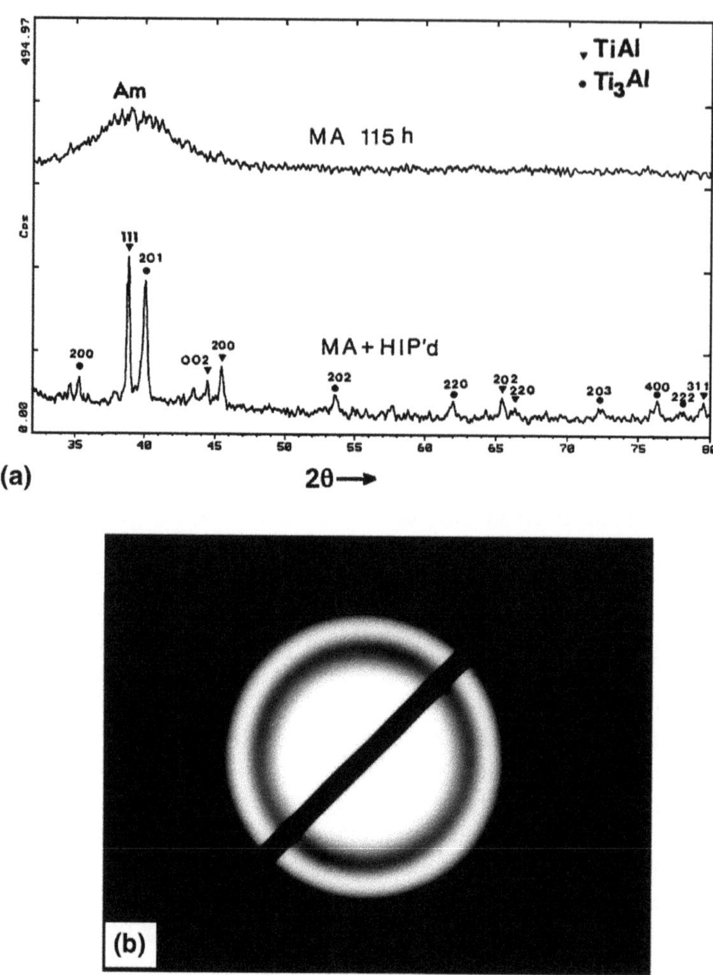

Figure 7.6 Typical (a) X-ray and (b) electron diffraction patterns from an amorphous phase obtained by MA.

scattering factors very close to each other. Consequently, the difference in their scattering factors and hence the diffracted intensity will be very small to be detected. Thus, it is desirable that XRD observations are confirmed by other techniques as well. For example, TEM studies can confirm the lack of contrast in the micrographs for a truly amorphous phase.

The appearance of a glass transition temperature during differential thermal analysis (DTA) or differential scanning calorimetry (DSC) studies is a clear and unambiguous indication of the presence of an amorphous (glassy) phase in the material studied. However, a glass transition may be obscured in many traditional metallic glass systems by the onset of crystallization and hence may not be observed in all cases. This is because the glass transition temperature and crystallization temperature are very close to each other. But, in the case of bulk metallic glasses, the glass transition and the crystallization temperatures could be separated by as much as 135 K [20,21]. DTA/DSC studies can show the presence of an exothermic peak on heating the sample indicating that crystallization of the amorphous phase has occurred. If the material is extremely fine grained (nanocrystalline, not amorphous), then only grain growth, driven by the decrease in grain boundary free energy, can occur in these alloys upon heating to high temperatures. Thus, in general, an exothermic peak in a constant heating–rate DSC plot may be due to (1) growth of a phase that nucleated at a lower temperature, or (2) nucleation and growth of a new phase. It is possible to separate these effects by monitoring the heat flow for a sample held at a constant temperature just below the onset of the exothermic peak. The shape of the heat flow signal indicates the nature of the transformation: nucleation and growth of a new phase gives a bell-shaped exothermic signal, while simple growth of a preexisting phase yields an exponentially decreasing exothermal signal. Other indications of grain growth are scanning peaks that are low and wide with a long high-temperature tail. Furthermore, these peaks shift to higher temperatures after preannealing [22,23]. Thus, one should be able to differentiate between amorphous and micro- or nanocrystalline (fine-grained) samples processed by MA techniques by using a combination of techniques such as microscopy, diffraction, and DTA/DSC methods.

7.5 MICROSTRUCTURAL FEATURES

The microstructural features of interest in both the as-synthesized powders and the consolidated condition are the sizes (of grains and second phase particles), shapes of these features, number and distribution of these phases, and their chemical compositions. Since the grain sizes of most mechanically alloyed/milled powders and consolidated products are small (often in the nanometer range), TEM is the most useful technique for observing microstructural features of mechanically alloyed powders. An added advantage of the TEM technique is that one could also simultaneously determine the crystal structure and chemical composition of these nanometer-sized individual phases. Furthermore, the volume fraction of the different phases can be determined using quantitative microscopy techniques. Details of interpretation of microstructural features that can be observed in TEM image may be found in standard textbooks (see, for example, Ref. 24).

Preparation of thin foil specimens for observing structural details in the TEM image is an involved and tedious process. Standard procedures are available to prepare electron-transparent thin-foil specimens from bulk samples, as obtained in a consolidated material [24,25]. However, special methods have been developed to prepare electron-transparent specimens from mechanically alloyed/milled powders to observe their microstructural features in the TEM image [26–29]. Since particulate materials are often too small to be prepared as a self-supporting specimen, their preparation incorporates the use of some form of embedding media to support the particles prior to the use of conventional thinning techniques. Typically, the as-milled powders are embedded in an epoxy, mechanically polished down to a thickness of about 20 μm, and then ion milled. Sometimes the milled powder can be dispersed in methanol using an ultrasonic bath; and then a drop of the particle suspension can be dried on a copper grid coated with holey carbon film. Some particles may be thin enough at the edges that one could observe the microstructural features without further preparation. Recently, focused ion beam milling has become a popular method to prepare thin-foil specimens from powders and other difficult-to-prepare specimens for electron microscopic observation [30,31]. In this method, a liquid metal (usually gallium) ion gun is used to cut trenches as small as 5 × 20 μm, and this small specimen is then milled to electron-transparent thickness and lifted out. This technique has come to be known as focused ion beam lift-out method. The greatest advantage of this technique is that electron-transparent specimens can be prepared from extremely small samples, including powders. Full details of this technique are given in Refs. 30 and 31.

Figure 7.7 shows a TEM micrograph of the Ti-45Al-2.4Si (at%) alloy powder obtained by hot isostatically pressing at a temperature of 875°C and a pressure of 200 MPa for 2 h. The MA powder was synthesized from a mixture of prealloyed Ti-48 at% Al and Ti_5Si_3 powders. The microstructure in the HIP compact shows a bimodal distribution of grain sizes—about 200-nm grains of γ-TiAl and about 100-nm particles of Ti_5Si_3, located mostly at the triple junctions of the γ-TiAl grains. The grain size was measured by the linear intercept method [32]. By measuring the mean grain size and the sizes of the individual phases, and also their frequency, their distributions have been plotted as shown in Figure 7.8. It may be noted that the overall and phase-specific grain size distributions show a nearly log-normal distribution [33].

The crystal structures and chemical compositions of the individual microcrystalline phases can also be determined using TEM techniques. Sometimes it is necessary to determine the chemistry of very thin surface layers (about 1–2 nm thick) of the powders or consolidated products, and for this techniques such as Auger electron spectroscopy (AES) and X-ray photoelectron spectroscopy (XPS) are extremely useful. Atom probe–field ion microscopy (AP-FIM) and three-dimensional atom probe (3D-AP) techniques have also been found to be very helpful in determining the homogeneity of the phases and their chemical compositions in very small areas [34]. As an example, from the contrast in the field ion micrographs recorded from a mechanically alloyed Cu-20 at% Co alloy, it could be concluded

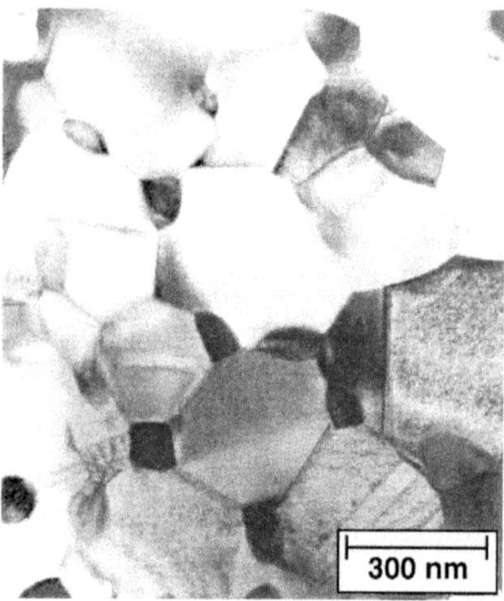

Figure 7.7 Transmission electron micrograph of MA Ti-45Al-2.4Si (at%) after hot isostatic pressing at 875°C.

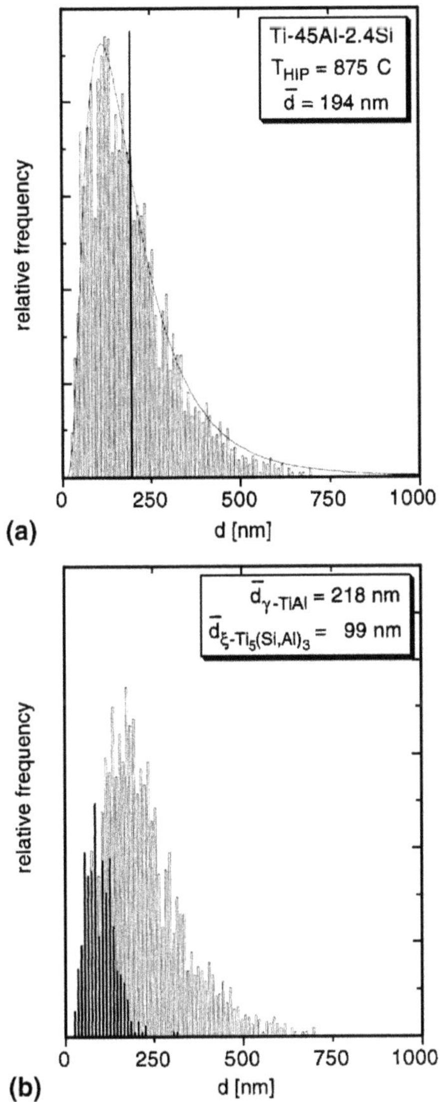

Figure 7.8 Grain size distribution in MA Ti-45Al-2.4Si (at%) after hot isostatic pressing at 875°C. (a) Overall distribution, showing log-normal distribution. (b) Phase-specific grain size distributions.

that the as-synthesized powder showed an inhomogeneous distribution of Co atoms in the solid solution [35]. Mössbauer spectroscopy is another common technique that has been extensively employed to determine the chemical and structural environment of iron atoms on a nearest-neighbor length scale, site occupancies of atoms, and magnetic characteristics of the phases in iron-based materials [36,37].

Experimental details and methods of interpreting the results obtained using the above-mentioned and other techniques may be found in standard handbooks and reference books [38,39]. Enzo [40] recently reviewed the existing issues concerning structural characterization of mechanically alloyed powders with special reference to amorphous and nanocrystalline alloys.

7.6 CRYSTALLITE SIZE AND LATTICE STRAIN

The crystallite size (size of the coherently diffracting domains, often equated with the grain size in mechanically alloyed/milled powders) and lattice strain are perhaps the most important parameters in mechanically alloyed/milled powders. The "grain" size, as obtained by XRD methods, is an average length of columns of the unit cells normal to the reflecting planes, and is also referred to as the column length. However, in subsequent discussions we will use the term grain size to denote the crystallite size/column length determined by the XRD methods.

The grain size of materials can also be determined by direct TEM methods. But the time-consuming sample preparation, the very small area investigated (and the doubt about whether it is representative of the whole sample studied), and the involved interpretation of the results can all be daunting and impractical if a large number of measurements must be made. On the other hand, sample preparation for XRD is simple and the information obtained is an average of a large number of grains. Therefore, XRD methods have been more commonly used to determine the grain size. Furthermore, it is possible to easily obtain the lattice strain data from the same analysis but it is more difficult by TEM methods. On the other hand, characterization of the defect structure, e.g., density and distribution of dislocations, presence of twins, stacking faults, antiphase boundaries, and so forth, can be done more conveniently and accurately by TEM methods. The grain size and lattice strain in mechanically alloyed/milled powders can be determined by measuring the width (breadth) of the XRD peaks and using standard procedures. This has been done in detail in a few cases [41,42]. However, the more common approach has been to calculate the grain size using the Scherrer formula, which does not take into account the lattice strain present in the powder. Consequently, this simple analysis underestimates the actual grain size.

XRD peaks are broadened due to (1) instrumental effects, (2) small particle size, and (3) lattice strain in the material. The presence of stacking faults also affects the positions and shapes of the peaks [10]. With increasing milling time, the particle size decreases and reaches a saturation value. Simultaneously, the lattice strain increases due to the generation of dislocations and other crystal defects. Both of these effects are shown in Figure 7.9 [43]. Consequently, the amount of peak broadening increases with milling time. However, beyond the time when the particle size reaches a saturation value, continued milling will not produce more dislocations (due to the difficulty of generating dislocations at very small grain sizes), but the existing dislocations will be rearranged and some will be annihilated. Therefore, the lattice strain shows a decrease in value.

Before analyzing the XRD patterns to estimate the grain size and lattice strain, it is necessary to first separate the α_1 and α_2 components in the peak, and then concentrate only on the α_1 component. The next step is to subtract the effect of instrumental broadening from the experimental pattern before the individual contributions of grain size and lattice strain could be separated. Since instrumental broadening is independent of the powder under consideration, its contribution is evaluated by recording a diffraction pattern from a well-annealed pure sample such as silicon or LaB_6, which will have a large grain size and no lattice strain. Subtraction of

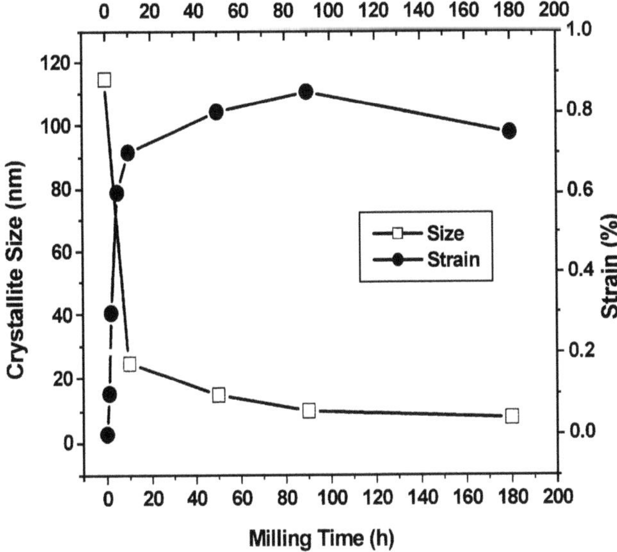

Figure 7.9 Typical plot of crystallite size and lattice strain vs. milling time during MA/MM. Note that the crystallite size decreases and lattice strain increases rapidly in the early stages of milling and then reaches a saturation value [43].

this instrumental broadening from the experimental profile is dependent on the assumption of the intrinsic peak profile (Cauchy, Gaussian, or combined), and the equations needed for this procedure can be found in the literature (see, for example, Ref. 42).

The individual contributions of grain size and lattice strain to total broadening can be separated using standard techniques such as the Williamson-Hall [44] or Warren-Averbach [45] approaches. These techniques have been subsequently modified to obtain grain size distributions [46,47]. Since details of the calculations can be found in some of the standard textbooks on XRD [10,12] and/or research papers [41,42,48–51], only a brief description is provided here. The suggested references can be consulted for full details and limitations of the techniques. Ungar [52] recently reviewed the meaning of size obtained from peak broadening in XRD patterns and compared the size values obtained by XRD methods with those obtained by direct microstructural studies, e.g., TEM.

The grain size of mechanically alloyed/milled powders decreases exponentially with time and reaches a saturation value when steady-state conditions have been achieved. As shown in Figure 7.9, the crystallite/grain size decreases rapidly in the early stages of milling and then more slowly reaching a few nanometers in a short milling time. The grain size then reaches a steady-state value and does not change even if the powder is milled for longer times. We shall later discuss the reasons for obtaining the small grain size and the minimal achievable grain size.

Most commonly the grain size is determined by measuring the full width at half maximal (FWHM) intensity of the Bragg diffraction peak and using the Scherrer formula:

$$B_t = \frac{0.9\lambda}{t \cos \theta} \tag{7.6}$$

where t is the crystallite size, λ is the wavelength of the X-radiation used, B_t is the peak width (in radians) due to particle size effect, and θ is the Bragg angle. This equation can give reasonably correct values only if appropriate corrections for instrumental and strain broadening have been made. However, if one is only interested in following the trend of change of grain size with milling conditions, this simple technique may be acceptable, even without the above-mentioned corrections. In other words, one assumes that the contribution of strain to peak broadening remains constant at all milling times, even though this is not true. The instrumental broadening is the same for all milling times and does not change.

While X-ray peak broadening due to small crystallite size is inversely proportional to cos θ, that due to lattice strain is proportional to tan θ. Thus, by combining these two equations one gets an equation for the total broadening (after subtracting the instrumental broadening) as:

$$B = \frac{0.9\lambda}{t \cos\theta} + \eta \tan\theta \tag{7.7}$$

where η is the strain. Note that B in the above equation is the peak width (in radians) after subtracting the peak width due to instrumental broadening from the experimentally recorded profile. On rearrangement, the above equation can be presented as:

$$B \cos\theta = \frac{0.9\lambda}{t} + \eta \sin\theta \tag{7.8}$$

Thus, when $B \cos\theta$ is plotted against $\sin\theta$, a straight line is obtained with the slope as η and the intercept as $0.9\lambda/t$ (Fig. 7.10). From these values, one could calculate the grain size t and the lattice strain η. It may be noted that the grain size is smaller the larger the intercept and that the strain is higher the larger the slope. A worked out example on how to determine the grain size and lattice strain in mechanically alloyed powders may be found in Ref. 53.

The grain sizes obtained by the indirect X-ray peak broadening studies, after incorporating the appropriate corrections, and the direct TEM techniques are expected to be the same, and this has been found to be true in some cases [54–58].

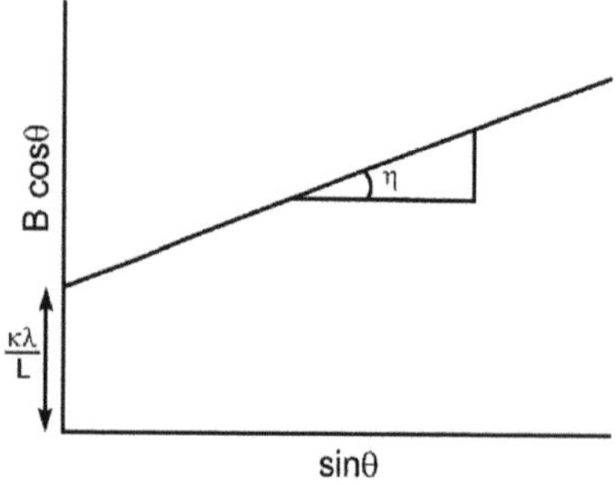

Figure 7.10 Plot of $B \cos\theta$ vs. $\sin\theta$ to determine the grain size and strain in mechanically alloyed powders. The intercept on the Y axis provides information on the grain size and the slope of the straight line gives the strain value.

But sometimes they do not match exactly. This is because sometimes researchers use the Scherrer equation without making the appropriate corrections. Furthermore, the Scherrer equation provides a volume-weighted average of the grain size and underestimates the grain size by 11–38% by neglecting strain broadening. The error in the Scherrer equation depends on the reflection used for the analysis, the amount of strain broadening, and the probability of stacking faults [45]. For example, the presence of stacking faults and twin boundaries are known to reduce the apparent grain sizes [59]. Nonlinearity of the $B \cos \theta$ vs. $\sin \theta$ is clear evidence of the presence of a large number of stacking faults in the crystalline phase. Stacking faults have been frequently observed in the mechanically alloyed/milled materials with face-centered cubic and hexagonal close-packed structures (see, for example, Refs. 60–62). It has been reported that the true grain sizes are an order of magnitude larger than the corresponding apparent values, calculated ignoring the presence of stacking faults [62]. Furthermore, stacking faults induce a decrease of the FWHM value for (111) reflections and an increase for (200) reflections. Therefore, the real grain size lies between these two calculated values [59]. It has been reported that the grain sizes determined by TEM methods are sometimes an order of magnitude larger than those determined by XRD methods. This difference could be due to the fact that TEM methods determine the true grain sizes (with high-angle grain boundaries) while the XRD methods determine the crystallite size; noting that the crystallite size is usually smaller than the grain size and that in extreme cases of very fine grain sizes, both could be the same size.

Comparing the grain size determinations by X-ray and TEM methods, Ungar [52] pointed out that the two measurements can be (1) identical to each other, (2) in good qualitative correlation, or (3) totally different. It was pointed out that the two measurements are in perfect agreement with each other when one is dealing with loose powder particles of submicrometer grain size or nanocrystalline bulk materials, i.e., when the particle size is within the sensitivity range of XRD peak broadening. On the other hand, the X-ray methods can give the average size of subgrains or dislocation cells when one deals with bulk materials containing strain or dislocations. TEM measurements give much larger values of the grain size when one is dealing with relatively larger grain sizes, beyond the general sensitivity of X-ray peak-broadening methods. It is useful to remember that whereas TEM techniques could be used to determine almost any crystallite/grain size, X-ray peak-broadening methods are most appropriate for grain sizes in the range 10–100 nm. For very large grain sizes, the intercept on the y axis becomes smaller and smaller, and therefore the magnitude of error increases. The general conclusion is that the grain size value calculated from the X-ray peak-broadening method is not much meaningful without knowing the microstructure of the sample and that these measurements should be supplemented with other techniques, such as TEM.

7.7 TRANSFORMATION BEHAVIOR

The mechanically alloyed/milled powders usually have very small grain sizes (reaching nanometer levels), and contain significant amount of lattice strain; in addition, the material is usually under far-from-equilibrium conditions containing metastable crystalline, quasi-crystalline, or amorphous phases. All of these effects, either alone or in combination, make the material highly metastable. Therefore, the transformation behavior of these powders to the equilibrium state by thermal treatments is of both scientific and technological importance. Scientifically, it is instructive to know whether transformations in mechanically alloyed/milled materials take place via the same transformation paths and mechanisms that occur in stable equilibrium phases or not. Technologically, it will be useful to know the maximal use temperature of the mechanically alloyed/milled material without any transformation occurring and thus losing the special attributes of this powder product. Some of the different important techniques available to study the transformation behavior of the mechanically alloyed/milled powders/products are described below.

One of the most useful techniques for studying transformation behavior, especially of metastable phases, is DTA or DSC. In this method, a small quantity of the mechanically alloyed/milled (usually metastable) powder is heated at a constant rate to high temperatures under vacuum or in an inert atmosphere. Since phase transformations involve either absorption or evolution of heat, one observes endothermic or exothermic peaks in the DTA/DSC scans as shown in Figure 7.11. Since the values of the peak onset temperature and peak areas depend on the position of the baseline, it is important to obtain an accurate baseline. This is normally done as follows: The sample is heated to the desired temperature when one observes the different peaks. The sample is then cooled back to the ambient temperature and then reheated to higher temperatures. The second plot could be used as the baseline or it could be subtracted from the first plot to obtain the accurate peak positions and areas.

It is true that one could observe endothermic or exothermic peaks during transformation from one equilibrium phase to another during heating (or cooling). For example, an endothermic peak will be observed at the time of melting of a material. But during cooling (or heating), the product phase will revert back to the parent phase. However, such reversible transformations do not occur during

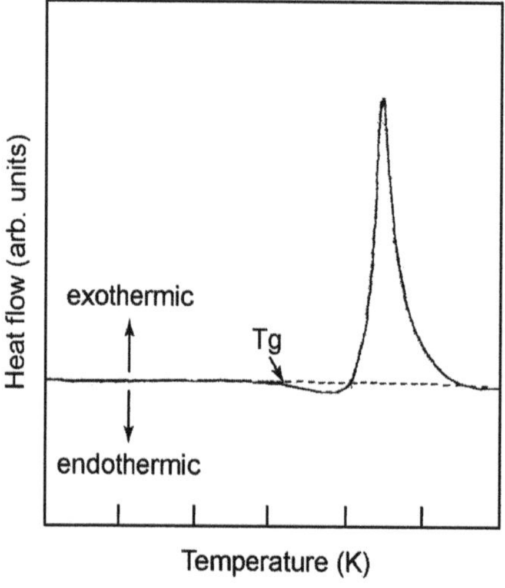

Figure 7.11 A schematic DSC curve showing the exothermic and endothermic peaks.

transformation of metastable phases such as amorphous phases, i.e., a peak of the opposite sign is not observed; in fact, there will be no peak at all. Furthermore, because metastable phases are always more energetic than the corresponding equilibrium phases, they often exhibit exothermic peaks in the DTA/DSC plots.

From the DTA/DSC scans, one can evaluate the temperatures at which phase transformations are occurring as well as the magnitudes of heat released or absorbed. For example, if one heats an amorphous alloy powder to higher and higher temperatures, one expects to observe a glass transition temperature T_g and a crystallization temperature T_x at a still higher temperature. A very broad and shallow peak is observed at relatively low temperatures if structural relaxation of the amorphous phase occurs. It is also possible that complete crystallization takes place in different stages; therefore, more than one exothermic peak is present in the DTA/DSC plot, each peak corresponding to a crystallization event. Figure 7.12 schematically represents these events of crystallization in an amorphous alloy. By quenching the sample from a temperature just above the DTA/DSC peak temperature, and recording the XRD patterns, one can investigate the structural changes that took place in the sample. TEM investigations can also be conducted to uncover the microstructural and crystal structure changes on a finer scale; in addition, compositional changes can be detected. Details of the crystallization behavior of metallic glasses produced by rapid solidification processing techniques have been reviewed [63–65]. Similar techniques and procedures could be used to study the crystallization behavior of amorphous alloys processed by MA/MM techniques. It may be pointed out, however, that there have not been many detailed crystallization studies of amorphous alloys synthesized by the MA/MM techniques [66].

The crystallization temperature is measured either as the temperature at which the exothermic peak starts or the temperature corresponding to the exothermic peak. Whatever is the value chosen, the crystallization temperature changes with the heating rate, and it increases with increasing heating rate. By

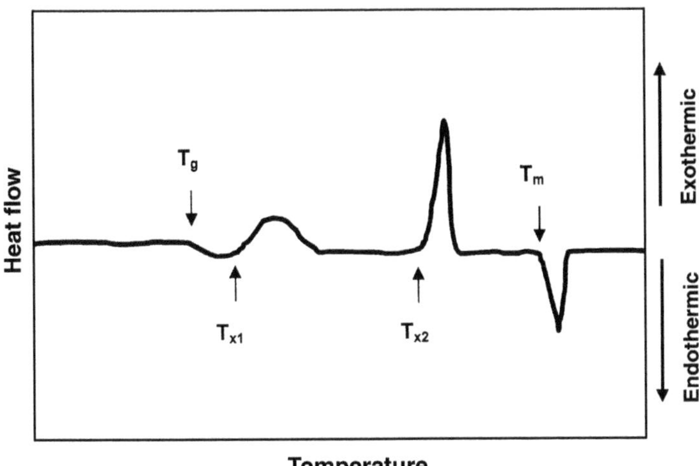

Figure 7.12 A schematic DSC curve depicting the different stages during crystallization of an amorphous phase.

measuring the peak temperatures T_p at different heating rates β, the activation energy for crystallization Q can be evaluated using the Kissinger equation:

$$\ln\left(\frac{\beta}{T_p^2}\right) = \left(-\frac{Q}{RT_p}\right) + A \tag{7.9}$$

where A is a constant and R is the universal gas constant. Thus, by plotting $(\beta/T_p^2$ against $1/T_p$ one obtains a straight line whose slope is $-Q/R$, from which the activation energy Q can be calculated [67]. It is also possible to evaluate the individual activation energies for the nucleation and growth stages of the transformation, from which information about the mechanism of transformation could be ascertained. Thus, with the combination of DTA/DSC and XRD/TEM techniques, it would be possible to obtain full information about the transformation temperatures, the number of stages in which the transformation is occurring, details about the product(s) of each individual transformation (crystal structure, microstructure, and chemical composition), and the activation energy (and also the atomic mechanism) for the transformation. The greatest advantage of the Kissinger method is that one could get the required data during continuous heating in a DSC, which is far more convenient to employ than the time-consuming isothermal analyses studies.

The Kissinger method may not be useful in all studies of decomposition. For example, metallic glasses may decompose by nucleation, growth, or a combination of both processes. In such a case, the decomposition is seldom described by first-order reaction kinetics, when the Kissinger method may not be applicable [68,69].

By recording the isothermal DSC scans at different temperatures, it is possible to study the kinetics of transformation by conducting the Johnson-Mehl-Avrami analysis. For example, during crystallization of an amorphous phase, this is done by calculating the volume fraction of the crystalline phase formed, $x(t)$, at time t and using the equation:

$$x(t) = 1 - \exp(-kt^n) \tag{7.10}$$

where k is a temperature-sensitive factor [$= k_o \exp(-Q/RT)$, where Q is the apparent activation energy and k_o a constant) and n is an exponent that reflects the nucleation rate and/or the growth mechanism. $x(t)$ is obtained by measuring the area under the peak of the isothermal DSC trace at different times and dividing that with the total area. Equation 7.10 can also be written as:

$$\ln[-\ln\{1 - x(t)\}] = \ln k + n \ln(t) \tag{7.11}$$

Thus, by plotting $\ln[-\ln(1 - x)]$ against $\ln(t)$, one obtains a straight line with the slope n. Such analyses have been done in some cases [70,71]. In such cases, when the first-order reaction kinetics are not obeyed, the analysis of the data according to the Johnson-Mehl-Avramai-Kolmogorov method will more accurately provide information about the nature of the transformation. Such an analysis was conducted by He and Courtney [72] for studying the decomposition behavior of mechanically alloyed W-Ni-Fe noncrystalline materials. However, if the reaction can be described as first order, the difference in the values of the activation energy may be small (typically about 10%).

7.8 GRAIN GROWTH STUDIES

Fine-grained materials can be considered to be in a metastable condition due to their small grain size and consequently large surface area, which provides a positive contribution to the free energy of the system. Their free energy can be minimized by increasing their grain size by high temperature annealing, i.e., by allowing grain growth to occur. A study of the grain growth behavior in fine-grained materials would aid the researchers in establishing the highest-use temperatures for consolidation or service. To study the grain growth behavior, the specimens are annealed at different temperatures and their grain sizes at each temperature are measured as a function of time. Figure 7.13 shows the variation of grain size with time on annealing at different temperatures for a Ti-47 at% Al-3 at% Cr powder obtained by MA and consolidated by hot isostatic pressing at 725°C [73]. If D_o is the initial grain size and D is the grain size on annealing for a time t at temperature T, then

$$D^n - D_o^n = Kt \tag{7.12}$$

where K is a constant and n is called the grain size exponent. In most cases, D_o is much smaller than D, and so the above equation, as a first approximation, may be represented as:

$$D^n = Kt \tag{7.13}$$

or

$$D = Kt^{1/n} \tag{7.14}$$

Thus, by plotting log D against log t, at a constant temperature (assuming that the initial grain size D_o is significantly smaller than the grain size D, and so it could be neglected), one obtains a straight line, with a slope of $1/n$. This value of n will be an indicator of the type of barriers to grain growth. But, sometimes it may so happen that the D_o value is not very small (or that the value of D is not substantially larger

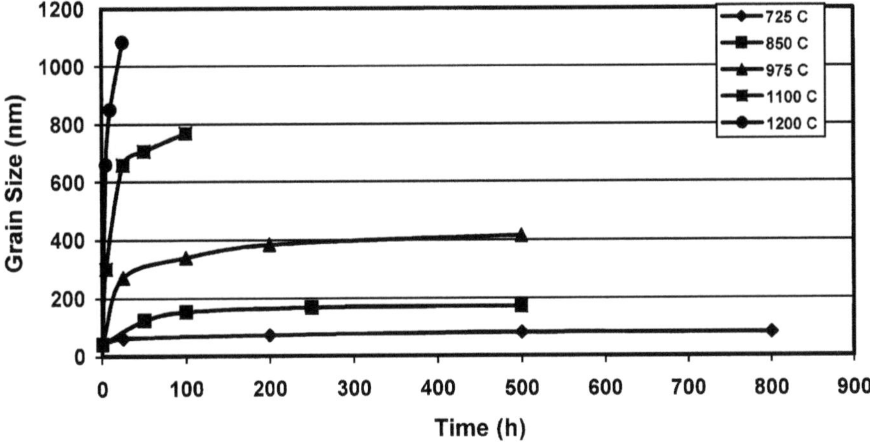

Figure 7.13 Variation of the average grain size with annealing time in mechanically alloyed Ti-47Al-3Cr (at%) powder consolidated at 725°C by hot isostatic pressing and annealed at the temperatures indicated.

than D_0 so that D_o could not be neglected). In such a case, the value of n could be calculated from the following Eq. (7.15), obtained by differentiating Eq. (7.12) with respect to time, t, and then taking logarithms on both sides:

$$\log \frac{dD(t)}{dt} = -(n-1)\log D(t) + \log K - \log n \qquad (7.15)$$

According to theoretical estimates, n is expected to be equal to 2, i.e., a plot of D^2 against t should be a straight line [74]. Most reported experimental work does not follow this trend, even though this is observed to be true in some cases when materials were annealed at very high homologous temperatures to study grain growth behavior or in extremely high-purity metals [75]. In reality, the grain size exponent has been found to vary from low values of 2 to as high as 10 [73], even though in some cases it has been reported to be unrealistically large, e.g., between 8.9 and 16.6, during annealing of mechanically alloyed γ-TiAl alloys [76]. The n values were even higher (between 12.1 and 44.3) when γ-TiAl alloys containing 10 vol% TiB_2 were investigated [76].

The value of K, as the intercept on the y axis, when log D is plotted against log t, measured at different temperatures, obeys the equation:

$$K = K_0 \exp(-Q/RT) \qquad (7.16)$$

where K_0 is another constant and R is the universal gas constant. By plotting ln K against $1/T$, one obtains a straight line, the slope of which corresponds to Q/R, from which Q, the activation energy for grain growth, can be calculated. Grain growth studies have been conducted using TEM, SEM, AP-FIM, XRD, electrical resistivity, and so forth. In fact, any property or technique that is sensitive to changes in grain size could be used to monitor the grain growth behavior.

In addition to the techniques mentioned above, some researchers have also used some specialized and advanced techniques to characterize the microstructure, crystal structure, and surface state/composition of the powders using AES, XPS, and other techniques. Details of most of the characterization techniques mentioned above may be found in standard reference books (see, for example, Refs. 38 and 39).

From the above description it is clear that a number of techniques are available to characterize the mechanically alloyed/milled materials to obtain information about the crystal structure and lattice parameters of the phase(s) produced, calculate the grain size and strain present, evaluate the microstructural features and compositions of the phases, and study the transformation behavior of the non-equilibrium phases. However, it is always beneficial and conclusive to confirm the results obtained by one technique with the aid of another.

REFERENCES

1. Groza, J. R. (1999). In: Suryanarayana, C., ed. *Non-equilibrium Processing of Materials*. Oxford, UK: Pergamon Press, pp. 345–372.
2. German, R. M. (1994). *Powder Metallurgy Science*. 2nd ed. Princeton, NJ: Metal Powder Industries Federation, pp. 28–78.
3. ASM Handbook. Vol. 7. (1998). *Powder Metal Technologies and Applications*. Materials Park, OH: ASM International, 1998.
4. Calka, A., Wexler, D. (2001). *Mater. Sci. For.* 360–362:301–310.

5. Harris, A. M., Schaffer, G. B., Page, N. W. (1996). *Scripta Metall. Mater.* 34:67–73.
6. Zoz, H., Ren, H., Reichardt, R., Benz, H. U. (1999). *P/M Sci. Tech. Briefs* 1:15–19.
7. Giesche, H. (1998). ASM Handbook. *Powder Metal Technologies and Applications*. Vol. 7. Materials Park, OH: ASM International, pp. 274–277.
8. Brunauer, S., Emmett, P. H., Teller, E. (1938). *J. Am. Chem. Soc.* 6:309.
9. Powder Diffraction File. Inorganic Phases Search Manual. Swarthmore, PA: International Center for Diffraction Data.
10. Suryanarayana, C., Norton, M. G. (1998). *X-Ray Diffraction: A Practical Approach*. New York: Plenum, pp. 237–249.
11. Suryanarayana, C. (2001). *Prog. Mater. Sci.* 46:1–184.
12. Cullity, B. D., Stock, S. R. (2001). *Elements of X-Ray Diffraction*. 3rd ed. Englewood Cliffs, NJ: Prentice Hall.
13. Charlot, F., Bernard, F., Gaffet, E., Klein, D., Niepce, J. C. (1999). *Acta Mater.* 47:619–629.
14. Gras, C., Bernsten, N., Bernard, F., Gaffet, E. (2002). *Intermetallics* 10:271–282.
15. Cargill, G. S. III (1975). *Solid State Phys.* 30:227–320.
16. Waseda, Y. (1980). *The Structure of Non-Crystalline Materials*. New York: McGraw-Hill.
17. Cocco, G. (1992). *Mater. Sci. For.* 88–90:703–710.
18. Hunt, J. A., Soletta, I., Battezzati, L., Cowlam, N., Cocco, G. (1993). *J. Alloys Compounds* 194:311–317.
19. Hunt, J. A., Rose, P., Li, M., Soletta, I., Cowlam, N., Cocco, G., Enzo, S., Battezzati, L. (1993). *Key Eng. Mater.* 81–83:115–120.
20. Inoue, A. (2000). *Acta Mater.* 48:279–306.
21. Hays, C., Kim, C. P., Johnson, W. L. (1999). *Appl. Phys. Lett.* 75:1089–1091.
22. Chen, L. C., Spaepen, F. (1988). *Nature* 336:366–367.
23. Chen, L. C., Spaepen, F. (1991). *J. Appl. Phys.* 69:679–688.
24. Williams, D. B., Carter, C. B. (1996). *Transmission Electron Microscopy*. New York: Plenum.
25. Thompson-Russell, K. C., Edington, J. W. (1977). *Electron Microscope Specimen Preparation Techniques in Materials Science*. Eindhoven, The Netherlands: NV Philips' Gloeilampenfabrieken.
26. Yang, S. Y., Nash, P., Bradley, S. (1987). *J. Mater. Sci. Lett.* 6:982–984.
27. Harris, A. M., Schaffer, G. B., Page, N. W. (1993). *J. Mater. Sci. Lett.* 12:160–161.
28. Huang, J. Y., He, A. Q., Wu, Y. K., Ye, H. Q. (1994). *J. Mater. Sci. Lett.* 13:1201–1203.
29. Zghal, S., Bhattacharya, P., Twesten, R., Wu, F., Bellon, P. (2002). *Mater. Sci. For.* 386–388, 165–174.
30. Prenitzer, B. I., Giannuzzi, L. A., Newman, K., Brown, S. R., Irwin, R. B., Shofner, T. L., Stevie, F. A. (1998). *Metall. Mater. Trans.* 29A:2399–2406.
31. Lomness, J. K., Giannuzzi, L. A., Hampton, M. D. (2001). *Microsc. Microanal.* 7:418–423.
32. Kurzydlowski, K. J., Ralph, B. (1995). *The Quantitative Description of the Microstructure of Materials*. Boca Raton, FL: CRC.
33. Bohn, R., Fanta, G., Klassen, T., Bormann, R. (2001). *J. Mater. Res.* 16:1850–1861.
34. Ivchenko, V. A., Wanderka, N., Czubayko, U., Naundorf, V., Yermakov, A.Ye, Uimin, M. A., Wollenberger, H. (2000). *Mater. Sci. For.* 343–346:709–714.
35. Ivchenko, V. A., Uimin, M. A., Yermakov, A. Ye, Korobeinikov, A. Yu. (1999). *Surf. Sci.* 440:420–428.
36. Le Caër, G., Delcroix, P., Foct, J. (1998). *Mater. Sci. For.* 269–272:409–418.
37. Yelsukov, E. P., Formin, V. M., Nemtsova, O. M., Shklyaev, D. A. (2002). *Mater. Sci. For.* 386–388:181–186.
38. Lifshin, E., ed. (1994). *Materials Science and Technology–A Comprehensive Treatment*. Vols. 2A and 2B. Weinheim, Germany: VCH.

39. Brundle, C. R., Evans, C. A. Jr., Wilson, S., Fitzpatrick, L. E., eds. (1992). *Encyclopedia of Materials Characterization*. Boston, MA: Butterworth-Heinemann.
40. Enzo, S. (1998). *Mater. Sci. For.* 269–272:363–372.
41. Pradhan, S. K., Chakraborty, T., Sen Gupta, S. P., Suryanarayana, C., Frefer, A., Froes, F. H. (1995). *Nanostructured Mater.* 5:53–61.
42. Tian, H. H., Atzmon, M. (1999). *Phil. Mag.* A79:1769–1786.
43. Lucks, I., Lamaparter, P., Mittemeijer, E. J. (2001). *Acta Mater.* 49:2419–2428.
44. Williamson, G. K., Hall, W. H. (1953). *Acta Met.* 1:22–31.
45. Warren, B. E., Averbach, B. L. (1950). *J. Appl. Phys.* 21:595.
46. Weismuller, J. (1996). In: Bourell, D. L., ed. *Synthesis and Processing of Nanocrystalline Powder*. Warrendale, PA: TMS, pp. 3–19.
47. Krill, C. E., Birringer, R. (1998). *Phil. Mag.* A77:621–640.
48. McCandlish, L. E., Seegopaul, P., Wu, L. (1997). In: Kneringer, G., Rödhammer, P., Wilhartitz, P., eds., Proc. 14th International Plansee Seminar. Vol. 4. Reutte, Tyrol, Austria: Plansee AG, pp. 363–375.
49. Lonnberg, B. (1994). *J. Mater. Sci.* 29:3224–3230.
50. Aymard, L., Dehahaye-Vidal, A., Portemer, F., Disma, F. (1996). *J. Alloys Compounds* 238:116–127.
51. Salimon, A. I., Korsunsky, A. M., Ivanov, A. N. (1999). *Mater. Sci. Eng.* A271:196–205.
52. Ungar, T. (2003). *Adv. Eng. Mater.* 5:323–329.
53. Suryanarayana, C., Norton, M. G. (1998). *X-Ray Diffraction: A Practical Approach*. New York: Plenum, pp. 207–221.
54. Malow, T. R., Koch, C. C. (1997). *Acta Mater.* 45:2177–2186.
55. Tsuzuki, T., McCormick, P. G. (2000). *Mater. Sci. For.* 343–346:383–388.
56. Liu, K. W., Mücklich, F., Pitschke, W., Birringer, R., Wetzig, K. (2001). *Z. Metallkde.* 92:924–930.
57. Ahn, J. H., Wang, G. X., Liu, H. K., Dou, S. X. (2001). *Mater. Sci. For.* 360–362:595–602.
58. Gonzalez, G., D'Angelo, L., Ochoa, J., D'Onofrio, J. (2001). *Mater. Sci. For.* 360–362:349–354.
59. Warren, B. E. (1990). *X-Ray Diffraction*. New York: Dover, p. 251.
60. Gayle, F. W., Biancaniello, F. S. (1995). *Nanostructured Mater.* 6:429–432.
61. Gialanella, S., Lutterotti, L. (1995). *Mater. Sci. For.* 179–181:59–64.
62. Tcherdyntsev, V. V., Kaloshkin, S. D., Tomilin, I. A., Shelekhov, E. V., Baldokhin, Y. V. (1999). *Z. Metallkde.* 90:747–752.
63. Scott, M. G. (1983). In: Luborsky, F. E., ed. *Amorphous Metallic Alloys*. London, UK: Butterworths, pp. 144–168.
64. Köster, U., Herold, U. (1981). In: Güntherodt, H. J., Beck, H. eds. *Glassy Metals I*. Berlin, Germany: Springer-Verlag, pp. 225–259.
65. Ranganathan, S., Suryanarayana, C. (1985). *Mater. Sci. For.* 3:173–185.
66. Sherif El-Eskandarany, M., Saida, J., Inoue, A. (2002). *Acta Mater.* 50:2725–2736.
67. Kissinger, H. E. (1957). *Anal. Chem.* 29:1702–1706.
68. Greer, A. L. (1982). *Acta Metall.* 30:171–192.
69. Henderson, D. W. (1979). *J. Non-Cryst. Solids* 30:301–315.
70. Baró, M. D., Suriñach, S., Malagaleda, J. (1993). In: deBarbadillo, J. J., et al. eds. *Mechanical Alloying for Structural Applications*. Materials Park, OH: ASM International, pp. 343–348.
71. Zhou, C. R., Lu, K., Xu, J. (2000). *Mater. Sci. For.* 343–346:116–122.
72. He, Z., Courtney, T. H. (2003). *Mater. Sci. Eng.* A346:141–148.
73. Senkov, O. N., Srisukhumbowornchai, N., Öveçoğlu, M. L., Froes, F. H. (1998). *J. Mater. Res.* 13:3399–3410.
74. Atkinson, H. V. (1988). *Acta Metall.* 36:469–491.
75. Hu, H., Rath, B. B. (1970). *Metall. Trans.* 1:3181–3184.
76. Kambara, M., Uenishi, K., Kobayashi, K. F. (2000). *J. Mater. Sci.* 35:2897–2905.

8

Temperature Rise During Milling

8.1 INTRODUCTION

The intense mechanical deformation experienced by powder particles during mechanical alloying (MA) results in the generation of a variety of crystal defects (dislocations, grain boundaries, stacking faults, antiphase boundaries, vacancies, and so forth). As mentioned earlier, the balance between cold welding and fracturing among the powder particles is responsible for structural changes and the alloying that occurs in them. However, crystal defects also play a major role in the alloying behavior, due to enhanced diffusivity of solute atoms in the presence of increased defect density. Furthermore, as mentioned in Chapter 5, a number of process variables during milling also affect the final constitution of the milled powder. Among these, the temperature experienced by the powder particles during milling is very important in determining the nature of the final powder product. This is again related to the increased diffusivity of solute atoms for alloying to occur.

It has been reported that more than 90% (different values have been quoted by different investigators) of the mechanical energy imparted to the powders during milling is transformed into heat [1] and this raises the temperature of the powder. Assuming that, under the given conditions of milling, an amorphous phase or a crystalline phase could form in the milled powder, the temperature experienced by the powder would determine the nature of the phase. If the temperature experienced by the powder is high, the associated higher diffusivity (higher atomic mobility) leads to processes resulting in recovery (and recrystallization). In such a case, a stable crystalline phase would form. This crystalline phase could be an intermetallic or a solid solution. On the other hand, if the temperature experienced by the powder is low, then defect recovery would be lower and an amorphous (or a nanocrystalline) phase would form. It has also been noted that formation of intermetallics becomes easier if they tend to be stable at higher temperatures (preferably, congruently melting), as indicated by the respective phase diagrams, suggesting that the milling temperature may have a strong influence on the nature of the phase formed [2].

8.2 TYPES OF TEMPERATURE EFFECTS

There could be two different kinds of temperature effects during MA. One is the local temperature pulse due to ball collisions. These local temperature pulses have a short duration, about 10^{-5} s, approximately the same as the collision time between balls. It is very difficult to experimentally measure these local temperature pulses. Therefore, they can only be estimated theoretically based on appropriate models encompassing the extent of plastic deformation the powder particles undergo and the efficiency of heat transfer [3]. These temperatures could be a few hundred degrees Celsius. The second kind is the overall temperature in a vial, and this could be estimated theoretically [1,4–8] or measured experimentally. Many experimental measurements have been reported over the years and the temperature was found to vary between 50°C and 150°C. The temperature of the milling container is usually measured by inserting thermocouples into the outer wall of the vial [9,10].

The temperature rise during milling is mainly due to ball-to-ball, ball-to-powder, and ball-to-wall collisions, and also due to frictional effects. The overall temperature rise of the powders during milling can be due to more than one reason. First, as mentioned above, the intense mechanical deformation due to kinetic energy of the grinding media (balls) raises the temperature of the powder. Thus, the higher the energy (milling speed, relative velocity of the balls, time of milling, size of the grinding balls, ball-to-powder weight ratio, etc.), the higher is the temperature rise.

Second, it is possible that exothermic processes occurring during the milling process cause the powder particles to ignite and generate additional heat [11]. It is possible to detect the onset of ignition during milling by recording the temperature of the vial. While the temperature of the vial increases gradually with milling time in the early stages of milling, a sudden increase in temperature is noted when an exothermic reaction occurs among the powder particles, i.e., ignition occurs in the particles.

In addition, one could externally change the temperature of the milling container to enable the powder to be milled at different temperatures. For example, one could mill the powders at low temperatures either by dripping liquid nitrogen or a mixture of liquid nitrogen and alcohol onto the outer surface of the milling container or by milling the powder in liquid nitrogen medium (cryomilling). On the other hand, elevated temperatures could be achieved by heating the vial, e.g., by wrapping it with an electrical heating tape. This intentional change of the vial temperature to mill the powder at different temperatures, as a process variable, was discussed in Chapter 5 and will be further discussed in the individual sections on solid solubility extensions, intermetallic phase formation, amorphous phases, and nanostructures. Therefore, this aspect will not be discussed again here.

If the temperature rise is due to both the intrinsic temperature rise (i.e., due to transfer of energy from the grinding media to the powders) and that due to ignition, then in practice it is difficult to separate these two factors into their individual components. This is because the temperature of the powder or of the milling container measured is probably due to a combination of these two factors. Another point to be remembered is that the *total* heat accumulated (and, consequently, the temperature experienced) by the vial, the powder, and the balls, due to intrinsic or external effects, remains constant during milling for a given set of milling conditions. This is due to the balance achieved between the temperature rise due to mechanical forces and/or exothermic reactions and the heat loss to the surroundings by conduction, radiation, and so forth. Therefore, when the mill is turned off, redistribution of the heat occurs between the balls and the vial. Experimentally, it has been observed that the temperature of the vial increases above the ambient after the mill is turned off, for a short duration of time, suggesting that heat transfer occurs from the powder and balls, which are at a higher temperature, to the vial which is at a lower temperature. Depending on the thermal conductivity of the vial material, attainment of thermal equilibrium may take a few minutes.

8.3 METHODS TO EVALUATE TEMPERATURE RISE

The temperature rise in powders during MA was evaluated in three different ways. One was to calculate the temperature rise based on some theoretical models by making some valid assumptions. The second method was to look at the microstructural changes and/or phase transformations that occur in the powders and estimate the approximate temperature the powder might have experienced. The most direct way is to measure the actual temperature rise experimentally, using thermocouples or other instruments. Results are available on all three methods of ascertaining the temperature increase. Koch published a comprehensive review on the temperature effects during mechanical alloying and milling [12].

Let us look at the different ways in which the temperature rise during MA has been estimated either theoretically or through microstructural means, as well as based on experimental measurements.

8.3.1 Methodology Behind the Theoretical Models

Several attempts have been made to estimate either the instantaneous temperature rise at the time of collision between two grinding media or the overall temperature rise during the milling operation. In brief, the investigators calculate the kinetic energy of the grinding medium and assume that a part of this kinetic energy is utilized in plastically deforming the powder particles. This fraction of the energy is transformed into heat, and so the temperature increase is calculated based on the thermal properties of the powder and the balls. The different models differ in the way the powder particles are assumed to become deformed (the nature of forces—impact, shear, or a combination of both), fraction of the energy utilized in plastically deforming the powder particles, and heat transfer conditions and efficiency of heat transfer.

Depending on the model used, the estimated temperature rises are different. For example, when Nb powder was impacted by an 8-mm steel ball at 6 m/s, the bulk temperature rise was estimated to be 48 K [4] or 130 K [13], depending on the assumptions made in the model. But much higher temperature rises have been predicted at the contact surface [3]. On the basis of the assumption that shearing of powder particles trapped between colliding grinding media increases the bulk temperature of the powder, Eckert et al. [14] estimated that the peak temperature for $Ni_{70}Zr_{30}$ particles could reach values between 130°C and 407°C, depending on the milling intensity. Based on a similar approach, and assuming that the temperature rise could be due to sliding friction, the effective temperature rise was estimated to be only between 10 °C and 40 °C for Ge-Si and Ni-Ti powders [4,5]. A temperature of 180 °C was estimated at the collision site for a Ni-Zr alloy [15], whereas temperatures as high as 668 °C for a ball velocity of 6 m/s and 1187 °C for a ball velocity of 8 m/s have been reported [3]. Let us now look at the details of the different models used to estimate the temperature rise during the milling operation.

8.3.2 Theoretical Models

One of the early models to calculate the temperature rise during milling of powders was due to Schwarz and Koch [4]. They assumed that the powder particles trapped between two colliding grinding media deform by localized shear and estimated the peak temperature attained by the powder particles during MA. Calculating the energy flux dissipated on the powder surface as $J = \sigma_n v_r$, where σ_n is the normal compressive stress developed by head-on collision of two balls (grinding media) and v_r is the relative velocity of the grinding balls before impact (assumed as 2 m/s), they estimated the maximum temperature rise, ΔT, as:

$$\Delta T = J\left(\frac{\Delta t}{\pi \kappa \rho c}\right)^{\frac{1}{2}} \tag{8.1}$$

where Δt is the time duration for which the normal stress lasts, κ is the thermal conductivity, ρ the density, and c the specific heat of the powder particles. Using the appropriate material constants for $Ni_{32}Ti_{68}$ and $Ni_{45}Nb_{55}$ alloys, the maximal temperature rise was calculated as 38 K for both the alloys [4]. However, application of this analysis to a Si-Ge powder mixture yielded a ΔT value of only 10 K for Ge and 7 K for Si [5]. A similar small temperature increase (4 K) was also estimated by assuming that the particles make contact with one another over a limited surface area and that sliding friction is the responsible feature for increase in temperature [5]. These models are very simplified approximations to the complex mechanics/physics occurring during milling of powder particles but yield results somewhat similar to those of the other more realistic models.

Davis et al. [7] tried to develop a mathematical model for the kinetics of milling in a SPEX shaker mill by recording the movement of grinding media in a transparent vial on videotape and inferring the ball movements based on classical mechanics and analytical geometry. They noted that the vast majority of ball impacts lead to energy dissipation of 10^{-3}–10^{-2} J during the collision events. This energy range corresponds to ball velocities of approximately 6 m/s (for 10^{-3} J) and 19 m/s (for 10^{-2} J), from which the normal energies due to head-on collisions and consequently the maximal temperature increase could be calculated. Equation (8.1) suggests that these ball velocities produce a temperature rise of 112 K and 350 K, respectively. Based on the statistical distribution of the number of impacts and their energies, the authors [7] concluded that after 15 min of milling only about 1% of the total impacts result in the expected temperature rise between 112 and 350 K.

Assuming that plastic deformation work during MA is entirely converted to an adiabatic temperature rise ΔT in the impacted powder volume, one can write:

$$V_c \int \sigma d\varepsilon = V_c c \, \Delta T \tag{8.2}$$

where V_c is the powder volume impacted, σ the flow stress, ε the strain, and c the specific heat. Using a low-temperature constitutive plasticity equation, $\sigma = \sigma_o + K\varepsilon^n$, which does not consider strain–rate effects, Maurice and Courtney [16] calculated the temperature rise as

$$\Delta T = c^{-1}\left(\sigma_0 \varepsilon_{\max} + \frac{K\varepsilon_{\max}^{n+1}}{n+1}\right) \tag{8.3}$$

Magini et al. [17,18] analyzed the energy transferred to the powders during milling in a planetary ball mill and a SPEX mill. The energy transferred or dissipated during collision, ΔE, is given as $K.d^3.\omega^2$, where d is the ball diameter and ω is the rotational velocity of the plate. For a constant rotational velocity of the plate, this expression simplifies to $\Delta E = K_1 d^3$. K and K_1 are constants incorporating the radius of the plate and velocity of the balls. Thus, as the ball diameter increases, the energy transferred to the powder increases with increasing plate velocity. Similarly, the energy transferred increases with increasing plate velocity for a constant ball diameter (Fig. 8.1). Assuming that the powder trapped in between the colliding balls can be approximated by the powder adhering to the ball surfaces, the maximum quantity of the trapped material, $Q_{\max}$, was calculated as

$$Q_{\max} = 2s\pi R_{\mathrm{h,max}}^2 \tag{8.4}$$

where s is the surface density of the adhering powder and $R_{\mathrm{h,max}}$ is the radius of contact at the maximum of compression. The energy transferred per unit mass is then given as

$$\frac{\Delta E}{Q_{\max}} = K_1 d\omega^{1.2} \tag{8.5}$$

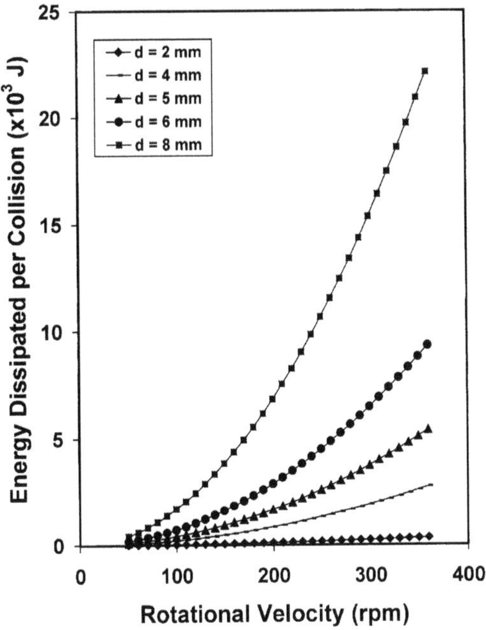

Figure 8.1 Energy transferred per collision as a function of plate velocity in a planetary ball mill [18].

Assuming that this energy results in an adiabatic temperature rise, ΔT can be given as

$$\Delta T = \frac{\Delta E}{Q_{\max} c} \tag{8.6}$$

where c is the specific heat of the powder material.

The models so far described estimate the bulk temperature of the powder. However, there have been some attempts [19] to explain that the local temperatures at the point of contact could be high enough to melt the powder particles. Furthermore, the local surface temperatures could be important in adhesion events among powder particles. Therefore, it would be instructive to calculate the maximal temperature rise at the time of contact between two grinding media.

Bhattacharya and Arzt [3] calculated the contact temperature of the powder compact's surfaces, considering that head-on collisions occur during collision between two grinding media. Therefore, the calculated value will be the upper bound of the temperature rise. In arriving at these values, the authors have assumed that (1) the time of impact, $\Delta\tau$, could be approximated using Hertz's theory of elastic impact, and (2) the energy flux at the compact surface is uniform over the entire contact area and is constant for times less than $\Delta\tau$. They have also assumed in their calculations that only a small fraction, β, of the kinetic energy $E = \frac{1}{2}mv^2$, where m is the mass and v is the relative velocity of the balls, is utilized in plastically deforming the powder, i.e., the expended plastic energy, $U_p = \beta E$. By taking the appropriate heat transfer conditions into consideration, they calculated the temperature rise, ΔT, as

$$\Delta T = \frac{2q_2\,\Delta\tau}{\rho c t_0} + \frac{q_2 t_0}{2\kappa}\left[\frac{1}{3} - \frac{2}{\pi^2}\sum_{n=1}^{\infty}\frac{(-1)^n}{n^2}\exp\left(-\frac{4n^2\pi^2\alpha\,\Delta\tau}{t_0^2}\right)\cos(n\pi)\right] \tag{8.7}$$

where

$$q_2 = \frac{\delta Q}{\pi r_0^2 \Delta\tau} \tag{8.8}$$

and Q represents the average quantity of heat arising from the total deformation process over a time interval $\Delta\tau$, δ is the fraction of heat flowing into the powder compact, κ is the thermal conductivity of the powder, t_0 is the thickness of the powder cylinder, r_0 is the radius of the powder compact, and α is the thermal diffusivity (Fig. 8.2). As mentioned earlier, this ΔT is the maximal temperature attained at the contact surface. Assuming that niobium powder is milled using stainless steel balls and that the powder compact has a thickness of $t_0 = 100$ μm and radius $r_0 = 263$ μm, the authors calculated the maximal temperature reached at the time of contact. These values are listed in Table 8.1. Even though the contact

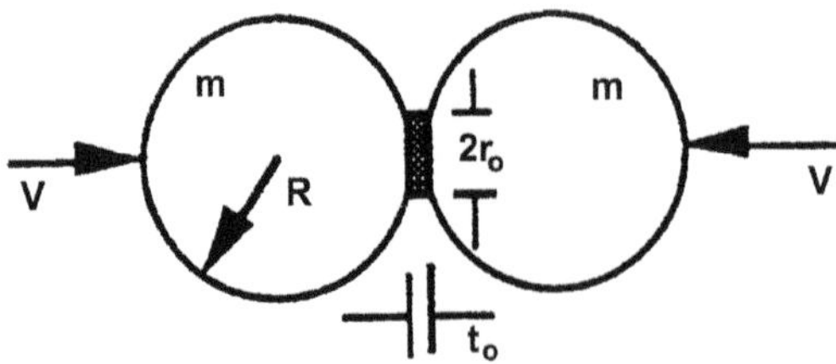

Figure 8.2 Geometrical parameters involving an impact between two grinding balls and the powder compact [3].

Table 8.1 Computed Values of the Contact Temperatures When Niobium Powder Is Milled Using Stainless Steel Balls[a]

	$T_c - T_0$ (°C)	
β	v = 6 m/s	v = 8 m/s
0.03	223	396
0.09	668	1187

β, fraction of the available energy utilized in plastically deforming the powder; T_c, contact (surface) temperature of the powder compact; T_0, ambient temperature; v, relative velocity of the grinding medium.
[a] Powder compact has a thickness of 100 μm and a radius of 263 μm [3].

temperatures calculated are quite high, they are surface temperatures and decay very fast with distance into the center of the compact. It was shown that the powder reaches room temperature at half the thickness of the compact, i.e., the middle of the compact is at room temperature. Furthermore, since the time scales for which these temperatures are maintained are short, the overall temperature experienced by the milled powder is low.

8.3.3 Observations Based on Microstructural/Phase Transformations

It has been noted above that there are many uncertainties in the measurement of either the bulk or instantaneous contact temperatures. These uncertainties include volume of the powder impacted during one collision, thickness and radius of the powder compact trapped between the balls, fraction of the kinetic energy of the grinding medium utilized in plastically deforming the powder, efficiency of heat transfer, and so forth. Because of these, the estimated powder temperatures could be substantially different from the actual values.

It is well known that blended elemental powders subjected to MA result in the formation of stable or metastable phases including solid solutions, quasi-crystalline, amorphous, and nanocrystalline phases. The phases formed may also be metastable in the sense that a new phase, which is stable only at higher temperatures and/or pressures, may appear at room temperature and atmospheric pressure. Furthermore, some type of microstructural and/or polymorphic changes may take place in the powder giving indication of the possible temperature the powder might have reached. Such microstructural features will be of use in evaluating the maximal temperatures reached during milling. Since phase transformations or microstructural changes are expected to occur at a critical temperature (depending on the kinetics, of course), observation of these changes indicates that the temperature attained by the powder is at least the critical temperature. It should be realized that these estimates are only indirect in nature. One must also exercise caution in interpreting these microstructural and/or phase changes since the intense plastic deformation can lead to a situation equivalent to high pressures and so the stability of phases under these conditions could be quite different from that under atmospheric pressure.

It has been known for some time that mechanically alloyed ductile material components exhibit features typical of cold working—deformation bands, slip lines [20], and high dislocation densities [21]. The presence of these features in the microstructure suggests that the milled material is still in the cold-worked condition and that recrystallization has not taken place. Unless the metal under consideration has a very low melting point, e.g., lead, tin, cadmium, indium, etc., the recrystallization temperature is much above room temperature. Therefore, it is possible to conclude that the temperature increase in the powder is not very much above room temperature but well below the recrystallization temperature.

The tempering response of a quenched Fe-1.2 wt% C steel during MA was studied by Davis et al. [7] to estimate the maximal temperature reached during MA. A steel sample quenched from the austenitic state to room temperature produces tetragonal martensite. On tempering the quenched steel, martensite decomposes to produce ε-carbide in the first stage of decomposition and cementite (Fe_3C) starts forming at higher temperatures. DSC studies indicated that ε-carbide starts forming at 418 K and cementite at 575 K [7]. By conducting separate annealing experiments at different temperatures, one could calculate the fraction of cementite formed as a result of decomposition of martensite. This helps in establishing a calibration curve relating the amount of cementite to the annealing temperature, which could be used as a baseline. Milling experiments were conducted on quenched steel (that contains only martensite) for different times, and the proportions of martensite and cementite in the milled powder were calculated using standard XRD procedures. The maximal temperature reached by the milled powder was then estimated from the above calibration curve. From such plots, it was estimated that the maximal temperature rise during milling was about 275°C [7]. It is, however, possible that introduction of point defects, dislocations, and other defects during milling could enhance carbon diffusivity, create heterogeneous nucleation sites, and allow precipitation of cementite to take place at lower temperatures. Thus, the above temperature may be assumed to be the upper limit of the temperature reached by the powder during milling.

Formation of an oxide phase or phase transformations occurring in materials during milling has also been utilized to indirectly estimate the temperature reached by the powder. During milling of Sb or Ga-Sb alloys it was noted that the antimony metal got oxidized and formed the high-temperature orthorhombic Sb_2O_3 phase. Since the Sb_2O_3 phase is stable only above 570°C, it was inferred that the local rise in temperature was at least 570°C [22,23]. Similarly, by observing the transformation of boehmite or gibbsite into α-alumina during milling, the temperature rise was estimated to be at least 1000°C [24,25]. Other estimated temperatures varied between 180°C for a Ni-Zr powder [15] and 590°C for an Al-Cu-Mn powder milled in a Fritsch Pulverisette 5 mill at an intensity of 9 [26]. The values of the estimated temperature rise (difference between the measured temperature and the ambient) are summarized in Table 8.2.

8.3.4 Experimental Observations

The temperature rises calculated and estimated above using theoretical models or microstructural changes are not accurate or reliable due to the assumptions made, some of which may not be valid. A major problem with these estimates is that one is not certain of the volume of the powder into which the heat is dissipated. Thus, the

Table 8.2 Temperature Rise Estimated from Microstructural and/or Phase Transformations During Mechanical Alloying/Milling

Alloy system	Type of mill	Temp. rise (°C)	Ref.
Al-Cu-Mn	Fritsch Pulverisette 5	590	26
Fe-1.2 wt% C	SPEX 8000	275	7
Ni-Al	Planetary ball mill	220	27
Ni-Zr	SPEX 8000	180	15
Sb or Sb-Ge	Centrifugal ball mill	>570	22,23
γ-AlOOH or $Al(OH)_3$	Fritsch Pulverisette 7	>1000	24,25

same amount of energy dissipated in a sufficiently small volume may produce a high temperature rise. Therefore, only experimental investigations may give reasonable and accurate values for the rise in temperature. In fact, only experimental observations will prove or disprove the different proposals for the formation of metastable phases in mechanically alloyed materials. For example, Ermakov et al. [19] proposed that the powder particles may melt during MA and that subsequent rapid solidification due to the cold powder surrounding it could result in the formation of amorphous phases. Furthermore, the actual temperature reached by the powder may also influence the nature of the phase formed and its kinetics of formation. It should be realized, though, that experimental measurements are not easy and that they measure only the bulk temperature of the vial, not the instantaneous temperature at the time of impact. Table 8.3 lists the experimentally measured temperature rises in different alloy systems. Here again, the temperature rise means the difference between the temperature measured and the ambient temperature. A number of

Table 8.3 Experimentally Measured Temperature Rise During Mechanical Alloying/Milling

Alloy system	Type of mill	Temp. rise (°C)	Ref.
—	SPEX 8000	50	12
Al-Cr	Conventional ball mill	90	28
Al-Mg	SPEX 8000	120	29
Al-Mg	Attritor	125	30
Co-Fe-Si-B	Vibrational ball mill	100	31
Cr-Cu	Vibrational ball mill	100	32
Fe-Al	Fritsch Pulverisette 5	80	33
$La_{2/3}Ca_{1/3}MnO_3$	Planetary ball mill	140	34
Mg-Ni	SPEX 8000	60	35
Nb-Si	SPEX 8000	<80	36
Ni-base superalloy	Attritor	<100–215	37
Ni-Ti	High-speed ball mill	172	38
Ta-Si	SPEX 8000	<80	36
Ti-C	Fritsch	70	39
V-Si	SPEX 8000	<80	36
$Zr_{70}Pd_{20}Ni_{10}$	Planetary ball mill	<100	40

investigators have measured the temperature of the vial during milling, but only a few typical results have been listed here.

The macroscopic temperature of the vial (or powder) has been measured by attaching thermocouples to the vial surfaces in some cases. A maximal temperature of 40–42°C was recorded [12,39] when the experiments were conducted with no grinding balls in the container; even with 13 balls in the SPEX mill, the temperature rise was noted to be only about 50°C [12]. On the other hand, when both the powder and the balls were loaded into the container and milling was done, the temperature of the container rose to about 70°C after approximately 6 h of milling [39]. Hence, it was concluded that most of the temperature rise comes from the motor and bearings.

Figure 8.3 shows the dependence of the vial temperature on milling time and vial contents during MA in a vibratory ball mill [1]. An approximately linear increase in temperature with milling time is noted up to 30 s of milling time, and a steady-state value is obtained after milling for about 20 min. This indicates that internal heating and external cooling are balanced out. Furthermore, as the ball-to-powder ratio increases, i.e., vial occupancy is higher, the temperature increases. This can be understood as due to the increased kinetic energy of the grinding medium and the consequent transfer of higher energy to the system.

The influence of ball diameter on the increase in temperature is shown in Figure 8.4 [41]. It may be noted that the temperature of the balls increases with the ball diameter. As the ball diameter increases, the weight of the ball increases; therefore, the kinetic energy ($E = \frac{1}{2}mv^2$) acquired by it is also higher. Consequently, the temperature attained is higher. However, when the ball diameter is very large, the relative velocity of the balls decreases, due to a decrease in the effective vial diameter,

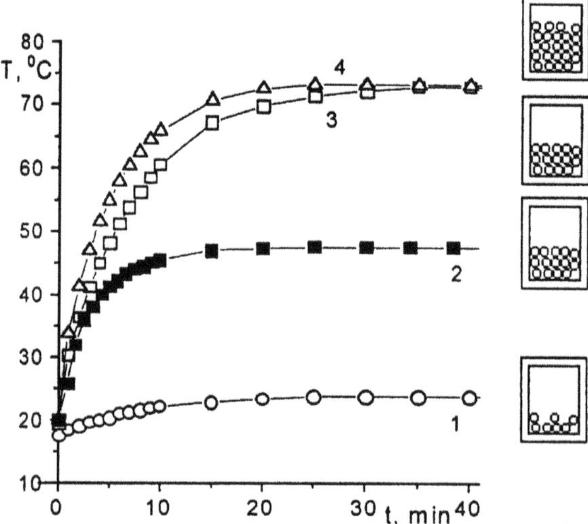

Figure 8.3 Measured external vial wall temperature as a function of milling time in a vibratory ball mill during mechanical alloying of Fe-Mn powders. The numbers on each curve correspond to the extent of filling in the vial, also indicated by the schematic figure adjacent to the curve. The occupancies of the vial are (1) 20%, (2) 45% with fan cooling of vial surface, (3) 45%, and (4) 70% [1].

Table 8.4 Comparison of the Measured and Calculated Temperature Values for Different Sizes of the Balls and a Vial Occupancy of 10–15% [42]

Size of the grinding ball (mm)	Calculated temp. (°C)	Measured temp. (°C)
3	400	160
5	480	455
9	610	520

i.e., the difference between the vial diameter and the ball diameter. This results in a decreased amount of kinetic energy and consequently the temperature attained is lower.

Shelekhov et al. [42] calculated the temperature rise during milling in a planetary ball mill (making assumptions similar to those in the earlier studies) and investigated the effect of occupancy of the vial as well as the size of the grinding medium. They noted that the experimentally measured values [43] are in good agreement with the values estimated from the model only for large balls. Table 8.4 compares the measured and estimated values for the different ball sizes. It was noted that the theoretical estimations of the milling temperatures were about 100°C higher than the experimental values.

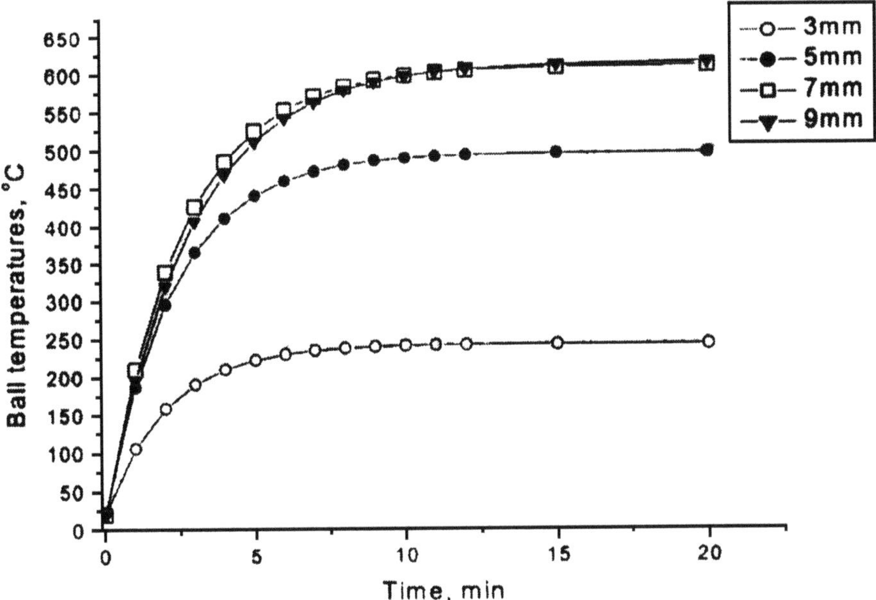

Figure 8.4 Measured temperature as a function of milling time for different sizes of the grinding balls in the absence of powder in an AGO-2 planetary ball mill. The mill conditions were: ball charge 200 g per vial, 1090 rpm for the planet carrier, 2220 rpm for the vials, and ball speed 8.24 ms^{-1} [41].

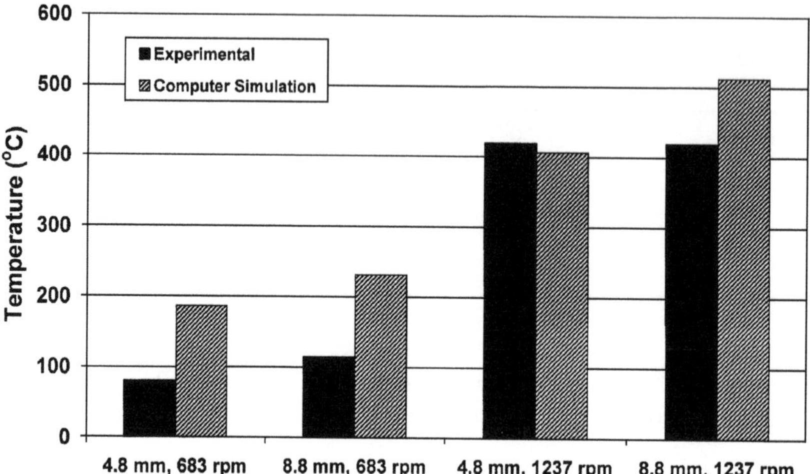

Figure 8.5 Experimental and calculated values of temperatures in a planetary ball mill for different sizes of the balls and rotation speeds. The calculated temperatures are the upper limits [1].

Pustov et al. [1] measured the temperature of the external wall of the milling container for different occupancies of the container by the powder and the balls. They noted that a steady-state condition was achieved when the internal heating and external cooling were equal. It was also noted that presence of the milled powder affected the rate of temperature rise and that the steady-state temperature was lower. This was explained as due to the higher thermal conductivity of the powder in relation to that of the normal vial atmosphere. A change in the ball-to-powder

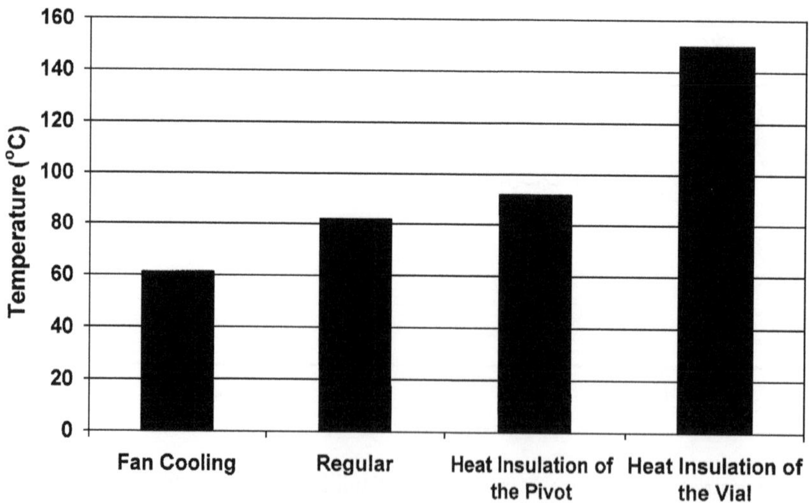

Figure 8.6 Measured milling temperatures in a vibrational ball mill under different cooling conditions [1].

weight ratio (BPR) did not cause any significant change either in the rate of temperature rise or in the final temperature attained. However, differences were noted between the estimated and measured temperatures for different ball sizes and also mill rotation speeds (Fig. 8.5). They had also measured the milling temperature under different types of external cooling conditions and noted that fan cooling produces the best results (Fig. 8.6).

8.4 TEMPERATURE OF THE BALLS

As mentioned earlier, the temperature of the powder (and the vial) increases due to the ball-to-ball, ball-to-powder, and ball-to-wall collisions. Since these collisions increase with milling time as well as with the size of the balls and the extent of vial filling, among other parameters, it is noted that the measured temperature increases with milling time and the extent of filling (Fig. 8.3). Since the balls make only point contacts with the vial, it is clear that the difference in the temperature between the balls and the vial could be significant.

A recent publication describes the measurement of the temperature of the balls during milling in a planetary ball mill and its influence on the milling behavior of Mg-Ni and Ni-Al alloys [41]. To allow accurate measurement of the temperature, cold water from a thermostat was circulated to cool the vial. Since the temperature of the vial remains reasonably constant for a short time after turning off the mill (say, up to about 30 s), the vial was transferred to a calorimeter within 30 s to measure its total heat content (H_t). By measuring the temperature change of the cooling water to calculate the dissipated energy and knowing the heat conductivity of the vial material, the temperature of the inner side of the vial walls, the average temperature of the vial, and the heat content during milling were calculated (H_v). The ball temperature was then calculated from the difference between H_t and H_v. These calculations were made for an AGO-2 planetary ball mill for four different sizes of hardened steel balls (3, 5, 7, and 9 mm in diameter), three different mill settings, and two different types of atmosphere. From Figure 8.4 it is clear that the ball temperature increases with milling time and reaches a steady-state value in about 20 min. But, the important point is that about 80% of the energy generated in the mill is consumed to heat the balls. Table 8.5 lists the milling conditions used and the steady-state temperatures recorded. From this it is clear that the temperature of the balls increases with the ball diameter and also with increasing milling time. The maximal temperature of the balls recorded is as high as 613°C under the most intense conditions of milling.

The above temperatures are for bare balls only, i.e., the vial did not contain any powder. When Mg and Ni powders corresponding to the composition Mg-33 at% Ni were milled with 7-mm-diameter balls for 10 min, the temperature was measured to be 360°C, as against 607°C, measured without the powder. The steady-state temperature was found to be different if a different powder was milled. It was found to be only 190°C, if Al_2O_3 powder was milled under similar conditions (Fig. 8.7). These results suggest that the ball temperature is lower in the presence of powder than only when bare balls are milled. It is possible that the powder forms coatings on the balls and the vial, or it could form large pellet-like agglomerates with a lamellar structure. It was noted that the ball temperature was lower when the powder formed agglomerates than when it formed coatings. The temperature was

Table 8.5 Mill Settings (for the AGO-2 Planetary Ball Mill) and the Ball Temperatures Under Steady-State Conditions (after 20 min of milling operation) [41]

	Rotation frequency, rpm (angular frequency, s^{-1})			Temp. (°C)			
			Ball	Ball diameter (mm)			
Step no.	Plate	Vials	speed (ms^{-1})	3	5	7	9
I	630 (66.0)	1290 (135.1)	4.79	130	261	304	308
II	890 (93.2)	1820 (190.6)	6.76	189	352	424	430
III	1090 (114.1)	2220 (232.5)	8.24	240	494	607	613

even lower when the powder was dispersed as small particles. This is because of the differences in the heat transfer efficiency in these three cases.

The nature of the gas and its pressure during milling also appear to play an important role. As shown in Figure 8.8, the steady-state ball temperature decreases with increasing gas pressure. It can also be noted that the temperature is much lower when helium is used instead of argon, and this effect is related to the thermal diffusivities of the two gases involved.

However, some investigators have reported very large temperature rises. The data available on the measured temperatures are summarized in Table 8.3. It may be

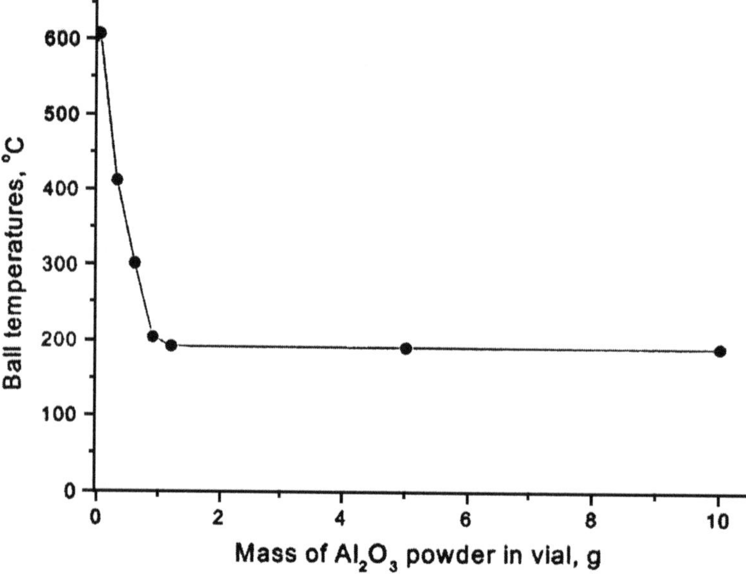

Figure 8.7 Steady-state ball temperature as function of the weight of Al_2O_3 powder in the vial during milling in an AGO-2 planetary ball mill. The milling conditions were: 7-mm-diameter balls, ball charge 200 g per vial, 1090 rpm for the planet carrier, 2220 rpm for the vials, and ball speed 8.24 ms^{-1} [41].

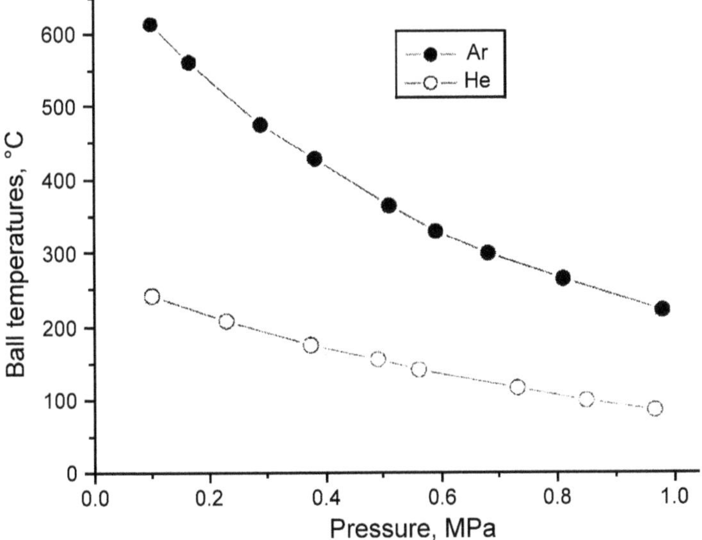

Figure 8.8 Steady-state ball temperatures as a function of the nature of the gas used and its pressure in the milling vial [41].

noted from the table that the maximal measured temperature is about 215°C, and more commonly it is about 100–120°C. It should be realized that this is the macroscopic temperature rise, even though it is recognized that local (microscopic) temperatures can be very high, often exceeding the melting points of some component metals.

8.5 METHODS TO MINIMIZE TEMPERATURE RISE

The above discussion clearly brings out the fact that there is always an increase in the temperature during milling of powders. The temperature reached depends on the nature of the mill, milling conditions, type of powder being milled, and heat transfer conditions. The temperature of the powder or the balls may be different, and the temperature could also differ from the theoretically calculated/estimated values. These higher temperatures normally enhance diffusion and thus accelerate the alloying process (see, for example, Ref. 1). However, in some cases this increased temperature may be a disadvantage by causing oxidation of the powder, crystallization of the amorphous phases produced, and so on. Therefore, one should be aware of the different ways in which the temperature rise could be minimized. The most normal practice is to use fan cooling, which has proved effective in minimizing the temperature increase. Another common practice is to mill the powder for a predetermined time, interrupt milling until the vial cools down to some reasonable temperature, and then resume milling. While using a SPEX mill, for example, one normally mills the powder for 30 min, waits for 15–30 min for the mill to cool down to close to room temperature, and repeats the sequence. However, if efficient fan or other means of cooling is provided, then one could mill the powder continuously, making sure that the container does not get too hot.

REFERENCES

1. Pustov, L. Yu., Kaloshkin, S. D., Tcherdyntsev, V. V., Tomilin, I. A., Shelekhov, E. V., Salimon, I. A. (2001). *Mater. Sci. For.* 360–362:373–378.
2. Kwon, Y.-S., Gerasimov, K. B., Lomovsky, O. I., Pavlov, S. V. (2003). *J. Alloys Compounds* 353:194–199.
3. Bhattacharya, A. K., Arzt, E. (1992). *Scripta Metall.* 27:749–754.
4. Schwarz, R. B., Koch, C. C. (1986). *Appl. Phys. Lett.* 49:146–148.
5. Davis, R. M., Koch, C. C. (1987). *Scripta Metall.* 21:305–310.
6. Courtney, T. H. (1995). *Mater. Trans. Jpn. Inst. Metals.* 36:110–122.
7. Davis, R. M., McDermott, B., Koch, C. C. (1988). *Metall. Trans.* 19A:2867–2874.
8. Courtney, T. H. (1994). *Rev. Part. Mater.* 2:63–116.
9. Koch, C. C., Pathak, D., Yamada, K. (1993). In: deBarbadillo, J. J, et al., eds. *Mechanical Alloying for Structural Applications.* Materials Park, OH: ASM International, pp. 205–212.
10. Klassen, T., Herr, U., Averback, R. S. (1997). *Acta Mater.* 45:2921–2930.
11. Takacs, L. (2002). *Prog. Mater. Sci.* 47:355–414.
12. Koch, C. C. (1994). *Int. J. Mechanochem. Mech. Alloy* 1:56–67.
13. Aikin, B. J. M., Courtney, T. H. (1993). *Metall. Trans.* 24A:647–657.
14. Eckert, J., Schultz, L., Hellstern, E., Urban, K. (1988). *J. Appl. Phys.* 64:3224–3228.
15. Schulz, R., Trudeau, M., Huot, J. Y., Van Neste, A. (1989). *Phys. Rev. Lett.* 62:2849–2852.
16. Maurice, D. R., Courtney, T. H. (1990). *Metall. Trans.* 21A:289–303.
17. Magini, M., Burgio, N., Iasonna, A., Martelli, S., Padella, F., Paradiso, E. (1993). *J. Mater. Synth. Proc.* 1:135–144.
18. Magini, M., Colella, C., Guo, W., Iasonna, A., Martelli, S., Padella, F. (1994). *Int. J. Mechanochem. Mech. Alloy* 1:14–25.
19. Ermakov, A. E., Yurchikov, E. E., Barinov, V. A. (1981). *Phys. Metals Metallogr.* 52(6):50–58.
20. Benjamin, J. S. (1976). *Sci. Am.* 234(5):40–48.
21. Schlump, W., Grewe, H. (1989). In: Arzt, E., Schultz, L., eds. *New Materials by Mechanical Alloying Techniques.* Oberursel, Germany: Deutsche Gesellschaft für Metallkunde, pp. 307–318.
22. Tonejc, A., Duževič, D., Tonejc, A. M. (1991). *Mater. Sci. Eng.* A134:1372–1375.
23. Tonejc, A., Tonejc, A. M., Duževič, D. (1991). *Scripta Metall. Mater.* 25:1111–1113.
24. Tonejc, A., Stubicar, M., Tonejc, A. M., Kosanović, K., Subotić, B., Smit, I. (1994). *J. Mater. Sci. Lett.* 13:519–520.
25. Tonejc, A., Tonejc, A. M., Bagović, D., Kosanović, C. (1994). *Mater. Sci. Eng.* A181/182:1227–1231.
26. Eckert, J., Schultz, L., Urban, K. (1990). *Z. Metallkde.* 81:862–868.
27. Mikhailenko, S. D., Kalinina, O. T., Dyunusov, A. K., Fasman, A. B., Ivanov, E., Golubkova, G. B. (1991). *Siberian J. Chem.* 5:93–104.
28. Kobayashi, K. F., Tachibana, N., Shingu, P. H. (1990). *J. Mater. Sci.* 25:3149–3154.
29. Zhang, D. L., Massalski, T. B., Paruchuri, M. R. (1994). *Metall. Mater. Trans.* 25A:73–79.
30. Cho, J. S., Kwun, S. I. (1993). In: Kim, N. J., ed. *Light Metals for Transportation Systems, Center for Advanced Aerospace Materials.* Pohang, South Korea: Pohang Univ of Sci & Tech, pp. 423–433.

31. Matyja, H., Oleszak, D., Latuch, J. (1992). *Mater. Sci. For.* 88–90:297–303.
32. Ogino, Y., Maruyama, S., Yamasaki, T. (1991). *J. Less Common Metals* 168:221–235.
33. Kuhrt, C., Schropf, H., Schultz, L., Arzt, E. (1993). In: deBarbadillo, J. J. et al., eds. *Mechanical Alloying for Structural Applications*. Materials Park, OH: ASM International, pp. 269–273.
34. Goya, G. F., Rechenberg, H. R., Ibarra, M. R. (2002). *Mater. Sci. For.* 386–388:433–438.
35. Ruggeri, S., Lenain, C., Roué, L., Liang, G., Huot, J., Schulz, R. (2002). *J. Alloys Compounds* 339:195–201.
36. Viswanadham, R. K., Mannan, S. K., Kumar, K. S. (1988). *Scripta Metall* 22:1011–1014.
37. Borzov, A. B., Kaputkin, Ya. E. (1993). In: deBarbadillo, J. J. et al., eds. *Mechanical Alloying for Structural Applications*. Materials Park, OH: ASM International, pp. S1–S4.
38. Qin, Y., Chen, L., Shen, H. (1997). *J. Alloys Compounds* 256:230–233.
39. Sherif El-Eskandarany, M. (1996). *Metall. Mater. Trans.* 27A:2374–2382.
40. Sherif El-Eskandarany, M., Saida, J., Inoue, A. (2003). *Metall. Mater. Trans.* 34A:893–898.
41. Kwon, Y.-S., Gerasimov, K. B., Yoon, S. K. (2002). *J Alloys Compounds* 346:276–281.
42. Shelekhov, E. V., Tcherdyntsev, V. V., Pustov, Yu. L., Kaloshkin, S. D., Tomilin, I. A. (2000). *Mater. Sci. For.* 343–346:603–608.
43. Gerasimov, K. B., Gusev, A. A., Kolpakov, V. V., Ivanov, E. (1991). *Siber. J. Chem.* 3:140–145.

9

Solid Solubility Extensions

9.1 INTRODUCTION

Solid solution alloys show many desirable properties such as increase in strength and elastic modulus, changes in density, and elimination of undesirable coarse second phases. Uniform precipitation of fine second-phase particles in a metallic matrix has been known for long to increase the strength and hardness of precipitation-hardening alloys [1]. The magnitude of improvement in the mechanical properties depends on interparticle spacing, and size and volume fraction of the precipitate or dispersoid. Since large-volume fractions of second-phase particles can be obtained during annealing/decomposition of highly supersaturated solid solutions, it is highly beneficial to synthesize supersaturated solid solutions, i.e., increase the solid solubility limits in alloy systems beyond the equilibrium limits. This leads to the synthesis of alloys with high strength and high strength-to-weight ratios.

9.2 HUME-ROTHERY RULES FOR SOLID SOLUTION FORMATION

Based on their pioneering work on alloying behavior of noble metal (Cu, Ag, and Au) alloy systems, Hume-Rothery and coworkers (see, for example, Ref. 2) have provided some empirical rules for the formation of primary substitutional solid solutions. It was shown that continuous series of substitutional solid solutions (complete solid solubility of one element in the other) may be obtained in binary alloy systems if the two component elements satisfy the following four empirical conditions.

1. *Atomic size factor*. If the atomic size of the solute atom does not differ by more than 14–15% from that of the solvent atom, then the size factor (defined as the ratio of the percentage difference in atomic sizes between the

solvent and solute atoms to the size of the solvent atom) is said to be favorable. In such a situation, the extent of primary solid solution may be unlimited. If the size factor differs by more than 15%, it is said to be unfavorable, and then the solid solubility is limited. Here the atomic sizes of the component elements are calculated based on the coordination number of 12.
2. *Electrochemical effect.* If the difference in the electronegativities (qualitatively, electronegativity is the power of an atom in a compound or solid solution to attract electrons to itself) of the solvent and solute atoms is large (i.e., one component is highly electropositive and the other highly electronegative), then stable intermetallic compounds form and consequently the extent of primary solid solubility is small. Therefore, for the formation of continuous series of solid solutions, the difference in electronegativity between the solvent and solute atoms should be small.
3. *Relative valency effect.* A metal of higher valency is more likely to dissolve in a metal of lower valency to a much larger extent than vice versa. Therefore, for complete solid solubility of one element in the other, the two elements should have the same valency.
4. *Crystal structure.* Formation of continuous series of solid solutions is possible only when the two components involved have the same crystal structure.

As examples, binary alloys of silver and gold, or copper and nickel, among others, satisfy the above four empirical rules and so form a continuous series of solid solutions. The equilibrium diagram is then said to be isomorphous. Even though copper and silver also satisfy the above rules, they do not form a continuous series of solid solutions; instead they form a eutectic system with limited primary solid solubilities. Therefore, it may be stated that the above Hume-Rothery rules are necessary but not sufficient to form extensive solid solutions in alloy systems. Consequently, the Hume-Rothery rules are sometimes referred to as negative rules. That is, if even one of the above rules is not satisfied, continuous series of solid solutions will not form. On the other hand, continuous series of solid solutions may not form even if all of the above rules are satisfied. There has been some discussion in recent years on the number and scientific basis of these "rules." It is now accepted that the relative valency effect rule does not work (about half the systems obey this rule and the other half do not), and therefore this rule has been abandoned [3].

9.3 FORMATION OF SUPERSATURATED SOLID SOLUTIONS

Solid solubility extensions, beyond the equilibrium values, have been achieved in many alloy systems by just quenching the alloys in the solid state from high temperatures. However, the extent of supersaturation is generally limited in such cases. Nevertheless, significant supersaturation of solute atoms has been achieved by nonequilibrium processing methods such as rapid solidification processing from the liquid state (RSP), vapor deposition, laser processing, sputtering, etc. [4]. However, the solid solubility extensions achieved by RSP methods have been most spectacular. Complete solid solubility of one element in the other has been reported to occur in systems such as Cu-Ag [5], Ni-Rh [6], GaSb-Ge [7], and other alloy systems [8], which, under equilibrium conditions, show only limited solid solubility of one component in the other. Note that the equilibrium solid solubility of Ag in Cu at the eutectic temperature of 779°C is 4.9

at% and that of Cu in Ag is 14.1 at%; these values are almost zero at room temperature. Supersaturated solid solutions in the entire composition range of the Cu-Ag binary system were obtained by rapidly solidifying the alloy melts at solidification rates of about 10^6 K/s. Even though complete solid solution formation may not be achieved in all rapidly solidified alloys, significant extensions of solid solubility limits have been reported in several alloy systems [8,9]. For example, the solid solubility of Fe in Al has been reported to increase by RSP up to 6 at% from the room temperature equilibrium value of only 0.025 at%.

Solid solubility extensions in rapidly solidified alloys have been rationalized using the concept of T_0, the temperature at which, for an alloy of a given composition, the free energies of the solid and liquid phases are equal [10–12]. In alloy systems such as Ag-Cu, where the two metals have the same crystal structure [face-centered cubic (fcc)] the T_0 vs. composition curve is continuous (Fig. 9.1). Accordingly, it is possible to obtain supersaturated solid solutions, either by quenching or other methods, if the alloy melt could be undercooled to below T_0 without the formation of any other phase. For a hypereutectic alloy this involves suppression of the formation of primary equilibrium β phase, of primary or secondary (intercellular or interdendritic) α-β eutectic, and of any alternatives involving nonequilibrium crystalline or amorphous phases. If the T_0 vs. composition curve is not continuous in some alloy systems, then undercooling the alloy melt to below T_0 is a minimal thermodynamic requirement.

Extensions of equilibrium solid solubility limits have also been achieved in mechanically alloyed powders. In fact, the technique of mechanical alloying (MA) has been shown to be capable of synthesizing both stable (equilibrium) solid solutions and

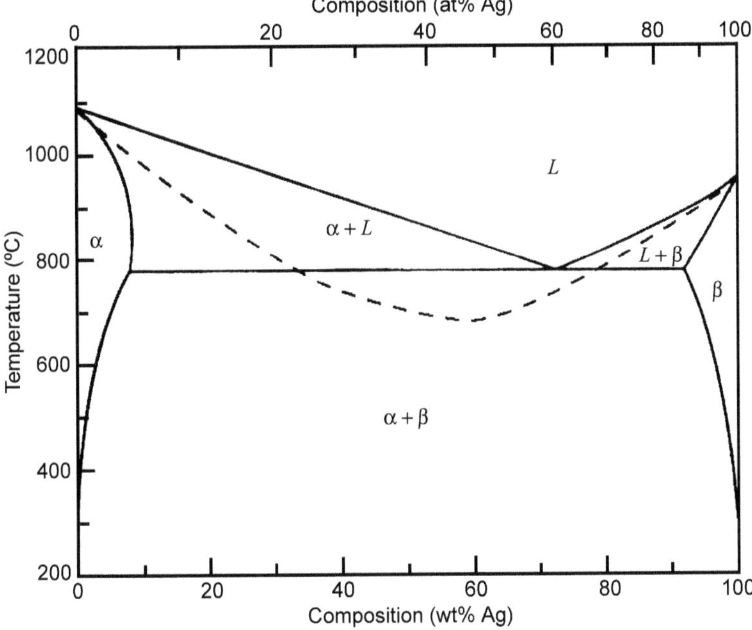

Figure 9.1 Variation of T_0 (denoted by the dotted line) with composition in the Cu-Ag system. The T_0 curve is continuous since both copper and silver have the same crystal structure (fcc).

metastable (nonequilibrium) supersaturated solid solutions starting from blended elemental powders in several binary and higher order systems. In the early years of MA research, observation of solid solubility extensions was not the primary objective; instead, formation of solid solutions was noted as a secondary result during amorphization of metal powder mixtures. Consequently, systematic studies were not conducted on the variation of solid solubility limits with the different process parameters. Results of such systematic investigations are being reported only in recent years.

9.4 MEASUREMENT OF SOLID SOLUBILITY LIMITS

Since the lattice parameter(s) of a solid solution phase is very sensitive to the solute content in the solid solution, measurement of lattice parameter(s) has been the most common method to evaluate solid solubility limits. Data regarding variation of lattice parameter(s) with solute content are available for many equilibrium solid solutions in standard reference books (see, for example, Ref. 13). Thus, if equilibrium solid solutions are formed, then the solute content in the solid solution can be estimated by calculating the lattice parameter of the solid solution from X-ray diffraction (XRD) patterns and estimating the solubility from a plot of lattice parameter vs. solute content, obtained from the literature.

The nature of the variation and the magnitude of change of lattice parameter of the solvent lattice with solute content is dependent on the relative atomic diameters of the solvent and solute atoms. The solvent lattice parameter increases with solute content if the solute atom has a larger atomic diameter than the solvent atom; otherwise it decreases. For small differences in atomic diameters, the change in the lattice parameter is also small.

XRD methods have been most commonly employed to determine lattice parameters of solid solutions. Solid solubility limits are then determined from the variation of lattice parameter(s) with solute content. Methods detailing the estimation of solubility levels may be found in standard textbooks on XRD [14,15]. In brief, the lattice parameter of the solid solution alloys with different compositions is calculated from XRD patterns and these values are plotted against solute content as shown in Figure 9.2. It may be noted that the lattice parameter increases (assuming that the atomic diameter of the solute is larger than that of the solvent) with solute content up to a certain composition, beyond which it remains constant. This composition limit, beyond which the lattice parameter remains constant, is the solid solubility limit at that temperature.

On mechanically alloying a blended elemental powder mixture of a known composition, let us say of solute content C_1, interdiffusion between the components takes place and alloying occurs. Assuming that right conditions exist and that a solid solution forms between the two elements, the lattice parameter of the solvent metal changes due to solute dissolution. The solute content in the solid solution is expected to increase with milling time as diffusion progresses. On continued milling, a saturation level is reached (when steady-state conditions of milling have been reached), beyond which no further extension of solid solubility occurs. A plot of lattice parameter vs. milling time indicates that the lattice parameter increases (assuming again that the solute atom has a larger atomic diameter than the solvent atom) with milling time until the maximal allowable solute is consumed to form the solid solution.

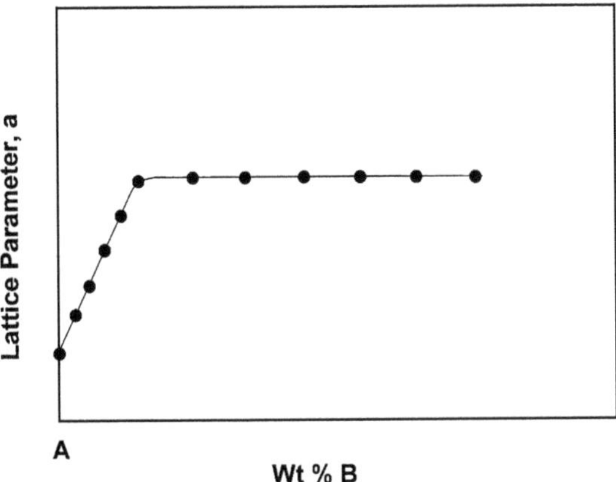

Figure 9.2 Variation of lattice parameter with solute content in a hypothetical alloy system where the solute has a larger atomic size than the solvent.

Beyond this time, the lattice parameter remains constant at a_1 (since the composition of the solid solution remains constant at C_1), as shown schematically in Figure 9.3. Repeating experiments like this with increasing solute content in the initial blended elemental powder mixture (e.g., with solute contents of $C_2, C_3, \ldots$, where $C_3 > C_2 > C_1$) results in lattice parameter saturations at $a_2, a_3, \ldots$, as shown schematically in Figure 9.4. Note that $a_3 > a_2 > a_1$. From these plots, the saturation lattice parameter values of the different blended elemental powder mixtures are plotted as lattice parameter vs. solute content as shown in Figure 9.5. In this plot, the lattice parameter increases with increasing solute content and reaches a saturation value beyond which the lattice parameter remains constant. This saturation value has been listed as the solid solubility level achieved by MA. For example, the solid solubility level is C_4 in the present example.

Note that Figure 9.5 is somewhat similar in trend to Figure 9.3. Whereas Figure 9.5 shows the variation of lattice parameter with solute content, Figure 9.3 shows the variation of lattice parameter with milling time for an alloy of a specific composition.

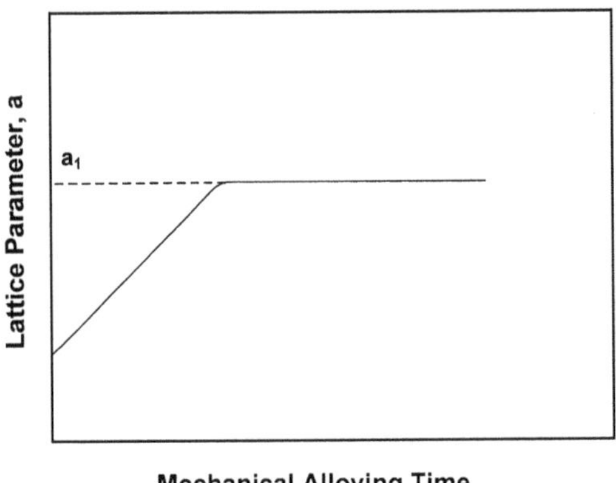

Figure 9.3 Increase in lattice parameter of the solid solution with mechanical alloying time.

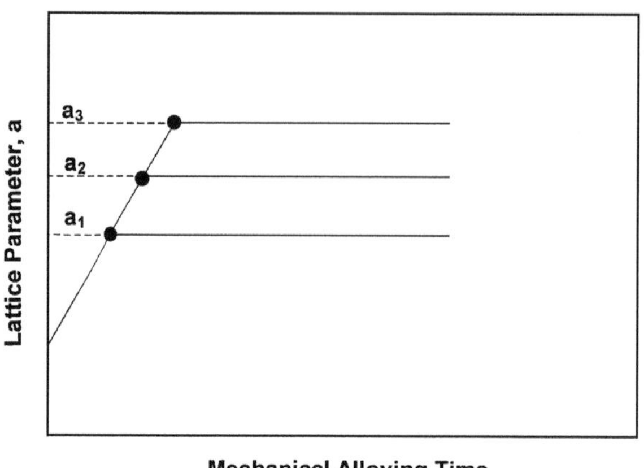

Figure 9.4 Variation of lattice parameter vs. mechanical alloying time for different powder blends with different initial solute contents.

If the two elements under consideration form continuous series of solid solutions, i.e., when the two component metals form an isomorphous system, the lattice parameter of one of the metals decreases and that of the other increases with milling time. Once steady-state conditions are established, the two lattice parameters merge and a homogeneous solid solution with the expected lattice parameter is formed. Figure 9.6 shows schematically the variation of the lattice parameter with milling time. Table 9.1 lists the solid solubility values obtained by mechanical alloying in several powder mixtures. Both room temperature and maximal equilibrium solid solubility values are also listed in the table to indicate the magnitude of increase of solubility achieved by MA. These equilibrium values have been taken from standard reference books (e.g., Ref. 292).

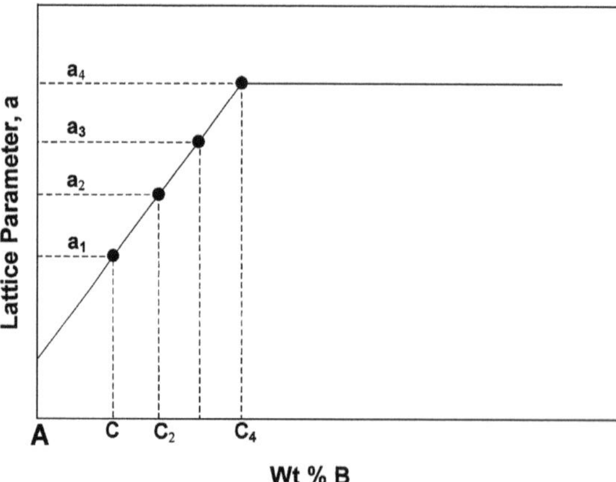

Figure 9.5 Variation of lattice parameter of the solid solution with solute content. The solid solubility limit can be determined by measuring the lattice parameter from X-ray diffraction patterns and using this master plot.

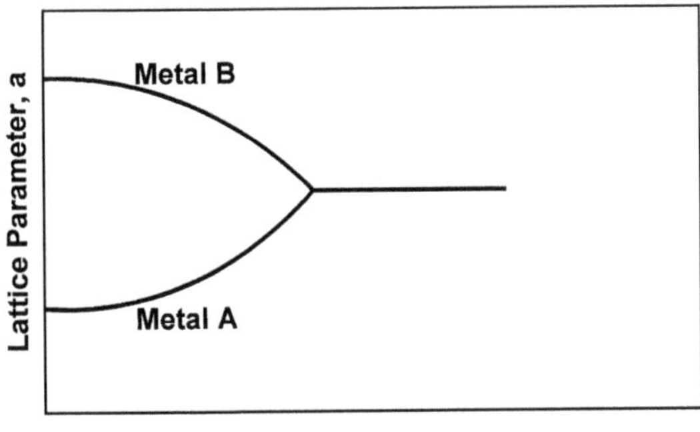

Figure 9.6 Schematic variation of the lattice parameter (solid solubility) of the individual component phases with mechanical alloying time in a binary alloy system that is expected to show complete solid solubility of one element in the other.

9.5 DIFFICULTIES IN SOLID SOLUBILITY DETERMINATION

As mentioned above, solid solubility limits have been generally determined from changes in lattice parameter values calculated from shifts in peak positions in the XRD patterns. Very frequently, absence of second-phase reflections in the XRD patterns has been inferred to indicate the absence of a second phase; therefore it was concluded that a homogeneous solid solution had formed. Both of these conclusions are fraught with problems; therefore, the values reported may not be always accurate and reliable. Hence, caution must be exercised in attaching too much importance to the solid solubility values reported in the literature. The difficulties associated with the limits of solid solubility determined this way in mechanically alloyed powders are discussed below.

1. The absence of diffraction peaks from the solute phase in XRD patterns is usually taken as proof of complete dissolution of the solute element in the solvent matrix; this has been interpreted as evidence for enhanced solid solubility values. It has been recently shown that solid solubility limits cannot be accurately determined only by noting the absence of solute peaks in XRD patterns. Kim et al. [293] reported that Ti reflections were absent in the XRD pattern recorded from an Al-20 wt% Ti powder mixture mechanically alloyed for 15 h, indicating that all the Ti atoms had dissolved in the aluminum matrix. But, transmission electron microscopy investigations on the same powder clearly showed the presence of nanometer-sized Ti particles dispersed in the aluminum matrix, suggesting that absence of XRD peaks from the solute phase may not be sufficient evidence for concluding that complete solid solubility has been achieved. Thus, it is desirable to use more than one technique to unambiguously determine the true solid solubility limits.

In some cases, the second component may become amorphous on milling, and in this case a very broad peak of low intensity will be present. For example, carbon, along with other components, was found to become amorphous on milling. If a number of peaks from the crystalline phase are also present, then the presence of the amorphous peak may not be noticed. In such a case, "disappearance" of the second-phase peaks in the XRD pattern may not be sufficient evidence to suggest achievement of complete solid solubility.

Table 9.1 Solid Solubility Limits (at%) Achieved by MA of Blended Elemental Powders[a]

Solvent	Solute	Equilibrium value		By MA	Ref.
		At RT	Max.		
Ag	Al	9.6	20.4	20	16
	Cu	0.0	14.1	100	17–19
	Gd	0.0	0.95	5	20
	Ni	0.0	0.0	4.3 (at RT) 36 to 44 at −195°C	21
	Pd	100	100	85[b]	22
Ag-15Sm	Fe			Fe:Sm 2:1	23
Al	Cr	0.0	0.37	5	24
	Cu	0.0	2.48	2.7	25
				0.2 (in Al-10Cu)	26
				5.6 (in Al-15Cu)	26, 27
	Cu-Fe	—	—	20Fe, 20Cu	28
	5Cu-5Zr			Partial, unspecified	29
	Fe	0.0	0.025	1.0	30
				2.0	31
				2.5	32
				4.5	33
				4–5	34
	Fe-V			45Fe, 5V	35
				25Fe, 25V	35
	Ge	0.0	2.0	14.1	27
	Mg	1.2	18.6	3.7 (in Al-10Mg)	26
				10.0 (in Al-20Mg)	26
				14.1 (in Al-30Mg)	26, 27
				13 (in Al-30Mg)	36
				18 (in Al-30Mg)	37
				24 (in Al-50Mg)	31
				40 (in Al-50Mg)	37
				23	38
	Mn	0.4	0.62	18.5	39
	Mn-Ti-Nb			8Mn, 24Ti, 1Nb	40
	Mo	0.0	0.06	0.4 (in Al-3Mo)	41
				1.3 (in Al-10Mo)	41
				2.4 (in Al-17Mo)	41
	Nb	0.0	0.065	Unspecified	42
				15	43
				25	44
	Ni	0.0	0.11	10	45
	Ni-Co	—	—	8Ni, 4Co	46
	Ru	0.0	0.008	14	47
	Si	0.0	1.5	0.7 (in Al-5Si)	26
				1.3 (in Al-10Si)	26
				4.5	26, 27
AlSb	InSb	—	—	100	48
Al	Ti	0.0	0.75	3.0	49
				4 (in Al–25Ti)	50
				10.4	51

Table 9.1 Continued

Solvent	Solute	Equilibrium value At RT	Equilibrium value Max.	By MA	Ref.
Al				12	52
				15–36	53
				20	54
				25	55–57
				35	58
				36	59
				50	60
	Ti-Cr	—	—	12Ti and 5–8Cr	61
	Ti-Fe	—	—	Unspecified	62
				12Ti and 3–5Fe	61
	Ti-Mn-Nb	—	—	23Ti, 8Mn, 2Nb	63
Al_2O_3	Fe_2O_3			25 mol %	64
Au_2Nb (PA)		9.0	57	33[b]	65
Bi	Sb	100	100	10[b]	66
Bi_2O_3	Nb_2O_5	—	—	100	67
	Y_2O_3	—	—	100	67
Cd	Zn	3.1	4.35	50	68
Co	C	0.0	4.1	6	69
	Cr	4.0	43.0	40[b]	70
	Cu	0.0	19.7	100	71–74
	Fe	100	100	10[b]	75
	Li_2O_3	—	—	Unspecified	76
	Mn	43.5	59.0	40[b]	70
	Ni	100	100	70[b]	77
	V	1.6	35.0	33[b]	70
	Zr	0.0	<1.0	5	70
Cr	Co	4.9	53.0	40[b]	70
	Cu	0.0	0.2	20	78
Cu	Ag	0.0	4.9	100	18
	Ag-Fe	—	—	10Ag, 20Fe	79
	Al	18.6	19.7	14[b]	80
				18[b]	25
	C	0.003	0.04 at 1373 K	25	81
				28.5	82
Cu	Co	0.0	8.0	50	83
				100	71–74, 84, 85
	Co-Fe			14Co, 6Fe	86
	Cr	0.0	0.9	50	87
	Fe	0.3	11.0	20	88
				30	17, 89, 90
				40	91
				50	92
				55	93, 94
				60	95–97
				65 (w/o ethanol)	98

Table 9.1 Continued

Solvent	Solute	Equilibrium value: At RT	Equilibrium value: Max.	By MA	Ref.
Cu				68 (w/ethanol)	98
				65	99, 100
	Hg	0.0	5.0	17	101
	Mn	100	100	35[b]	84
	Nb	0.1	0.1	3	102
	Ni	100	100	13[b]	103
				100	104
	Sb	1.1	5.8	3.7[b]	105
	Sn	0.0	9.1	Unspecified	106, 107
				9.8	108, 109
	Ti	0.9	8.0	0.6 (in Cu-10Ti)	110
				2.8 (in Cu-20Ti)	110
				3.7 (in Cu-30Ti)	110
				9.4 (in Cu-40Ti)	110
	V	0.0	0.1	Unspecified	84
	W	0.0	0.97	Unspecified	111
	Zn	32.0	38.3	44	112
Cu-29.7Zn (PA)	C			38.5	113
Fe	Ag	0.0	0.0	10	114
				Unspecified	115
	Ag-Cu	—	—	10Ag, 10Cu	79
	Al	20.8	45.0	7.9[b]	116
				26.7[b]	117
				20[b]	118
				25[b]	119–121
				25[b]	122
				27[b]	123
				40[b]	124, 125
				50	30, 126–128
				50	129
				50	35, 130
	Al-Ti			25Al, 5–15Ti	122
				33.3Al, 33.3Ti	131
	Al-Ti-C			25Al, 25Ti, 25C	132
	B	0.0	0.0	4.0	133
	B-Cu	—	—	13B, 1Cu	134
	B-Si	—	—	5Si, 4B	135
	C	0.01	0.11	4	136, 137
				5, but mentioned as 1.3 in Ref. 139	138
				33 (hcp, corresponding to high-pressure form)	140
	Co	100	100	20[b]	75
	Cr	100	100	46.5[b]	141
				50[b]	142

Table 9.1 Continued

Solvent	Solute	Equilibrium value At RT	Equilibrium value Max.	By MA	Ref.
Fe				70[b]	143
	Cu	7.2	15.0	5 to 20	144
				13.5[b]	145
				15	93, 94, 99, 100
				20	91, 97
				20	146
				25	96
				30	95
				40	90, 147
				18 (w/ethanol)	98
				24.5 (w/o ethanol)	98
	Cu-Zr			46.5Cu, 7Zr	148, 149
	Ge	10.0	17.5	25	121, 150, 151
				30	152
	Hf	0.0	0.51	10	153
	Hf-B			7Hf, 3B	154
	Mg	0.0	0.0	20	155
	Mn	3.0	100	10[b]	156
				24[b]	157
	Mn-Si			24Mn, 6Si	157
	Nb-B			7Nb, 3B	154
				6Nb, 12B (w/o PCA)	158
	Ni	2.0	4.7/100	36	159
				40	160
	Ni-Zr-B			7–28Ni, 10Zr, 20B	161
	Pb	0.0	0.0	Unspecified	162
	Rh	19.0	100	50 (fcc)[b]	163
	Sb	1.5	5.0	2.1[b]	164
	Si	3.0	19.5	Unspecified	165
				5[b]	133
				12[b]	166
				24.5	167
				26	168
				25	121, 169, 170
				30	171
				35	152
	Si-Nb	—	—	23.75Si, 5Nb	170
	Sn	0.0	9.2	20	172, 173
				22	152
				25	118, 174, 175
				30	172, 176
	Ta	0.3	2.5	40	177
	Tb	0.0	0.0	36	178
	W	0.0	14.3	Unspecified	179, 180
				20	181, 182

Table 9.1 Continued

Solvent	Solute	Equilibrium value At RT	Equilibrium value Max.	By MA	Ref.
Fe				30	183
	Zn	0.0	42.0	25[b]	121, 174
	Zr	0.0	6.5	5[b]	184–187
				10	153
				14	134
	Zr-B			7Zr, 3B	154
	Zr-B	—	—	7Zr, 7B	134
	Zr-B-Cu	—	—	7Zr, 6B, 1Cu	134
	Zr-Cu	—	—	7Zr, 7Cu	134
				13Zr, 1Cu	134
				11Zr, 3Cu	134
$Fe_{0.5}Cu_{0.5}$ (PA)	Zr			11	188
$Fe_{60}Mn_{10}Al_{30-x}B_x$ (PA)				Up to $x = 6$	189
Fe_2O_3	Cr_2O_3			100	190
α-Fe_2O_3	SnO_2			10 ± 3 mol % (in 15 mol % SnO_2)	191
				20 ± 4 mol % (in 29 mol % SnO_2)	191
FeS (PA)	Al	—	—	7Al in Fe	192
	Mn	—	—	Unspecified Mn	192
Ge	Si	100	100	89[b]	193
	Sn	0.48	1.1	24	194
Mg	Al	0.9	11.0	2.2 (in Mg-10Al)	26
				3.7 (in Mg-15Al)	26, 27
	Fe			5	155
	Ti	0.0	0.0	6	195, 196
α-Mn	Co	4.0	38.0	50	70
	Fe	30.0	35.0	5–7[b]	197
				28[b]	198
Mo	Al	0.0	20.5	Unspecified	41
Nb	Al	5.9	21.5	15.4[b]	199
				25	200–202
				33.3	203
				60	43, 44
Nb_3Al (PA)	—	—	—	25	204
Nb_3Au (PA)	—	19.0	54.0	25[b]	65, 205
Nb	Be	9.1	16.0	Unspecified	206
	Cu			20	78
	Ge	2.8	9.2	8.6[b]	207
	Ni	1.2	5.0	10	208
	Si	0.0	3.5	Unspecified	209
Ni	Ag	0.0	3.0	5	210
				6.6	21
	Al	7.5	21.2	0.5[b]	135

Table 9.1 Continued

Solvent	Solute	Equilibrium value At RT	Max.	By MA	Ref.
Ni				25	34, 150, 211–214
				27	215
				35	216
	Al-Nb			28–40Al, 10Nb	217
	Al-Ti-C			40–45Al, 5Ti, 5C	218
Ni_3Al (PA)	—	7.5	21.2	25	219
NiAl (PA)			39 to 55 Al	33–62 Al	220
				40–65 Al	215
Ni	Al-Nb-C			18–23Al, 5Nb, 5C	221
	C	0.0	7.4	9.0 (in Ni–10 C)	69
				12.0 (in Ni–15 C)	69
	Co	100	100	60[b]	104
				70[b]	77
	Cr	22.6	50.0	50	34
	Fe	3.8	100	40[b]	179
	Ga	10.8	24.3	50	222
	Ge	9.6	12.0	25	150
	In	0.0	14.5	13[b]	222
	Mg	0.0	0.0	10	223
	Mo	17.0	28.0	21[b]	224
				23[b]	225
	Nb	3.2	14.0	10[b]	208
				15	226
	Ru	4.6	34.5	31[b]	227
	Si	10.0	15.8	24	228
				25	150
Ni_3Si (PA)				25	229
Ni	Sn	0.0	10.6	17.5	222
	Ta	3.0	17.2	30	230
Ni	Ti	6.0	13.9	6[b]	231
				28	232
				8.9 (in Ni-10Ti)	110
				13.8 (in Ni-20Ti)	110
				18.3 (in Ni-25Ti)	110
Ni_3V (PA)	—	15.0	43.0	25[b]	233
Ni_2V (PA)	—			33[b]	233
Ni	W	12.5	17.5	14[b]	34
				23	234
				28	235
Pd	Ag	100	100	95[b]	22
	Y	7.6	13.0	12[b]	236
Pd_3Zr (PA)	—	12.0	16.2	25	237
PTFE	PE	—	—	Unspecified	238
Ru	Al	0.6	4.0	14	47
				Unspecified	239, 240

Table 9.1 Continued

Solvent	Solute	Equilibrium value		By MA	Ref.
		At RT	Max.		
Ru	Ni	6.4	50.0	53	227
Si	Ge	100	100	28[b]	241
				50[b]	242
	Sn	0.0	0.0	Unspecified	243
Ta	Al	0.0	4.0	33	244
				50	245
	Cu	0.0	0.0	30	246
	Zr-Al-Ni-Cu			10Zr-10Al-10Ni-15Cu	247
Ti	Al	11.0	47.5	10[b]	248
				16[b]	249, 250
				25[b]	55, 58, 204, 251–253
				28[b]	62
				≥30[b]	249
				33[b]	254
				50	58, 255
				60	55, 56
Ti-49Al (PA)	—	—	—	49	55, 56
Ti	Al-Nb	—	—	Unspecified	256, 257
				24Al, 11Nb (hcp)	202
	Al-Nb	—	—	22Al, 23 Nb (bcc)	258, 259
	Al-Ni	—	—	6 Al, 6 Ni	260
	Cu	0.0	1.6 (α), 13.5 (β)	8	78
				10	261
	Fe	0.0	0.04 (α), 22 (β)	Unspecified	262
	Mg	0.0	0.0	3.6	263
				6	264–266
				16.3	267
				24	268
				30	195, 196
				38	269
				50	196
	Ni	0.0	0.2 (α), 10.0 (β)	2[b]	270
	Si	0.0	0.5 (α), 3.5 (β)	8	271
				37.5	203
TiH_2	Fe			Unspecified	262
V	Co	7.3	22.0	Unspecified	84
				40	70
V_3Ga (PA)	—	9.2	41.0	25	272
V_3Ni (PA)	—	2.0	24.0	25	233
W	Al	2.0	2.0	86	273
	Fe	2.0	2.6	3	182

Table 9.1 Continued

Solvent	Solute	Equilibrium value		By MA	Ref.
		At RT	Max.		
W	Fe			3.4	274
				20	181
				40	275
	Ni-Fe	—	—	Unspecified	276
	Re	—	—	24.8	277
Yb	Ce	0.0	1.0	Unspecified	278
Zr	Al	1.0	26.0	Unspecified	279, 280
				10 (in the temperature range 173–573 K	281
				12.5 (at RT)[b]	282
				8.0 (at 300°C)	282
				15[b]	283–285
				17.5[b]	286
				50	280
	Co	1.0	5.0	4[b]	70
	Ni	0.0	3.0	7	187, 287
ZrO_2 + 8mol% Y_2O_3 (8YSZ)	Al_2O_3	—	—	1.15 wt %	288
ZrO_2	TiO_2			44 to 60 mol %	289
	ZrN	—	—	Unspecified	290
Zr_2O_3	Al_2O_3	—	—	Unspecified	291
	CoO	—	—	16.7 mol %	291
	Fe_2O_3	—	—	16.7 mol %	291
	Y_2O_3	—	—	16.7 mol %	291

RT Room temperature.
[a] In some cases, prealloyed powders have been used, and this has been indicated by mentioning PA next to the solvent component.
[b] In these cases, the composition of the alloy investigated is mentioned and is not necessarily the maximal solid solubility achieved by MA.

Huang et al. [135] analyzed the solid solubility data on 10 alloy systems of equiatomic composition processed by MA. Solid solubility extensions have been reported in these alloys by noting the absence of second-phase reflections. The authors have noted that XRD peaks associated with one of the elements disappear during milling while those of the other persist. Based on the results of these 10 alloy systems, they have concluded that the element with the higher atomic number persisted in the XRD pattern, while the one with the lower atomic number did not. Since the X-ray atomic scattering factor of an element is directly proportional to its atomic number, the element with the lower atomic number, and hence lower atomic scattering factor, has a lower intensity and hence with continued milling it tends to disappear. This argument assumes that atomic scattering factor is the major factor contributing to the intensity of a reflection, and this is not far from true. Accordingly, in the Al-Mo system, Mo (Z = 42) peaks persist while Al (Z = 13) peaks disappear. Similarly, in the Ni-Ru system, Ru (Z = 44) peaks persist while those of Ni (Z= 28) disappear.

2. Due to the mechanical deformation introduced into the powder, particle and crystallite size refinement occurs and lattice strain increases. These two effects cause broadening of XRD peaks and a consequent decrease in peak heights. In fact, formation of nanocrystalline solid solutions can significantly broaden the XRD peaks. Furthermore, if the weight fraction of the second component is small, the corresponding peaks may even be "absent." Thus, the disappearance of Al reflections in the XRD pattern of a mechanically alloyed Ni-Al powder mixture has been attributed to crystallite refinement to the level of about 5 nm, even though a solid solution has not actually formed.

Even if a reflection is present in the diffraction pattern, its broadening makes the location of the peak position difficult and consequently the calculated lattice parameters may not be accurate.

3. Another difficulty associated with small particle sizes is the level of detectability by the XRD technique. It has been shown that while 2 wt% of Ti can be easily detected if the particle size is in the range of 26–38 μm, one requires about 25 wt% Ti if the particle size is in the range of 0.05–1.0 μm. Thus, a much larger amount of material is required when the particle size is in the submicrometer range [293]. Mechanically alloyed materials often have particle sizes in the latter range of less than 1 μm, and consequently the presence of small amounts of second phases cannot be easily detected by XRD techniques.

4. There is also an inherent difficulty in determining solid solubility levels using the technique of XRD. It is very difficult to differentiate between XRD patterns from specimens containing a perfectly homogeneous solid solution and a finely decomposed mixture of two phases, e.g., formed during spinodal decomposition. This is because in both the situations, a single-phase XRD pattern and a Vegard law dependence of the lattice parameters is expected to occur. It was shown [294] that single-phase fcc structures were observed in the XRD patterns from codeposited Cu-Co thin films up to 65 at% Co content, suggesting that the solid solubility of Co in Cu was increased up to 65 at%. However, magnetoresistance measurements on these films pointed to the existence of superparamagnetic Co-rich clusters when the Co content was higher than 12 at%. Thus, conventional XRD techniques may not reveal information about the homogeneity of a solid solution; complementary techniques should be utilized to confirm these observations. However, complete solid solubility of Co in Cu was achieved by MA [71], as confirmed by both XRD and differential scanning calorimetry techniques.

5. Very often, solid solubility levels are determined by extrapolating the data of lattice parameter vs. solute content obtained either for equilibrium solid solutions or for metastable supersaturated solid solutions obtained by other nonequilibrium processing techniques. If such data were not available, then it could be assumed that Vegard's law [linear variation of lattice parameter (or atomic diameter or atomic volume) with solute content] is valid and that the lattice parameter varies linearly from one pure component to the other. Extrapolations can be erroneous, especially if the solid solubility is small under equilibrium or metastable conditions. Furthermore, assumption of the applicability of Vegard's law may not always be valid. The extent of deviation from linearity of lattice parameter(s) with solute content may be significant when one is dealing with pairs of metals that have different crystal structures. This problem is even more serious when the components involved have noncubic structures.

6. Heavy deformation (cold working) introduces stacking faults on the {111} planes in fcc metals and alloys. Stacking faults can also be introduced in the basal

$\{0001\}$ or prismatic $\{10\bar{1}0\}$ planes of alloys with hexagonal close-packed structures. These faults can cause anomalous hkl-dependent peak broadening in the XRD patterns [255,295]. The effective crystallite size (D_{eff}) measured from a reflection is due to the combined effects of true average crystallite size (D_{true}) and the "effective stacking fault diameter" (D_{SF}) according to the relation:

$$\frac{1}{D_{\text{eff}}} = \frac{1}{D_{\text{true}}} + \frac{1}{D_{\text{SF}}} \tag{9.1}$$

The presence of stacking faults thus leads to a refined "effective" crystallite size. The density of stacking faults is dependent on alloy composition and it has been shown that it varied from one in 133 atomic planes for a mechanically alloyed Cu-5 at% Co powder to one in every 23 planes for a Cu-90 at% Co alloy powder. In other words, the calculated crystallite size, ignoring the contribution of stacking faults and assuming that peak broadening is only due to the small crystallite size, grossly underestimates the true crystallite size. The true crystallite size was reported to be 3–10 times the apparent crystallite size in mechanically alloyed Cu-Co alloys [295]. The contribution of stacking faults to broadening and/or shift of peak positions is very important in alloys with low to moderate stacking fault energy, where a high density of stacking faults can be expected.

The peak shift caused by the presence of stacking faults can further complicate the situation in that the effects of solid solubility and stacking faults can cause the peak shifts to occur in opposite ways.

7. The technique of MA is often known to introduce impurities (substitutional impurities like iron, chromium, and nickel from the steel milling container as well as the grinding medium and interstitial impurities like nitrogen, oxygen, and carbon from the steel vial, milling atmosphere, and/or process control agent) and their amounts increase with milling time (see Chapter 15 for further details about levels of contamination in mechanically alloyed powders). The interstitial impurities dilate the lattice and substitutional impurities can dilate or contract the lattice depending on the relative atomic sizes of these elements with respect to the solvent atom. The possible presence of impurities and their influence on lattice parameter(s) of the solvent lattice must be taken into account when determining the real solid solubility values. An accurate chemical analysis for the impurity content and the effect of these impurities on the magnitude of peak shift should be evaluated before solid solubility extensions can be accurately reported.

8. It has also been noted in several instances that solid solubility levels increase with increasing initial solute content in the powder mix (see Table 9.1). This has been reported in alloy systems such as Al-Cu [26,27], Al-Mg [26,37], Al-Mo [41], Al-Si [26], Cu-Ti [110], Mg-Al [26,27], Ni-C [69], and Ni-Ti [110], among others. This was a true observation since milling was carried out to the steady-state condition for each alloy composition and this result was reported as reproducible by several investigators. It has also been reported that different solid solubility levels could be achieved by milling the powder at different temperatures. The solid solubility extensions appear to be higher when the powders were milled at lower temperatures [21,282]. Thus, it is difficult to list *one* specific solid solubility value for a given alloy system, at least in some cases, under the metastable conditions obtained during MA. These observations can be further complicated by variations in the nature of the mill, intensity of milling, milling atmosphere, and so forth.

Even though all the above-listed factors have not been taken into account in all of the cases, solid solubility extensions, beyond the equilibrium values have been reported in a number of alloy systems (Table 9.1).

In some cases *equilibrium* solid solutions were synthesized starting from blended elemental powders. In those alloy systems, which exhibit complete solid solubility under equilibrium conditions (isomorphous systems), the lower values indicated in Table 9.1 represent the composition of the alloy investigated and not the true maximal solid solubility level that can be achieved by MA. A similar comment may also be made with respect to other solid solutions that dissolve a large amount of the solute, but not in the whole composition range, under equilibrium conditions. One of the interesting observations made is that supersaturated solid solutions could be synthesized by MA even in immiscible systems that show a positive heat of mixing and hence do not have any solid solubility under equilibrium conditions.

The formation of solid solutions (both equilibrium and metastable) during MA may be attributed to the effects of severe plastic deformation. As mentioned earlier, plastic deformation refines the particle sizes and reduces the crystallite sizes of both the elemental powders. This process produces additional interfaces between the two elemental powders and increases the grain boundary area. The decreased particle size and consequent increase in the density of interfaces reduces the diffusion distances between particles and facilitates faster diffusion between the two elements. Since heavy plastic deformation introduces a variety of crystal defects and their density increases with the extent of deformation, more diffusion occurs along short circuit paths such as dislocations and grain boundaries. Diffusion is further aided by a local rise in temperature (see Chapter 8 for further details). The combined effect of all these effects permits sufficient diffusion to occur in the interfacial regions of the ultrafine grains to form a random solid solution in which the component elements are alloyed on an atomic level.

9.6 EFFECT OF PROCESS VARIABLES

Mechanical alloying is a process in which a number of process parameters can be varied. The solid solubility levels attained will be different when these variables are altered, thus affording a means of achieving the desired solute content in the solid solution and hence the properties. Since alloying is expected to occur due to particle/crystallite size reduction and increased defect density (grain boundaries, vacancies, etc.) and these effects become more significant with the intensity of deformation, the different process parameters should be taken into consideration while determining the solid solubility levels. Furthermore, the nature of the mill and milling conditions should to be accurately specified during reporting of solid solubility values.

The effect of different process variables on the extent of solid solution formation has been investigated. The important variables investigated include temperature of milling, nature and amount of process control agent (PCA), and initial composition of the powder blend, among others.

9.6.1 Milling Temperature

The effect of milling temperature seems to be an important factor in determining the extent of solid solution formation. Blended elemental powder mixtures were milled at different temperatures ranging form liquid nitrogen temperature up to very high temperatures of a few hundred degrees Celsius. While low temperatures were achieved by dripping liquid nitrogen or liquid nitrogen and alcohol mixture on the milling container, high temperatures were obtained by heating the milling container with an electrical heating tape. Even though cryomilling has recently become popular to achieve certain desirable properties in mechanically alloyed powder products, there are no reports on solid solubility investigations on cryomilled alloys.

It has been shown that a steady-state condition is achieved and optimal solid solubility of one element in the other is obtained when a balance occurs between mixing and decomposition reactions [296–298]. During milling, continuous shearing across the interfaces of the two component metals A and B results in an increase of the A/B interfacial area and this results in intermixing of the two components. This may be considered as "forced" or ballistic mixing under nonequilibrium conditions, and the diffusion flux for this case can be designated as J_B. On the other hand, thermally activated jumps of vacancies (due to temperature rise in the powder) increase thermal diffusion and favor decomposition of the supersaturated solid solution into the equilibrium phase mixture. The diffusional flux in this case may be designated as J_{th}. Under conditions of MA, a balance between these two atomic fluxes must be considered to decide the effectiveness in forming a solid solution. Thus, a forcing parameter, γ, has been introduced [299] as the ratio between the forced and thermally activated jump frequencies:

$$\gamma = \Gamma_b / \Gamma_{th} \tag{9.2}$$

where Γ_b is the atomic jump frequency caused by external forcing, which helps in the intimate intermixing of the components, and Γ_{th} is the atomic jump frequency due to thermal diffusion, that aids in the equilibration of the system (favors phase decomposition and ordering). A dynamic equilibrium is maintained between these two competing processes. For zero or low values of γ, thermally activated diffusion will drive the system to its equilibrium state. On the other hand, for infinitely large values of γ, the microstructure is continuously refined until a random solid solution is obtained. Usually the value of γ is neither of these two extremes but intermediate. Ma et al. [298] estimated the value of γ to be about 3 for a Cu-50 at% Fe alloy.

Klassen et al. [300,301] ball milled an Ag-50 at% Cu powder mixture to various steady-state conditions at different temperatures ranging from 85 to 473 K and noted that a homogeneous fcc solid solution was obtained after milling at subambient temperatures. On the other hand, a fully decomposed two-phase mixture was obtained after milling at 473 K. Coexistence of an equiatomic solid solution phase and decomposed terminal solid solutions was reported at intermediate temperatures of milling. Klassen et al. also reported that decomposition of the supersaturated solid solution was only mildly temperature dependent and that excess vacancies induced by shear were responsible for the decomposition of the supersaturated solid solution. In the Ni-Ag system an increased solid solubility level was also achieved when the powders were milled at low temperatures. The solid solubility of Ni in Ag was reported to be 4.3 at% when the powders were milled at room temperature, while it increased to 36–44 at% on milling the powders at –195°C [21]. But, in the Zr-Al system, the maximal solid solubility achieved was 12.5 at% Al in Zr when milling was conducted at room temperature, whereas it was only 8.0 at% Al at 300°C [282].

However, Sheng et al. [281], reported that the solid solubility of Al in Zr continued to be constant at 10 at% at all the milling temperatures investigated (173–573 K). This is the only report where milling temperature did not alter the solid solubility level.

Differences in alloying behavior were observed on mechanically alloying different alloy systems that are characterized by a large positive heat of mixing (ΔH_m), i.e., they do not alloy under equilibrium conditions. Cu-Fe and Ag-Fe are two such immiscible systems. While extensive increase in solid solubility levels was reported in the Cu-Fe system under the metastable conditions of MA, solid solution formation was not reported in the Ag-Fe system. Atomic level mixing and formation of solid solutions in the Cu-Fe system, and phase separation in the Ag-Fe system (on milling both at room temperature or liquid nitrogen temperature), were explained by Ma et al. [298] on the basis of forced mixing and the γ parameter referred to above. They considered an effective temperature T' defined as $T' = T(1 + \gamma)$, where T is the temperature at which milling is conducted and γ is the forcing parameter defined in Eq. (9.2). Since milling introduces a variety of crystal defects such as dislocations, grain boundaries, point defects, interphase interfaces, and so forth, the enthalpy associated with these milling-induced defects can be considered as ΔH_b. This is expected to decrease the value of ΔH_m, the enthalpy of mixing of the solution with respect to elemental states. Thus, the effective enthalpy of mixing,

$$\Delta H = \Delta H_m - \Delta H_b \tag{9.3}$$

has the same effect as increasing the value of T'. That means the system behaves as if it were in a high-temperature high-entropy state, i.e., in a metastable state. The ΔH_m for Cu-50 at% Fe was measured as 11 kJ/mol and ΔH_b was estimated as about 6 kJ/mol. Thus, the value of T' was calculated as 2400 K, i.e., the state of the milled Cu-Fe mixture is equivalent to that at about 2400 K. Since Cu and Fe are miscible in the liquid state at this elevated temperature, alloying is expected to occur under the nonequilibrium conditions as obtained during MA.

In contrast, Ag-Fe alloys have a large positive heat of mixing (>25.5 kJ/mol). Accordingly Ag-Fe alloys continue to be immiscible even in the liquid state at these very high temperatures. Consequently, the decreased ΔH value and increased T' cannot cause alloying in this system.

9.6.2 Process Control Agent

The use of a PCA also appears to affect the maximal achieved solid solubility limit, although the results reported are not consistent. For example, Gaffet et al. [98] reported that the solid solubility of Fe in copper was increased to 68 at% Fe when ethanol was used as a PCA, while it was only 65 at% Fe without the use of ethanol. Contrary to this, they reported that the solubility decreased with the use of ethanol as a PCA on the Fe-rich side of the Cu-Fe phase diagram. Accordingly, the solid solubility was 18 at% Cu in Fe when ethanol was used and it was 24.5 at% Cu in Fe without ethanol [98]. These two effects appear to be contradictory, but no explanation for this has been offered.

Kaloshkin et al. [302] reported that the solid solubility levels achieved were higher when the Fe-Cu alloy powders were milled in an atmosphere containing oxygen. Yavari and Desré [303] suspected that the solid solubilities were higher in oxygen-containing alloys because the presence of oxygen results in a negative heat of mixing of the multicomponent mixture. Another possible reason for this could be that oxygen will act like a PCA and promotes fracturing of the particles and helps in the formation of very fine-grained (nanostructured) material. This greatly helps in better alloying and achieving higher solid solubility levels. In principle, it may be expected that the oxygen present in the PCAs may also have a similar effect. However, the amount of oxygen released by the decomposition of the PCA during milling is relatively small and so may not have a significant effect. In fact, it is surprising that the solid solubility levels achieved were lower in some cases when the powders were milled with a PCA.

9.6.3 Starting Composition of the Powder Blend

An interesting observation that has been made in mechanically alloyed powders is that the solid solubility limit achieved increased with increasing solute content in the starting blended elemental powder mixture. This has been observed in several aluminum-, copper-, and nickel-based alloys, among others (see Table 9.1). Such a situation may possibly arise during rapid solidification of metallic melts when the solute concentration at the dendrite tip can be quite high and is determined by the operative solute partition ratio. But since there is no melting and consequently no dendrite formation involved in the MA process, such a mechanism may not be applicable here. Another possibility is that the local overheating may change the solid solubility limit achieved; higher temperatures lead to lower limits of solid solubility. However, it is most unlikely that the temperature rise is substantially different in alloy powder mixes with different initial solute contents; therefore, this explanation also is not satisfactory. Furthermore, increased temperatures normally decrease the solid solubility levels due to the enhanced equilibration effects.

The increased solid solubility level with increasing initial solute content may be tentatively explained on the basis that concentration gradients between different regions in the powder sample are steeper at higher initial solute contents and that these are expected to result in increased diffusion and consequently in higher solid solubility levels. This variation may arise possibly due to kinetic considerations only since thermodynamically the maximal solubility that can be achieved is fixed, under the given conditions of mixing, for a given alloy system. Calka et al.[37] noted that the solid solubility of Mg in Al was 40 at% in the Al-50 at% Mg powder blend and only 18 at% Mg in the Al-30 at% Mg powder blend. They had explained that this may be due to the higher driving force for solid-state reaction caused by an increased value of the negative heat of mixing for the Al-50 at% Mg powder blend.

The influence of other process parameters also was investigated in some cases. It has been reported that milling of blended elemental Cu-Cr powders in an air-free argon atmosphere resulted in the formation of a bcc solid solution [304]. However, when the powder was milled in an air-containing argon atmosphere, the solid solution was not formed; instead, an amorphous phase formed. A similar result was reported for the powders milled in a nitrogen atmosphere. The difference was explained on the basis that the presence of air (or nitrogen) enables refinement of the grain structure to nanometer levels, which then further facilitates formation of the amorphous phase.

Gonzalez et al. [117] reported that the kinetics of alloying (time required for the formation of solid solutions) in the Fe-Al system was dependent on the intensity of milling. They noted that the Fe-Al solid solution formed in 1 h at high milling intensities, whereas alloying was not complete even after 5 h at lower milling intensities.

9.7 MECHANISMS OF SOLID SOLUBILITY EXTENSION

During the early stages of milling the powder particles undergo severe plastic deformation resulting in repeated cold welding, shearing and fracturing (due to the increased hardness of the powder particles), and rewelding. In the case of ductile materials, this process leads to the formation of lamellar structures of alternate layers of the constituent metals. Interdiffusion then takes place across the interfaces due to the reduced diffusion distances and increased diffusivity aided by creation of lattice defects and increase in temperature. However, the individual lamellae will be heterogeneous, i.e., they have varying levels of the individual components. With continued milling the interlamellar spacing decreases, compositional inhomogeneity is reduced, and the interlamellar spacing decreases to such a fine level that one would not be able to observe the lamellar structure under an optical microscope. At this stage, the composition is homogeneous on an atomic scale, all the powder particles have the same composition, and true alloying occurs.

Neither the actual mechanism(s) for the formation nor the limits for solid solubility extensions in alloy systems obtained by MA have been well investigated. The general concept has been to explain the achieved solid solubility limit based on the common tangent approach mentioned in Chapter 1. Accordingly, the "metastable" equilibrium between the terminal solid solution and the amorphous or a metastable intermediate phase determines the extent of supersaturation that can be obtained. Schwarz et al. [232] suggested that the increased solid solubility of Ti in Ni in mechanically alloyed Ni-Ti powder mixtures was due to the metastable equilibrium between the α-Ni solid solution and the Ni-Ti amorphous phase as opposed to the stable equilibrium between the α-Ni and Ni_3Ti phases. Grigorieva et al. [222] suggested that the solid solubility can be extended up to the position of the intermetallic compound in the phase diagram if the base metal and the intermetallic have similar crystal structures. Figure 9.7a shows that the maximal solid solubility achieved under equilibrium conditions is obtained by drawing the common tangent between the free energy minima for the α-solid solution and the equilibrium Ni_3Ti intermetallic phase. But under conditions of MA, since an amorphous phase has formed in this system, the common tangent between the extended solid solution and the amorphous phase determines the solid solubility level (Fig. 9.7b). That is, the extent of supersaturation is limited to the composition where the amorphous (or the intermediate) phase has the lowest free energy. A similar conclusion was drawn by other investigators [110]. However, it has been noted in recent years that significant solid solubility extensions can be achieved in some alloy systems (e.g., Ti-Mg) even when an amorphous phase did not form in those alloy systems [263]. Thus, alternative explanations are needed to explain the formation of solid solutions and their limits of solid solubility.

Suryanarayana and Froes [264] suggested that formation of supersaturated solid solutions is closely related to the formation of fine-grained structures, reaching down to nanometer levels, during MA. The heavy mechanical deformation introduces a variety of structural defects into the powder particles, and these produce local stresses

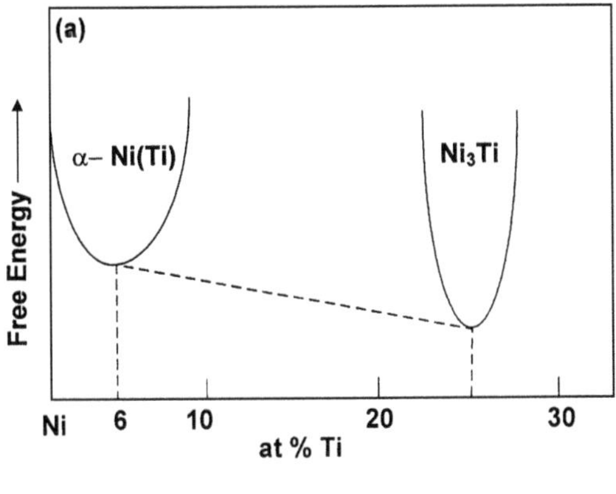

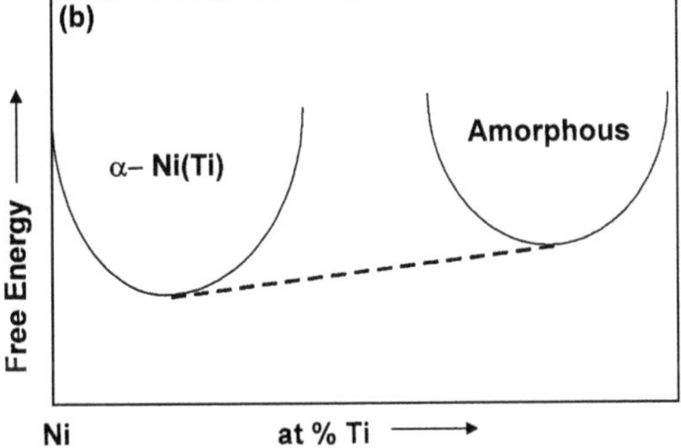

Figure 9.7 (a) Common tangent approach to establish the maximal solid solubility of Ti in Ni. The point of contact of the common tangent with the free energy curve of the α-Ni(Ti) solid solution phase drawn between the free energy curves of the α-Ni(Ti) and Ni_3Ti intermetallic determines the maximal equilibrium solid solubility of Ti in Ni. (b) Under nonequilibrium conditions the maximal solid solubility of Ti in Ni is obtained by drawing a common tangent between the free energy curves of the α-Ni(Ti) solid solution and the amorphous phases.

around the grain boundaries. This causes the nanocrystals to be in a higher energy state compared with the powder mixture of blended elemental powders. Added to this, a large volume fraction of atoms resides in grain boundaries in nanocrystalline materials. Both of these factors are expected to enhance diffusion and consequently the solid solubility levels are higher in systems that contain crystallites/grains with nanocrystalline dimensions. In support of this argument it was shown that the solid solubility of Mg in Ti is nearly zero when the grain size of the titanium phase is on the order of micrometers, whereas significant solubility extension (about 3 at%) was observed when the titanium grain size was on the order of nanometers. Similar results have also been reported by a number of other investigators [72,172,305]. For example, Huang et al. [72] investigated the effect of particle/crystallite size on solid solubility levels obtained in mechanically alloyed Co-Cu powders. The particle/grain sizes were measured using transmission electron microscopy techniques and the chemical composition of the particles was obtained by energy-dispersive X-ray analysis of the powders. They reported that the solute content was higher the smaller the particle/

crystallite size in their mechanically alloyed Co-Cu powders. The cobalt content was on an average 6 at% when the crystallite size was 25–30 nm, and it increased to 20 at% for a crystallite size of 12–15 nm, and to 24 at% for a crystallite size of 5–8 nm. Similarly, Yelsukov et al. [172] noted that to achieve a solid solubility of 30 at% Sn in Fe, reduction of grain size to nanometer levels was important and that the role of grain boundaries was critical in this process.

In comparison to a coarse-grained material, the volume fraction of grain boundaries in a nanocrystalline structure is much higher [306]. Furthermore, diffusivity of the component atoms is increased due to the large amount of structural defects (such as dislocations and grain boundaries) and the local stresses in such a material [306]. Under these conditions, solute atoms substitute the solvent atoms in the grain boundaries and vice versa. Rapid diffusion of atoms from one grain into the other leads to quick homogenization and results in the formation of solid solutions. Veltl et al. [307] suggested that the energy stored in the grain boundaries of nanocrystalline materials serves as a driving force for the formation of a solid solution. It was shown earlier [308] that a substantial amount of enthalpy can be stored in nanocrystalline materials due to the large grain boundary area. It was also suggested that the enhanced solid solubility might be due to the high dislocation density produced during milling [93,309].

Yavari et al. [310–312] proposed another mechanism based on the assumption that upon deformation elemental fragments with small tip radii are formed and a small fraction of the material is expected to be as crystallites 2 nm or less in diameter. In these cases the capillary pressure forces atoms at the tips of the fragments to dissolve. This process continues to full dissolution with increased milling time, due to generation of such small particles by necking at the tips of larger ones [310]. A similar mechanism was also suggested by Gente et al. [71]. They suggested that the chemical enthalpy associated with the interface between the elemental components can enhance the free energy of a composite above that of the related solid solution, thus providing a driving force for alloying in systems with a positive free energy of mixing. The main difference between the models proposed by Yavari et al. [310–312] and Gente et al. [71] is that the former assumes incoherent interfaces between elemental components whereas the latter suggests formation of composites with coherent elemental domains before alloying occurs. The experimental information presently available is not sufficient to indicate one model over the other.

Based on their experimental results, Yelsukov et al. [172,173] suggested that in alloy systems that have limited solid solubility of one element in the other, formation of supersaturated solid solutions is likely to be preceded by intermetallic formation. Taking the Fe-Sn system as an example, they showed that a mixture of α-Fe and $FeSn_2$ intermetallic had formed in the early stages (1–2 h) of milling the blended elemental powders. The proportion of the intermetallic formed peaked at about 40% with milling time noting that the proportion of the intermetallic was higher the higher the Sn content in the starting powder mixture. All of the powder mixture transformed into a supersaturated solid solution at later stages of milling (at about 5 h) according to the reaction:

$$\alpha\text{-Fe} + \text{FeSn}_2 \rightarrow \alpha\text{-Fe(Sn)} \tag{9.4}$$

Assuming that segregation of Sn to grain boundaries is the mechanism by which solid solubility increases, the authors showed that the grain size of the solid solution phase plays an important role in determining the solid solubility level. Accordingly, it was concluded that reduction of the grain size to nanometer level is critical to achievement of high solubility levels.

Formation of (supersaturated) solid solutions by MA is more noteworthy in liquid- and solid-immiscible systems than in solid-miscible systems. Gente et al. [71], Huang et al. [72,85], and Jiang et al. [313] reported formation of extensive solid solutions in the Cu-Co and Fe-Cu systems.

The mechanism responsible for enhanced solid solubility in systems with positive heat of mixing is not clear. Bellon and Averback [296] suggested that alloying in such systems may be a result of the γ factor effect. That is, the frequency of continuous shearing of atomic planes during milling (Γ_b) may be dominant over the thermal equilibration effects (Γ_{th}). It has also been suggested that interdiffusion between adjacent elemental domains, when their dimensions are reduced to a few nanometers, may be responsible for alloying to occur [71,312]. Under these conditions, it was suggested that sufficient thermodynamic driving force (interfacial energy) for dissolution is generated and the nanomixture may then behave as a system with a negative heat of mixing. Ma et al. [298] comment that neither of these two approaches alone appears adequate to explain solid solution formation observed in these systems. The thermodynamic argument alone is sufficient only when the domain sizes are reduced to 1 nm or less. However, high-resolution transmission electron microscopy investigations have shown that only about 1% or so of the crystallites have such small dimensions. Furthermore, it has also been shown that complete solid solution could form even when the domain sizes are as large as 8 nm [72]. On the other hand, the limited probability of the kinetic shearing-mixing process alone may also not achieve the desired solid solution formation. In fact, efficient shearing of atomic planes may become difficult when the crystallite is very small because such small crystallites may not support any dislocations. The shearing deformation is controlled by grain boundary mechanisms. Thus, the achievement of solid solubility in alloy systems with positive heat of mixing may not be due to just one single mechanism but rather to a combination of mechanisms.

Schwarz [314] proposed another model based on dislocation kinetics and solute-dislocation interactions. Atomically clean interfaces are produced during MA by the fracture of powder particles. The solute-dislocation interaction in an edge dislocation occurs through elastic locking, and the interaction energy for this locking is proportional to the misfit parameter, δ.

$$\delta = \frac{1}{a}\left(\frac{da}{dc}\right) \tag{9.5}$$

where a is the lattice parameter of the solvent and da/dc is the rate of change in the lattice parameter with the solute concentration c. Thus, the interaction is strongest when the misfit parameter is large, i.e., the atomic volumes of the solvent and solute are substantially different. Because of this interaction, solute atoms larger (smaller) than the solvent are attracted to the expanded (contracted) lattice regions near the cores of the edge dislocations. Thus, the solute atoms penetrate into the cores of edge dislocations in the solvent lattice. Since the solute atoms have an attractive interaction with the dislocations, the solute concentration in the core of the dislocations attains values largely in excess of the equilibrium solubilities. The next time the particle is trapped between the grinding media, it experiences an applied stress much higher than the flow stress of either of the two constituents. As a result, the dislocations start moving out through glide and reach another point in the powder. However, the solute atoms cannot diffuse along with the fast-moving dislocations and consequently are left behind, in a state of supersaturtion.

9.8 SOLID SOLUBILITY PLOTS

As mentioned at the beginning of this chapter, Hume-Rothery rules [2] suggest that the metallic radii of the solvent and solute should not differ by more than 15% to achieve good solid solubility of one element in the other. This was observed to be true in mechanically alloyed Co–transition metal alloys [70]. To further confirm whether a similar situation could be obtained in other alloy systems processed by MA, Bansal et al. [121] studied the alloying behavior of powder blends of Fe with 25 at% Al, As, Ge, In, Sb, Si, Sn, and Zn. They concluded that significant solid solubility (5 at%) could be achieved by MA even when the difference in the atomic radii between the constituent elements was about 30%. Stated differently, if the atomic radius difference was less than 15%, solubilities of up to 25 at% could be achieved; at larger differences in the atomic radii, the solid solubility was lower.

Figures 9.8–9.10 compare the equilibrium room temperature solid solubilities of different solute elements in aluminum, copper, and iron as the solvents with the extended solubilities obtained by MA. The data are taken from Table 9.1. It may be noted that the room temperature solid solubility in most of these systems is virtually zero and that MA has yielded significant increases in the solid solubility limits of most of the elements. To rationalize the solid solubility increases in mechanically alloyed powders in terms of the Hume-Rothery rules (small difference in atomic size; similar electronegativities, valencies, and crystal structures of the two components involved), the extended solid solubility values of different solute elements in copper are plotted against atomic radius in Figure 9.11. The ±15% limits of the atomic radius of copper

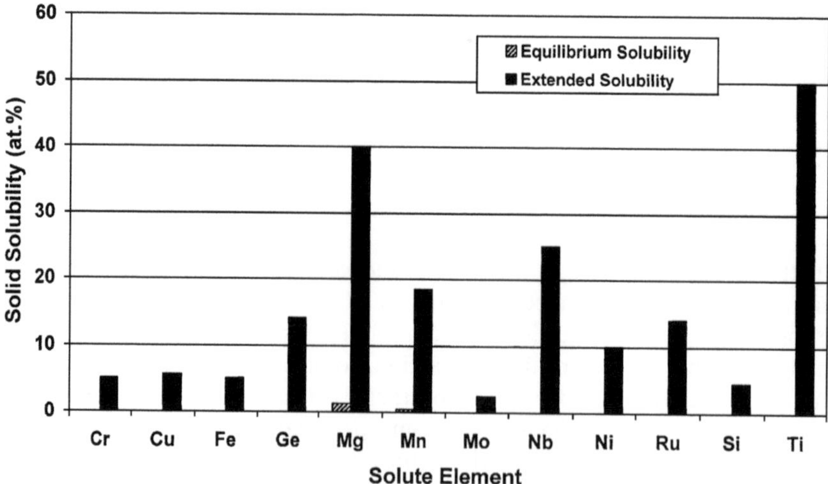

Figure 9.8 Comparison of equilibrium and extended solid solubility limits of different solute elements in aluminum achieved by MA.

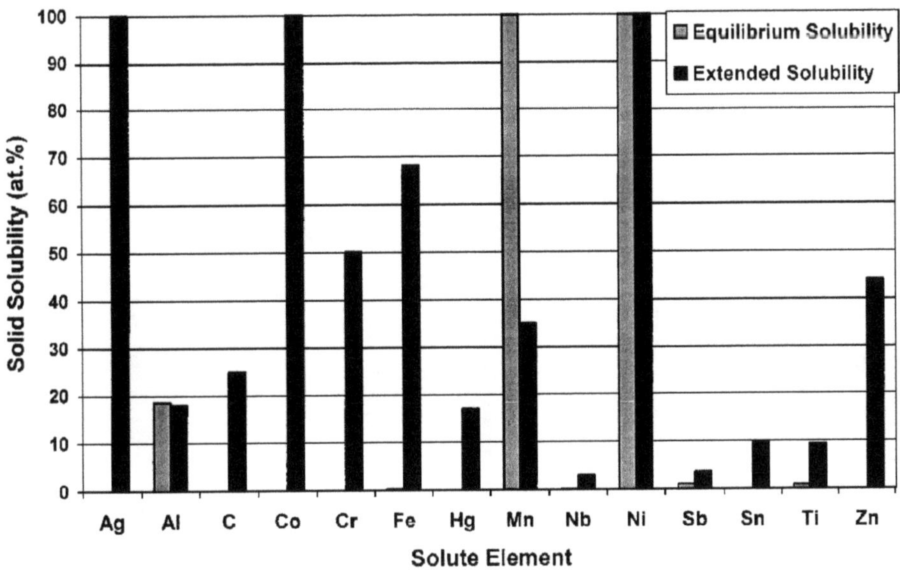

Figure 9.9 Comparison of equilibrium and extended solid solubility limits of different solute elements in copper achieved by MA.

are also indicated in Figure 9.11. It is interesting that in all cases in which the solid solubility limit has been considerably increased by MA, the atomic radius of the solute is well within the 15% limit suggested by Hume-Rothery. In some cases, significant increases in solid solubility limits have been observed even when the atomic size difference is on the border line (e.g., Cu-Hg). Figure 9.11 also makes it clear that the solid solubility increase is very small when the difference in the atomic sizes between

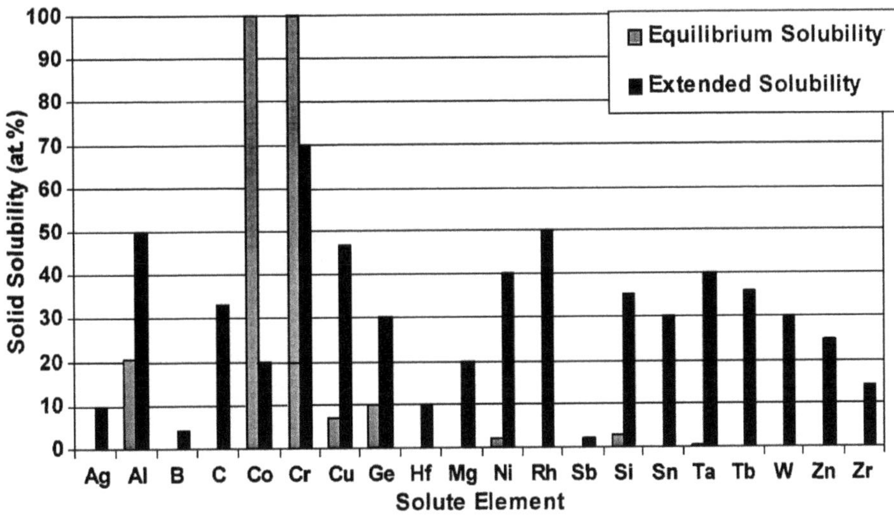

Figure 9.10 Comparison of equilibrium and extended solid solubility limits of different solute elements in iron achieved by MA.

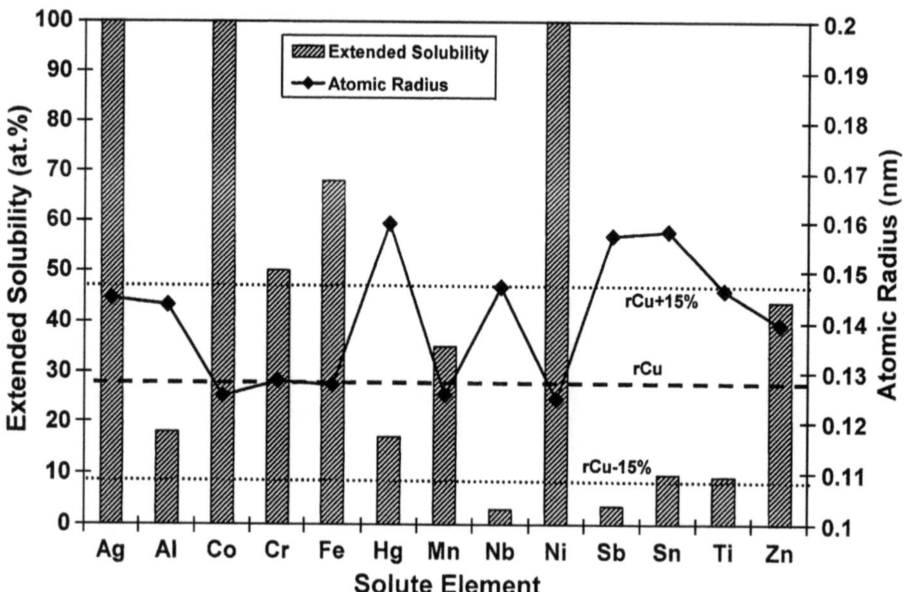

Figure 9.11 Extended solid solubility limits of different solute elements in copper achieved by MA of blended elemental powder mixtures plotted against atomic radius of the solute element.

the solvent and the solute is very large (e.g., Cu-Nb and Cu-Sb). However, it should, be noted that the equilibrium solid solubility of niobium in copper is only 0.1 at%, and MA increased it to about 3 at%. Thus, if the ratio of the extended solid solubility achieved by MA and the equilibrium solid solubility is plotted, then the result is even more striking. Since it is not possible to easily plot the infinite value of the ratio of extended to equilibrium solid solubility limits, the reciprocal of this value is plotted in Figure 9.12. In this graph, it may be noted that the ratios are always less than 1; i.e., the solid solubilities are always extended by MA. If complete solid solubility of one element in the other has been achieved under both equilibrium and mechanically alloyed conditions, then the ratio is 1. (Even if complete solid solubility is not achieved, but the same values are obtained in both the conditions, then also the ratio will be 1. However, such situations have not been reported in mechanically alloyed powders). A value of zero here corresponds to a situation whereby the equilibrium solid solubility is zero and extension has been achieved by MA. Irrespective of the extension achieved, the ratio is always zero. Thus, one notes from Figure 9.12 that substantial increases in solid solubility have been obtained for Ag, C, Co, Cr, Hg, Sn, V, and W. The actual extended solid solubility values may be obtained from Table 9.1.

This analysis confirms that the rules applicable for solid solution formation under equilibrium conditions are also valid under nonequilibrium conditions. It may be further noted that the maximal expected solubility limits, not observed under equilibrium conditions when the Hume-Rothery rules are satisfied, have been realized under nonequilibrium conditions. The best example for this is the Cu-Ag system, where the Hume-Rothery rules predict complete mutual solid solubility of Cu and Ag whereas only a limited solid solubility is observed under equilibrium conditions.

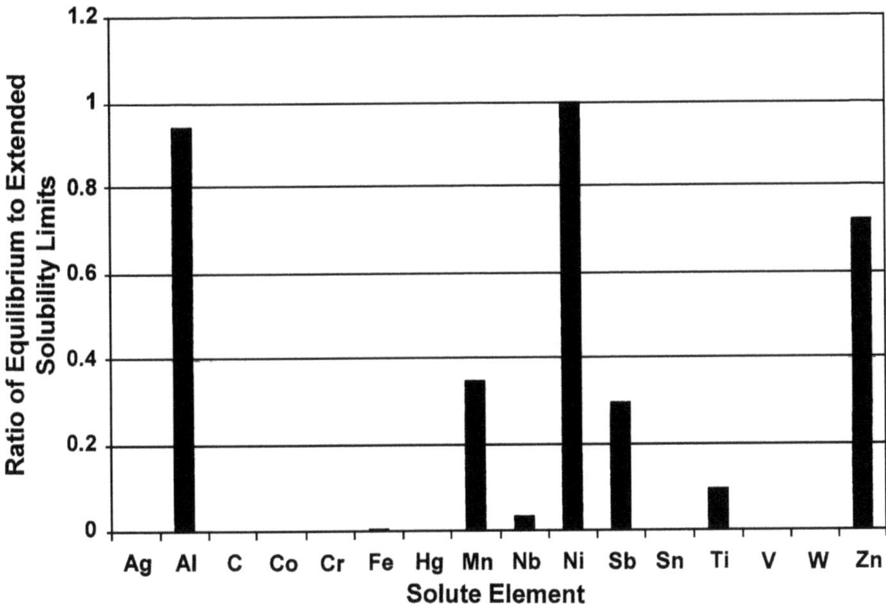

Figure 9.12 Ratio of equilibrium to extended solid solubility levels in different mechanically alloyed copper-based alloy systems.

Complete solid solubility of Cu in Ag and Ag in Cu has been achieved under nonequilibrium conditions obtained during MA. In fact, complete solid solubility of one component in the other is obtained in a number of other alloy systems also processed by MA (Table 9.2).

In addition to the atomic size, Darken and Gurry [315] included the electronegativity factor and developed plots (referred to as Darken-Gurry plots) to predict solid solubility limits in alloy systems. In this case, the electronegativity is plotted against the

Table 9.2 Alloy Systems Exhibiting Complete Solid Solubility in the Mechanically Alloyed Condition

		Solid solubility (at%)		
Solvent	Solute	Maximum under equilibrium conditions	Extended	Ref.
Ag	Cu	14.1	100	17–19
AlSb	InSb	—	100	48
Bi_2O_3	Nb_2O_5	—	100	67
Bi_2O_3	Y_2O_3	—	100	67
Co	Cu	19.7	100	71–74
Cu	Ag	4.9	100	18
Cu	Co	8.0	100	71–74,84,85
Fe_2O_3	Cr_2O_3	—	100	190

metallic radius corresponding to a coordination number of 12, and an ellipse is inscribed with an atomic size difference of ±15% and an electronegativity difference of ±0.3. Elements lying within the ellipse are expected to show high solid solubilities, and those outside the ellipse are expected to show limited solid solubilities.

Figure 9.13 shows the traditional Darken-Gurry plot, and it can be clearly seen that all the elements lying within the ellipse have a very high solid solubility in copper. Among the elements outside the ellipse, Ag is soluble up to 100% in copper under the nonequilibrium conditions of MA. Hg also has a reasonably high solubility of 17 at%, and Al 18 at%. The other elements (Nb, Sb, Sn, and Ti) have solubility of less than 10 at%.

Girgis suggested [316] that to determine the solid solubility of one element in the other, one could construct

1. A small ellipse, with the element in the center, with ±0.2 electronegativity difference on the *y* axis and ±8% atomic radius difference on the *x* axis,
2. A bigger ellipse, with ±0.4 electronegativity difference on the *y* axis and ±15% atomic radius difference on the *x* axis.

The solid solution alloys can then be divided into three groups: (1) those for which the second element has high solubility (more than 15%) lie inside the small ellipse; (2) those for which the second element has low solubility lie between the two ellipses; and (3) those in which the second element tends to be almost insoluble are outside both the ellipses. Figure 9.14 shows the modified Darken-Gurry plot for mechanically alloyed copper-based powder mixtures. Here it may be noted that solid solubility extensions are high (>50 at%), when the elements are lying within the

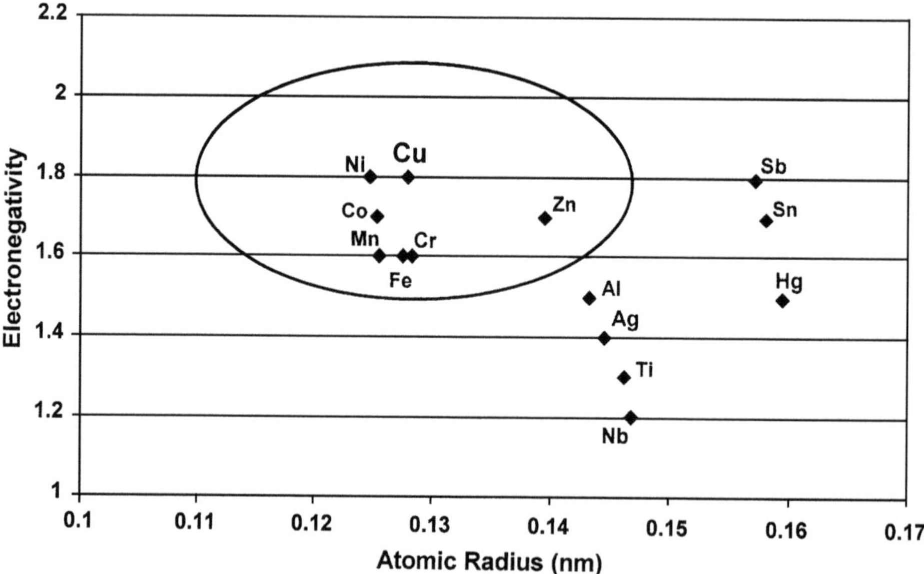

Figure 9.13 Darken-Gurry (electronegativity vs. atomic size) plot for mechanically alloyed copper-based powder mixtures. Elements lying inside the ellipse drawn with ±15% atomic size and ±0.3 unit of electronegativity are expected to show high solid solubilities.

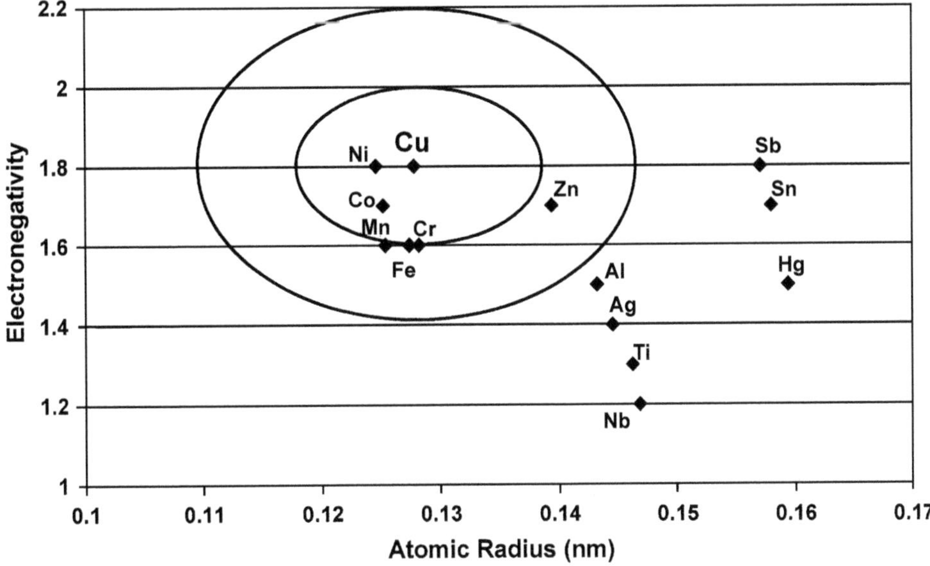

Figure 9.14 Modified Darken-Gurry plot for mechanically alloyed copper-based powder mixtures. Two ellipses are drawn to determine high solid solubilities. The inner ellipse is drawn with ±8% atomic size and ±0.2 unit of electronegativity of copper, while the outer ellipse is drawn with ±15% atomic size and ±0.4 unit of electronegativity. Elements lying inside the inner ellipse are expected to show very high solid solubilities and those outside the outer ellipse are least soluble.

smaller ellipse and reasonably extensive (35–50 at%) when they are lying in-between the two ellipses. However, the complete solubility of Ag in Cu and the reasonably large solubilities of Hg (17 at%) and Al (18 at%) are not predicted by this plot since these two elements lie outside the larger ellipse. From these plots it appears that neither the traditional nor modified Darken-Gurry plots account for the extended solubilities in alloy systems achieved by MA.

9.9 COMPARISON BETWEEN MECHANICAL ALLOYING AND RAPID SOLIDIFICATION PROCESSING

The absolute limits of solid solubility extension in alloys processed by rapid solidification processing (RSP) have been determined from thermodynamic considerations using the concept of T_0 temperature at which, for a given composition, the solid and liquid phases have the same free energy. Supersaturated solid solutions could be obtained only when the liquid was undercooled to a temperature below T_0 [11]. Even though MA also is a nonequilibrium processing technique, there have been no attempts so far to rationalize the available experimental data of solid solubility extensions. Further, the maximal possible limits of solid solubility extensions have not been defined theoretically. However, since no liquid phase is involved during MA, it is doubtful if the same criterion (of T_0) as in the RSP case could be used to rationalize solid solubility extensions achieved during MA.

Table 9.3 Comparison of Solid Solubility Limits (at%) Achieved by Rapid Solidification Processing (RSP) and Mechanical Alloying (MA)

		Equilibrium value		Extended value	
Solvent	Solute	At RT	Max.	By MA	By RSP
Ag	Cu	0.0	14.0	100	100
	Gd	0.0	0.95	5.0	5.0–6.0
Al	Cr	0.0	0.37	5	6
	Cu	0.0	2.48	5.6	18
	Fe	0.0	0.025	4.5	6.0
	Ge	0.0	2.0	14.1	40
	Mg	1.2	18.6	40	40
	Mn	0.4	0.62	18.5	9
	Mo	0.0	0.06	2.4	2.4
	Nb	0.0	0.065	25	2.4
	Ni	0.0	0.11	10	7.7
	Ru	0.0	0.008	14	4.5
	Si	0.0	1.5	4.5	16
	Ti	0.0	0.75	50	2.0
AlSb	InSb	—	—	100	—
Cd	Zn	3.1	4.35	50	35
Co	Cu	0.0	19.7	100	25
	Zr	0.0	< 1.0	5	1.5
Cu	Ag	0.0	4.9	100	100
	Co			100	
	Cr	0.0	0.0	50	4.5
	Fe	0.3	11.0	68	20
	V	0.0	0.1	Unspecified	1.0
Fe	B	0.0	0.0	4.0	4.3
	Cu	7.2	15.0	40	15
	Ge	10.0	17.5	25	25
	Sn	0.0	9.2	25	20.0
	W	0.0	14.3	Unspecified	20.8
Mg	Al	0.9	11.0	3.7	21.6
Nb	Al	5.9	21.5	60	25
Ni	C	0.0	7.4	12	8.2
	Ge	9.6	12.0	25	22
	Nb	3.2	14.0	15	15.1
	Si	10.0	15.8	24	20
	Sn	0.0	10.6	17.5	17.0
	Ta	3.0	17.2	30	16.6
	V	15.0	43.0	25	51.0
Ti	Si	0.0	3.5	37.5	6.0

The mechanisms by which supersaturated solid solutions form by MA and RSP techniques are different; even then comparisons have been frequently made, with no specific conclusions. Table 9.3 compares the available data. For example, solid solutions have been formed in the whole composition range in the Cu-Ag system by both the techniques. Rapid solidification processing produced continuous series of solid solutions in many other alloy systems [8], but MA did not obtain similar results. Solid solutions have been obtained in the full composition range in the AlSb-InSb and Cu-Co systems by MA but not by RSP. Very few examples are reported in the literature where complete solid solubility has been achieved in the same alloy system by both the MA and RSP techniques. Cu-Ag is the only such system. It will be useful to conduct investigations on other alloy systems to find out whether similar results are obtained by both techniques. The levels of solid solubility achieved are also different in different systems by these two techniques.

It was frequently observed that the extent of supersaturation in rapidly solidified alloys increased with increasing rate of solidification [4,8–10,317]. Thus, the solubility extension is higher in an alloy system when the liquid was melt spun at about 10^5–10^6 K/s than when the melt was atomized at 10^2–10^3 K/s. But a similar situation has not been reported in mechanically alloyed powders. For example, one could investigate the solid solubility extensions as a function of milling intensity, energy input, ball-to-powder weight ratio, or other parameters. In fact, taking all these variables into consideration one could define an effective "quench" rate or milling efficiency. It would be then worthwhile to investigate if the solid solubility extensions are higher at higher milling efficiencies in the MA case also.

The limits of solid solubility extensions achieved by MA have also been compared with the values obtained by other methods. Figure 9.15 shows schematically

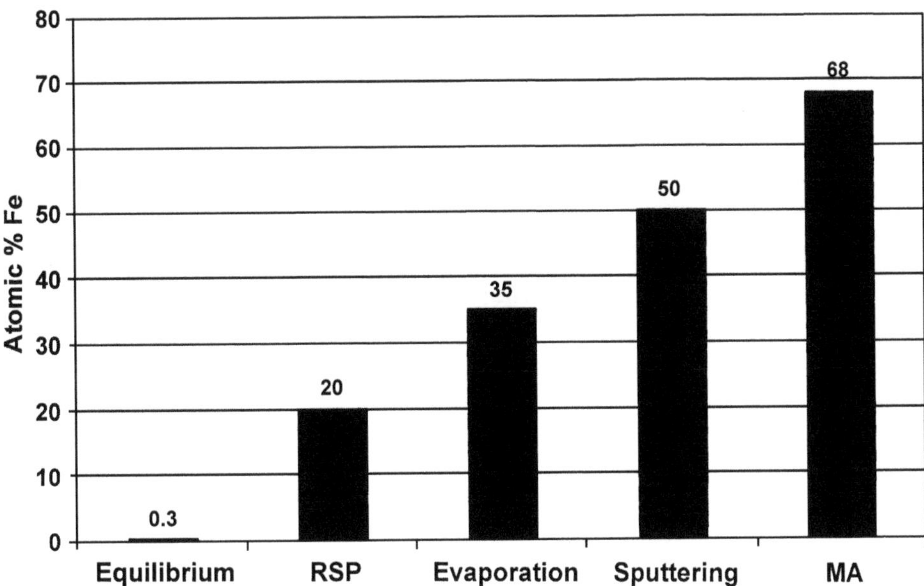

Figure 9.15 Solid solubility limits of Fe in Cu obtained by different nonequilibrium processing techniques.

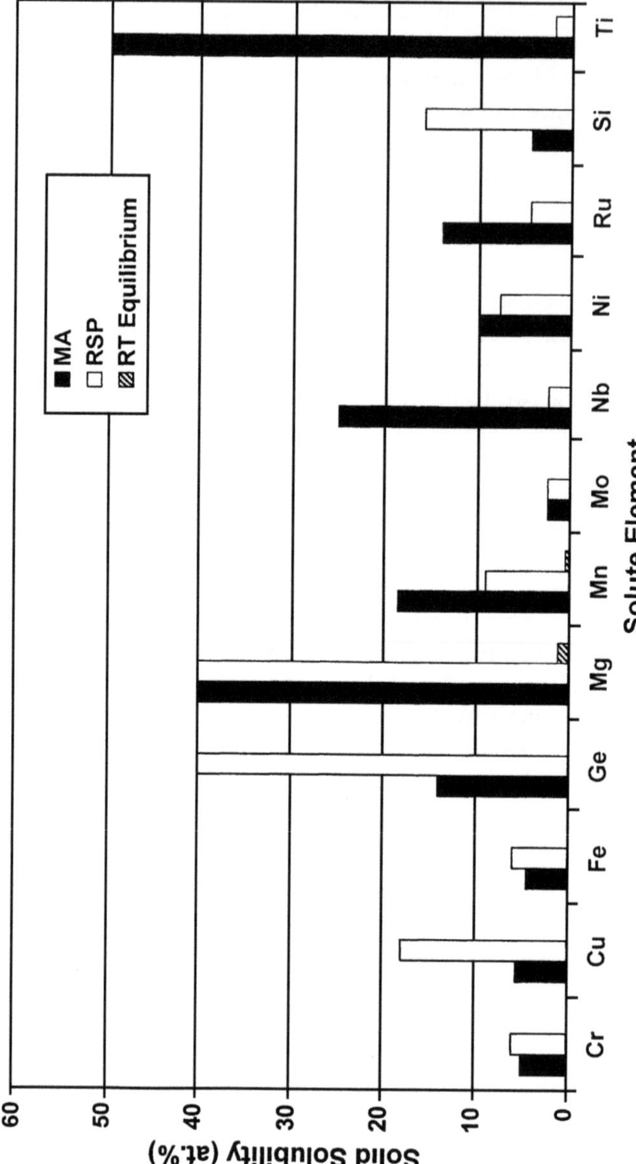

Figure 9.16 Comparison of room temperature equilibrium solid solubility of solute elements in Al and those obtained by mechanical alloying (MA) and rapid solidification processing (RSP) techniques.

the room temperature solid solubility of Fe in Cu obtained by different nonequilibrium processing methods. It may be noted that the highest solid solubility extension was achieved in the mechanically alloyed powders.

It was shown in Chapter 1 that the maximal departure from equilibrium obtainable by MA is higher than that by RSP. Therefore, one expects larger solid solubility extensions during MA than during RSP. But the available results do not always confirm this. As can be noted from Table 9.3 and Figure 9.16, one can make a general observation that the maximal solid solubility extension achieved by MA is generally higher than that by RSP except in the cases of Cu, Ge, and Si as solute elements. This appears to be even more impressive when the room temperature equilibrium solid solubility is very small or almost zero. This is understandable because MA is a near–room temperature process and therefore the room temperature solid solubility is more important than the maximal solid solubility at the higher temperatures; the latter could be important in RSP studies.

REFERENCES

1. Wilm, A. (1911). *Metallurgie* 8:225; Kelly, A., Nicholson, R. B. (1963). *Prog. Mater. Sci.* 10:149–391.
2. Hume-Rothery, W., Smallman, R. E., Haworth, C. W. (1969). *The Structure of Metals and Alloys*. London, UK: Institute of Metals.
3. Massalski, T. B. (2000). In: Turchi, P. E. A., Shull, R. D., Gonis, A., eds. *The Science of Alloys for the 21st Century*. Warrendale, PA: TMS, pp. 55–70.
4. Suryanarayana, C., ed. *Non-Equilibrium Processing of Materials*. Oxford, UK: Pergamon.
5. Duwez, P., Willens, R. H., Klement, W. Jr. (1960). *J. Appl. Phys.* 31:1136–1137.
6. Luo, H. L., Duwez, P. (1964). *J. Less Common Metals* 6:248–249.
7. Duwez, P., Willens, R. H., Klement, W. Jr. (1960). *J. Appl. Phys.* 31:1500.
8. Anantharaman, T. R., Suryanarayana, C. (1971). *J. Mater. Sci.* 6:1111–1135.
9. Anantharaman, T. R., Ramachandrarao, P., Suryanarayana, C., Lele, S., Chattopadhyay, K. (1977). *Trans. Indian Inst. Metals* 30:423–448.
10. Jones, H. (1982). *Rapid Solidification of Metals and Alloys. Monograph No. 8*. London, UK: Institution of Metallurgists.
11. Boettinger, W. J., Perepezko, J. H. (1993). In: Liebermann, H. H., ed. *Rapidly Solidified Alloys: Processes, Structures, Properties, Applications*. New York: Marcel Dekker, pp. 17–78.
12. Suryanarayana, C. (2002). In: Buschow, K. H. J., Cahn, R. W., Flemings, M. C., Kramer, E. J. Mahajan, S., eds. *Encyclopedia of Materials: Science and Technology—Updates*. Oxford, UK: Pergamon.
13. Pearson, W. B. (1967). *A Handbook of Lattice Spacings and Structures of Metals and Alloys*. Oxford, UK: Pergamon.
14. Suryanarayana, C., Norton, M. G. (1998). *X-Ray Diffraction: A Practical Approach*. New York: Plenum.
15. Cullity, B. D., Stock, S. R. (2001). *Elements of X-Ray Diffraction*. 3rd ed. Englewood Cliffs, NJ: Prentice Hall.
16. Paruchuri, M. R., Zhang, D. L., Massalski, T. B. (1994). *Mater. Sci. Eng.* A174:119–125.
17. Shingu, P. H., Ishihara, K. N., Uenishi, K., Kuyama, J., Huang, B., Nasu, S. (1990). In: Clauer, A. H., deBarbadillo, J. J., eds. *Solid State Powder Processing*. Warrendale, PA: TMS, pp. 21–34.

18. Uenishi, K., Kobayashi, K. F., Ishihara, K. N., Shingu, P. H. (1991). *Mater. Sci. Eng.* A134:1342–1345.
19. Najafabadi, R., Srolovitz, D. J., Ma, E., Atzmon, M. (1993). *J. Appl. Phys.* 74:3144–3149.
20. Tang, J. (1996). *Mater. Sci. For.* 225–227:477–482.
21. Xu, J., Herr, U., Klassen, T., Averback, R. S. (1996). *J. Appl. Phys.* 79:3935–3945.
22. Aymard, L., Dehahaye-Vidal, A., Portemer, F., Disma, F. (1996). *J. Alloys Compounds* 238:116–127.
23. Jianu, A., Kuncser, V., Nicula, R., Röhlsberger, R., Stir, M., Burkel, E. (2001). *Mater. Sci. For.* 360–362:589–594.
24. Kobayashi, K. F., Tachibana, N., Shingu, P. H. (1990). *J. Mater. Sci.* 25:3149–3154.
25. Li, F., Ishihara, K. N., Shingu, P. H. (1991). *Metall. Trans.* 22A:2849–2854.
26. Clark, C. R., Suryanarayana, C., Froes, F. H. (1995). In: Phillips, M., Porter, J., compilers. *Advances in Powder Metallurgy and Particulate Materials—1995*. Part I. Princeton, NJ: Metal Powder Industries Federation, pp. 135–143.
27. Clark, C. R., Suryanarayana, C., Froes, F. H. (1995). In: Froes, F. H., Suryanarayana, C., Ward-Close, C. M., eds. *Synthesis and Processing of Lightweight Metallic Materials*. Warrendale, PA: TMS, pp. 175–182.
28. Kim, H. G., Myung, W-N., Sumiyama, K., Suzuki, K. (2001). *J. Alloys Compounds* 322:214–219.
29. García Escorial, A., Torralba, M., Caruana, G., Cardoso, K. R., Lieblich, M. (1998). *Mater. Sci. For.* 269–272:169–174.
30. Huang, B., Ishihara, K. N., Shingu, P. H. (1997). *Mater. Sci. Eng.* A231:72–79.
31. Fadeeva, V. I., Leonov, A. V. (1992). *Mater. Sci. For.* 88–90:481–488.
32. Zhou, F., Lück, R., Lu, K., Rühle, M. (2001). *Z. Metallkde.* 92:675–681.
33. Mukhopadhyay, D. K., Suryanarayana, C., Froes, F. H. (1995). *Metall. Mater. Trans.* 26A:1939–1946.
34. Polkin, I. S., Kaputkin, E. Ya, Borzov, A. B. (1990). In: Froes, F. H., deBarbadillo, J. J., eds. *Structural Applications of Mechanical Alloying*. Materials Park, OH: ASM International, pp. 251–256.
35. Fadeeva, V. I., Portnoy, V. K., Baldokhin, Yu V., Kotchetov, G. A., Matyja, H. (1998). *Nanostructured Mater.* 12:625–628.
36. Radlinski, A. P., Calka, A., Ninham, B. W., Kaczmarek, W. A. (1991). *Mater. Sci. Eng.* A134:1346–1349.
37. Calka, A., Kaczmarek, W., Williams, J. S. (1993). *J. Mater. Sci.* 28:15–18.
38. Zhang, D. L., Massalski, T. B., Paruchuri, M. R. (1994). *Metall. Mater. Trans.* 25A:73–79.
39. Suryanarayana, C., Sundaresan, R. (1991). *Mater. Sci. Eng.* A131:237–242.
40. Sun, J., Zhang, J. X., Fu, Y. Y., Hu, G. X. (2002). *Mater. Sci. Eng.* A329–331:703–707.
41. Zdujic, M. V., Kobayashi, K. F., Shingu, P. H. (1991). *J. Mater. Sci.* 26:5502–5508.
42. Peng, Z., Suryanarayana, C., Froes, F. H. (1992). *Scripta Metall. Mater.* 27:475–480.
43. Peng, Z., Suryanarayana, C., Froes, F. H. (1996). *Metall. Mater. Trans.* 27A:41–48.
44. Peng, Z., Suryanarayana, C., Froes, F. H. (1993). In: deBarbadillo, J. J., et al., eds. *Mechanical Alloying for Structural Applications*. Materials Park, OH: ASM International, pp. 335–341.
45. Shingu, P. H. (1991). In: Hirano, K., et al., eds. *Science and Engineering of Light Metals*. Tokyo: Japan Inst. Light Metals, pp. 677–684.
46. Benameur, T., Inoue, A., Masumoto, T. (1994). *Nanostructured Mater.* 4:303–322.
47. Xu, Y., Makhlouf, S. A., Ivanov, E., Wakoh, K., Sumiyama, K., Suzuki, K. (1994). *Nanostructured Mater.* 4:437–444.
48. Uenishi, K., Kobayashi, K. F., Ishihara, K. N., Shingu, P. H. (1992). *Mater. Sci. For.* 88–90:453–458.

49. Cocco, G., Soletta, I., Battezzati, L., Baricco, M., Enzo, S. (1990). *Phil. Mag.* B61:473–486.
50. Murty, B. S., Naik, M. D., Mohan Rao, M., Ranganathan, S. (1992). *Mater. For.* 16:19–26.
51. Abe, S., Saji, S., Hori, S. (1990). *J. Jpn. Inst. Metals* 54:895–902.
52. Saji, S., Neishi, Y., Araki, H., Minamino, Y., Yamane, T. (1995). *Metall. Mater. Trans.* 26A:1305–1307.
53. Fan, G. J., Gao, W. N., Quan, M. X., Hu, Z. Q. (1995). *Mater. Lett.* 23:33–37.
54. Miki, M., Yamasaki, T., Ogino, Y. (1993). *Mater. Trans. Jpn. Inst. Metals* 34:952–959.
55. Oehring, M., Yan, Z. H., Klassen, T., Bormann, R. (1992). *Phys. Stat. Sol.* (a)131: 671–689.
56. Oehring, M., Klassen, T., Bormann, R. (1993). *J. Mater. Res.* 8:2819–2829.
57. Fan, G. J., Quan, M. X., Hu, Z. Q. (1995). *Scripta Metall. Mater.* 33:377–381.
58. Leonov, A. V., Szewczak, E., Gladilina, O. E., Matyja, H., Fadeeva, V. I. (1997). *Mater. Sci. For.* 235–238:67–72.
59. Fan, G. J., Quan, M. X., Hu, Z. Q. (1995). *Scripta Metall. Mater.* 32:247–252.
60. Fadeeva, V. I., Leonov, A. V., Szewczak, E., Matyja, H. (1998). *Mater. Sci. Eng.* A242:230–234.
61. Saji, S., Araki, H., Hashimoto, K., Murata, E. (1996). *Mater. Trans. Jpn. Inst. Metals* 37:1061–1066.
62. Fan, G. J., Quan, M. X., Hu, Z. Q. (1995). *J. Mater. Sci.* 30:4847–4851.
63. Che, X., Wang, Q., Hu, G. (1995). *Scripta Metall. Mater.* 33:2019–2023.
64. Jiang, J. Z., Mørup, S., Linderoth, S. (1996). *Mater. Sci. For.* 225–227:489–496.
65. Di, L. M., Bakker, H., Bárczy, P., Gácsi, Z. (1993). *Acta Metall. Mater.* 41:2923–2932.
66. Martin-Lopez, R., Zandona, M., Scherrer, H. (1996). *J. Mater. Sci. Lett.* 15:16–18.
67. Esaka, T., Takai, S., Nishimura, N. (1989). *Denki Kagaku* 64:1012.
68. Mukhopadhyay, D. K., Suryanarayana, C., Froes, F. H. (1994). *Scripta Metall. Mater.* 30:133–137.
69. Tanaka, T., Ishihara, K. N., Shingu, P. H. (1992). *Metall. Trans.* 23A:2431–2435.
70. Eckert, J., Schultz, L., Urban, K. (1990). *J. Less Common Metals* 166:293–302.
71. Gente, C., Oehring, M., Bormann, R. (1993). *Phys. Rev.* B48:13244–13252.
72. Huang, J. Y., Wu, Y. K., He, A. Q., Ye, H. Q. (1994). *Nanostructured Mater.* 4:293–302.
73. Cabañas-Moreno, J. G., Dorantes, H., López-Hirata, V. M., Calderon, H. A., Hallen-Lopez, J. M. (1995). *Mater. Sci. For.* 179–181:243–248.
74. Arce Estrada, E. M., Díaz De la Torre, S., López Hirata, V. M., Cabañas Moreno, J. G. (1996). *Mater. Sci. For.* 225–227:807–812.
75. Kuhrt, Ch, Schultz, L. (1992). *J. Appl. Phys* 71:1896–1900.
76. Antolini, E., Daturi, M., Ferretti, M. (1996). *J. Mater. Sci. Lett* 15:416–418.
77. Aymard, L., Dumont, B., Viau, G. (1996). *J. Alloys Compounds* 242:108–113.
78. Shen, T. D., Koch, C. C. (1996). *Acta Mater.* 44:753–761.
79. Uenishi, K., Kobayashi, K. F., Ishihara, K. N., Shingu, P. H. (1992). *Mater. Sci. For.* 88–90:459–466.
80. Ying, D. Y., Zhang, D. L. (2000). *Mater. Sci. Eng.* A286:152–156.
81. Yamane, T., Okubo, H., Hisayuki, K., Oki, N., Konishi, M., Komatsu, M., Minamino, Y., Koizumi, Y., Kiritani, M., Kim, S. J. (2001). *J. Mater. Sci. Lett.* 20:259–260.
82. Saji, S., Kadokura, T., Anada, H., Notoya, K. (1998). *Mater. Trans. Jpn. Inst. Metals* 39:778–781.
83. Diaz-Barriga-Arceo, L., Orozco, E., Tsuchiya, K., Umemoto, M., López-Hirata, V. M. (2002). *Mater. Sci. For.* 386–388:147–152.
84. Baricco, M., Battezzati, L., Enzo, S., Soletta, I., Cocco, G. (1993). *Spectrochim. Acta* A49:1331–1344.
85. Huang, J. Y., Yu, Y. D., Wu, Y. K., Li, D. X., Ye, H. Q. (1997). *J. Mater. Res.* 12:936–946.

86. Galdeano, S., Chaffron, L., Mathon, M. H., Vincent, E., De Novion, C. H. (2001). *Mater. Sci. For.* 360–362:367–372.
87. Ogino, Y., Murayama, S., Yamasaki, T. (1991). *J. Less Common Metals* 168:221–235.
88. Barro, M. J., Navarro, E., Agudo, P., Hernando, A., Crespo, P., Garcia Escorial, A. (1997). *Mater. Sci. For.* 235–238:553–558.
89. Enzo, S., Mulas, G., Frattini, R., Principi, G., Gupta, R., Cooper, R., Cowlam, N. (1997). *Mater. Sci. For.* 235–238:529–534.
90. Ueda, Y., Ikeda, S., Mori, Y., Zaman, H. (1996). *Mater. Sci. Eng.* A217–218:371–375.
91. Eckert, J., Holzer, J. C., Krill, C. E. III, Johnson, W. L. (1992). *J. Mater. Res.* 7:1980–1983.
92. Macri, P. P., Rose, P., Frattini, R., Enzo, S., Principi, G., Hu, W. X., Cowlam, N. (1994). *J. Appl. Phys.* 76:4061–4067.
93. Eckert, J., Holzer, J. C., Johnson, W. L. (1993). *J. Appl. Phys.* 73:131–141.
94. Eckert, J., Holzer, J. C., Krill, C. E. III, Johnson, W. L. (1993). *J. Appl. Phys.* 73:2794–2802.
95. Uenishi, K., Kobayashi, K. F., Nasu, S., Hatano, H., Ishihara, K. N., Shingu, P. H. (1992). *Z. Metallkde.* 83:132–135.
96. Ma, E., Atzmon, M., Pinkerton, F. E. (1993). *J. Appl. Phys.* 74:955–962.
97. Qi, M., Zhu, M., Yang, D. Z. (1994). *J. Mater. Sci. Lett.* 13:966–968.
98. Gaffet, E., Harmelin, M., Faudot, F. (1993). *J. Alloys Compounds* 194:23–30.
99. Huang, J. Y., Wu, Y. K., Hu, K. Y., Meng, X. M. (1993). *Acta Metall. Sinica* B29:60–63.
100. Huang, J. Y., He, A. Q., Wu, Y. K. (1994). *Nanostructured Mater.* 4:1–10.
101. Ivanov, E. (1992). *Mater. Sci. For.* 88–90:475–480.
102. Benghalem, A., Morris, D. G. (1992). *Scripta Metall. Mater.* 27:739–744.
103. Kim, K. J., Sumiyama, K., Suzuki, K. (1994). *J. Non-Cryst Solids* 168:232–240.
104. Shen, T. D., Koch, C. C. (1995). *Mater. Sci. For.* 179–181:17–24.
105. Kis-Varga, M., Beke, D. L. (1996). *Mater. Sci. For.* 225–227:465–470.
106. Ivanov, E. (1993). In: deBarbadillo, J. J., et al., eds. *Mechanical Alloying for Structural Applications*. Materials Park, OH: ASM International, pp. 171–176.
107. Yazenko, S. P., Hayak, V. G., Filipov, V. A., Ivanov, E., Grigorieva, T. F. USSR Patent 4725973, March 22, 1990.
108. Ivanov, E., Patton, V., Grigorieva, T. F. (1996). *Mater. Sci. For.* 225–227:575–580.
109. Ivanov, E. Y., Grigorieva, T. F. (1997). *Solid State Ionics* 101–103:235–241.
110. Murty, B. S., Mohan Rao, M., Ranganathan, S. (1993). *Nanostructured Mater.* 3:459–467.
111. Gaffet, E., Louison, C., Harmelin, M., Faudot, F. (1991). *Mater. Sci. Eng.* A134:1380–1384.
112. Pabi, S. K., Joardar, J., Murty, B. S. (1996). *J. Mater. Sci.* 31:3207–3211.
113. Yamane, T., Okubo, H., Hisayuki, K., Oki, N., Konishi, M., Minamino, Y., Koizumi, Y., Kiritani, M., Komatsu, M., Kim, S. J. (2001). *Metall. Mater. Trans.* 32A:1861–1862.
114. Kuyama, J., Ishihara, K. N., Shingu, P. H. (1992). *Mater. Sci. For.* 88–90:521–528.
115. Yoshioka, T., Yasuda, M., Miyamura, H., Kikuchi, S., Tokumitsu, K. (2002). *Mater. Sci. For.* 386–388:503–508.
116. Jiang, H. G., Perez, R. J., Lau, M. L., Lavernia, E. J. (1997). *J. Mater. Res.* 12:1429–1432.
117. Gonzalez, G., D'Angelo, L., Ochoa, J., Lara, B., Rodriguez, E. (2002). *Mater. Sci. For.* 386–388:159–164.
118. Nasu, S., Imaoka, S., Morimoto, S., Tanimoto, H., Huang, B., Tanaka, T., Kuyama, J., Ishihara, K. N., Shingu, P. H. (1992). *Mater. Sci. For.* 88–90:569–576.

119. Bonetti, E., Scipione, G., Valdre, G., Cocco, G., Frattini, R., Macri, P. P. (1993). *J. Appl. Phys.* 74:2053–2057.
120. Bonetti, E., Valdre, G., Enzo, S., Cocco, G., Soletta, I. (1993). *Nanostructured Mater.* 2:369–375.
121. Bansal, C., Gao, Z. Q., Hong, L. B., Fultz, B. (1994). *J. Appl. Phys.* 76:5961–5966.
122. Zhu, S. M., Iwasaki, K. (1999). *Mater. Sci. Eng.* A270:170–177.
123. Gonzalez, G., D'Angelo, L., Ochoa, J., Bonyuet, D. (2002). *Mater. Sci. For.* 386–388:153–158.
124. Wolski, K., Le Caër, G., Delcroix, P., Fillit, R., Thévenot, F., Le Coze, J. (1996). *Mater. Sci. Eng.* A207:97–104.
125. Gialanella, S. (1995). *Intermetallics* 3:73–76.
126. Kuhrt, C., Schropf, H., Schultz, L., Arzt, E. (1993). In: deBarbadillo, J. J., et al., eds. *Mechanical Alloying for Structural Applications*. Materials Park, OH: ASM International, pp. 269–273.
127. Liu, Z. G., Guo, J. T., He, L. L., Hu, Z. Q. (1994). *Nanostructured Mater.* 4:787–794.
128. Schropf, H., Kuhrt, C., Arzt, E., Schultz, L. (1994). *Scripta Metall. Mater.* 30:1569–1574.
129. Oleszak, D., Pękała, M., Jartych, E., Żurawicz, J. K. (1998). *Mater. Sci. For.* 269–272:643–648.
130. Portnoy, V. K., Leonov, A. V., Fadeeva, V. I., Matyja, H. (1998). *Mater. Sci. For.* 269–272:69–74.
131. Krasnowski, M., Matyja, H. (2001). *Mater. Sci. For.* 360–362:433–438.
132. Krasnowski, M., Matyja, H. (2000). *Mater. Sci. For.* 343–346:302–307.
133. Perez, R. J., Huang, B. L., Crawford, P. J., Sharif, A. A., Lavernia, E. J. (1996). *Nanostructured Mater.* 7:47–56.
134. Stiller, C., Eckert, J., Roth, S., Schäfer, R., Klement, U., Schultz, L. (1996). *Mater. Sci. For.* 225–227:695–700.
135. Huang, B. L., Perez, R. J., Lavernia, E. J., Luton, M. J. (1996). *Nanostructured Mater.* 7:67–79.
136. Umemoto, M., Liu, Z. G., Hao, X. J., Masuyama, K., Tsuchiya, K. (2001). *Mater. Sci. For.* 360–362:167–174.
137. Djahanbakhsh, M., Lojkowski, W., Bürkle, G., Baumann, G., Ivanisenko, Yu V., Valiev, R. Z., Fecht, H. J. (2001). *Mater. Sci. For.* 360–362:175–182.
138. Rochman, N. T., Kawamoto, K., Sueyoshi, H., Nakamura, Y., Nishida, T. (1999). *J. Mater. Proc. Technol.* 89–90:367–372.
139. Rochman, N. T., Kuramoto, S., Fujimoto, R., Sueyoshi, H. (2003). *J. Mater. Proc. Technol.* 138:41–46.
140. Tokumitsu, K., Umemoto, M. (2002). *Mater. Sci. For.* 386–388:479–484.
141. Yang, H., Di, L. M., Bakker, H. (1993). *Intermetallics* 1:29–33.
142. Koyano, T., Mizutani, U., Okamoto, H. (1995). *J. Mater. Sci. Lett.* 14:1237–1240.
143. Koyano, T., Takizawa, T., Fukunaga, T., Mizutani, U., Kamizuru, S., Kita, E., Tasaki, A. (1993). *J. Appl. Phys.* 73:429–433.
144. Tokumitsu, K. (1999). *Mater. Sci. For.* 312–314:557–562.
145. Tcherdyntsev, V. V., Kaloshkin, S. D., Tomilin, I. A., Shelekhov, E. V., Serdyukov, V. N. (2001). *Mater. Sci. For.* 360–362:361–366.
146. Tokumitsu, K. (1998). *Mater. Sci. For.* 269–272:467–472.
147. He, L., Allard, L. F., Breder, K., Ma, E. (2000). *J. Mater. Res.* 15:904.
148. Crespo, P., Hernando, A., García-Escorial, A., Castaño, F. J., Multinger, M. (2000). *Mater. Sci. For.* 343–346:793–799.
149. Multinger, M., Hernando, A., Crespo, P., Rivero, G., García-Escorial, A., Stiller, C., Schultz, L., Eckert, J. (2000). *Mater. Sci. For.* 343–346:800–805.
150. Wang, X. M., Aoki, K., Masumoto, T. (1996). *Mater. Sci. For.* 225–227:423–428.
151. Fultz, B., Gao, Z. Q., Hamdeh, H. H., Oliver, S. A. (1994). *Phys. Rev.* B49:6312–6315.

152. Cabrera, A. F., Sánchez, F. H., Mendoza Zélis, L. (1999). *Mater. Sci. For.* 312–314:85–90.
153. Kataoka, N., Suzuki, K., Inoue, A., Masumoto, T. (1991). *J. Mater. Sci.* 26:4621–4625.
154. Chiriac, H., Moga, A., Urse, M., Hison, C. (2000). *Mater. Sci. For.* 343–346:806–811.
155. Hightower, A., Fultz, B., Bowman, R. C. Jr. (1997). *J. Alloys Compounds* 252:238–244.
156. Tcherdyntsev, V. V., Kaloshkin, S. D., Tomilin, I. A., Shelekhov, E. V., Baldokhin, Yu V. (1999). *Z. Metallkde.* 90:747–752.
157. Liu, T., Zhao, Z. T., Ma, R. Z., Hu, T. D., Xie, Y. N. (1999). *Mater. Sci. Eng.* A271: 8–13.
158. Caamaño, Z., Perez, G., Zamora, L. E., Suriñach, S., Muñoz, J. S., Baró, M. D. (2001). *J. Non-Cryst. Solids* 287:15–19.
159. Hays, V., Marchand, R., Saindrenan, G., Gaffet, E. (1996). *Nanostructured Mater.* 7:411–420.
160. Xia, S. K., Scorzelli, R. B., Souza Azevedo, I., Baggio-Saitovich, E., Takeuchi, Y. (1996). *Mater. Sci. For.* 225–227:453–458.
161. Liu, Y. J., Chang, I. T. H. (2002). *Acta Mater.* 50:2747–2760.
162. Nunes, E., Passamani, E. C., Larica, C., Freitas, J. C. C., Takeuchi, A. Y. (2002). *J. Alloys Compounds* 345:116–122.
163. Navarro, E., Fiorani, D., Yavari, A. R., Rosenberg, M., Multinger, M., Hernando, A., Caciuffo, R., Rinaldi, D., Gialanella, S. (1998). *Mater. Sci. For.* 269–272:133–138.
164. Kis-Varga, M., Beke, D. L., Mészáros, S., Vad, K., Kerekes, Gy, Daróczi, L. (1998). *Mater. Sci. For.* 269–272:961–966.
165. Garcia-Escoral, A., Adeva, P., Cristina, M. C., Martin, A., Carmona, F., Cebollada, F., Martin, V. E., Leonato, M., Gonzalez, J. M. (1991). *Mater. Sci. Eng.* A134:1394–1397.
166. Kohmoto, O., Yamaguchi, N., Mori, T. (1994). *J. Mater. Sci.* 29:3221–3223.
167. Yelsukov, E. P., Fomin, V. M., Nemtsova, O. M., Shklyaev, D. A. (2002). *Mater. Sci. For.* 386–388:181–186.
168. Abdellaoui, M., Barradi, T., Gaffet, E. (1992). *J. Physique IV* 2:C3-73–C3-78.
169. Gao, Z., Fultz, B. (1993). *Nanostructured Mater.* 2:231–240.
170. Zhou, T., Zhong, J., Xu, J., Yu, Z., Gu, G., Wang, D., Huang, H., Du, Y., Wang, J., Jiang, Y. (1996). *J. Mag. Mag. Mater.* 164:219–224.
171. Gupta, A., Dhuri, P. (2001). *Mater. Sci. Eng. A* 304–306:394–398.
172. Yelsukov, E. P., Dorofeev, G. A., Barinov, V. A., Grigorieva, T. F., Boldyrev, V. V. (1998). *Mater. Sci. For.* 269–272:151–156.
173. Dorofeev, G. A., Yelsukov, E. P., Ulyanov, A. L., Konygin, G. N. (2000). *Mater. Sci. For.* 343–346:585–590.
174. Hong, L. B., Bansal, C., Fultz, B. (1994). *Nanostructured Mater.* 4:949–956.
175. Nasu, S., Shingu, P. H., Ishihara, K. N., Fujita, F. E. (1990). *Hyperfine Interactions* 55:1043–1050.
176. Cabrera, A. F., Sánchez, F. H., Mendoza-Zélis, L. (1995). *Mater. Sci. For.* 179–181:231–236.
177. Schlump, W., Grewe, H. (1989). In: Arzt, E., Schultz, L., eds. *New Materials by Mechanical Alloying Techniques*. Oberursel, Germany: Deutsche Gesellschaft für Metallkunde, pp. 307–318.
178. Ruuskanen, P. (1998). *Mater. Sci. For.* 269–272:139–144.
179. Oleszak, D., Jachimowicz, M., Matyja, H. (1995). *Mater. Sci. For.* 179–181:215–218.
180. El-Eskandarany, M. S., Sumiyama, K., Suzuki, K. (1997). *Acta Mater.* 45:1175–1187.
181. Bai, H. Y., Michaelsen, C., Sinkler, W., Bormann, R. (1997). *Mater. Sci. For.* 235–238:361–366.
182. Pękała, K., Oleszak, D., Jartych, E., Żurawicz, J. K. (2000). *Mater. Sci. For.* 343–346:825–829.
183. Herr, U., Samwer, K. (1992). *Nanostructured Mater.* 1:515–521.
184. Hellstern, E., Schultz, L. (1986). *Appl. Phys. Lett.* 49:1163–1165.
185. Michaelsen, C., Hellstern, E. (1987). *J. Appl. Phys.* 62:117–119.

186. Hellstern, E., Schultz, L. (1988). *Mater. Sci. Eng.* 97:39–42.
187. Schultz, L. (1988). *Mater. Sci. Eng.* 97:15–23.
188. Crespo, P., Marin, P., Agudo, P., Alocén, M. C., Hernando, A., Garcia-Escorial, A., Eckert, J., Roth, S., Schultz, L. (2002). *Mater. Sci. For.* 386–388:175–180.
189. Rico, M. M., Sort, J., Suriñach, S., Muñoz, J. S., Greneche, J. M., Pérez Alcázar, G. A., Baró, M. D. (2002). *Mater. Sci. For.* 386–388:497–502.
190. Michel, D., Mazerolles, L., Chichery, E. (1998). *Mater. Sci. For.* 269–272:99–104.
191. Jiang, J. Z., Lin, R., Mørup, S. (1998). *Mater. Sci. For.* 269–272:449–454.
192. Matteazzi, P., Le Caër, G. (1992). *Mater. Sci. Eng.* A156:229–237.
193. Gaffet, E., Faudot, F., Harmelin, M. (1991). *Mater. Sci. Eng.* A149:85–94.
194. Boolchand, P., Koch, C. C. (1992). *J. Mater. Res.* 7:2876–2883.
195. Hida, M., Asai, K., Takemoto, Y., Sakakibara, A. (1996). *Mater. Trans. Jpn. Inst. Metals* 37:1679–1683.
196. Hida, M., Asai, K., Takemoto, Y., Sakakibara, A. (1997). *Mater. Sci. For.* 235–238:187–192.
197. Zhukov, A. P., Ivanov, S. A., Nudelman, M. A., Ponomarev, B. K., Kaloshkin, S. D., Shatov, A. A. (1993). *J. Appl. Phys.* 73:6414.
198. Ziller, T., Le Caër, G., Delcroix, P. (1999). *Mater. Sci. For.* 312–314:33–42.
199. Tracy, M. J., Groza, J. R. (1992). *Nanostructured Mater.* 1:369–378.
200. Hellstern, E., Schultz, L., Bormann, R., Lee, D. (1988). *Appl. Phys. Lett.* 53:1399–1401.
201. Groza, J. R., Tracy, M. J. H. (1993). In: deBarbadillo, J. J., et al., eds. *Mechanical Alloying for Structural Applications*. Materials Park, OH: ASM International, pp. 327–334.
202. Goodwin, P. S., Ward-Close, C. M. (1995). In: Froes, F. H., ed. *P/M in Aerospace, Defense, and Demanding Applications*. Princeton, NJ: Metal Powder Industries Federation, pp. 89–95.
203. Oleszak, D., Burzynska-Szyszko, M., Matyja, H. (1993). *J. Mater. Sci. Lett.* 12:3–5.
204. Oehring, M., Bormann, R. (1990). *J. Physique* 51:C4-169–C4-174.
205. Di, L. M., Bakker, H. (1991). *J. Phys. C Condens. Matter* 3:9319–9326.
206. Chou, T. C., Nieh, T. G., Wadsworth, J. (1992). *Scripta Metall. Mater.* 27:881–886.
207. Kenik, E. A., Bayuzick, R. J., Kim, M. S., Koch, C. C. (1987). *Scripta Metall.* 21:1137–1142.
208. Lee, P. Y., Koch, C. C. (1987). *J. Non-Cryst. Solids* 94:88–100.
209. Lou, T., Fan, G., Ding, B., Hu, Z. (1997). *J. Mater. Res.* 12:1172–1175.
210. Zghal, S., Bhattacharya, P., Twesten, R., Wu, F., Bellon, P. (2002). *Mater. Sci. For.* 386–388:165–174.
211. Itsukaichi, T., Shiga, S., Masuyama, K., Umemoto, M., Okane, I. (1992). *Mater. Sci. For.* 88–90:631–638.
212. Cardellini, F., Cleri, F., Mazzone, G., Montone, A., Rosato, V. (1993). *J. Mater. Res.* 8:2504–2509.
213. Lu, L., Lai, M. O., Zhang, S. (1994). *Mater. Design* 15:79–86.
214. Guerero-Paz, J., Robles-Hernández, F. C., Martínez-Sánchez, R., Hernández-Silvam, D., Jaramillo-Vigueras, D. (2001). *Mater. Sci. For.* 360–362:317–322.
215. Ivanov, E., Grigorieva, T. F., Golubkova, G. V., Boldyrev, V. V., Fasman, A. B., Mikhailenko, S. D., Kalinina, O. T. (1988). *Mater. Lett.* 7:51–54.
216. Portnoy, V. K., Blinov, A. M., Tomilin, I. A., Kuznetsov, V. N., Kulik, T. (2002). *J. Alloys Compounds* 336:196–201.
217. Krivoroutchko, K., Kulik, T., Fadeeva, V. I., Portnoy, V. K. (2002). *J. Alloys Compounds* 333:225–230.
218. Krivoroutchko, K., Portnoy, V. K., Matyja, H., Fadeeva, V. I. (2000). *Mater. Sci. For.* 343–346:308–313.
219. Gialanella, S., Delorenzo, R., Marino, F., Guella, M. (1995). *Intermetallics* 3:1–8.

220. Boldyrev, V. V., Golubkova, G. V., Grigorieva, T. F., Ivanov, E., Kalinina, O. T., Mihailenko, S. D., Fasman, A. B. (1987). *Doklady Akad. Nauk SSSR* 297:1181–1184.
221. Krivoroutchko, K., Kulik, T., Matyja, H., Fadeeva, V. I., Portnoy, V. K. (2001). *Mater. Sci. For.* 360–362:385–390.
222. Grigorieva, T. F., Barinova, A. P., Boldyrev, V. V., Ivanov, E. Y. (1996). *Mater. Sci. For.* 225–227:417–422.
223. Yang, Q. M., Lei, Y. Q., Wu, J., Qang, Q. D., Lu, G. L., Chen, L. S. (1993). *Key Eng. Mater.* 81–83:169–173.
224. Oleszak, D., Portnoy, V. K., Matyja, H. (1999). *Mater. Sci. For.* 312–314:345–350.
225. Huot, J. Y., Trudeau, M. L., Schulz, R. (1991). *J. Electrochem. Soc.* 138:1316–1321.
226. Portnoy, V. K., Fadeeva, V. I., Zaviyalova, I. N. (1995). *J. Alloys Compounds* 224:159–161.
227. Van Neste, A., Lamarre, A., Trudeau, M. L., Schulz, R. (1992). *J. Mater. Res.* 7:2412–2417.
228. Jang, J. S. C., Tsau, C. H., Chen, W. D., Lee, P. Y. (1998). *J. Mater. Sci.* 33:265–270.
229. Cho, Y. S., Koch, C. C. (1993). *J. Alloys Compounds* 194:287–294.
230. Lee, P. Y., Chen, T. R. (1994). *J. Mater. Sci. Lett.* 13:888–890.
231. Battezzati, L., Cocco, G., Schiffini, L., Enzo, S. (1988). *Mater. Sci. Eng.* 97:121–124.
232. Schwarz, R. B., Petrich, R. R., Saw, C. K. (1985). *J. Non-Cryst. Solids* 76:281–302.
233. Yang, H., Bakker, H. (1993). In: deBarbadillo, J. J., et al., eds. *Mechanical Alloying for Structural Applications.* Materials Park, OH: ASM International, pp. 401–408.
234. Mi, S., Courtney, T. H. (1998). *Scripta Mater.* 38:637–644.
235. Zbiral, J., Jangg, G., Korb, G. (1992). *Mater. Sci. For.* 88–90:19–26.
236. Bryden, K. J., Ying, J. Y. (1996). *Mater. Sci. For.* 225–227:895–900.
237. Katona, G. L., Kis-Varga, M., Beke, D. L. (2002). *Mater. Sci. For.* 386–388:193–198.
238. Ishida, T. (1994). *J. Mater. Sci. Lett.* 13:623–628.
239. Liu, K. W., Mücklich, F., Birringer, R. (2001). *Intermetallics* 9:81–88.
240. Liu, K. W., Mücklich, F. (2002). *Mater. Sci. Eng.* A329–331:112–117.
241. Davis, R. M., McDermott, B., Koch, C. C. (1988). *Metall. Trans.* A19:2867–2874.
242. Davis, R. M., Koch, C. C. (1987). *Scripta Metall.* 21:305–310.
243. Gaffet, E., Harmelin, M. (1990). In: Froes, F. H., deBarbadillo, J. J., eds. *Structural Applications of Mechanical Alloying.* Materials Park, OH: ASM International, pp. 257–264.
244. Sherif El-Eskandarany, M. (1996). *Metall. Mater. Trans.* 27A:3267–3278.
245. El-Eskandarany, M. S. (1994). *J. Alloys Compounds* 203:117–126.
246. Liu, L., Chu, Z. Q., Dong, Y. D. (1992). *J. Alloys Compounds* 186:217–221.
247. Sherif El-Eskandarany, M., Zhang, W., Inoue, A. (2003). *J. Alloys Compounds* 350:222–231.
248. Suryanarayana, C., Chen, G. H., Frefer, A., Froes, F. H. (1992). *Mater. Sci. Eng.* A158:93–101.
249. Burgio, N., Guo, W., Martelli, S., Magini, M., Padella, F., Soletta, I. (1990). In: Froes, F. H., deBarbadillo, J. J., eds. *Structural Applications of Mechanical Alloying.* Materials Park, OH: ASM International, pp. 175–183.
250. Guo, W., Martelli, S., Burgio, N., Magini, M., Padella, F., Paradiso, E., Soletta, I. (1991). *J. Mater. Sci.* 26:6190–6196.
251. Bonetti, E., Cocco, G., Enzo, S., Valdre, G. (1990). *Mater. Sci. Tech.* 6:1258–1262.
252. Bonetti, E., Valdre, G., Enzo, S., Cocco, G. (1993). *J. Alloys Compounds* 194:331–338.
253. Szewczak, E., Paszula, J., Leonov, A. V., Matyja, H. (1997). *Mater. Sci. Eng. A* 226–228:115–118.
254. Suryanarayana, C., Chen, G. H., Froes, F. H. (1992). *Scripta Metall. Mater.* 26:1727–1732.

255. Fadeeva, V. I., Leonov, A. V., Szewczak, E., Matyja, H. (1998). *Mater. Sci. Eng.* A242:230–234.
256. Chen, G. H., Suryanarayana, C., Froes, F. H. (1991). *Scripta Metall. Mater.* 25:2537–2540.
257. Suryanarayana, C., Chen, G. H., Froes, F. H. (1992). In: Froes, F. H., et al., eds. *Advancements in Synthesis and Processing*. Covina, CA: SAMPE, pp. M671–M683.
258. Bououdina, M., Guo, Z. X. (2002). *Mater. Sci. Eng.* A322:210–222.
259. Bououdina, M., Luklinska, Z., Guo, Z. X. (2001). *Mater. Sci. For.* 360–362:421–426.
260. Itsukaichi, T., Norimatsu, T., Umemoto, M., Okane, I., Wu, B-Y. (1992). In: Tamura, I., ed. *Heat and Surface '92*. Tokyo, Japan: Japan Tech. Info. Center, pp. 305–308.
261. Abe, Y. R., Johnson, W. L. (1992). *Mater. Sci. For.* 88–90:513–520.
262. Novakova, A. A., Agladze, O. V., Tarasov, B. P., Sidorova, G. V., Andrievsky, R. A. (1998). *Mater. Sci. For.* 269–272:127–132.
263. Zhou, E., Suryanarayana, C., Froes, F. H. (1995). *Mater. Lett.* 23:27–31.
264. Suryanarayana, C., Froes, F. H. (1990). *J. Mater. Res.* 5:1880–1886.
265. Sundaresan, R., Froes, F. H. (1989). *Key Eng. Mater.* 29–31:199–206.
266. Sundaresan, R., Froes, F. H. (1989). In: Arzt, E., Schultz, L., eds. *New Materials by Mechanical Alloying Techniques*. Oberursel, Germany: Deutsche Gesellschaft für Metallkunde, pp. 243–262.
267. Wilkes, D. M. J., Goodwin, P. S., Ward-Close, C. M., Bagnall, K., Steeds, J. W. (1996). *Mater. Lett.* 27:47–52.
268. Sun, F., Froes, F. H. (2002). *J. Alloys Compounds* 340:220–225.
269. Wilkes, D. M. J., Goodwin, P. S., Ward-Close, C. M., Bagnall, K., Steeds, J. (1996). In: Bormann, R., et al., eds. *Metastable Phases and Microstructures*. Vol. 400. Pittsburgh, PA: *Mater. Res. Soc.* , 267–274.
270. Skakov, Yu. A., Edneral, N. V., Frolov, E. V., Povolovzki, J. A. (1995). *Mater. Sci. For.* 179–181:33–38.
271. Yan, Z. H., Oehring, M., Bormann, R. (1992). *J. Appl. Phys.* 72:2478–2487.
272. Di, L. M., Bakker, H. (1991). *J. Phys. C: Condens. Matter* 3:3427–3432.
273. Tang, H. G., Ma, X. F., Zhao, W., Yan, X. W., Hong, R. J. (2002). *J. Alloys Compounds* 347:228–230.
274. Shen, T. D., Wang, K. Y., Quan, M. X., Wang, J. T. (1992). *Mater. Sci. For.* 88–90:391–397.
275. Herr, U., Samwer, K. (1992). *Nanostructured Mater.* 1:515–521.
276. Öveçoğlu, M. L., Özkal, B., Suryanarayana, C. (1996). *J. Mater. Res.* 11:673–682.
277. Ivanov, E., Suryanarayana, C., Bryskin, B. D. (1998). *Mater. Sci. Eng.* A251:255–261.
278. Ivanov, E., Sumiyama, K., Yamauchi, H., Suzuki, K. (1993). In: deBarbadillo, J. J., et al., eds. *Mechanical Alloying for Structural Applications*. Materials Park, OH: ASM International, pp. 409–413.
279. Biswas, A., Dey, G. K., Haq, A. J., Bose, D. K., Banerjee, S. (1996). *J. Mater. Res.* 11:599–607.
280. Chen, G. H., Suryanarayana, C., Froes, F. H. (1993). In: deBarbadillo, J. J., et al., eds. *Mechanical Alloying for Structural Applications*. Materials Park, OH: ASM International, OH, pp. 367–375.
281. Sheng, H. W., Lu, K., Ma, E. (1999). *J. Appl. Phys.* 85:6400–6407.
282. Fu, Z., Johnson, W. L. (1993). *Nanostructured Mater.* 3:175–180.
283. Fecht, H. J., Han, G., Fu, Z., Johnson, W. L. (1990). *J. Appl. Phys.* 67:1744–1748.
284. Ma, E., Atzmon, M. (1991). *Phys. Rev. Lett.* 67:1126–1129.
285. Ma, E., Atzmon, M. (1993). *J. Alloys Compounds* 194:235–244.
286. Ma, E., Atzmon, M. (1992). *Mater. Sci. For.* 88–90:467–474.
287. Schultz, L., Hellstern, E., Thoma, A. (1987). *Europhys. Lett.* 3:921–926.

288. Kwon, N. H., Kim, G. H., Song, H. S., Lee, H. L. (2001). *Mater. Sci. Eng.* A299:185–194.
289. Kong, L. B., Ma, J., Zhu, W., Tan, O. K. (2002). *J. Alloys Compounds* 335:290–296.
290. Katamura, J., Yamamoto, T., Qin, X., Sakuma, T. (1996). *J. Mater. Sci. Lett.* 15:36–37.
291. Tonejc, A. M., Tonejc, A. (1996). *Mater. Sci. For.* 225–227:497–502.
292. Massalski, T. B., ed. *Binary Alloy Phase Diagrams.* 2nd ed. Materials Park, OH: ASM International.
293. Kim, H. S., Suhr, D. S., Kim, G. H., Kum, D. W. (1996). *Metals and Materials (Korea)* 2:15–21.
294. Michaelsen, C. (1995). *Phil. Mag.* A72:813–828.
295. Gayle, F. W., Biancaniello, F. S. (1995). *Nanostructured Mater.* 6:429–432.
296. Bellon, P., Averback, R. S. (1995). *Phys. Rev. Lett.* 74:1819–1822.
297. Martin, G. (1984). *Phys. Rev.* B30:1424–1436.
298. Ma, E., He, J. H., Schilling, P. J. (1997). *Phys. Rev.* B55:5542–5545.
299. Pochet, P., Tominez, E., Chaffron, L., Martin, G. (1995). *Phys. Rev.* B52:4006–4016.
300. Klassen, T., Herr, U., Averback, R. S. (1997). *Acta Mater.* 45:2921–2930.
301. Klassen, T., Herr, U., Averback, R. S. (1996). In: Bormann, R., et al., eds. *Metastable Phases and Microstructures.* Vol. 400. Pittsburgh, Pa: *Mater. Res. Soc.*: pp. 25–30.
302. Kaloshkin, S. D., Tomlin, I. A., Andrianov, G. A., Baldokhin, U. V., Shelekhov, E. V. (1997). *Mater. Sci. For.* 235–238:565–570.
303. Yavari, A. R., Desré, P. (1992). *Mater. Sci. For.* 88–90:43–50.
304. Ogino, Y., Yamasaki, T., Murayama, S., Sakai, R. (1990). *J. Non-Cryst Solids* 117/118:737–740.
305. Sui, H. X., Zhu, M., Qi, M., Li, G. B., Yang, D. Z. (1992). *J. Appl. Phys.* 71:2945–2949.
306. Suryanarayana, C. (1995). *Int. Mater. Rev.* 40:41–64.
307. Veltl, G., Scholz, B., Kunze, H.-D. (1991). *Mater. Sci. Eng.* A134:1410–1413.
308. Fecht, H. J., Hellstern, E., Fu, Z., Johnson, W. L. (1990). *Metall. Trans.* 21A:2333–2337.
309. Desré, P. J. (1994). *Nanostructured Mater.* 4:957–963.
310. Yavari, A. R., Desré, P. J., Benamuer, T. (1992). *Phys. Rev. Lett.* 68:2235–2238.
311. Yavari, A. R. (1994). *Mater. Sci. Eng.* A179/180:20–26.
312. Drbohlav, O., Yavari, A. R. (1995). *Acta Mater.* 43:1799–1809.
313. Jiang, J. Z., Gente, C., Bormann, R. (1998). *Mater. Sci. Eng.* A242:268–277.
314. Schwarz, R. B. (1998). *Mater. Sci. For.* 269–272:665–674.
315. Darken, L. S., Gurry, R. W. (1953). *Physical Chemistry of Metals.* New York: McGraw-Hill.
316. Girgis, K. (1983). In: Cahn, R. W., Haasen, P., eds., *Physical Metallurgy.* 3rd ed. Amsterdam, The Netherlands: Elsevier, pp. 219–270.
317. Liebermann, H. H., ed. (1993). *Rapidly Solidified Alloys: Processes, Structures, Properties, Applications.* New York: Marcel Dekker.

10

Synthesis of Intermetallics

10.1 INTRODUCTION

Intermetallic compounds (or intermediate phases, or intermetallic phases, or intermetallics for short) consist of metals in approximately stoichiometric ratios. They usually crystallize with a structure other than those of its components and possess ordered crystal structures. An alloy is considered to be in the ordered state if two or more sublattices are required to describe its crystal structure. The driving force for ordering in a binary alloy is the greater attraction between unlike neighbors, i.e., the strength of the A-B bonds in an alloy made up of components A and B is greater than that of the A-A or B-B bonds that they replace on ordering. Thus, unlike atoms are favored over like ones as nearest neighbors. If the ordering energy is low, then as the temperature increases, the entropic tendency for disorder dominates and causes the lattice to disorder, as, for example, in β-CuZn with the B2 structure. Such types of phases have been designated as reversibly ordered intermetallics [1], i.e., it is possible to go from the disordered to the ordered state and vice versa. For other metallic combinations, the ordering energy is high enough that the ordered structure is retained up to the melting temperature. Such materials may be simply referred to as intermetallics or as irreversibly ordered (or permanently ordered) compounds. A majority of the intermetallics fall under the permanently ordered category.

The ordered nature of intermetallics leads to an attractive combination of properties such as high strength, increased stiffness, creep resistance, low-temperature damage tolerance, and excellent environmental resistance. Furthermore, these properties could be retained up to reasonably high temperatures. These attractive elevated-temperature properties are due to the reduced dislocation motion (since pairs of dislocations—superdislocations—must move together to retain the ordered nature of the lattice) and low diffusivities. Hence, many of these intermetallics and intermetallic-based alloys have been identified as useful structural materials. Table 10.1 lists the properties of some commercially useful intermetallics. The reader is advised to consult some of the excellent reviews, monographs, and conference proceedings on intermetallics for a full and detailed description of their structures, properties, and applications [2–9].

In spite of the attractive combination of properties listed above, intermetallics suffer from low ambient temperature ductility and fracture toughness, which is also associated with the reduced dislocation activity, and this precludes large-scale industrial applications of intermetallics. Hence, several attempts have been made in recent years to alleviate this problem. The common routes adopted to improve the room temperature ductility of intermetallics include:

1. Reduction in grain size
2. Disordering of the lattice, to improve the dislocation motion (superdislocations do not exist in disordered lattices and therefore only single dislocations need to move for deformation to occur), and
3. Modifying the crystal structure of the alloy phase into a more symmetrical, e.g., cubic, so that the minimal required number of slip systems for plastic deformation are available.

Mechanical alloying (MA) can achieve all the above effects simultaneously. Therefore, this processing technique has been extensively employed to synthesize intermetallics and study their mechanical behavior. The process of MA will also lead to disordering of the intermetallics. However, in this chapter we concentrate on the synthesis of intermetallics by MA. The aspect of disordering of intermetallics by MA

Table 10.1 Crystal Structure, Physical Properties, and Room Temperature Mechanical Properties of Some Important Intermetallics

Intermetallic	Crystal structure	Pearson symbol	Density (g/cm^3)	Melting temp. (°C)	Fracture mode	DBTT (°C)	Young's modulus (GPa)
Ni_3Al	$L1_2$, fcc	cP4	7.50	1400	IG[a]	—	200
Ni_3Si	$L1_2$, fcc	cP4	7.30	1140	IG	—	
Fe_3Al	DO_3, bcc	cF16	6.72	1540	TG[b]	—	
TiAl	$L1_0$, tetragonal	tP4	3.91	1460	TG	800–1000	173
Ti_3Al	DO_{19}, hexagonal	hP8	4.20	1600	TG	—	140
FeAl	B2, bcc	cP2	5.56	1300	IG+TG	—	261
NiAl	B2, bcc	cP2	5.86	1640	IG+TG	150–250	184
$MoSi_2$	$C11_b$, tetragonal	tI6	6.24	2020	IG+TG	1200	
Ti_5Si_3	$D8_b$		4.32	2130	TG	700–900	
Al_3Ti	DO_{22}	tI8	3.20	1350	TG	—	216

[a] Intergranular.
[b] Transgranular.

will be covered in Chapter 11, and the synthesis of intermetallics by mechanochemical processing will be discussed in detail in Chapter 14.

The types of intermetallics synthesized by MA include both quasi-crystalline and crystalline phases. Both equilibrium and metastable phases have been synthesized in the latter category, and these also include the disordered and ordered phases.

10.2 QUASI-CRYSTALLINE PHASES

Quasi-crystalline phases are metallic phases, which were first reported in a rapidly solidified Al-Mn alloy in 1984 [10]. These phases possess a unique crystallographic structure characterized by long-range aperiodic order and crystallographically forbidden rotational symmetries. Accordingly, they exhibit 5-fold (icosahedral), 7-fold, 8-fold, 10-fold (decagonal), 12-fold, 15-fold, etc., rotational symmetries and quasi-periodic translational order. In the crystallographic hierarchy, quasi-crystals occupy a position intermediate between perfect crystals and amorphous phases.

Electron diffraction patterns from quasi-crystals show sharp diffraction spots; thus, it could be concluded that they exhibit ordering in the lattice and that they are not amorphous. A characteristic feature of diffraction patterns from these phases is that the neighboring pairs of spots along any direction perpendicular to the 5-fold and 10-fold rotation axes are aperiodic (Fig. 10.1), i.e., the diffraction spots are separated by a distance corresponding to the golden mean (i.e., $(1 + \sqrt{5})/2 = 1.618$). While the icosahedral quasi-crystals exhibit quasi-periodicity along all three directions, the

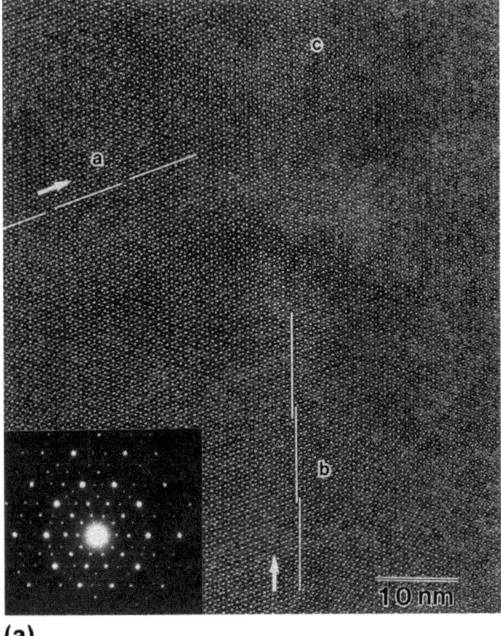

(a)

Figure 10.1 High-resolution electron micrographs and electron diffraction patterns from (a) Al-Mn-Si icosahedral quasi-crystal and (b) Al-Mn decagonal quasi-crystal. The incident beam is parallel to the fivefold and tenfold symmetry axes in (a) and (b), respectively.

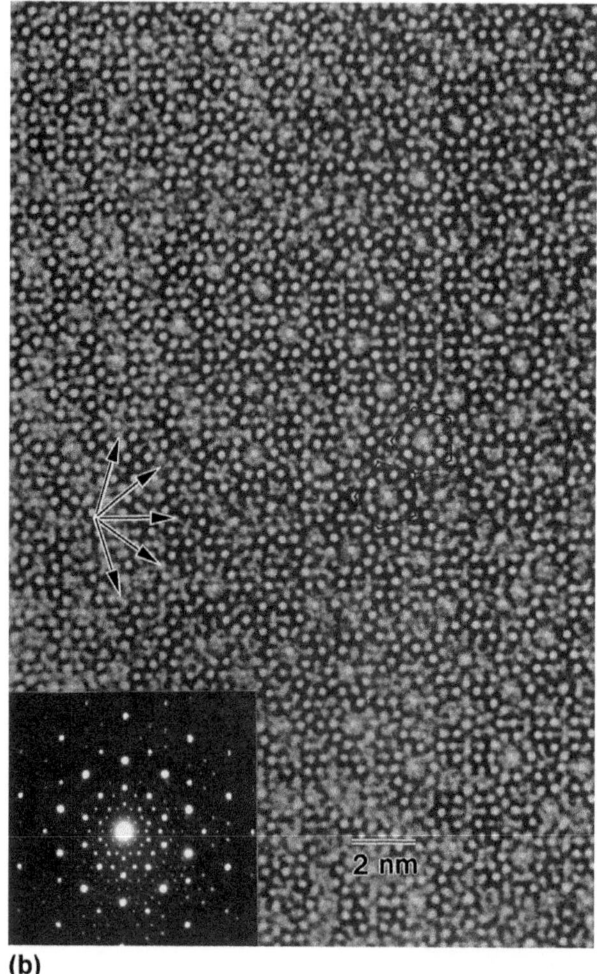

(b)

Figure 10.1 Continued.

decagonal quasi-crystals show quasi-periodicity only along two directions and exhibit perfect periodicity along the third direction. Accordingly, electron diffraction patterns recorded from a decagonal phase in a direction perpendicular to the quasi-periodic planes (i.e., along the 10-fold symmetry axis) show periodicity. Thus, there is no difference in the diffraction patterns from crystals and decagonal quasi-crystals when viewed along the 10-fold symmetry axis. This is why the discovery of quasi-crystals was narrowly missed by 6 years in 1978 [11].

Even though some thermodynamically stable quasi-crystals (e.g., Al-Cu-Fe, Al-Li-Cu, Al-Cu-Ru, and Zn-Mg-RE where RE = Y, Gd, Tb, Dy, Ho, Er) have been discovered [12], a great majority of the quasi-crystalline phases synthesized by non-equilibrium processing methods, such as rapid solidification processing or MA, are metastable in nature. Quasi-crystals have been shown to have high strength, reduced friction coefficient, and interesting electrical properties. Some of the potential applications for quasi-crystals are as catalysts, materials for thermopower generation, hydrogen storage, solar light absorption, and as reinforcements in metal-matrix composites. An interesting and commercially available application is to reinforce nano-icosahedral particles in a maraging steel used for different parts in electric shavers. Another commercial application of quasi-crystalline materials is as coatings for the production of scratch-resistant films (e.g., on frying pans) offering a lowered adhesion to food or to some polymers. The structure, properties, and applications of quasi-crystalline phases have been discussed in several earlier reviews (see, for example, Refs. [13–18]).

10.2.1 General Features

The first report of synthesis of quasi-crystalline phases by MA appeared in 1989 [19,20]. Even today only a limited number of reports exist on the synthesis and many fewer on the transformation behavior of these quasi-crystalline phases. The quasi-crystalline phases synthesized to date have been summarized in Table 10.2 along with the milling conditions (mostly using a planetary ball mill) employed to synthesize them. It should be noted that *all* the quasi-crystalline phases synthesized so far are of the icosahedral type (having the 5-fold rotational symmetry) and that there are no reports of synthesis of phases with other symmetries; a number of the latter category have been synthesized by other methods such as rapid solidification from the melt [13,14]. The icosahedral phases synthesized by MA are similar to those produced by rapid solidification processing.

Unambiguous identification of quasi-crystalline phases by X-ray diffraction (XRD) techniques has not been easy. For example, in most quasi-crystalline phases, a number of closely spaced diffraction peaks occur in the XRD patterns. Furthermore, broadening of the diffraction peaks in mechanically alloyed materials due to decreased particle/crystallite size and increased lattice strain may hinder a clear identification of the phases produced and also a proper distinction between the different types of quasi-crystalline phases (icosahedral, decagonal, etc.). This could be one reason why phases other than those with the icosahedral symmetry have not been obtained or reported with MA. Confirmatory transmission electron microscopy (TEM) studies are necessary and useful for the correct identification of these phases. Even while using a transmission electron microscope, it is difficult to orient the small powder particles along the 10-fold symmetry axis, even if a decagonal phase was produced.

As noted above, the peaks in the XRD patterns representing the icosahedral phase in mechanically alloyed powders are usually broad. Annealing of the powders to relieve the strain and increasing the crystallite size causes sharpening of the diffraction peaks and enables a clear identification of the phase [41,43,44]. However, care must be taken to ensure that the annealing temperature is not high enough to transform the metastable quasi-crystalline phase to the equilibrium phase(s).

Quasi-crystalline phases have been synthesized either directly by MA or on subsequent annealing of the amorphous phase produced by MA. Sometimes annealing the nanocrystalline phase obtained by MA has also produced a quasi-crystalline phase [21]. Results are available to support these different routes. At present, quasi-crystalline phases have been synthesized by MA only in those alloy compositions and systems, at which these phases were produced by other nonequilibrium processing techniques, most notably rapid solidification processing. But it is not known if MA

Table 10.2 Quasicrystalline Phases Formed by Mechanical Alloying of Blended Elemental Powders

Alloy/composition	Mill	Ball-to-powder ratio	Milling intensity	Milling time (h)	Ref.
$Al_{65}Cu_{20}Co_{15}$	Fritsch P5	15:1	—	55[a]	21
Al-Cu-Cr	Fritsch P5	15:1	9	—	21
$Al_{70}Cu_{20}Fe_{10}$	Fritsch P5	10:1	—	40	22
$Al_{65}Cu_{20}Fe_{15}$	Fritsch P7/ AGO–2U planetary ball mill	15:1	7	15[b]	23–25
$Al_{70-x}(Cu,Fe)_{30+x}$	Fritsch P5	10:1	—	20–30	26
$Al_{65}Cu_{20}Mn_{15}$	Fritsch P5	15:1	7	90	19, 21, 27
$Al_{40}Cu_{10}Mn_{25}Ge_{25}$	Fritsch	20:1	450–650 rpm	66	28
$Al_{65}Cu_{20}Ru_{15}$	—	—	—	—	29
$Al_{70}Cu_{12}Ru_{18}$	—	—	—	—	29
$Al_{75}Cu_{15}V_{10}$	Fritsch P5/P7	15:1	9	c	21, 30
$Al_{92}Mn_6Ce_2$	Fritsch P5	13:1	300 rpm	700	31
$Al_{91}Mn_7Fe_2$	Fritsch P5	13:1	300 rpm	700	31
$Al_{50}Mn_{20}Ge_{30}$	Fritsch	20:1	45–650 rpm	47	32
$Al_{50}Mn_{20}Si_{20}Ge_{10}$	Fritsch	20:1	45–650 rpm	80	32
$Al_{75}Ni_{10}Fe_{15}$	—	—	—	d	33
$Al_{70}Pd_{20}Mn_{10}$	Fritsch P7	15:1	7	30	34
$Mg_{32}Cu_8Al_4$	Planetary	—	900 rpm	—	19, 20, 35, 36
Mg-Al-Pd	Planetary mill	40:1	150G	4	37
$Mg_{32}(Al,Zn)_{49}$	Planetary mill	—	—	3 min	20, 35, 36, 38, 39
	Fritsch	—	600 m/s^2	10 min	40
$Mg_3Zn_{5-x}Al_x$ (x = 2–4)	Planetary mill	—	900 rpm	—	19, 36
$Mg_{44}Al_{15}Zn_{41}$	Planetary ball mill	—	600 m/s^2	30 min	41
$Ti_{56}Ni_{18}Fe_{10}Si_{16}$	SPEX 8000	6:1	—	e	42
$Ti_{45}Zr_{38}Ni_{17}$	Fritsch P7	8:1	9.5	f	43
$Ti_{45}Zr_{38}Ni_{17}$	AGO-2 planetary centrifugal mill	20:1	400 m/s^2	2	44

[a] MA for 55 h to a nanocrystalline phase followed by annealing for 1 h at 600°C.
[b] MA + anneal at 550°C.
[c] MA for 20 h to an amorphous phase followed by annealing at 450°C.
[d] MA for 400 h followed by annealing for 20 h at 800°C.
[e] MA for 30 h to an amorphous state followed by annealing for 30 min at 750°C.
[f] MA for 20 h to an amorphous phase followed by heating to 828 K in the differential scanning cobrimeter at 5°C/min.

could produce quasi-crystalline phases at other compositions and/or in alloy systems and if any special conditions must be satisfied to form a quasi-crystalline phase by MA.

The mechanism for the formation of quasi-crystalline phases in mechanically alloyed powders has not been investigated in detail, but it could be expected that the general features of alloying observed during MA will also be applicable here. Accordingly, the repeated cold welding, fracturing, and rewelding of powder particles results in the mutual dispersion of the constituent metal powder particles. The action of mechanical pulses and local heating raises the temperature of the powder and facilitates diffusion. During this process it is possible that icosahedral clusters, which are the basic elements of the icosahedral quasi-crystalline structure, form. Growth of these clusters could result in the formation of the icosahedral structure. Konstanchuk et al. [44] suggested that the icosahedral clusters could first build the amorphous structure, which on subsequent milling could transform into the icosahedral phase, as a result of crystallization of the amorphous phase.

Since formation of the quasi-crystalline alloy phase, either directly or as a result of crystallization of the amorphous phase, requires diffusion, the magnitude of the interdiffusion coefficient determines the time required for formation of the icosahedral phase. Thus, it was noted that formation of the icosahedral phase in the Mg-Al-Zn system requires only a few minutes (3–10) [20,38–40] whereas it requires about 2 h in the Ti-Zr-Ni system [44], related to the lower diffusivities in the higher melting Ti-Zr-Ni system.

10.2.2 Effect of Process Variables

There have been only a few studies investigating the effect of process variables on the formation of quasi-crystalline phases. Eckert et al. [27] studied the effect of milling intensity in the Fritsch Pulverisette 5 mill on the nature of the phase formed in a mechanically alloyed $Al_{65}Cu_{20}Mn_{15}$ powder blend. Four different intensities representing the settings of 3, 5, 7, and 9 corresponding to calculated velocities of the hitting balls of 2.5, 3.6, 4.7, and 5.8 m/s were employed. They reported that the nature of the phase synthesized was different depending on the intensity of milling. It was noted that an amorphous phase formed when the blended elemental powder was milled at an intensity of 5, and an equilibrium intermetallic phase formed at an intensity of 9. At the intermediate intensity of 7, a quasi-crystalline phase was formed. This was explained on the basis of the temperature rise during milling. The estimated maximal peak temperatures were 247°C, 407°C, and 590°C at the milling intensities of 5, 7, and 9, respectively. The temperature reached at the milling intensity of 9 is above the quasi-crystalline-to-crystalline phase transformation temperature of the quasi-crystalline phase produced and therefore an equilibrium intermetallic phase formed under this condition of milling. This observation confirms that "soft" milling conditions, i.e., lower ball-to-powder weight ratios, lower milling intensities, and so forth, favor the formation of metastable phases—both amorphous and quasi-crystalline [45–47]. Yet another conclusion that can be drawn from the above observation is that an amorphous phase is farthest from equilibrium whereas the quasi-crystalline phase is intermediate in departure from equilibrium between an amorphous phase and an equilibrium crystalline phase.

Salimon et al. [25] reported that the presence of nitrogen during milling of Al-Cu-Fe powder mixture produced the equilibrium cubic β phase instead of the quasi-crystalline phase. The quasi-crystalline phase was reported to form in the absence of a nitrogen environment. This was rationalized by the fact that the highly reactive aluminum bonded with nitrogen and consequently the sample composition shifted away from that required for the formation of a quasi-crystalline phase.

As in the case of amorphization of intermetallics on milling [48,49] (to be discussed later), amorphization has also been reported to occur on milling of the quasi-crystalline phases. Accordingly, it has been reported that the $Al_{65}Cu_{20}Fe_{15}$ quasi-crystalline phase becomes amorphous (as detected by XRD methods) after milling for 300 h in a QM-1SP planetary ball mill at a BPR of 10:1 [50]. Similar results were reported by Nasu et al. [23]. However, TEM investigations have shown the presence of a few quasi-crystalline phase particles with a size of 10–15 nm in the amorphous matrix. Furthermore, there are also reports of formation of a quasi-crystalline phase on annealing an amorphous phase [30,42]. This is somewhat similar to the mechanical crystallization behavior observed when an amorphous alloy was found to crystallize on mechanical milling [51]. This aspect also will be discussed in chapter 12.

Apart from an early curiosity about synthesizing quasi-crystalline phases by a route other than rapid solidification processing and the possibility of producing large amounts of powders for subsequent consolidation to form bulk alloys, interest in the study of quasi-crystalline phases synthesized by MA is now decreasing. The discovery of stable (equilibrium) quasi-crystals in recent years [12,14,16,17] has further lessened interest in the synthesis and characterization of quasi-crystals by MA.

10.3 CRYSTALLINE INTERMEDIATE PHASES

An important group of intermediate phases is the so-called Hume-Rothery phases or electron compounds. They crystallize with the same structure, if they possess the same average number of valence electrons per atom, or valence electron concentration (VEC). It has been shown that for VEC values of 3/2, 21/13, and 7/4 the crystal structures adopted are usually body-centered cubic (bcc), complex cubic, and hexagonal close packed (hcp), respectively. However, under certain circumstances, phases with VEC = 3/2 can take up the β-Mn and hcp structures also. Because these three groups of phases have their prototypes in the Cu-Zn system, they are traditionally referred to as β-brass, γ-brass, and ε-brass structures, respectively.

The equilibrium Hume-Rothery electron compounds—β-brass, γ-brass, and ε-brass phases—were synthesized by MA starting from blended elemental pure Cu and Zn powders mixed in the correct proportions [52–54]. Similarly, a number of other intermetallic phases have been synthesized starting from blended elemental powder mixtures. Even though the mechanism of formation of intermetallics by MA is not the same in every case, it seems to occur by a self-propagating high-temperature synthesis (SHS, commonly known as combustion) reaction, at least in some cases. In this category, there are also examples where the combustion took place only after "interrupted milling," i.e., milling of the powders for a given length of time, aging the powder at room temperature after stopping the milling, and then resuming the milling operation. An NiAl compound was synthesized in this way [55]. Similar results were also reported in other systems [56]; in fact, these results are not significantly different from those obtained in the conventional SHS process [57], except that in this case mechanical activation of the solid plays an important role.

The type of crystalline intermediate phases synthesized by MA includes both the metastable (nonequilibrium) and stable (equilibrium) phases.

10.3.1 Metastable Crystalline Intermediate Phases

The available results show that only a few *metastable* crystalline intermediate phases have been synthesized by MA. This is in stark contrast to the very large number of metastable intermediate phases reported to occur in rapidly solidified alloys in the early years of the development of the technique [58,59]. A metastable deformation-induced martensite phase has been reported to form in mechanically alloyed Cu-Zn powders [52,54]. It is surprising that not many such deformation-induced metastable phases have been reported in mechanically alloyed powder mixtures, since MA involves heavy plastic deformation of metal powder particles. This observation should also be viewed in the light of the fact that there are several instances in which metastable phases are formed by cold working; a good example is the formation of deformation-induced metastable ε (hcp) martensite in austenitic stainless steels. The synthesis of a bct-Fe phase by prolonged processing of iron powder in a nitrogen atmosphere has also been reported [60,61]. Recently, a metastable crystalline phase with rhombohedral crystal structure was reported to form in mechanically alloyed Al-Ge powders [62,63]. Two metastable phases designated β′ (Ni_3Sn) and β″ (Ni_3Sn_2) have been detected in mechanically alloyed Ni-Sn alloys at 25 and 40 at% Sn compositions, respectively [64]. The nature of the phases and the conditions under which these metastable phases formed in mechanically alloyed powder mixtures are summarized in Table 10.3. It is interesting that a number of metastable ordered phases are also formed directly on MA.

If a phase that is stable at high temperatures and/or high pressures is retained by MA at room temperature and atmospheric pressure, then also it is considered metastable. However, such phases are listed separately, under the category of polymorphic phase transformations, in Table 10.4.

As mentioned in Chapter 8, the room temperature formation of phases stable at high temperatures during milling has been used as an indicator to estimate the maximal temperature reached during milling. Accordingly, formation of the high-temperature orthorhombic Sb_2O_3 phase during milling of Sb and Sb-Ga alloys at room temperature was taken to suggest that the maximal temperature reached was at least 570°C, the temperature above which this phase is stable [145,146]. Thus, the high-temperature orthorhombic Sb_2O_3 phase at room temperature is metastable in nature.

The pure metal cobalt is known to undergo a martensitic transformation on cold working; the room temperature hcp form transforms to the elevated temperature fcc form. Both the pure metal Co and Co-based alloys have been subjected to MA, and the reversible hcp ↔ fcc martensitic transformation has been reported to occur in all the cases. However, the end product of mechanical milling has been the fcc form in some cases [99,101,103] and the hcp form in others [100,101]. In some instances, the fcc phase has been shown to transform to the hcp form after milling for a short period of time and then again to the fcc form after prolonged milling. Since the fcc → hcp transformation and the reverse transformation are known to occur via the formation of stacking faults, the nature and density of stacking faults present in the mechanically alloyed powder sample might have been responsible for the specific phase present in

Table 10.3 Metastable Phases Formed in Mechanically Alloyed Powder Mixtures

Alloy	Mill	BPR	Time (h)	Phase	Ref.
Al-Cu	Ball mill	90:1	400	m-Bcc	65
Al-Ge	SPEX 8000	—	120	Rhombohedral γ_1	62
Al-Ge	Fritsch P5	10:1	100	Rhombohedral	63
Al-Hf	SPEX 8000	—	20	$L1_2$ Al_3Hf	66, 67
Al-Hf-Fe	SPEX 8000	10:1	8	$L1_2$ $(Al,Fe)_3Hf$	68
Al-Hf-Ni	SPEX 8000	10:1	10	$L1_2$ $(Al,Ni)_3Hf$	68
Al-Mn	SPEX 8000	10:1	[a]	Fcc	69
Al–25Nb				Al_3Nb + m-Nb_2Al	70
Al-Ti	Fritsch P5	10:1	100	$L1_2$ Al_3Ti	66, 71
Al-Zr	Fritsch P5	10:1	20	$L1_2$ Al_3Zr	66, 67
Al-Zr-Fe	SPEX 8000	10:1	20	$L1_2$ $(Al,Fe)_3Zr$	72, 73
Al-Zr-Ni	SPEX 8000	10:1	20	$L1_2$ $(Al,Ni)_3Zr$	72, 73
Co-C	Ball mill	100:1	500	m-Co_3C	74
Co-25Ti	Fritsch P5	17:1	24	m-Co_3Ti	75–77
Cu-In-Ga-Se	Fritsch P5	10:1	2	Cubic $CuIn_{0.7}Ga_{0.3}Se_2$	45
Cu-Zn	SPEX 8000	5:1	0.5	Deformation martensite	52
Cu-Zn	Fritsch P5	10:1	—	Martensite	54
Fe-B	Fritsch	50:1	3	m-Fe_2B	78
Mg-Al-Ti-B	Fritsch P5	20:1	MA/40 h + extrusion at 400°C	m-Ti_3B_4	79
Mg-Sn	SPEX 8000	10:1	12	Orthorhombic	80
Nb-Ge	SPEX 8000	—	5	Fcc	81
Ni-Al	SPEX 8000	6:1	20	m-Ni_3Al	55
Ni-Al				m-NiAl	82
Ni-C	Ball mill	100:1	500	m-Ni_3C	74
Ni-Sn	SPEX 8000	—	1	β' and β''	64
Ni-Sn	Fritsch P5	20:1	150	NiSn	83
Ni_3Sn_4	Fritsch P5	20:1	150	NiSn	83
Sn-S (1:1.85)	URF-AGO2 planetary ball mill	—	1.5	4H-SnS_2	84, 85
Te-Ag	SPEX 8000	8:1	50	π phase	86
Ti	Super Misuni NEV MA-8	6:1	100	m-Ti_2N	87
Ti-Al	Fritsch P5	10:1	10	Fcc γ-TiAl	88, 89
Ti-Al	Fritsch P5	10:1	25	Fcc γ-TiAl	90
TiH_2-22Al-23Nb	Fritsch P5	10:1	10	Bcc-$(Nb,Ti)H_2$	91
Ti-Si	Fritsch P5	10:1	—	m-$TiSi_2$ (C49)	92, 93
Ti-Si	Ball mill	—	180	Bcc	94
Zr-Al	Attritor	10:1	25	m-Zr_3Al (DO_{19})	95

[a] MA/12 h plus aged for 30 days at room temperature.

the sample. Several other high-temperature polymorphs have also been reported to form at room temperature by MA. Some of the available results on these polymorphic transformations are summarized in Table 10.4.

The particle size during powder processing is known to have a significant influence on the phase constitution. Pure ZrO_2 is known to exhibit the phenomenon of polymorphism. At room temperature it exists in the monoclinic structure and at temperatures greater than 1170°C it transforms to the tetragonal form. The tetragonal form transforms to the cubic form above 2370°C and the cubic form is stable up to the melting point. However, it has been shown that a decrease in crystallite size could affect the structural stability, and this has been explained on the basis of known thermodynamic arguments. Garvie [147] explained the stabilization of tetragonal ZrO_2 at room temperature due to the change in the surface energy conditions. Assuming that the tetragonal and monoclinic forms of ZrO_2 are represented by the symbols t and m, respectively, the following thermodynamic equations could be written for the monoclininc and tetragonal forms of ZrO_2:

$$\Delta G_{\mathrm{m}} = \Delta G_{\mathrm{v(m)}} + A_{\mathrm{m}}\gamma_{\mathrm{m}} \tag{10.1}$$

and

$$\Delta G_{\mathrm{t}} = \Delta G_{\mathrm{v(t)}} + A_{\mathrm{t}}\gamma_{\mathrm{t}} \tag{10.2}$$

where ΔG is the free energy, ΔG_{v} is the volume change in free energy, A is the surface area, and γ is the surface energy per unit area. The change in free energy during transformation from the monoclinic to the tetragonal form of ZrO_2 can be written as:

$$\Delta G_{\mathrm{m}\to\mathrm{t}} = \Delta G_{\mathrm{t}} - \Delta G_{\mathrm{m}} = \left\{\Delta G_{\mathrm{v(t)}} - \Delta G_{\mathrm{v(m)}}\right\} + \{A_{\mathrm{t}}\gamma_{\mathrm{t}} - A_{\mathrm{m}}\gamma_{\mathrm{m}}\} \tag{10.3}$$

Since the monoclinic phase is the stable form at room temperature, $\Delta G_{\mathrm{v(t)}} - \Delta G_{\mathrm{v(m)}}$ is positive. But since the tetragonal form has a lower specific surface energy than the monoclinic form, $A_{\mathrm{t}}\gamma_{\mathrm{t}} - A_{\mathrm{m}}\gamma_{\mathrm{m}}$ is negative. Furthermore, the surface area increases as the particle size decreases. Therefore, the $A_{\mathrm{t}}\gamma_{\mathrm{t}} - A_{\mathrm{m}}\gamma_{\mathrm{m}}$ term becomes increasingly negative for the formation of the tetragonal form of ZrO_2. Consequently, at some critical value of the grain (crystallite) size, $\Delta G_{\mathrm{m}\to\mathrm{t}}$ becomes zero. Based on the available thermodynamic data, this critical grain size has been calculated to be about 30 nm. Thus, below about 30 nm the tetragonal phase gets stabilized even at room temperature. Sometimes even the cubic form of ZrO_2 has been shown to be stable at room temperature due to a reduction in particle size. Several investigators have observed these changes experimentally [107,135,141–144].

10.3.2 High-Pressure Phases

It has been suggested that very high pressures are generated during milling of powders; according to some estimates, the pressures could be of the order of 6 GPa [148,149]. Therefore, the high pressures developed during milling should be sufficient to stabilize the high-pressure polymorphs of phases at atmospheric pressure. The high-pressure polymorphs of Dy_2S_3 and Y_2S_3 have been reported to be retained on MA at atmospheric pressure [107]. Similarly, high-pressure phases of other chalcogenides, such as $Cu_{2-x}S$ and $CuSe_2$, were also synthesized by MA at atmospheric pressure [106]. A high-pressure cubic polymorph of the equilibrium tetragonal $Cu(In_{0.7}Ga_{0.3})Se_2$ phase was also reported to form in mechanically alloyed Cu-In-Ga-Se powders [45].

Table 10.4 Polymorphic Phase Transformations in Metals and Alloys by Mechanical Alloying/Mechanical Milling

Starting phase	Mill/MA time	Final phase	Comments	Ref.
γ-AlOOH (boehmite)	Fritsch P7/2.5 h	α-Al_2O_3 (corundum)		96
$Al(OH)_3$ (gibbsite)	Fritsch P7/7 h	α-Al_2O_3 (corundum)		96
γ-Al_2O_3	56 h	α-Al_2O_3		97
γ-Al_2O_3	Fritsch P9/2 h	δ-Al_2O_3		98
δ-Al_2O_3	Fritsch P9/2 h	θ-Al_2O_3		98
Co (fcc + hcp)	20 h	Co(hcp)	BPR 10:1	99
Co (fcc + hcp)	20 h	Co(hcp + fcc)	BPR 20:1	99
Co (fcc + hcp)	20 h	Co(fcc)	BPR 30:1	99
Co (fcc)		Co (hcp)		100
Co (fcc)	5 h	Co (hcp)		101
Co (hcp)	2 h	Co (fcc)	HT phase	101
	15 h	Co (fcc)	HT phase	102
	20 h	Co (fcc)	HT phase	103
Co_3Sn_2 (ordered orthorhombic)	48 h	Co_3Sn_2 (disordered hexagonal)	HT phase	104
Cr_2O_3 (hexagonal)	18 h	Cr_2O_3 (cubic)		105
$Cu(In_{0.7}Ga_{0.3})Se_2$				45
$Cu_{2-x}S$				106
$CuSe_2$				106
Dy_2O_3 C type (bixbyte)	—	Monoclinic Dy_2O_3 B type		107
γ-Dy_2S_3	—			108
Er_2O_3 C-type (bixbyte)	—	Monoclinic Er_2O_3 B-type		107
β-FeB (HT form)	Fritsch P7/62 h	α-FeB (LT form)	LT phase	109, 110
Fe_2B (bct)	Fritsch	Fe_2B (orthorhombic)		111
γ′-Fe_4N (fcc)	—	ε-Fe_4N (hexagonal)	HT phase	112
α-Fe_2O_3	SPEX 8000/70 h	Fe_3O_4		113
α-Fe_2O_3	SPEX 8000	Fe_3O_4		114
α-Fe_2O_3	SPEX 8000	Fe_3O_4→FeO		115
α-Fe_2O_3	Wet milling with water for 96 h	Fe_3O_4 (fcc)		116
γ-Fe_2O_3	Oscillating mill/40 min	α-Fe_2O_3		117
Fe_3O_4	SPEX 8000/70 h	α-Fe_2O_3		118
FeS		Hexagonal		119
α-$FeSi_2$	Shaker mill/50 h	β-$FeSi_2$		120
GeO_2	—		Rutile type	107
Iron anhydrous ammonia	—		HT phase	121
α-$MgCl_2$	96 h	δ-$MgCl_2$		122
α-$MoSi_2$	—	β-$MoSi_2$	Due to reduction in particle size	123

Table 10.4 Continued

Starting phase	Mill/MA time	Final phase	Comments	Ref.
Mg-33.3 Sn	Mg_2Sn (hexagonal)	Cubic → hexagonal	HP phase	124
Mg(+Si)-33.3Sn		Mg_2Sn (hexagonal) + Si	HP phase	125
Nb	Fritsch P5	Bcc → fcc	—	126
Nd	MA/10 h	Dhcp → fcc	HP phase[a]	127, 128
Ni-5,10Si	Fritsch P5/50 h	Fcc → hcp		129
Ni_3Sn_2 (orthorhombic)	40 h	Ni_3Sn_2 (B8-type hexagonal)	HT phase	130
PbO	—			115
SiC	WL-1 planetary ball mill/60 h	6H → 3C (α → β)		131, 132
TaN (CoSn-type)	Fritsch P7/1 h	TaN (WC type)		133
TiO_2	—		HP phase	134
TiO_2 (anatase)	Uni-ball mill/140 h	TiO_2 (rutile)	HT Phase	135
TiO_2 (anatase)	—	TiO_2 (rutile)		107
TiO_2 (anatase)	SPEX 8000	Orthorhombic TiO_2(II)	HP phase	136
TiO_2 (anatase)	SPEX 8000	Monoclinic TiO_2(B)		136
TiO_2 (anatase)	Fritsch P7/15 min	TiO_2 II		137
Y_2O_3 (C-type)	—	Y_2O_3 (B-type)		107
γ-Y_2S_3	—			108
Yb (divalent)	—	Yb (trivalent)		138, 139
Yb_2O_3 (C-type)	—	Yb_2O_3 (B-type)		107
$Zn_{10}Fe_3$ (τ)	—	Zn_7Fe (δ)		140
$Zn_{21}Fe_5$ ($τ_1$)	—	$Zn_{13}Fe$ (ξ)		140
ZrO_2 monoclinic	—	CaF_2-type ZrO_2		141
ZrO_2 monoclinic (baddeleyite)	Uni-Ball-Mill/70 h	Tetragonal or cubic ZrO_2		135, 142
ZrO_2 monoclinic (baddeleyite)	—	Cubic ZrO_2 (fluorite type)		107
ZrO_2 monoclinic	SPEX 8000/40 h	Orthorhombic ZrO_2		143
ZrO_2 (tetragonal)	—	ZrO_2 (cubic)	HT phase Due to reduction in particle size	144

LT Low temperature; HT, high-temperature; HP, high pressure.
[a] Due to powder contamination by iron, oxygen, etc.

Table 10.5 Formation of Equilibrium Intermetallic Phases by Mechanical Alloying

System	Phase	Structure	Comments	Ref.
Ag-Al	ξ	Hcp, A3 (hP2)		152
Ag-Al	μ	Complex cubic, A13 (cP20)		152
Ag-33.3Cu-33.3Se	AgCuSe			153
Ag-33.3S	Ag_2S			153
Ag-33.3Se	Ag_2Se			153
Ag-34.5Te	$Ag_{1.9}Te$			153
Ag-40Te	Ag_3Te_2			153
Al-40 to 50Co	AlCo	B2	MA/38 h	154
Al-40Co	AlCo	B2		155
Al-31.5Co	β–CoAl	B2		156
Al- 18.2 to 50Co	β′-AlCo			101
Al-20Cu-15Fe	FeAl (partially disordered)			25
	Al_2Cu	C16		25
Al-Fe	Al_6Fe		MA + 300°C	157
Al-50 to 60Fe	AlFe	B2	MA/38 h	154
Al-50Fe	AlFe	A2, a = 0.2907 nm		158
Al-25,29Fe	Al_5Fe_2			159, 160
Al-20,25Fe	Al_3Fe			161
Al-12.5Fe	Al_3Fe		MA/12 h + 500°C	161, 162
Al-27.4Fe-28.7C	AlFe	B2	MA/24 h + 475°C	163
Al-45Fe-5Ni	AlFe	B2		164
Al-25Fe-25Ni	Al(Ni,Fe)	B2		164
Al-25Fe-25Ti–25C	AlFe	B2, a = 0.288 nm		165
Al-25Fe-25V	Al_7Cu_2Fe	Bcc	MA + anneal at 625 K	158
Al-45Fe- 5V	Al_7Cu_2Fe	Bcc	MA + anneal at 625 K	158
Al-25Ge-10Fe	Al_4Ge_2Fe	Tetragonal, a = 0.6417, c = 0.9521 nm		166
Al-Mg	Al_3Mg_2	γ		167
	$Al_{12}Mg_{17}$	A12		167
Al-Mn	Al_6Mn	Orthorhombic	MA + extrusion	69, 168
Al-3Mo	$Al_{12}Mo$		MA + HT	169, 170
	Al_5Mo			169, 170
	Al_4Mo			169, 170
	Al_8Mo_3			169, 170
Al-Nb	Al_3Nb			171–176
Al-25Nb	Al_3Nb + m-Nb_2Al			70
Al-Nb-Ti	Nb_2Al			175
Al-50Ni	AlNi	B2	MA/38 h	154
Al-35 to 50Ni	NiAl	B2		55, 177, 178

Table 10.5 Continued

System	Phase	Structure	Comments	Ref.
Al-35Ni	Al_3Ni_2		MA/38 h	154
Al-25Ni	Al_3Ni			55
Al-8Ni-4Co	$Al_9(Co,Ni)_2$			179
Al-40 to 49Ni-2 to 20Cr	B2	Cubic		180
Al-Ni-Fe	B2			181
Al-40 to 49Ni-2 to 20Fe	B2			180
Al-39Ni-22Ti	Disordered NiAl	Cubic		182
Al-40Ru	RuAl	B2		183
Al-Sb	AlSb			184
Al-30Si-15Fe-5Ni	$(Al,Si)_7Ni_3$		10 h, partial	185
Al-Ti	Al_3Ti		50 h	186, 187
Al-25$TiH_{1.924}$	Al_3Ti		MA/10 h + 620°C/7 days	188
Al-Zr	AlZr	Orthorhombic		189, 190
Al-Zr	Al_3Zr_2	Orthorhombic		189, 190
Al_2O_3–TiO_2	β-Al_2TiO_5	β phase	MA/15 min + 1340°C	191
BaO-TiO_2	$BaTiO_3$			192
Be-Nb	$NbBe_{12}$		MA/72 h + 1000°C/4 h	193
Be-Nb	Nb_2Be_{17}		MA/72 h + 1000°C/4 h	193
CaO-TiO_2	$CaTiO_3$			192
Co	CoO		Milling in atmospheric air	194
Co	Co_3O_4		Milling in atmospheric air	194
Co-30 to 60Sn	Co_3Sn_2 CoSn $CoSn_2$			195
Co-25Ti	Co_3Ti	Bcc	Cyclic Am → Cryst → Am	75
Co-50Ti	CoTi	Bcc	Cyclic Am → Cryst → Am	196
Cr-33Nb	Cr_2Nb			197
Cu-25In-50S	$CuInS_2$			153
Cu-25In-50Se	$CuInSe_2$			153
Cu-25In-50Te	$CuInTe_2$			153
Cu-30 to 70Al	Cu_9Al_4			198
Cu-11Ni-18P	Cu_3P, Ni_3P			199
Cu-20P	Cu_3P			200
Cu-33S	$Cu_{2-x}S$			149
Cu-25 to 75Sb	η-Cu_2Sb			201
Cu-66.7Se	$CuSe_2$			106
Cu-Sn	Cu_6Sn_5 Cu_3Sn			202
Cu-46.8, 49.3Zn	β-CuZn			52

Table 10.5 Continued

System	Phase	Structure	Comments	Ref.
CuY-33.3 mol % CuBa	$YBa_2Cu_3O_{7-\delta}$		MA + anneal in oxygen atm.	203
Cu-Zn	β	Bcc		52,54,204
	γ	Cubic		
	ε	Hexagonal		
Fe-10 to 30Al	FeAl	B2	MA/48 h + 1150°C/4 h	205
Fe-25Al	Fe_3Al		MA/4 h (SS) + 550 K	206
Fe-25Al	FeAl			207
Fe-40Al	FeAl (disordered)			208
Fe-50Al	FeAl		MA/500 h + 1200 K	209
Fe-Al	FeAl		MA [Fe(Al)] + 700°C	210
Fe-40Al-10V-10C	FeAl	B2, a = 0.2907 nm	MA/16 h + heated to 720°C in the DSC	211
Fe-50Co	FeCo	Disordered		212, 213
Fe-Cr-S (1:2:4)	$FeCr_2S_4$	Cubic, Fd3m	Partial	214
Fe-S	FeS			215
Fe-Sn	Fe_3Sn_2			216
	FeSn			216
	$FeSn_2$			216
Fe-32Sn	$FeSn_2$		MA/4 h; partial	217
Fe–40 to 50Tb	Fe_2Tb			218
Fe-Ti	FeTi			219
Fe-33W	Fe_2W			220
Fe-33Zr	Fe_2Zr			221
Fe_3O_4-Nd_2O_3	$NdFeO_3$			222
α-Fe_2O_3 + CuO	$CuFe_2O_4$	Small amount		114
α-Fe_2O_3 + MgO	$MgFe_2O_4$	Small amount		114
α-Fe_2O_3-$MnCO_3$	$MnFe_2O_4$			223
α-Fe_2O_3-$MnCO_3$-ZnO	$Zn_{0.25}Mn_{0.75}Fe_2O4$			223
α-Fe_2O_3 + NiO	$NiFe_2O_4$	Small amount		114
α-Fe_2O_3 + ZnO	$ZnFe_2O_4$			114
Fe_2O_3-33.3ZnO	$ZnFe_2O_4$		MA/18 min + 1100 K/400 min	224
α-Fe_2O_3-ZnO	$ZnFe_2O_4$			223
$Fe(OH)_3$-MnO_2-ZnO	$MnFe_2O_4$			223
$Fe(OH)_3$-MnO_2	$Zn_{0.25}Mn_{0.75}Fe_2O_4$			223
Hf-50Co	HfCo	Primitive cubic, a = 0.3165 nm	MA/5 h	225
Hf-50Ru	HfRu	Cubic, B2		226

Table 10.5 Continued

System	Phase	Structure	Comments	Ref.
In + $(NH_2)_2CO.H_2O_2$	In_2O_3		MA/20 min + anneal 473–573 K	227
In-Sb	InSb			184
La_2O_3 + Li_2CO_3 + TiO_2	$La_{(2/3)-x}Li_{3x}TiO_3$		MA/120 h	228
Mg	MgH_2		H_2 atm/47.5 h	229
	β-MgH_2	Tetragonal	25 h in H_2 atm	230
	γ-MgH_2	Orthorhombic	35 h in H_2 atm	230
Mg	MgH_2		Partial after 5 h	231
Mg-50Ag	MgAg	B2		232
Mg-Ge	Mg_2Ge	Cubic, a = 0.6445 nm		80
MgH_2 + Co	Mg_2CoH_5	Hexagonal		233
Mg-50La	MgLa	B2		232
Mg-50Ni	Mg_2Ni + $MgNi_2$		Due to crystallization of amorphous Mg-Ni phase	234
Mg-33.3Ni	Mg_2Ni	Hexagonal,	>10 h	235
Mg-50Pd	MgPd	B2		232
Mg-Sn	Mg_2Sn	Cubic, a = 0.6761 nm		80
Mg-33.3Sn	Mg_2Sn	Cubic		124
Mg-33Sn	Mg_2Sn		Horizontal ball mill, 20:1, MA/100h	236
MgO-TiO_2	$MgTiO_3$			192
Mn-47Al	β-MnAl	B2		237
Mn-As	MnAs	Orthorhombic		238
Mn-As	MnAs	Orthorhombic	MA + hot pressing + anneal at 773 K/0.5 h	239
Mn-50Bi	MnBi			149
Nb-Al	Nb_3Al		MA + 800–950 °C	240
Nb-15.4Al	Nb_3Al		MA + 1150 K	241
Nb-74Al	Nb_2Al		200 h	174
	$NbAl_3$		300 h	
Nb-25Al	Nb_3Al		MA + 1500 K/30 MPa/3–4 min	242
Nb-33.3 to 70Al	σ-Nb_2Al + $NbAl_3$		MA + 825°C/2 h	171
Nb-25Al	Nb_3Al		MA + 825°C/2 h	171
Nb-25Ge	Nb_3Ge, Nb_5Ge_3, δ-$NbGe_2$			81
Nb	$NbH_{1.01}$	Cubic, a = 0.34391 nm	In hydrogen atmosphere	243

Table 10.5 Continued

System	Phase	Structure	Comments	Ref.
Nb-S	$Nb_{1-x}S$			215
Nb-Sn	Nb_3Sn		MA + 750–850°C	240
Ni	NiO		Milling in atmospheric air	194
Ni-Al	B2-NiAl			210
Ni-Al	Al_3Ni (ordered) NiAl (ordered) Ni_3Al(disordered)			244
Ni-Al	NiAl		MA for <1 h	245
Ni-15 to 60Al	NiAl		Between 39 and 60 Al	246
Ni-24Al	$L1_2$-Ni_3Al			247, 248
Ni-25Al	Disordered Ni_3Al			249
Ni-25Al	Ni_3Al			250
Ni-49Al	NiAl			248
Ni-50Al	B2-NiAl			210, 251–254
Ni-60 to 65Al	NiAl			177
Ni-63Al	NiAl(disordered)			255
Ni-65Al	Ni_2Al_3 NiAl		17 min 125 min	255, 256
Ni-70Al	Ni_2Al_3			255
Ni-75Al	$NiAl_3$, Ni_2Al_3			256
Ni-50Al-25Fe	β′-(Fe,Ni)Al			257
Ni-7.5Al-17.5Nb	$Ni_3Al + Ni_3Nb$			248
Ni-38.7Al-19.3Nb	NiAl + NiAlNb			248
Ni-25 to 30Al-20 to 25Ti	β′-Ni(Al,Ti)			181, 258
$Ni_{50-x}Al_{50-x}M_{2x}$ (M = Cr, Fe)	NiAl		Partially ordered up to $x = 8$ and completely ordered at $x = 10$	254
Ni-43Mo	γ	fcc, a = 1.1098 nm	MA/63 h + 720°C	259
Ni-40S	Ni_3S_2			215, 260
Ni-Sn	Ni_3Sn_4		MA/150 h + anneal 1000°C/2h	83
Ni-40Ti	Ni_3Ti			261, 262
Pb-Te	PbTe	Cubic, NaCl type	2 min	263
PbO + MgO + Nb_2O_5	$Pb(Mg_{1/3}Nb_{2/3})O_3$		MA/20 h	264, 265
(PbO + MgO + Nb_2O_5) + 10 to 35 mol % TiO_2		Perovskite structure	MA/20 h	265

Table 10.5 Continued

System	Phase	Structure	Comments	Ref.
PbO-TiO_2	$PbTiO_3$			266
PbO_2-TiO	$PbTiO_3$			266
Pd-50Al	PdAl			267
Ru-50Al	RuAl	B2	MA/25 h or MA/2 h + 733°C	268, 269
Ru-50Al-10 to 25Ir	(Ru,Ir)Al	B2	MA/6 h	270
Sb-15Fe	Sb_2Fe		MA/400 h; partial	271
Sb-57.1Zn	Sb_3Zn_4		MA/16 h + anneal at 805 K	272
Sm-Fe	$SmFe_2$		MA/7 h + 500°C/2 h	273
Sm-Fe	Sm_2Fe_{17}		MA/7 h + 800°C/2 h	273, 274
Sm-57.2Se	Sm_3Se_4			150
Sm-60 Se	Sm_2Se_3			150
Sm-50 Te	SmTe			150
Sn-66.7S	SnS_2			215
SrO-TiO_2	$SrTiO_3$			192
Ti-60Al	Fcc γ-TiAl			275, 276
Ti-50, 55Al	γ-TiAl		Partial, MA/10 h + 615°C/1 week	277
Ti-50Al	Ti_3Al			278
Ti-50Al	Disordered TiAl			279
Ti-50Al	TiAl		MA/650h + 800°C/2h	280
Ti-48Al	TiAl		MA + 1000°C/1h hot press + 1200C/4 h	281
Ti-45Al	α_2-Ti_3Al	DO_{19}	MA/200 h + VHP 1173 K/1 h	282
Ti-24Al	α_2-Ti_3Al	DO_{19}	MA/3 h + 630°C/4 weeks	283
Ti-24Al-11Nb	B2	Cubic		284
Ti-25Al-25Nb	B2	Cubic		285, 286
Ti-28.5Al-23.9Nb	B2	Cubic		285, 286
Ti-37.5Al-12.5Nb	B2	Cubic		286
Ti-42.4Fe	TiFe	B2	Annealing of the amorphous phase at 750°C/0.5 h	287
Ti-Ni	Ti_2Ni		1 min	288
	TiNi	B2	5 min	288
Ti-25Ru-25Fe	Ti_2RuFe			289
Ti-50V	TiV	Bcc		290
	$TiVH_{4.7}$	Fcc		290
$TiH_{1.924}$-50Al	TiAl	Fct	MA/16.5 h + 620°C/7 days	188
TiH_2 + Al_3Ti	γ-TiAl		Partial	291

Table 10.5 Continued

System	Phase	Structure	Comments	Ref.
TiO_2-$Ba(OH)_2.8H_2O$	$BaTiO_3$			292
Y_2O_3-$BaCO_3$-CuO	$Y_2Cu_2O_5$			293
Y_2O_3-CuO	$Y_2Cu_2O_5$			293
Y_2O_3-BaO_2-CuO	Y_2BaCuO_5			293
Zn-10 wt% Al-22wt%Cu	η and ε phases			294
Zn-15Cu	ε phase		Milling for 1 h	204
	γ phase		After long milling	204
Zn-Fe	Fe_3Zn			295
Zn-50S	ZnS			215, 260
Zr	$ZrH_{1.66}$			229
ZrO_2+SiO_2, $0.5H_2O$	$ZrSiO_4$		MA/15 min + 1200°C/2 h	191
$ZrO_2.4.3H_2O + SiO_2$	$ZrSiO_4$		MA/15 min + 1200°C/2 h	191
$ZrO_2 + TiO_2$	$ZrTiO_4$	Partial	MA/15 min	191
$ZrO_2 + TiO_2$	$ZrTiO_4$	Partial	MA/15 min	191
$ZrO_2.nH_2O$ + $TiO_2.nH_2O$	$ZrTiO_4$		Mechanical activation to form an amorphous phase and subsequent annealing at 600°C	191

Sen et al. [136] synthesized the high-pressure orthorhombic form of TiO_2(II) at atmospheric pressure by MA at room temperature. Alonso et al. [128] reported that they synthesized the high-pressure fcc polymorphs of lanthanide metals Dy, Gd, Nd, and Sm by MA of the powders for about 24 h in a SPEX mill using a hardened steel vial and alloy steel balls. However, they later recognized [150] that the phases produced were the NaCl-type interstitial phases formed by reaction of these reactive metals with oxygen, nitrogen, and hydrogen during milling in a poorly sealed vial. As an example, the milled Nd powder contained 16.2 at% oxygen, 22.3 at% nitrogen, and 10.3 at% hydrogen. Thus, it is not clear whether the high-pressure phases of pure metals, especially the reactive ones, can be synthesized by MA. But it appears that high-pressure compound phases of oxides, chalcogenides, and so forth can certainly be synthesized by MA, since contamination is not likely to play any role in the formation of these less reactive compounds.

The above observation reminds us that extreme caution should be exercised in identifying metastable crystalline phases produced by MA. During MA processing, the powder picks up appreciable quantities of interstitial elements like oxygen, nitrogen, hydrogen, and carbon (from the ambient atmosphere and/or the process control agents) as well as iron and chromium (from the grinding media and/or the mill walls). These metallic and nonmetallic impurities can have a very significant effect on the stability of the phases. In addition, these interstitials can combine with the metallic elements and form interstitial compounds. Thus, it is possible to misinterpret these interstitial phases (e.g., carbides, nitrides, hydrides, oxides, oxynitrides) as new intermetallic phases. An accurate chemical analysis of the final resultant powder can, in many cases, clear up this confusion [151].

10.3.3 Equilibrium Crystalline Phases

MA has been used for the synthesis of several intermetallic phases in both the disordered and ordered conditions. In many cases the intermetallics were synthesized directly by MA, but in others an additional heat treatment was required after MA to form the intermetallic. The intermetallics synthesized include aluminides (mostly based on titanium, nickel, and iron), borides, carbides, nitrides, silicides, composites, and some exotic varieties. Table 10.5 presents a complete listing of the intermetallics that have been synthesized by MA. As a typical example we will now describe the phase evolution in mechanically alloyed blended elemental Ti-Al powder mixtures.

The Ti-Al binary phase diagram (Fig. 10.2) [296] features three intermetallics: Ti_3Al (α_2), TiAl (γ), and $TiAl_3$. All these intermetallics are light weight, have high specific strengths, high elastic moduli, good corrosion resistance, and retain sufficient strength at elevated temperatures. The two titanium-rich titanium aluminides—α_2 (Ti_3Al) and γ (TiAl)—are attractive candidate materials for potential applications in advanced aerospace engine and airframe components [296,297]. However, they suffer from low room-temperature ductility, thus preventing their easy fabrication and widespread applications. The technique of MA has been utilized extensively to synthesize these intermetallics in a nanocrystalline and disordered state with a view to improving their room temperature ductility. Even though actual improvement in ductility is only partially realized in nanostructured intermetallics, whether produced by MA or other methods such as electrodeposition or severe plastic deformation [298–304], its realization in the mechanically alloyed nanostructured intermetallics, including the titanium aluminides, is yet to be achieved. However, a vast amount of information has been generated on phase evolution in these alloys.

Figure 10.3 shows a series of XRD patterns indicating the phase evolution as a function of milling time in a Ti-24 at% Al powder mixture [283]. The as-mixed

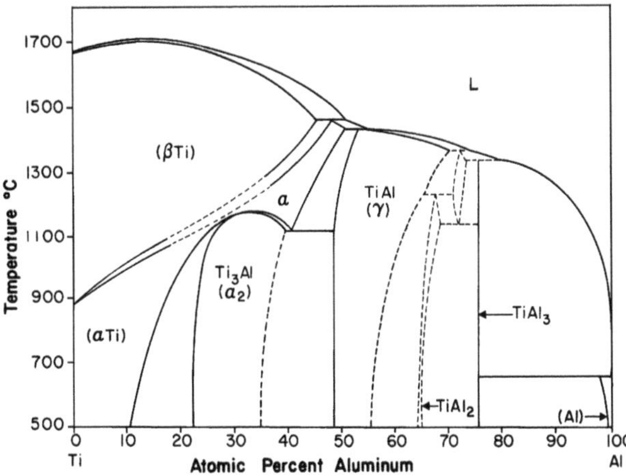

Figure 10.2 Equilibrium phase diagram of the Ti-Al system [296].

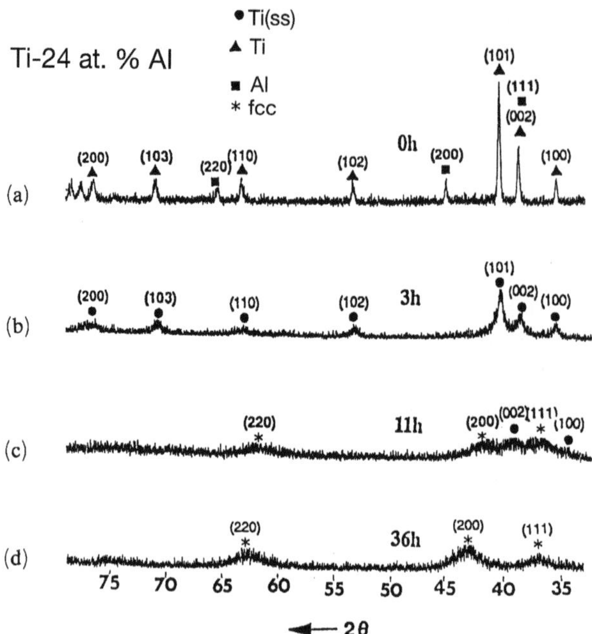

Figure 10.3 X-ray diffraction patterns of the blended elemental Ti-24 at% Al powder mixture mechanically alloyed for different times. (a) As-mixed powder, (b) milled for 3 h showing the formation of Ti(Al) solid solution, (c) milled for 11 h showing the coexistence of Ti(Al) solid solution, amorphous, and fcc phases, and (d) milled for 36 h showing the homogeneous fcc phase [283].

elemental blend (Fig. 10.3a) shows the presence of both elemental titanium and aluminum phases, without any alloying occurring between them. On milling for 3 h, one could detect the formation of a solid solution of Al in Ti, as evidenced by the absence of the Al peaks in the XRD pattern (Fig. 10.3b) and a shift in the Ti peak positions. On continued milling for 11 h, the powder started losing crystallinity, as shown by the presence of a broad peak in the XRD pattern, indicative of formation of an amorphous phase (Fig. 10.3c). Some crystalline peaks are also superimposed on this peak, suggesting the coexistence of the amorphous and some crystalline phases. Based on the peak positions, it could be concluded that the crystalline phases present are the Ti(Al) solid solution and another crystalline phase. Further milling to 36 h produced an fcc phase (Fig. 10.3d), which has been identified as (Ti,Al)N, a contaminant phase [151]. Thus, the phase formation sequence in this powder mixture can be represented as

$$\begin{aligned} &\text{BE Ti} + \text{Al} \rightarrow \text{Ti(Al) solid solution} \rightarrow \text{amorphous phase} \\ &\qquad \rightarrow \text{fcc (contaminant) phase} \end{aligned} \tag{10.4}$$

If the milling were to be stopped at an appropriate time, the powder would not get seriously contaminated, and one could end up with either a solid solution or an amorphous phase. Even under these conditions, if proper precautions are not taken, the powder would be somewhat contaminated, especially when reactive powders like titanium and zirconium are being milled. See Chapter 15 for a discussion of powder contamination during MA.

Most of the investigators reported a similar sequence of events in the phase evolution on MA of Ti-Al powder blends, with the difference that the time required for the formation of different phases is different depending on the process variables [type of mill, ball-to-powder weight ratio (BPR), use or absence of process control agents (PCAs), etc.]. As an example, the times required for the formation of different phases [the Ti solid solution, Ti_{ss}, the cubic B2/bcc phase, and the amorphous (Am) phase] in the mechanically alloyed ternary Ti-Al-Nb powders milled in different atmospheres is presented in Figure 10.4. Milling was conducted in helium, argon, nitrogen, and air atmospheres by filling the milling vial with the appropriate atmosphere before milling had started. Even though the sequence of phase formation was the same in all atmospheres, the times required for the formation of the different phases were found to be different. For example, in the Ti-25 at% Al-25 at% Nb powder mixture, it took about 25 h to form the cubic B2 phase in a protective helium atmosphere, while it took only about 10–11 h in air or nitrogen atmosphere [305]. This suggests that contamination could perhaps play an important role in phase formation during MA, at least in reactive alloys. Furthermore, depending on the purity of the milling atmosphere, the contaminant phase is minimized or completely avoided. The extent of supersaturation has also been reported to be different depending on the degree of contamination (see Table 9.1 for the extent of supersaturation achieved by different investigators). An exhaustive summary of the structural evolution in mechanically alloyed Ti-Al alloys, including the presence of a contaminant phase, is presented in Ref. 151.

In all the investigations on the Ti-Al system it has been reported that either a supersaturated solid solution or an amorphous phase has formed on MA, and that neither of the intermetallic phases—α_2-Ti_3Al and γ-TiAl—could be synthesized directly by milling. The desired intermetallic was obtained in the appropriate composition range, only after a suitable heat treatment of the mechanically alloyed powder. For example, annealing of the mechanically alloyed Ti-24 at% Al powder for 4 weeks at 903 K produced the α_2-Ti_3Al phase [283]. Similarly, the γ-TiAl phase could be produced only after annealing the as-milled amorphous powder for 168 h at 888 K

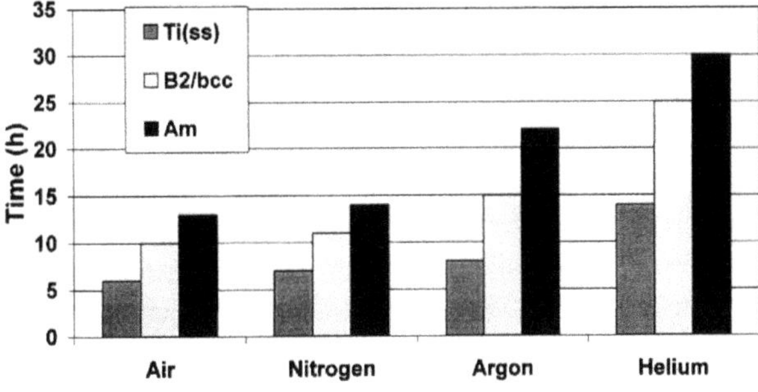

Figure 10.4 Times required for the formation of different phases in blended elemental Ti-25Al-25Nb (at%) powder mixture milled in different atmospheres [305].

[277,306]. These temperature and time combinations required for producing the intermetallics could vary considerably depending on the powder particle/grain size. However, the third intermetallic, i.e., the $TiAl_3$ phase, could be synthesized in a metastable $L1_2$ structure directly by MA [66,71].

The times required to form a particular phase depend on the initial concentration of the solute in the powder mixture. If the solute content is much less than the exact stoichiometry, the expected phase (although the proportion of the phase in the mixture of phases is less) would form at much longer milling times. For example, while investigating the formation of the β′-AlCo phase in Al-Co alloys, Sui et al. [101] noted that it required 10 h to form the β′-AlCo phase in an Al-50 at% Co powder mixture. On the other hand, it required 30 h in Al-28.6 at% Co, 70 h in Al-23.5 at% Co, and 200 h in Al-18.2 at% Co mixtures to form the β′-AlCo phase under identical milling conditions. Thus, it is clear that the times required for the formation of the β′-AlCo phase increase as one deviates more from the stoichiometric composition with the times being much longer at larger deviations from stoichiometry. This is easy to understand because formation of a particular phase requires diffusion and equilibration, both requiring time at a given temperature. The times can be probably reduced if the temperature at which milling is conducted is increased. However, there have not been any investigations reported to confirm this hypothesis. The time required to form a phase can also be substantially reduced if the BPR is increased. This has been shown to be true in several cases but is likely to lead to increased levels of contamination.

The sequence of phase formation during MA of blended elemental powders appears to be related to the diffusivity of the predominantly mobile species. Thus, if in an alloy system A-B, the diffusivity of A is higher than that of B, i.e., $D_A > D_B$, then the first phase to form will be rich in the B component. Accordingly, it was reported [307] that in the Ni-Ti system, the first phase to form was $NiTi_2$, since $D_{Ni} \gg D_{Ti}$. Similarly, in the Al-Ti system, the first phase to form was $AlTi_3$ because $D_{Al} > D_{Ti}$, thus confirming the above hypothesis. A similar situation was also reported for the formation of intermetallic phases in the Cu-Zn system [54]. It has been shown that nucleation is an important factor in determining the first phase to be formed. This has been predicted on the basis of the classical nucleation theory and the interfacial area of the particles obtained during MA [308].

Schaffer [309] showed that mixtures of intermetallic phases can be produced via MA by choosing the appropriate compositions. If the two phases are in equal proportion, then it has been noted that grain growth was considerably hindered. Thus, by having approximately equal proportions of the NiAl and Ni_3Al phases in a nanocrystalline state, it was shown that grain growth could be slowed down up to at least 500°C.

In majority of the cases, equilibrium crystalline phases have been synthesized at the compositions at which they were expected to be present under equilibrium conditions.However, in some cases the intermetallics could be synthesized at compositions away from the equilibrium homogeneity range, i.e., the homogeneity range of the intermetallics has been extended. For example, Sui et al. [101] reported that the β-AlCo phase formed at a composition corresponding to Co-70 at% Al, while the equilibrium β-AlCo phase was stable between 44.3 and 53.7 at% Al. A similar result was reported by Guerrero-Paz et al. [310], who had reported the formation of only the β-CoAl phase in the Co-68.5 at% Al alloy. Similarly, Singh et al. [311] reported that the homogeneity ranges of the β (Al_3Mg_2) and γ ($Al_{12}Mg_{17}$) phases in the Mg-Al system could be significantly extended. While the β phase was expected to be present homogeneously between 59.7 and 61.5 at% Al under equilibrium conditions, it was found to be present homogeneously at even a low value of 50 at% Al in the mechanically alloyed powders. Similarly, the γ phase was expected to be stable between 39.5 and 55 at% Al, but its homogeneity range could be extended down to 30 at% Al in the mechanically alloyed powders.

As was mentioned in Chapter 5, a number of process variables affect the nature and kinetics of the intermetallic phase formed during MA. It was reported by Gauthier et al. [312,313] that the Nb_3Al intermetallic could be synthesized by a combination of MA and another processing step. Accordingly, they had mechanically alloyed blended elemental powders of Nb and Al for a short time to obtain a nanoscale distribution of the reactants (without the formation of any product phase). This nanoscale mixture was subsequently cold compacted and subjected to either field activation (simultaneous application of pressure, 56–84 MPa, and current, 1500–1650 A) [312], or self-propagating high-temperature synthesis reaction [313]. The combination of these processes led to the formation of the intermetallic in a short time (3–6 min) as against a much longer period required when only conventional MA was done without combining it with other treatments.

10.4 ROLE OF HYDROGEN AS A TEMPORARY ALLOYING ELEMENT

Hydrogen can dissolve in titanium and its alloys up to 8.3 at% in the α phase and up to 60.3 at% in the β phase. Since reaction of hydrogen with titanium is reversible [314], hydrogen can be easily removed from the alloy by vacuum annealing. Thus, hydrogen can be used as a temporary alloying element to improve the performance of the alloys (by bringing about phase transformations), and this process has come to be known as thermochemical processing (TCP). It has been shown earlier that TCP could enhance the processability and mechanical behavior of titanium alloys, mostly through grain refinement via a phase transformation [315]. The actual processing consists of adding hydrogen to the alloy by holding the material at a relatively high temperature (usually in the β-phase field to dissolve more hydrogen) in a hydrogen environment. The presence of hydrogen then allows the titanium alloy to be processed at lower stresses/lower temperatures due to the presence of an increased amount of the β phase. This alloy could then be heat treated to produce novel microstructures with enhanced mechanical behavior after hydrogen removal. The hydrogen is simply removed by a vacuum annealing treatment to levels below which no detrimental effects occur. Consequently, thermomechanically processed titanium aluminides exhibit improved strength, ductility, and fatigue resistance (because hydrogen allows the material to behave as a eutectoid former system).

As mentioned earlier, MA of alloy powders containing titanium results in the formation of a contaminant fcc TiN phase. The use of a PCA can also introduce contamination into the powder; therefore, avoiding the use of a PCA could also reduce contamination of the powder. Since a PCA is used mainly to minimize excessive cold welding and obtain a balance between fracturing and cold welding of powder particles, nonusage of the PCA may result in excessive cold welding and formation of larger particles; true alloying may also not occur. This problem can be overcome by using intermetallics or other compounds, which are inherently brittle (reduced ductility), and therefore excessive cold welding between particles is not of a serious concern. Accordingly, Suryanarayana et al. [291] synthesized the γ-TiAl phase by mixing titanium hydride (instead of pure titanium) and the Al_3Ti intermetallic (instead of pure aluminum) powders in the proper proportion, according to the relation:

$$Al_3Ti + 2TiH_2 \rightarrow 3TiAl + 2H_2 \uparrow \tag{10.5}$$

On mechanically alloying the above powder mixture for 52 h in a SPEX mill, they obtained 55 vol% of the γ-TiAl phase; this value was increased to 95% on hot isostatic pressing of the mechanically alloyed powder at 1023 K and 275 MPa for 5 h. Based on the success of synthesizing the γ-TiAl phase, this procedure has been subsequently extended to synthesize the Ti_3Al, TiAl, and Al_3Ti phases by milling titanium hydride and aluminum powders in the ratios of 3:1, 1:1, and 1:3, respectively [188,316]. However, in these cases the respective intermetallics could not be synthesized directly by MA; probably related to the ductile aluminum being present in the powder mixture instead of the brittle Al_3Ti compound in the earlier investigation. A subsequent annealing treatment at 893 K for 7 days was necessary to produce the TiAl and Al_3Ti phases. However, the Ti_3Al phase could not be synthesized even after the heat treatment due to the very large amount of hydrogen content present in the powder. That is, hydrogen was not sufficiently removed by vacuum annealing. The Ti_3Al phase could, however, be synthesized after dehydrogenation of the mechanically alloyed powder at 873 K for 22 h and then annealing this powder at 893 K for 7 days. The major advantage of using titanium hydride instead of titanium in the powder mixture is that PCAs need not be used during milling, and this avoids or minimizes contamination of the powder.

Bououdina and Guo [91,317] investigated the powder yield and particle size refinement on MA of Ti-22Al-23Nb or TiH_2-22Al-23Nb (at%) powder mixtures in a Fritsch P5 planetary ball mill using stainless balls and a ball-to-powder weight ratio of 10:1. They have noted that the total free-powder yield (the amount of powder that did not stick to the container walls and therefore could be recovered) did not change with milling time up to 20 h and remained constant at a high value of 80% when the hydride powder was used. On the other hand, when titanium powder was used, the yield decreased from about 70% at a milling time of 2 h to about 20% for a milling time of 20 h. On continued milling the yield decreased more rapidly in the case of titanium hydride, and eventually the yield was the same (about 10%) whether the hydride or pure titanium powder was used. This is shown schematically in Figure 10.5. It should be noted, however, that the powder stuck to the walls of the vial could be easily removed after milling when the hydride powder was used.

It was also noted [317] that the powder particles were much finer and some had even reached nanometer levels when the titanium hydride powder was used instead of pure titanium powder. This can be understood in terms of the hard and brittle nature of the hydride particles, which could be easily fractured during the milling process. It was also claimed that the use of hydride powder significantly lowered the levels of oxygen and nitrogen contamination in the milled powder and consequently accelerated the amorphization process.

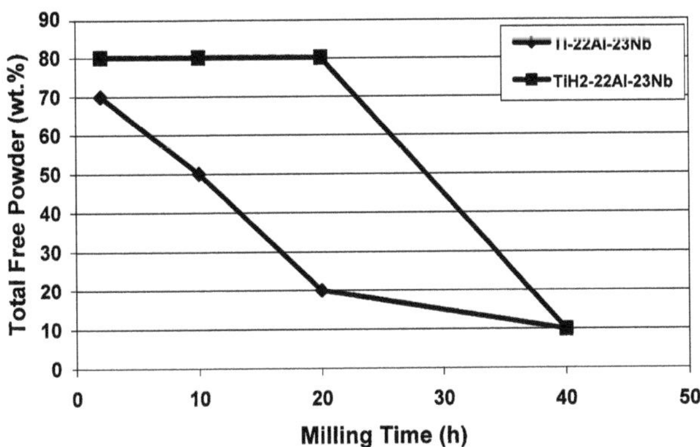

Figure 10.5 Change in the weight of free powder with milling time for the blended elemental Ti-22Al-23Nb and TiH_2-22Al-23Nb powder mixtures [91].

10.5 COMBUSTION REACTIONS DURING MA

There is a considerable amount of literature on the synthesis of nickel and iron aluminides by MA. Several intermetallic phases, such as Ni_3Al, NiAl, and Al_3Ni, have been synthesized by MA in the appropriate composition ranges of the Ni-Al system. An interesting observation is that the equiatomic NiAl phase is produced via a combustion synthesis [also known as the self-propagating high-temperature synthesis (SHS)] [55]. In this and other similar systems, the combustion reaction took place only after "interrupted milling," i.e., milling of the powders for a given length of time, aging the powder at room temperature after stopping the milling, and then resuming the milling operation. For example, in the Al-Ni system, an intimate mixture of Al and Ni phases was detected after milling the blended elemental Al-Ni mixture for 2 h in a SPEX mill. If the milling was stopped and the powder stored at room temperature for 30 min, and then milling resumed, it was noted that the NiAl phase formed just after 1 min of milling due to the occurrence of an explosive reaction [55]. Similar reactions were also observed in the synthesis of $MoSi_2$ [123,318–321], $Ti_{50}Ni_{20}C_{30}$ [322], $NbSi_2$ [323], TiB_2 [324], $PbTiO_3$ [266], and some other compounds. However, in the case of synthesis of $NbSi_2$ phase, it was noted that interruption during milling was not required for the abrupt reaction to take place [323].

In the MA of blended elemental powder mixtures, which undergo the SHS reaction, it has been observed that milling of the powder mix produces only a fine and intimate mixture of the two or more metal powders involved. No reaction has been found to occur at this stage. However, on continued milling, an explosive reaction occurs and the compound suddenly forms. This has been the sequence of events in the formation of compounds by the SHS reaction. (In fact, it has been noted that obtaining a homogeneous mixture of the constituent powder particles is a prerequisite to achieve easy alloying. Subsequent treatments such as SHS or annealing at some high temperature seem to make alloy formation easier.) The times at which the explosive reaction occurs depends on the condition of the powder mixture, e.g., the fineness of the powders. Refining the powder particle size, by premilling the powder before mixing the individual particles or by milling the powders at high intensities or large BPR values, has been reported to decrease the times at which the explosive reaction occurs [325]. This has been attributed to the improved homogeneity of the powder mixture. It has also been recently reported that the time for the initiation of the explosive reaction can be delayed by 20–30 min by the addition of ternary elements such as Ti and Fe to an Al-Ni mixture (Fig. 10.6) [181]. This is similar to the delayed reaction observed by the addition of diluents during the SHS reaction.

It has been known that the SHS reactions take place when the material is heated fast. Therefore, one could assume that a large BPR value could lead to the occurrence of the SHS reaction during milling. However, it was reported that lower BPR and short times induce the SHS reaction in Al and Zr powder mixtures and not the higher BPR values [326].

It was also reported [327] that the Al_3Ni_2 phase formed by a combustion reaction in an Al-35 at% Ni powder mixture only when heavy grinding balls, a high BPR, and long milling times were employed. For example, the spontaneous reaction occurred only when 6 balls weighing 8.5 g each, a BPR of 10:1, and a time longer than 6.9 ks (1 h and 55 min) were used. Lower BPR of 5:1 required longer periods, i.e., 11.88 ks (3 h and 18 min). On the other hand, when the lighter grinding balls (weighing 1.04 or 0.15 g each) were used, the spontaneous combustion reaction did not take place at a BPR of 5:1 even after 32.4 ks (9 h). In fact, no alloying took place even when 333 grinding balls of 0.15 g weight were used. Only partial formation of the Al_3Ni_2 phase occurred when 48 balls of 1.04 g were used for 18 ks (5 h); at shorter milling times, no alloying was reported to occur. It is possible that it is not the mass of the grinding balls but the rate of energy transfer that is important in inducing the SHS reactions. Thus, it appears that the conditions for the occurrence of the spontaneous combustion reaction could be different in different alloy systems and compositions.

Explosive reactions were also reported to occur when the grinding container was opened to the atmosphere soon after milling was stopped [328]. This can be attributed to the energy released by the oxidation of the component metal powders. On the other hand, the absence of an explosive reaction when the Ni-Al powders were milled in air [255] suggests that the elements were perhaps being continuously oxidized. Thus, the

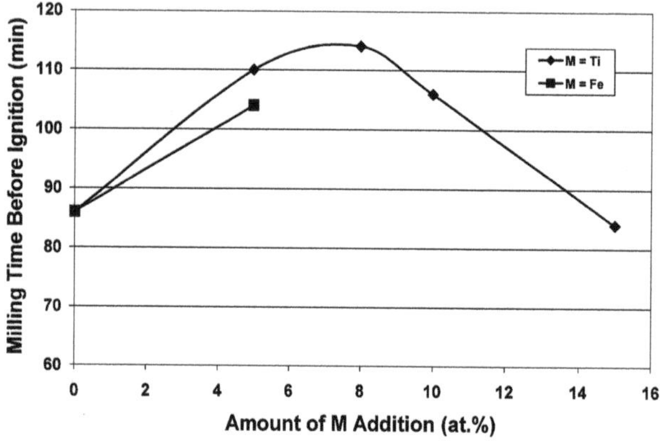

Figure 10.6 Effect of Ti and Fe addition on the time for initiation of the exothermic reaction during the formation of the NiAl phase [181].

slow diffusion between the oxide-coated components would have resulted in reduced kinetics and prevented the explosive reaction from occurring. Reactions similar to the explosive type, termed "discontinuous additive mixing," were also reported to occur, resulting in the formation of ordered intermetallics in the Ni-Al system [244]. In these cases, it has been reported that the reactions occurred only when the component powder particles were refined to a minimal size. At this stage, an SHS-type reaction occurs and the intermetallic forms. The crystallite size of the intermetallic at the time of its formation was found to be the sum of the crystallite sizes of the individual components before the reaction (Fig. 10.7). This minimal critical crystallite size below which only the explosive-type reaction occurs was found to increase with a decrease in the enthalpy of formation of the ordered intermetallic. It has also been reported that a gradual diffusive mode of mixing occurs once the product is disordered, e.g., $AlNi_3$ or AlNi(Cr) [329].

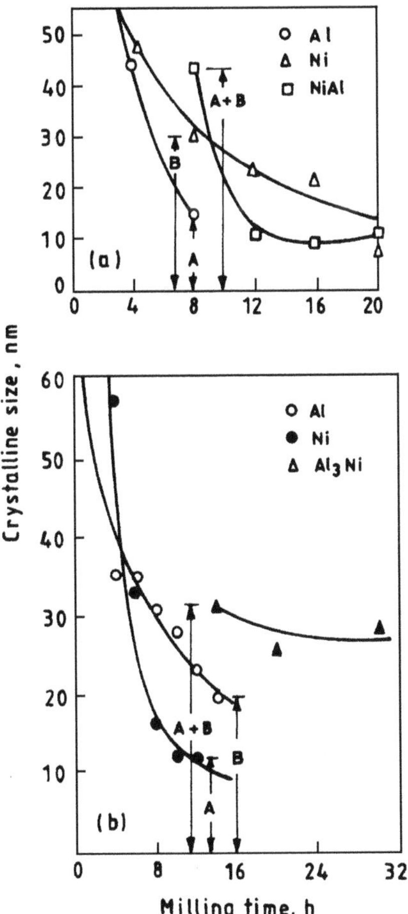

Figure 10.7 Discontinuous additive mixing during the formation of (a) AlNi and (b) Al_3Ni ordered intermetallic phases by MA. Note that the crystallite size of the intermetallic at the time of its formation is the sum of the individual crystallite sizes of the components [244].

It has also been reported in some cases that intermixing among the powder particles takes place continuously and metastable phases form prior to formation of the equilibrium phases. In other cases, a series of other equilibrium phases form before the actual expected phase forms. For example, the α-$MoSi_2$ phase formed when the Mo-Si powder mixture was milled in a planetary ball mill for 6 min, while the β-$MoSi_2$ phase formed after milling for 40 min [123]. Similarly, the Ni_2Al_3 phase formed after milling for 17 min in an Ni-65Al powder and the NiAl phase after 125 min; both phases were formed by exothermic reaction [256].

10.6 CYCLIC PHASE TRANSFORMATIONS

As will be shown later, amorphous phase formation has been achieved in a number of alloy systems by both MA and mechanical milling (MM) techniques. In most cases the amorphous phase was found to exist in a metastable condition at room temperature. In other cases it has been reported that the amorphous phases formed by milling have transformed to a crystalline phase, i.e., the amorphous phase has crystallized, on continued milling (or after thermal annealing in most cases. Here we will ignore the report of the formation of a contaminant crystalline phase in the milled Ti-Al alloy powders beyond the formation of an amorphous phase as an example of formation of a crystalline phase on prolonged milling). An interesting observation is that an amorphous phase can be crystallized to produce a crystalline intermetallic phase on milling, and this is referred to as mechanical crystallization [51]. On the other hand, an intermetallic or a solid solution or a blended elemental powder mixture can be made to undergo a crystalline → amorphous transformation on milling, and this is sometimes referred to as mechanical disordering or amorphization. These two processes, with fundamentally different mechanisms, are opposite in direction from a thermodynamic point of view. There have been few reports of combining these two phenomena (amorphization and crystallization) in the same metal or alloy, and even fewer on the cyclic behavior of these transformations.

It has been reported that the cyclic crystalline → amorphous → crystalline transformation occurs in the Co-25 at% Ti and Co-50 at% Ti powder mixtures [75–77,196]. When the Co and Ti powders in the ratio of 3:1 were milled in a Fritsch Pulverisette P5 mill, an amorphous phase had formed after milling for 3 h. On continued milling for 24 h, the amorphous phase had crystallized to a metastable bcc-Co_3Ti phase. This Co_3Ti phase transformed into an amorphous phase on further milling to 48 h. Furthermore, it was shown that the amorphous and crystalline phases could be produced alternately on continued milling as shown in Figure 10.8 [75,77]. Since the bcc-Co_3Ti phase is ferromagnetic in nature and the amorphous phase is not, the authors could follow the cyclic crystalline → amorphous → crystalline transformation by measuring the magnetization behavior of the mechanically milled powder. This shows a cyclic behavior with time as represented in Figure 10.9 [196]. The rapid decrease of magnetization during the first couple of hours indicates a significant decrease in the volume fraction of the pure Co particles and the formation of an amorphous phase. On continued milling the amorphous phase crystallizes into the bcc-Co_3Ti phase, which is ferromagnetic; therefore, the magnetization increases. Thus, the cyclic transformation between the amorphous and crystalline phases leads to a cyclic decrease and increase in the magnetization of the powder, respectively.

On annealing the amorphous phase obtained by MA for 3 h in the Co-25 at% Ti alloy at 1200 K, it had transformed to the equilibrium fcc-Co_3Ti phase [77].

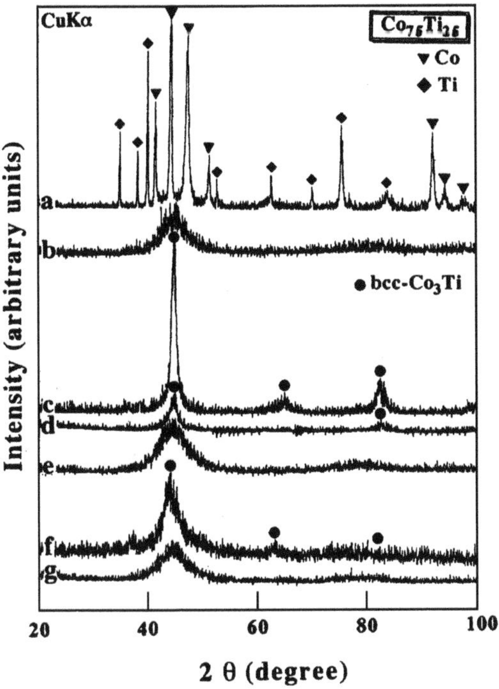

Figure 10.8 X-ray diffraction patterns of Co-25 at% Ti powder milled for (a) 0, (b) 3, (c) 24, (d) 48, (e) 100, (f) 150, and (g) 200 h. Note the cyclic transformation of amorphous and crystalline compounds [75].

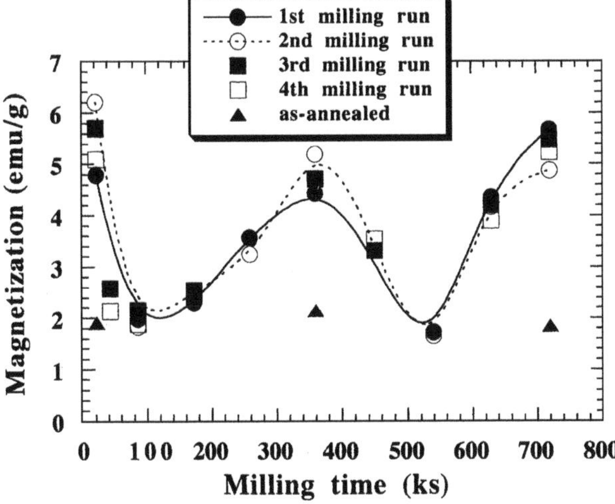

Figure 10.9 Dependence of the room temperature magnetization behavior of mechanically milled Co-50 at% Ti alloy powder as a function of the milling time. While the bcc-CoTi phase is ferromagnetic, the amorphous phase is not. The symbols in the figure represent the values for three different runs showing the reproducibility of the result [196].

Furthermore, the metastable bcc-Co_3Ti phase formed on milling the Co-25 at% Ti powder for 24 h had also been transformed to the equilibrium fcc-Co_3Ti phase on annealing at a high temperature of 1589 K. Thus, it appears that both the amorphous and metastable bcc-Co_3Ti phases transformed to the equilibrium fcc-Co_3Ti phase on annealing at a sufficiently high temperature, but only the metastable phases were found to form during MA. This complete set of transformations that take place on milling in the Co-25 at% Ti alloy powder is represented schematically in Figure 10.10 [77].

A similar cyclic crystalline → amorphous → crystalline phase formation was also reported in a mechanically alloyed Co-50 at% Ti powder mixture [196]. An amorphous phase had formed on MA of a Co-50 at% Ti blended elemental powder mixture for 6 h, which transformed into the equilibrium crystalline CoTi phase with a bcc structure on further milling to 24 h. This cyclic amorphous and crystalline phase formation was reported to occur repeatedly on continued milling of the powder mixture. The sequence of phase formation in this alloy could be represented as

$$\text{Co} + \text{Ti} \xrightarrow{6\ \text{h}} \text{amorphous} \xrightarrow{24\ \text{h}} \text{bcc-CoTi} \xrightarrow{100\ \text{h}} \text{amorphous} \xrightarrow{150\ \text{h}} \text{bcc-CoTi} \xrightarrow{200\ \text{h}} \text{amorphous} \tag{10.6}$$

Even though reasons for the occurrence of the cyclic crystalline → amorphous → crystalline transformation are not clear, a possible cause could be either the contam-

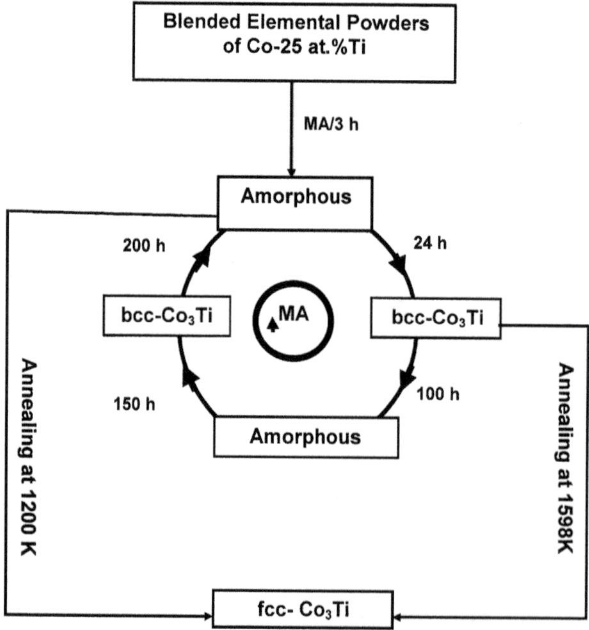

Figure 10.10 Schematic illustration of the amorphous → crystalline → amorphous transformation during MA of Co-25 at% Ti powders. Formation of the equilibrium fcc-Co_3Ti phase on annealing either the amorphous or metastable bcc-Co_3Ti phases externally is also indicated [77].

ination of the powder or an increase in the powder temperature during milling. It has been shown earlier that a crystalline phase had formed on continued milling after the formation of the amorphous phase in some alloy systems, explained as due to severe contamination of the milled powder [151]. If the crystalline phase formed were due to contamination, then subsequent milling would not transform that phase to the amorphous state. Furthermore, high-temperature annealing also would not change the constitution of the powder. However, the present authors [77,196] have shown that both the metallic and interstitial contamination of the powder was much less than 1 at% for each of these in both the alloy compositions (Fig. 10.11). Therefore it is unlikely that formation of the crystalline phase after formation of the amorphous phase is due to powder contamination. Furthermore, if the crystalline phase formed were to be the contamination phase, it would not have transformed to the equilibrium phase on annealing.

Another possibility for the formation of the crystalline phase from the amorphous phase on continued milling is an excessive rise of the powder temperature during milling. This possibility also was discounted by the authors who ensured that the Co-25 at% Ti powder milled for 3 h (amorphous phase) and 24 h (bcc Co_3Ti phase) did not show any phase transformation even on annealing at 700 K, a temperature much higher than that obtained during milling.

The explanation offered by these authors for such cyclic transformations is as follows: Introduction of crystal defects during MA raises the strain energy of the system and destroys the long-range periodic structure resulting in the formation of an amorphous phase. On continued milling the frictional forces between the grinding medium or between the grinding medium and the vial surface generate some heat and this relaxes the amorphous phase into forming a metastable crystalline phase. But formation of a metastable phase was not reported in the Co-50 at% Ti powder mixture. Introduction of additional defects into the crystalline phase during continued milling is supposed to have destabilized the amorphous phase by mechanical disorder-

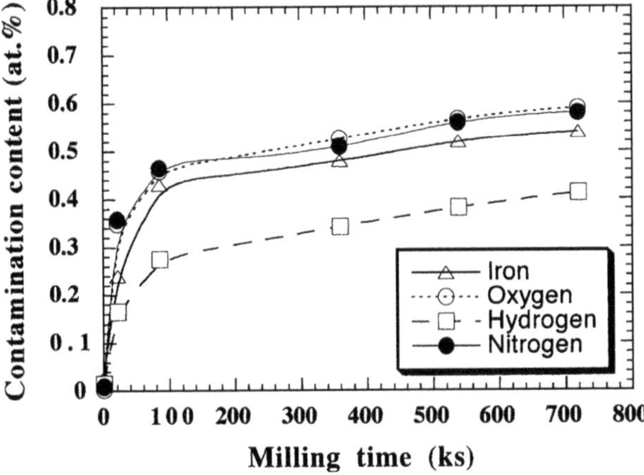

Figure 10.11 Increase in the extent of contamination of iron, oxygen, hydrogen, and nitrogen as a function of milling time in the Co-50 at% Ti powder. Note that the highest content for each of the contaminants is less than 1 at%. A similar behavior is also reported for the Co-25 at% powder [196].

ing. However, a more convincing explanation is needed to satisfactorily explain the crystalline → amorphous → crystalline observations.

As noted above, a metastable intermediate phase did not form in the Co-50 at% Ti powder. Furthermore, it is not easy to understand the repetitive cyclic transformation, like the one reported in the Co-Ti powder mixtures, on the basis of the above explanations. Thus, alternative explanations for this phenomenon are needed.

Guo and Lu [330] have also reported that amorphous → nanocrystalline → amorphous transformations could be observed in elemental selenium. Amorphous Se, obtained by melt quenching, was milled in a vibrating ball mill for 30 min, when diffraction peaks of trigonal Se were detected. At later stages of milling only nanocrystalline (about 15 nm) trigonal Se was detected. On continued milling, the amorphous Se phase started appearing again at 5 h, and the material was completely amorphous at 10 h. It should be noted that these authors have not continued these experiments to demonstrate that crystallization of the amorphous phase takes place on further milling.

During MA/MM two different processes can occur. First, the free energy of the system can be raised by the accumulation of different types of crystal defects. Second, the free energy can be lowered by activating a phase transformation to a more stable state (such as crystallization). When the first process is predominant during MA/MM, the material transforms into a highly defective state and eventually becomes amorphous. On the other hand, when the second process dominates, a crystalline phase would become more stable. Thus, the actual phase under the given processing conditions is determined by the interplay between these two competing processes—accumulation of crystal defects and activation of a phase transformation.

It is of interest to note that all the component metals, i.e., Co, Ti, and Se, exhibit allotropic transformations. For example, Co exhibits the fcc → hcp, Ti the hcp → bcc, and Se the trigonal → monoclinic polymorphic transformations, under equilibrium conditions. Such cyclic transformations have not yet been reported to occur in other alloy systems. Therefore, it would be instructive to investigate whether such cyclic transformations could be observed in other alloy systems not containing elements that exhibit polymorphic transformations under equilibrium conditions.

10.7 FORMATION OF ORDERED INTERMETALLICS

Even though intermetallics synthesized by MA include both the ordered and disordered types, it is not surprising that MA produces mostly the disordered phases. This is because MA involves heavy deformation, which is known to destroy long-range ordering in the lattice [48,331,332; see also Chapter 11]. Among the large number of intermetallic phases synthesized by MA (Table 10.5) both the disordered and ordered types have been reported, even though the proportion of the disordered variety is much higher than the ordered type.

In some instances, ordered intermetallics have been found to form directly on MA. This has been shown to be particularly true in Al-rich Al-transition metal systems. Some examples are Al_5Fe_2 [159,160], Al_3Hf [68], and $(Al,X)_3Hf$ where X = Fe or Ni [68], Al_3Nb [333], Al_3Zr [66–68,72], and $(Al,X)_3Zr$ where X = Fe or Ni [72,73].

Reasons for the formation of ordered intermetallics have not been investigated in detail. It may be assumed that a phase will exist either in the ordered or disordered condition depending on the balance between atomic disordering introduced by MA and the thermally activated reordering. The reordering is caused by the difference in energy between the ordered and disordered states. Thus, if this difference in energy is small the alloy will exist in the disordered state whereas if it is large the alloy will be in the ordered state. It has been shown that the NiAl compound was produced in the ordered state by MA of the blended elemental powder, and upon MM, the ordered NiAl compound continued to be in the ordered state without getting disordered. However, the FeAl phase was found to be in the disordered state both in the as-produced condition by MA and on mechanically milling of the ordered compound [251]. The ordering energy is related and scales up with the enthalpy of formation (ΔH_f). The ΔH_f values for NiAl and FeAl are 72 and 25 kJ/mol, respectively [251], confirming the above argument.

Superconducting compounds such as Nb_3Sn [240], $YBa_2Cu_3O_{7-\delta}$ [203,293,334–336], MgB_2 [337], and others [338–341], including YNi_2B_2C [342], have also been prepared by MA. Thus, the capabilities of MA in synthesizing a variety of intermetallic phases appear unlimited.

10.8 TRANSFORMATION BEHAVIOR OF METASTABLE INTERMETALLIC PHASES

Unlike in rapidly solidified alloys, systematic studies on the transformation behavior of metastable crystalline phases to the equilibrium phases have not been undertaken in detail in the mechanically alloyed powders. Only a few reports are available. For example, the transformation behavior of the $L1_2$-type Al_3Ti alloy phase produced in Al-25 at% Ti alloys was studied by annealing the powder at different temperatures [66]. The as-milled powder contained the $L1_2$ intermetallic and the Al(Ti) solid solution phases. Annealing of this two-phase mixture at 673 K for 1 h produced another metastable DO_{23} phase. The equilibrium DO_{22} phase was obtained only after annealing the as-milled powder for 1 h at temperatures higher than 800°C [66] or 1000°C [280]. The $Al_{24}Ti_8$ phase was produced on annealing the as-milled powder at 800°C [280]. Thus, the transformation sequence of the metastable $L1_2$ phase in the mechanically alloyed Al_3Ti powders may be represented as

$$\begin{aligned}\text{Metastable } L1_2 \text{ structure} &\rightarrow \text{metastable } DO_{23} \text{ structure}\\ &\rightarrow \text{equilibrium } DO_{22} \text{ structure}\end{aligned} \tag{10.7}$$

Figure 10.12 presents the crystal structures of the $L1_2$, DO_{23}, and DO_{22} phases. It may be clearly visualized that these crystal structures are closely related to each other. While the $L1_2$ phase has a cubic structure, the DO_{23} and DO_{22} phases have tetragonal structures whose lattice parameters for the Al_3Ti composition are listed in Table 10.6. From the relation between the lattice parameters of the different crystal structures, it may be seen that

$$a_{DO_{23}} \cong a_{L1_2};\ c_{DO_{23}} \cong 4\ a_{L1_2}, \tag{10.8}$$

and

$$a_{DO_{22}} \cong a_{L1_2};\ c_{DO_{22}} \cong 2\ a_{L1_2} \tag{10.9}$$

Consequently, it should be relatively easy to transform one crystal structure to the other under nonequilibrium processing conditions such as those obtained during MA.

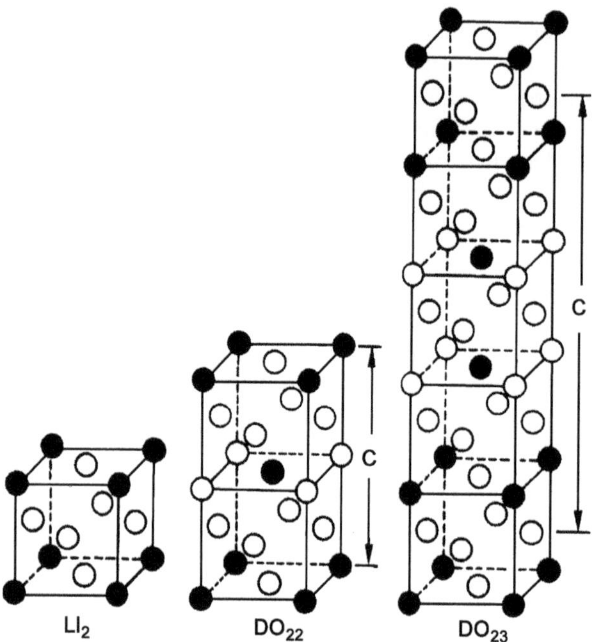

Figure 10.12 Crystal structures of the $L1_2$, DO_{23}, and DO_{22} phases.

Sun et al. [343] investigated the transformation behavior of the metastable phases produced in a mechanically milled $Al_{67}Mn_8Ti_{24}Nb_1$ alloy. The starting alloy had the ordered $L1_2$ and DO_{22} phases, which transformed into the disordered fcc supersaturated solid solution on milling in a planetary ball mill for 15 h. On continued milling this supersaturated solid solution transformed into an amorphous phase at 60 h. Crystallization occurred on annealing the amorphous alloy. A disordered fcc phase (without any superlattice peaks) started forming on annealing the amorphous alloy for 30 min at 420 °C, which transformed to the ordered $L1_2$ structure on annealing for 30 min at 650 °C. Simultaneously, diffraction peaks of the equilibrium DO_{22} phase also started appearing. Fully ordered $L1_2$ structure with plenty of DO_{22} precipitates was obtained on annealing the sample at 850 °C. Even though the temperatures at which these transformations occurred were lower during hot-stage electron microscopy

Table 10.6 Lattice Parameters of the Different Polytypes of the Al_3Ti Intermetallic Compound [66]

Phase	Crystal structure	Pearson symbol	Lattice parameters		
			a (nm)	c (nm)	c/a
$L1_2$	Cubic	cP4	0.3967	—	—
DO_{22}	Tetragonal	tI8	0.3851	0.8611	2.236
DO_{23}	Tetragonal	tI16	0.3890	1.6822	4.324

experiments, the sequence of transformation was identical. Thus, the sequence of phase transformation could be represented as:

$$
\begin{aligned}
&Al_{67}Mn_8Ti_{24}Nb_1(L1_2 + DO_{22})\\
&\rightarrow MA/16h \rightarrow \text{supersaturated solidsolution}\\
&\rightarrow MA/60\ h \rightarrow \text{amorphous phase}\\
&\rightarrow \text{annealing at } 420^\circ C \rightarrow \text{disordered fcc solid solution}\\
&\rightarrow \text{annealing at } 650^\circ C \rightarrow L1_2 \text{ phase with a small amount of } DO_{22}\\
&\rightarrow \text{annealing at } 850^\circ C \rightarrow L1_2 \text{ phase with a large amount of } DO_{22} \text{ precipitates}
\end{aligned}
\tag{10.10}
$$

The sequence of phase transformations in mechanically alloyed powders of Al_3Ti appears to be related to the lattice stability calculations of Carlsson and Meschter [344], who predicted that the stability of the different phases of Al_3Ti composition increases in the sequence reported above. That is, the DO_{22} is the most stable structure, the DO_{23} less stable, and the $L1_2$ the least stable under equilibrium conditions. The stability of these phases also appears to be related to the lattice mismatches. The lattice mismatch between the Al solid solution and the different structures of Al_3Ti is calculated using the equation suggested by Zedalis and Fine [345]:

$$
\delta = \frac{1}{3}\left[2\left(1 - \frac{a(Al_3X)}{a(Al)}\right) + \left(\frac{c'(Al_3X)}{a(Al)} - 1\right)\right] \tag{10.11}
$$

where δ is the overall mismatch, a is the base lattice parameter, $c' = c/4$ for the DO_{23} structure, and $c' = c/2$ for the DO_{22} structure. The δ values calculated were 2.0, 3.9, and 5.4, respectively for the $L1_2$, DO_{23}, and DO_{22} structures. Since coherency could be maintained when the δ values are small, and also because it is easier to nucleate a coherent precipitate, it has been suggested that the lower coherency is responsible for the formation of the $L1_2$ phase in the mechanically alloyed materials.

The Ti-Al binary phase diagram suggests that the γ-TiAl phase is ordered up to the melting point, i.e., it is a permanently ordered intermetallic (or irreversibly ordered) phase. A phase with an fcc structure is theoretically expected to form if the γ-TiAl phase (with the fct structure) were to be disordered, and it was calculated that this hypothetical fcc phase should have a lattice parameter of $a = 0.402$ nm. There have been some reports on the possible synthesis of a disordered γ-TiAl (fcc) phase by MA of blended elemental powders [89,90,276,346,347]. But a detailed analysis of the XRD patterns and a calculation of the expected intensities for the fcc phase reflections clearly show that the fcc phase, formed in mechanically alloyed alloy powders of near Ti-50 at% Al composition and interpreted as the disordered form of the γ-TiAl phase, was the contaminant nitride phase referred to earlier [151]. A fine structure analysis of the contaminant phase using a combination of techniques (XRD, TEM, laser Raman spectroscopy, and X-ray photoelectron spectroscopy) also indicated that the fcc phase was nanocrystalline TiN [348]. It was suggested that MA under ambient atmosphere results in the dynamic and rapid incorporation of excess nitrogen atoms, far beyond the solid solubility limits, and leads to the formation of an amorphous phase. This

Table 10.7 Crystal Structure Data of the Hypothetical γ-TiAl Fcc Phase and the Possible Contaminant Phases in Mechanically Alloyed Ti-Al and Ti-Al-Nb Powder Alloys

Compound	Pearson symbol	Structure type	Space group	Lattice parameter (nm)
γ-TiAl	cF4	—	—	0.4020
δ-TiN	cF8	NaCl	Fm3m	0.4242
τ-TiO	cF8	NaCl	Fm3m	0.4177
TiC	cF8	NaCl	Fm3m	0.4328
AlN	cF8	NaCl	Fm3m	0.4120
NbN	cF8	NaCl	Fm3m	0.4390

amorphous phase exhibits a nitride-like interaction with the titanium atoms. Prolonged milling eventually led to the formation of a nanocrystalline TiN phase.

Whereas the disordered form of γ-TiAl is expected to have an fcc structure with 4 atoms per unit cell (Pearson symbol: cF4), the contaminant phases have fcc structures with 8 atoms per unit cell (Pearson symbol: cF8). Accordingly, the intensity ratios of the 111 and 200 peaks in the XRD patterns of these two structures are reversed. That is, the 111 peak has a higher intensity than the 200 peak in the cF4 structure, while the 200 peak has a much higher intensity than the 111 peak in the cF8 structure. The possible confusion between the contaminant phases and the disordered form of γ-TiAl arises because the contaminant phases also have fcc structures with lattice parameters close to that expected for the disordered γ-TiAl phase (Table 10.7). Furthermore, it should be noted that the disordered forms of ordered phases have been synthesized by MA only when that phase exists in the phase diagram at higher temperatures, i.e., the phase is reversibly ordered (see Chapter 11 for full details on this aspect). Since this is not the situation in the Ti-Al system, the disordered form of the γ-TiAl phase was not expected to be produced by MA.

10.9 REFRACTORY COMPOUNDS

Refractory compounds based on metals and metalloids have desirable properties such as extremely high hardness, high-temperature stability, high thermal conductivity, and good corrosion resistance. For example, silicides could be used as potential structural materials in the field of microelectronics and electrical technology [349]. In particular, $MoSi_2$ has received considerable attention as a material for high-temperature applications. Its properties include a desirable combination of high melting point (2020°C), high Young's modulus (440 GPa), high oxidation resistance in air or in a combustion gas environment, and a relatively low density (6.25 g/cm^3) [350]. The refractory compounds, especially the carbides and nitrides, are usually synthesized on a commercial scale by direct reaction of the metal with the appropriate ambient at high temperatures and pressures. This process is carried out under isothermal conditions and the end product is often heterogeneous with some unreacted metal being present. Thus, the conventional methods of preparing refractory compounds are time consuming, expensive, and involved. Mechanical alloying has been a one-step successful route in synthesizing these compounds in powder form in an inexpensive and faster way.

Tables 10.8, 10.9, 10.10, and 10.11, respectively, list the borides, carbides, nitrides, and silicides synthesized by MA.

It is clear from the data presented in these tables that the refractory compounds could be synthesized by milling the metal powder along with required nonmetalloid source. For example, the boride phases are synthesized by milling the metal powders in the presence of boron or a boron source. Sometimes the boride phase can be synthesized in situ and be dispersed in the metallic matrix. As an example, the TiB_2 powder produced by MA can be successfully used in the development of dispersion-strengthened materials, e.g., Al-TiB_2 composites [452], since these powder particles have nanometer dimensions. It is also possible to produce TiB_2 dispersions *in situ* by milling blended elemental mixtures of Ti, Al, and B [453,454].

The boride phases are not always directly produced by MA. In some cases, such as in the Fe-B system, MA of iron and boron powder mixture produces either an amorphous phase or a solid solution [353] depending on the milling conditions. Heat treatment of these powders then precipitates the boride phases such as Fe_2B and FeB. However, Ruuskanen and Heczko [354] reported that they were able to directly synthesize FeB by MA.

Detailed investigations were also conducted on the synthesis of carbide phases by MA. Whereas some carbide phases were synthesized directly by MA, e.g., Al_4C_3, Fe_3C, Fe_7C_3, etc., an appropriate heat treatment was necessary in other cases. Some phase transformations were also observed on continued milling and/or heat treatment. For example, during milling of Fe-25 at% C powders it was noted that the Fe_3C phase was produced after 5 h, whereas a mixture of ε-Fe_2C and χ-$Fe_{2.5}C$ phases was

Table 10.8 Synthesis of Boride Phases by Mechanical Alloying

System	Phase	Comments	Ref.
Co-33B	Co_2B	MA (Am) + 500°C	351
Co-33 to 50B	t-Co_2B		352
Co-20 to 50B	t-Co_2B	MA (Am) + HT	102
Fe-20B	Fe_2B	MA/400h (α-Fe) + 800°C/2 h	353
Fe-20 to 30B	Fe_2B		354
Fe-50B	FeB		354
Fe-50B	FeB Fe_3B		78
Fe-50B	FeB	200 h milling under impact followed by 30 min of spark milling	355
Nb-50B	NbB		197, 356
Nb-67B	NbB_2		356
Ni-B	Ni_3B		357
Ti-B	TiB_2		324
Ti-66B	TiB_2		324, 358
Ti-67B	TiB_2	MA/1 h	359
TiO_2-B	TiB_2	MA/25 h + 1050°C	360
ZrO_2 + B	ZrB_2	MA/20 h + 100°C	360

Table 10.9 Synthesis of Carbide Phases by Mechanical Alloying

System	Phase	Comments	Ref.
Al-C	Al_4C_3		361
Al-27.4Fe-28.7C	$AlFe_3C_{0.5}$	MA/24 h + 475°C	163
Al-25Fe-25Ti-25C	TiC	MA/3 h	362
Co-C	Co_3C		361, 363
Cr-C	Cr_3C_2		361, 364–366
	Cr_7C_3		365, 366
Cr-40 C	Cr_3C_2	20 h in a SPEX mill + anneal at 800°C/70 min	366
Cr-30 C	Cr_7C_3	Milled for 192 h in a vibratory mill and 4 h in a SPEX mill	366
Cr-21.5 C	$Cr_{23}C_6$	MA + 800°C/2 h	364, 365
Fe-40Al-10V-10C	V_4C_3 (small amount)	MA/16 h + heated to 720°C in the DSC	367
Fe-17 to 25C	Fe_3C		368
Fe- 29 to 70C	Fe_7C_3		368
Fe-C	Fe_3C		363
Fe-C	Fe_3C		361
	Fe_5C_2		
	Fe_2C		
	Fe_7C_3		
Fe-25 to 33C	Fe_3C	20 h in Fe-25C 15 h in Fe-29C 12.5 h in Fe-33C	369
Fe-25 C	Fe_3C	5 h	370
Fe-25 C	Fe_3C	MA/100 h + anneal 1173 K for 15 min	371
Fe_3C (+Cr, Mn, V, Mo)	Fe_3C	MA/100 h + anneal 1173 K for 15 min	372
Fe-60 to 70C	Fe_7C_3		373
Fe-50C	Fe_7C_3	210 h	374
	Fe_3C	MA/210 h + 500°C/15 min	374
Fe-29C	Fe_3C	15 h	375
Fe-29C	HCP phase	20 h	375
Fe-25C	ε-Fe_2C and χ $Fe_{2.5}C$	10 h	370
Fe-25C	Fe_3C	20 h	375
Fe-25C	HCP phase	40 h	375
Fe-25C	Fe_3C	140 h	376
	Fe_7C_3	285 h	
Mn-Al-C			237
Mn-C	Mn_3C		361
Mo-C	MoC and MoC_x $(x < 0.5)$		361
Nb-C	$NbC_{0.75}$ and $NbC_x(x < 0.5)$		361

Table 10.9 Continued

System	Phase	Comments	Ref.
Nb + graphite or hexane	NbC		377
Ni-40Al-5Nb-5C	NiAl	B2, a = 0.2865 nm; MA/60 h	378
	NbC		
Ni-C	Ni_3C		361, 363
Re-C	ReC		361
Si-C	β-SiC	MA/300 h	361, 379–381
Si-C	SiC	>10 h	382
Ti-45Al	Ti_2AlC	MA/200 h + VHP 1173 K/1 h	282
Ti-Al-V-C	(Ti,Al,V)C		361
Ti-C	TiC		361, 381–384
Ti-graphite	TiC	MA/5 h	385
Ti-graphite	TiC	MA/19.5 h	386
Ti-C_{60} (C_{70})	TiC	MA/3 h	385
Ti-50C	TiC	MA/25 min	387
V-C	VC and $VC_x (x < 0.5)$		361
V-30C	V_2C	Only at high-energy milling	388
V-30C	V_2C	Low-energy milling for 700 h + anneal	389
V-50C	VC	Only at high-energy milling	388
V-50C	VC	Low-energy milling for 700 h + anneal	389
V-50C	VC	High energy	390
	VC	Medium energy + anneal	390
	V_2C	Low-energy MA (nano V + am. C) + anneal	390
	V + VC	Low-energy MA (nano V + am. C) + anneal	390
W-50C	WC		391
W-C-Co	WC	>100 h	392
W-C	FeW_3C and $(Fe,W)_6C$		361
W-activated C	WC	MA/310 h + 1000°C/1 h	393
WO_3-Mg-C	WC		394
Y-Ni-B-C	YNi_2B_2C		395
Y + 2Ni + 2B + C	YNi_2B_2C		396
Y-Pd-Pt-B-C	$Y(Pd_{1-x}Pt_x)_2B_2C$ $(0.1 \leq x \leq 0.7)$	MA + 1273 K/10 min	397
Zr-C	ZrC		361

Table 10.10 Synthesis of Nitride Phases by Mechanical Alloying

System	Phase	Comments	Ref.
Al	AlN	In a nitrogen atmosphere	398
B	BN	MA/140 h in a nitrogen atm + 800°C/1 h	398
B	BN	In a nitrogen or ammonia atmosphere	390
B	BN (hexagonal, nanotubes)	In anhydrous ammonia. MA/150 h + 1300°C/6 h	399
Cr	CrN, Cr_2N	MA/300 h + 800°C/1 h	398
Cu	Cu_3N	MA/60 h in N_2 atm	398
Fe	ε-Iron nitride	In NH_3 atm., HT phase	400
Fe	ξ- Fe_2N		401
Fe	Fe_4N		401
Fe	ε-Fe_3N		402
Fe	Fe_3N	MA of Fe with piperazine ($H_{10}C_4N_2$) or pyrazine ($H_4C_4N_2$)	403
Ga	GaN	In ammonia atm/300 h	404
Hf	HfN		405
Hf	HfN, cF8, a = 0.453 nm	MA/4 h in a nitrogen flux	406
Mg	Mg_3N_2	N_2 atm/60 h	398
Mo	Mo_2N	MA/300 h + 800°C/1 h; N_2 atm	398
	Mo_2N	Nanostr. + 520°C/72 h; N_2 atm	390
	MoN	Nanostr. + 800°C/1 h; N_2 atm	390
Mo	Mo_2N	NH_3 atm/280 h	390
Nb	NbN		407
Si	α-Si_3N_4		390, 398, 408
Si	α-Si_3N_4	MA + annealing at 1000°C	409
Si	α-Si_3N_4	MA/168 h at RT or MA/48 h at 100°C in anhydrous ammonia + anneal at 1100°C	410
Si	α-Si_3N_4 (partial)	In ammonia atm MA/160 h + anneal 1200°C	411
Si		MA in nitrogen atm + anneal at 900°C/4 h	412
Ta	TaN	In a nitrogen atm; MA + 580°C/72 h	390
Ta	TaN	In an ammonia atm., MA + 900°C/24 h	390
Ta	Ta_2N	In an ammonia atm., MA + 450°C/72 h	390
Ta-Al	(Ta,Al)N and TaN	In a nitrogen atm.	413
Ti	m-Ti_2N	Super Misuni NEV MA-8/6:1/100 h	87
Ti	TiN and Ti_2N	High- or medium-energy milling Low-energy milling	390
Ti	TiN	In N_2 atm	229, 414–420
Ti	Ti_2N	In N_2 atm	401
Ti	Ti(O,N)		421
Ti-5Al	Ti(O,N)		421
Ti-25Al	(Ti,Al)N		419, 420, 422
Ti-50Al	(Ti,Al)N	In N_2 atm	419
Ti-50V	$TiVN_{0.16}$		290
Ti-50Zr	$(Ti,Zr)N_{0.9}$		423
V	VN		398
V-30Cu	$(V_{0.7}Cu_{0.3})N$	In N_2 atm	424

Table 10.10 Continued

System	Phase	Comments	Ref.
W	WN and W_2N	In N_2 atm, MA/300 h + 800°C/1 h	398
W	WN and W_2N	In NH_3 atm, MA (nano) + 1000°C/1 h	390
Zr	ZrN	N_2 atm	398, 425
Zr	ZrN	N_2 atm	390, 425
Zr	ZrN	NH_3 atm	426

produced after 10 h [370]. Similarly, Wang et al. [374] reported that the Fe_7C_3 phase was produced on milling an Fe-50C powder mixture for 210 h in a Uni-Ball mill. On annealing the above-milled powder for 15 min at 500°C, it transformed into the Fe_3C phase.

As mentioned in Chapter 5, it is possible to change the intensity of milling in a Uni-Ball mill by changing the position of the magnets. Thus, one can have high energy (HE), medium energy (ME), or low energy (LE). Calka et al. [390] milled a V-50 at% C mixture under these three different conditions and observed differences in the nature of the final product. At the highest energy of milling, the equiatomic VC was directly produced. At intermediate milling energy, a nanostructured phase was produced, which on annealing transformed to the VC phase. At the lowest energy of milling a mixture of amorphous carbon and nanostructured vanadium phases was produced, which on heat treatment transformed to the V_2C phase. At slightly higher energies than the lowest energy, the mixture of amorphous carbon and nanostructured vanadium transformed to a mixture of vanadium and VC on annealing. Similarly, direct formation of TiN was observed only when milling was conducted at high energies and not at lower energies; low-energy milling led to the formation of the Ti_2N phase [390], again confirming that "soft" milling conditions result in the formation of metastable phases.

Several nitride phases have been synthesized by milling pure metals either in a nitrogen or ammonia atmosphere. These are listed in Table 10.10. Whereas a majority of the nitride phases have formed directly by MA, in some cases an additional heat treatment was necessary. For example, nitrides of Al, B, Cu, Fe, Ga, Mg, Nb, Ta, Ti, V, and Zr could be produced directly by MA. In the case of molybdenum, it was noted that milling in a nitrogen atmosphere produced a nanocrystalline phase, which on annealing for 72 h at 520°C produced a mixture of the Mo and Mo_2N phases. The same as-milled nanocrystalline phase transformed to a mixture of Mo, MoN, and Mo_2N on annealing at 800°C for 1 h [390]. A similar situation was reported in the synthesis of tungsten nitrides by MA [390].

Milling of reactive metals like titanium and zirconium in improperly sealed containers always produced the nitride phases. Since a minimal amount of nitrogen is required for the formation of the nitride phase, the nitride phase does not form in the initial stages. Often the nitride contamination phase forms after the formation of an amorphous phase in the titanium and zirconium base alloy systems. The reader is advised to consult Ref. 151 for a critical discussion of the nitride phase formation in mechanically alloyed titanium alloys.

Table 10.11 Synthesis of Silicide Phases by Mechanical Alloying

System	Phase	Structure	Comments	Ref.
Cr-Si	Cr_3Si		MA/35 h (Cr-Si) + 900°C/4 h	427
Cr-67Si	$CrSi_2$			356
Fe-25Al-25Si	Fe(Al,Si)	B2, a = 0.2826 nm	MA/3.5 h	428
	FeSi	B20, a = 0.4490 nm		
Fe-Si	α-$FeSi_2$		MA + 300°C	429, 430
Fe-Si	β-$FeSi_2$		500 h	431
Fe-Si	β-$FeSi_2$		MA/500 h + 973 K/200 h	432
Fe-Si	α-$FeSi_2$ FeSi			430
Fe-24.5Si	Fe_3Si	DO_3, ordered	MA/8 h + anneal 1173 K	433
Fe-25 to 50Si	FeSi			434
Fe-40Si	FeSi			354
Mg-Si	Mg_2Si	Cubic, a = 0.6359 nm		80, 435, 436
Mn-63 to 65Si	$MnSi_{1.73}$		MA/50 h + anneal at 700 K	437
Mo 25Si	Mo_5Si_3			438
Mo-37Si	Mo_5Si_3			439
Mo-37.5Si	Mo_5Si_3			115
	β-$MoSi_2$		MA + 500°C	
Mo-38Si	Mo_5Si_3			318, 438
Mo-67Si	α-$MoSi_2$		SHS reaction occurred	318, 438
Mo-66.7Si	α-$MoSi_2$		MA/6 min	123
	β-$MoSi_2$		MA/40 min	
Mo-66.7Si	$MoSi_2$		MA/210 h	440
Mo-66.7Si	t-$MoSi_2$		SHS at ≥3 h	441, 442
Mo-66.7Si	$MoSi_2$		MA + 760°C/24 h	443
Mo-67Si	$MoSi_2$(α and β)			439
Mo-80Si	$MoSi_2$(α and β)			318, 438
Mo-Si	α-$MoSi_2$		SHS; 14 h 15 min	318
	Mo_5Si_3		100 h	
Nb-16Si	Nb_3Si			444
Nb-35Si	α-Nb_5Si_3			445
Nb-45 to 59Si	$NbSi_2$ + β-Nb_5Si_3			446
Nb-37.5Si	Nb_5Si_3 (α, β, and γ types)			445, 446
Nb-66.7 Si	$NbSi_2$		SHS partly; reaction takes place abruptly after 65 min	197, 323, 356, 446
Ni-Si	Ni_5Si_2 and Ni_2Si		200 h	447
Ni-25 Si	Ni_3Si_2			439

Table 10.11 Continued

System	Phase	Structure	Comments	Ref.
Ni-33 Si	Ni_2Si			439
Ni-50 Si	NiSi			439
Ni-50 Si	NiSi		MA + 700°C/1 h	94
Ni-50 Si	NiSi		Milling for 400 h under impact followed by 30 min of spark milling	355
Ni-67 Si	$NiSi_2$	Fcc		439
Pd-19 Si	Pd_9Si_2 and Pd_3Si			448
Pd-20 Si	Pd_3Si			448
Ta-66.7 Si	$TaSi_2$			446
Ti-Si	m-$TiSi_2$ (C 49)		Fritsch P5/10:1	92, 93
Ti-Si		BCC	Ball mill/180 h	94
Ti-67 Si	$TiSi_2$		m-bcc + 800°C/30 min	94
Ti-50 Si	TiSi			325
Ti-44.4 Si	Ti_5Si_4			325
Ti-37.5 Si	Ti_5Si_3			325
Ti-37.5 Si	Ti_5Si_3			278, 449
Ti-37 Si	Ti_5Si_3		MA/180 h (Am) + 800°C/10 min	94
Ti-Si	Ti_5Si_3, Ti_5Si_4			450
Ti-Si	TiSi, Ti_5Si_3, Ti_5Si_4			325
V-25 Si	V_3Si + V_5Si_3			47
V-37.5 Si	V_5Si_3			446
V-66.7 Si	VSi_2			446
W-Si	WSi_2		MA(W + Si) + HIP 1200°C/100MPa/3h	364, 449
W-66.7 Si	WSi_2		MA/210 h	319
W-66.7 Si	WSi_2 + W			451

Synthesis of transition metal silicides by MA has attracted a great deal of attention in recent times. Among the different silicides synthesized, $MoSi_2$ has been the subject of many investigations because it was found to form in different ways depending on the milling conditions.

REFERENCES

1. Cahn, R. W. (1989). *Metals Mater. Proc.* 1:1–19.
2. Fleischer, R. L., Dimiduk, D. M., Lipsitt, H. A. (1989). *Annu. Rev. Mater. Sci.* 19:231–263.
3. Stoloff, N. S., Sikka, V. K., eds. (1996). *Physical Metallurgy and Processing of Intermetallic Compounds.* New York: Chapman & Hall.
4. Sauthoff, G. (1996). In: Matucha, K. H., ed. *Structure and Properties of Nonferrous Alloys. Vol. 8 of Materials Science and Technology—A Comprehensive Treatment.* Weinheim, Germany: VCH, pp. 643–803.
5. George, E. P., Yamaguchi, M., Kumar, K. S., Liu, C. T. (1994). *Annu. Rev. Mater. Sci.* 24:409–451.

6. Westbrook, J. H., Fleischer, R. L., eds. *Intermetallic Compounds: Principles and Practice*, Vol. 1: Principles (1995), Vol. 2: Practice (1995), and Vol. 3: Progress (2002). Chichester, UK: Wiley.
7. Yamaguchi, M., Umakoshi, Y. (1990). *Prog. Mater. Sci.* 34:1–148.
8. The Materials Research Society Conference Series on High Temperature Ordered Intermetallic Alloys I, II, III, etc. These proceedings have been published as MRS Proceedings I (vol. 39, 1985), II (vol. 81, 1987), III (vol. 133, 1989), IV (vol. 213, 1991), V (vol. 288, 1993), VI (vol. 364, 1995), VII (vol. 460, 1997), VIII (vol. 552, 1999), IX (vol. 646, 2001).
9. Proceedings of the series of conferences on Structural and Functional Intermetallics, I (1989), II (1991), III (1994), IV (1997), and V (2000). These proceedings have been published. The first was published as a separate book (High Temperature Aluminides and Intermetallics, edited by S. H. Whang, C. T. Liu, D. P. Pope, and J. O. Stiegler, Warrendale, PA: TMS, 1989) and the subsequent ones as special issues of the journal Mater. Sci. Eng. A (III, vols. 192–193, 1995; IV, vols. 239–240, 1997; V, vols. 329–331, 2002).
10. Schechtman, D., Blech, I., Gratias, D., Cahn, J. W. (1984). *Phys. Rev. Lett.* 53:1951–1953.
11. Sastry, G. V. S., Suryanarayana, C., Van Sande, M., Van Tendeloo, G. (1978). *Mater. Res. Bull.* 13:1065–1070.
12. Tsai, A. P., Inoue, A., Masumoto, T. (1987). *Jpn. J. Appl. Phys.* 26:L1505–1507.
13. Suryanarayana, C., Jones, H. (1988). *Int. J. Rapid Solid.* 3:253–293.
14. Kelton, K. F. (1993). *Int. Mater. Rev.* 38:105–137.
15. Series of articles on different aspects of quasicrystals in MRS *Bull.* 22(11):34–72.
16. Dubois, J. -M., Thiel, P. A., Tsai, A. P., Urban, K., eds. (1999). *Quasicrystals: Preparation, Properties and Applications, MRS Symp. Proc. Vol. 553*. Warrendale, PA: Mater. Res. Soc.
17. Suck, J. -B., Schreiber, M., Häussler, P. (2002). *Quasicrystals: An Introduction to the Structure, Physical Properties, and Applications*. New York: Springer-Verlag (Vol. 55 of Springer Series in Materials Science).
18. Dubois, J. -M. (2001). *J. Phys. C: Condens. Matter* 13:7753–7762.
19. Eckert, J., Schultz, L., Urban, K. (1989). *Appl. Phys. Lett.* 55:117–119.
20. Ivanov, E., Konstanchuk, I. G., Bokhonov, B. D., Boldyrev, V. V. (1989). *Reactiv. Solids* 7:167–172.
21. Eckert, J. (1992). *Mater. Sci. For.* 88–90:679–686.
22. Barua, P., Srinivas, V., Murty, B. S. (2000). *Phil. Mag.* A80:1207–1217.
23. Nasu, S., Miglierini, M., Ishihara, K. N., Shingu, P. H. (1992). *J. Phys. Soc. Jpn.* 61: 3766–3772.
24. Asahi, N., Maki, T., Matsumoto, S., Sawai, T. (1994). *Mater. Sci. Eng.* A181/182:841–844.
25. Salimon, A. I., Korsunsky, A. M., Kaloshkin, S. D., Tcherdyntsev, V. V., Shelekhov, E. V., Sviridova, T. A. (2001). *Mater. Sci. For.* 360–362:137–142.
26. Barua, P., Murty, B. S., Srinivas, V. (2001). *Mater. Sci. Eng.* A304–306:863–866.
27. Eckert, J., Schultz, L., Urban, K. (1990). *Z. Metallkde.* 81:862–868.
28. Bahadur, D., Singh, K., Roy, M. (1992). *Mater. Sci. Eng.* A154:79–84.
29. Politis, C., Krauss, W., Leitz, H., Schommers, W. (1989). *Mod. Phys. Lett.* 3:615–618.
30. Asahi, N., Noguchi, S., Matsumura, K. (1994). *Mater. Sci. Eng.* A181/182:819–822.
31. Schurack, F., Börner, I., Eckert, J., Schultz, L. (1999). *Mater. Sci. For.* 312–314:49–54.
32. Roy, M., Singh, K., Bahadur, D. (1992). *J. Mater. Sci. Lett.* 11:858–861.
33. Meyer, M., Mendoza-Zélis, L., Sánchez, F. H. (1995). *Mater. Sci. For.* 179–181:177–182.
34. Asahi, N. (1997). *Mater. Sci. Eng.* A226:67–69.
35. Ivanov, E. Yu., Konstanchuk, I. G., Bokhonov, B. D., Boldyrev, V. V. (1989). *Doklady Phys. Chem.* 304(1–3):82–85.
36. Ivanov, E., Bokhonov, B., Konstanchuk, I. (1991). *J. Jpn. Soc. Powder Powder Metall.* 38:903–905.

37. Takeuchi, T., Yamada, Y., Fukunaga, T., Mizutani, U. (1994). *Mater. Sci. Eng.* A181/182:828–832.
38. Ivanov, E., Bokhonov, B., Konstanchuk, I. (1991). *J. Mater. Sci.* 26:1409–1411.
39. Ivanov, E., Bokhonov, B., Konstanchuk, I., Boldyrev, V. V. (1993). In: deBarbadillo, J. J., et al., eds. *Mechanical Alloying for Structural Applications*. Materials Park, OH: ASM International, pp. 421–424.
40. Bokhonov, B., Konstanchuk, I., Ivanov, E., Boldyrev, V. V. (1992). *J. Alloys Compounds* 187:207–214.
41. Bokhonov, B. B., Ivanov, E. Y., Tolochko, B. P., Sharaphutdinov, M. P. (2000). *Mater. Sci. Eng.* A278:236–241.
42. Bahadur, D., Dunlap, R. A., Foldeaki, M. (1996). *J. Alloys Compounds* 240:278–284.
43. Takasaki, A., Han, C. H., Furuya, Y., Kelton, K. F. (2002). *Phil. Mag. Lett.* 82:353–361.
44. Konstanchuk, I. G., Ivanov, E. Yu., Bokhonov, B. B., Boldyrev, V. V. (2001). *J. Alloys Compounds* 319:290–295.
45. Suryanarayana, C., Ivanov, E., Noufi, R., Contreras, M. A., Moore, J. J. (1999). *J. Mater. Res.* 14:377–383.
46. Gerasimov, K. B., Gusev, A. A., Ivanov, E. Y., Boldyrev, V. V. (1991). *J. Mater. Sci.* 26:2495–2500.
47. Liu, L., Casadio, S., Magini, M., Nannetti, C. A., Qin, Y., Zheng, K. (1997). *Mater. Sci. For.* 235–238:163–168.
48. Bakker, H., Zhou, G. F., Yang, H. (1995). *Prog. Mater. Sci.* 39:159–241.
49. Suryanarayana, C. (2001). *Prog. Mater. Sci.* 46:1–184.
50. Zhang, F. X., Wang, W. K. (1996). *J. Alloys Compounds* 240:256–260.
51. Schulz, R., Trudeau, M. L., Dussault, D., Van Neste, A. (1990). In: Yavari, A. R., Desré, P., eds. *Multilayer Amorphisation by Solid State Reaction and Mechanical Alloying. J. Physique* 51(C4)Suppl 14:259–264.
52. McDermott, B. T., Koch, C. C. (1986). *Scripta Metall.* 20:669–672.
53. Martelli, S., Mazzone, G., Scaglione, S., Vittori, M. (1988). *J. Less Common Metals* 145:261–270.
54. Pabi, S. K., Joardar, J., Murty, B. S. (1996). *J. Mater. Sci.* 31:3207–3211.
55. Atzmon, M. (1990). *Phys. Rev. Lett.* 64:487–490.
56. Takacs, L. (2002). *Prog. Mater. Sci.* 47:355–414.
57. Moore, J. J., Feng, H. J. (1995). *Prog. Mater. Sci.* 39:243–316.
58. Anantharaman, T. R., Suryanarayana, C. (1971). *J. Mater. Sci.* 6:1111–1135.
59. Anantharaman, T. R., Ramachandrarao, P., Suryanarayana, C., Lele, S., Chattopadhyay, K. (1977). *Trans. Indian Inst. Metals* 30:423–448.
60. Rawers, J. C., Doan, R. C. (1994). *Metall. Mater. Trans.* 25A:381–388.
61. Rawers, J., Govier, D., Cook, D. (1995). *Scripta Metall. Mater.* 32:1319–1324.
62. Yvon, P. J., Schwarz, R. B. (1993). In: deBarbadillo, J. J., et al., eds. *Mechanical Alloying for Structural Applications*. Materials Park, OH: ASM International, pp. 421–424.
63. Chattopadhyay, K., Wang, X.-M., Aoki, K., Masumoto, T. (1996). *J. Alloys Compounds* 232:224–231.
64. Martelli, S., Mazzone, G., Vittori-Antisari, M. (1991). *J. Mater. Res.* 6:499–504.
65. Li, F., Ishihara, K. N., Shingu, P. H. (1991). *Metall. Trans.* 22A:2849–2854.
66. Srinivasan, S., Desch, P. B., Schwarz, R. B. (1991). *Scripta Metall. Mater.* 25:2513–2516.
67. Schwarz, R. B., Desch, P. B., Srinivasan, S. (1993). In: deBarbadillo, J. J., et al., eds. *Mechanical Alloying for Structural Applications*. Materials Park, OH: ASM International, pp. 227–235.
68. Li, W., Suryanarayana, C., Froes, F. H. (1995). In: Phillips, M. A., Porter, J., eds. *Advances in Powder Metallurgy and Particulate Materials. Part I*. Princeton, NJ: Metal Powder Industries Federation, pp. 145–157.
69. Suryanarayana, C., Sundaresan, R. (1991). *Mater. Sci. Eng.* A131:237–242.

70. Lee, K. M., Yang, H. U., Lee, J. S., Kim, S. S., Ahn, I. S., Park, M. W. (2000). *Mater. Sci. For.* 343–346:314–319.
71. Klassen, T., Oehring, M., Bormann, R. (1994). *J. Mater. Res.* 9:47–52.
72. Suryanarayana, C., Li, W., Froes, F. H. (1994). *Scripta Metall. Mater.* 31:1465–1470.
73. Li, W., Suryanarayana, C., Froes, F. H. (1995). In: Froes, F. H., Suryanarayana, C., Ward-Close, C. M., eds. *Synthesis/Processing of Lightweight Metallic Materials*. Warrendale, PA: TMS, pp. 203–213.
74. Tanaka, T., Ishihara, K. N., Shingu, P. H. (1992). *Metall. Trans.* 23A:2431–2435.
75. Sherif El-Eskandarany, M., Aoki, K., Sumiyama, K., Suzuki, K. (1997). *Appl. Phys. Lett.* 70:1679–1681.
76. Aoki, K., Sherif El-Eskandarany, M., Sumiyama, K., Suzuki, K. (1998). *Mater. Sci. For.* 269–272:119–126.
77. Sherif El-Eskandarany, M., Aoki, K., Sumiyama, K., Suzuki, K. (2002). *Acta Mater.* 50:1113–1123.
78. Barinov, V. A., Tsurin, V. A., Elsukov, E. P., Ovechkin, L. V., Dorofeev, G. A., Ermakov, A. E. (1992). *Phys. Metals Metallogr.* 74(4):412–415.
79. Lu, L., Lai, M. O., Toh, Y. H., Froyen, L. (2002). *Mater. Sci. Eng.* A334:163–172.
80. Clark, C. R., Wright, C., Suryanarayana, C., Baburaj, E. G., Froes, F. H. (1997). *Mater. Lett.* 33:71–75.
81. Kenik, E. A., Bayuzick, R. J., Kim, M. S., Koch, C. C. (1987). *Scripta Metall.* 21: 1137–1142.
82. Cardellini, F., Mazzone, G., Montone, A., Vittori-Antisari, M. (1994). *Acta Metall. Mater.* 42:2445–2451.
83. Ahn, J. H., Wang, G. X., Liu, H. K., Dou, S. X. (2001). *Mater. Sci. For.* 360–362:595–602.
84. Baláž, P., Ohtani, T. (2000). *Mater. Sci. For.* 343–346:389–392.
85. Baláž, P., Takacs, L., Ohtani, T., Mack, D. E., Boldižárová, E., Soika, V., Achimovičovă, M. (2002). *J. Alloys Compounds* 337:76–82.
86. Chitralekha, J., Raviprasad, K., Gopal, E. S. R., Chattopadhyay, K. (1995). *J. Mater. Res.* 10:1897–1904.
87. Sherif El-Eskandararany, M., Sumiyama, K., Aoki, K., Suzuki, K. (1992). *J. Mater. Res.* 7:888–893.
88. Burgio, N., Guo, W., Martelli, S., Magini, M., Padella, F., Soletta, I. (1990). In: Froes, F. H., deBarbadillo, J. J., eds. *Structural Applications of Mechanical Alloying*. Materials Park, OH: ASM International, pp. 175–183.
89. Guo, W., Martelli, S., Burgio, N., Magini, M., Padella, F., Paradiso, E., Soletta, I. (1991). *J. Mater. Sci.* 26:6190–6196.
90. Guo, W., Iasonna, A., Magini, M., Martelli, S., Padella, F. (1994). *J. Mater. Sci.* 29: 2436–2444.
91. Bououdina, M., Guo, Z. X. (2002). *Mater. Sci. Eng.* A332:210–222.
92. Oehring, M., Yan, Z. H., Klassen, T., Bormann, R. (1992). *Phys. Stat. Sol.* (a)131: 671–689.
93. Yan, Z. H., Oehring, M., Bormann, R. (1992). *J. Appl. Phys.* 72:2478–2487.
94. Radlinski, A. P., Calka, A. (1991). *Mater. Sci. Eng.* A134:1376–1379.
95. Biswas, A., Dey, G. K., Haq, A. J., Bose, D. K., Banerjee, S. (1996). *J. Mater. Res.* 11:599–607.
96. Tonejc, A., Stubicar, M., Tonejc, A. M., Kosanović, K., Subotić, B., Smit, I. (1994). *J. Mater. Sci. Lett.* 13:519–520.
97. Jiang, J. Z., Mørup, S., Linderoth, S. (1996). *Mater. Sci. For.* 225–227:489–496.
98. Kostić, E., Kiss, Š., Bošković, S., Zec, S. (1997). *Powder Tech.* 91:49–54.
99. Huang, J. Y., Wu, Y. K., Ye, H. Q. (1996). *Acta Mater.* 44:1201–1209.
100. Huang, J. Y., Wu, Y. D., Wu, Y. K., Li, D. X., He, Y. Q. (1997). *J. Mater. Res.* 12:936–946.
101. Sui, H. X., Zhu, M., Qi, M., Li, G. B., Yang, D. Z. (1992). *J. Appl. Phys.* 71:2945–2949.

102. Corrias, A., Ennas, G., Marongiu, M., Musinu, A., Paschina, G. (1993). *J. Mater. Res.* 8:1327–1333.
103. Huang, J. Y., Wu, Y. K., He, A. Q., Ye, H. Q. (1994). *Nanostructured Mater.* 4:293–302.
104. Di, L. M., Zhou, G. F., Bakker, H. (1993). *Phys. Rev.* B47:4890–4895.
105. Takacs, L. (1993). *Nanostructured Mater.* 2:241–249.
106. Ohtani, T., Motoki, M., Koh, K., Ohshima, K. (1995). *Mater. Res. Bull.* 30:1495–1504.
107. Gaffet, E., Michel, D., Mazerolles, L., Berthet, P. (1997). *Mater. Sci. For.* 235–238:103–108.
108. Han, S. H., Gshneidner, K. A., Beaudry, B. J. (1991). *Scripta Metall. Mater.* 25:295–298.
109. Barinov, V. A., Dorofeev, G. A., Ovechkin, L. V., Elsukov, E. P., Ermakov, A. E. (1991). *Phys. Stat. Sol.* (a)123:527–534.
110. Balogh, J., Bujdoso, L., Faigel, Gy, Granasy, L., Kemeny, T., Vincze, I., Szabo, S., Bakker, H. (1993). *Nanostructured Mater.* 2:11–18.
111. Barinov, V. A., Dorofeev, G. A., Ovechkin, L. V., Elsukov, E. P., Ermakov, A. E. (1992). *Phys. Metals Metallogr.* 73(1):93–96.
112. Foct, J., de Figueiredo, R. S. (1994). *Nanostructured Mater.* 4:685–697.
113. Matteazzi, P., Le Caër, G. (1991). *Mater. Sci. Eng.* A149:135–142.
114. Jiang, J. Z., Gerward, L., Mørup, S. (1999). *Mater. Sci. For.* 312–314:115–120.
115. Kosmac, T., Courtney, T. H. (1992). *J. Mater. Res.* 7:1519–1525.
116. Wu, E., Campbell, S. J., Kaczmarek, W. A., Hoffmann, M., Kennedy, S. J., Studer, A. J. (1999). *Mater. Sci. For.* 312–314:121–126.
117. Torres Sanchez, R. M. (1996). *J. Mater. Sci. Lett.* 15:461–462.
118. Sorescu, M. (1998). *J. Mater. Sci. Lett.* 17:1059–1061.
119. Baláž, P., Takacs, L., Jiang, J. Z., Soika, V., Luxová, M. (2002). *Mater. Sci. For.* 386–388:257–262.
120. Tokumitsu, K., Wada, M. (2002). *Mater. Sci. For.* 386–388:473–478.
121. Chen, Y., Williams, J. S., Wang, G. M. (1996). *J. Appl. Phys.* 79:3956–3962.
122. Bart, J. C. J. (1993). *J. Mater. Sci.* 28:278–284.
123. Bokhonov, B. B., Konstanchuk, I. G., Boldyrev, V. V. (1995). *J. Alloys Compounds* 218:190–196.
124. Urretavizcaya, G., Meyer, G. O. (2002). *J. Alloys Compounds* 339:211–215.
125. Schilz, J., Riffel, M., Pixius, K., Meyer, H. J. (1999). *Powder Technol.* 105:149–154.
126. Chattopadhyay, P. P., Pabi, S. K., Manna, I. (2001). *Mater. Sci. Eng.* A304–306:424–428.
127. Khelifati, G., Le Breton, J. M., Teillet, J. (2000). *Mater. Sci. For.* 343–346:344–350.
128. Alonso, T., Liu, Y., Parks, T. C., McCormick, P. G. (1991). *Scripta Metall. Mater.* 25:1607–1610.
129. Datta, M. K., Pabi, S. K., Murty, B. S. (2000). *J. Mater. Res.* 15:1429–1432.
130. Zhou, G. F., Di, L. M., Bakker, H. (1993). *J. Appl. Phys.* 73:1521–1527.
131. Yang, X. Y., Wu, Y. K., Ye, H. Q. (2000). *Phil. Mag. Lett.* 80:333–339.
132. Yang, X. Y., Wu, Y. K., Ye, H. Q. (2001). *J. Mater. Sci. Lett.* 20:1517–1518.
133. Mashimo, T., Tashiro, S. (1994). *J. Mater. Sci. Lett.* 13:174–176.
134. Chaudhuri, J., Ram, M. L., Sarkar, B. K. (1994). *J. Mater. Sci.* 29:3484–3488.
135. Millet, P., Hwang, T. (1996). *J. Mater. Sci.* 31:351–355.
136. Sen, S., Ram, M. L., Roy, S., Sarkar, B. K. (1999). *J. Mater. Res.* 14:841–848.
137. Girot, T., Devaux, X., Begin-Colin, S., Le Caër, G., Mocellin, A. (2001). *Phil. Mag.* A81:489–499.
138. Ivanov, E., Sumiyama, K., Yamauchi, H., Suzuki, K. (1993). *Solid State Ionics* 59: 175–177.
139. Ivanov, E., Sumiyama, K., Yamauchi, H., Suzuki, K. (1993). *J. Alloys Compounds* 192:251–252.
140. Liu, Z. T., Uwakweh, O. N. C. (1996). *J. Mater. Res.* 11:1665–1672.
141. Michel, D., Faudot, F., Gaffet, E., Mazerolles, L. (1993). *Rev. Metall.-CIT* 219–225.

142. Tonejc, A. M., Tonejc, A. (1996). *Mater. Sci. For.* 225–227:497–502.
143. Qi, M., Fecht, H. J. (1998). *Mater. Sci. For.* 269–272:187–191.
144. Bailey, J. E. (1972). *J. Br. Ceram. Soc.* 71:25.
145. Tonejc, A., Dužević, D., Tonejc, A. M. (1991). *Mater. Sci. Eng.* A134:1372–1375.
146. Tonejc, A., Tonejc, A. M., Dužević, D. (1991). *Scripta Metall. Mater.* 25:1111–1113.
147. Garvie, R. J. (1965). *J. Phys. Chem.* 69:1238.
148. Maurice, D. R., Courtney, T. H. (1990). *Metall. Trans.* 21A:289–303.
149. Davis, R. M., McDermott, B., Koch, C. C. (1988). *Metall. Trans.* A19:2867–2874.
150. Alonso, T., Liu, Y., Parks, T. C., McCormick, P. G. (1992). *Scripta Metall. Mater.* 26:1931–1932.
151. Suryanarayana, C. (1995). *Intermetallics* 3:153–160.
152. Paruchuri, M. R., Zhang, D. L., Massalski, T. B. (1994). *Mater. Sci. Eng.* A174:119–125.
153. Ohtani, T., Maruyama, K., Ohshima, K. (1997). *Mater. Res. Bull.* 32:343–350.
154. Angeles-Islas, J. (2002). *Mater. Sci. For.* 386–388:217–222.
155. Makhlouf, S. A., Ivanov, E., Sumiyama, K., Suzuki, K. (1992). *J. Alloys Compounds* 189:117–121.
156. Guerro-Paz, J., Robles-Hernández, F. C., Martínez-Sánchez, R., Hernández-Silva, F. D., Jaramillio-Viguera, D. (2001). *Mater. Sci. For.* 360–362:317–322.
157. Polkin, I. S., Kaputkin, E.Ya, Borzov, A. B. (1990). In: Froes, F. H., deBarbadillo, J. J., eds. *Structural Applications of Mechanical Alloying.* Materials Park, OH: ASM International, pp. 251–256.
158. Fadeeva, V. I., Portnoy, V. K., Baldokhin, Y. V., Kotchetov, G. A., Matyja, H. (1999). *Nanostructured Mater.* 12:625–628.
159. Cardellini, F., Contini, V., Mazzone, G., Montone, A. (1997). *Phil. Mag.* B76:629–638.
160. Mukhopadhyay, D. K., Suryanarayana, C., Froes, F. H. (1994). *Scripta Metall. Mater.* 31:333–338.
161. Morris, M. A., Morris, D. G. (1992). *Mater. Sci. For.* 88–90:529–535.
162. Morris, M. A., Morris, D. G. (1991). *Mater. Sci. Eng.* A136:59–70.
163. Zhang, D. L., Adam, G., Ammundsen, B. (2002). *J. Alloys Compounds* 340:226–230.
164. Portnoy, V. K., Leonov, A. V., Fadeeva, V. I., Matyja, H. (1998). *Mater. Sci. For.* 269–272:69–74.
165. Krasnowski, M., Matyja, H. (2000). *Mater. Sci. For.* 343–346:302–307.
166. Burzynska-Szyszko, M., Fadeeva, V. I., Matyja, H. (1997). *Mater. Sci. For.* 235–238: 97–102.
167. Zhang, D. L., Massalski, T. B., Paruchuri, M. R. (1994). *Metall. Mater. Trans.* 25A: 73–79.
168. LeBrun, P., Froyen, L., Delaey, L. (1990). In: Froes, F. H., deBarbadillo, J. J., eds. *Structural Applications of Mechanical Alloying.* Materials Park, OH: ASM International, pp. 155–161.
169. Zdujić, M., Kobayashi, K. F., Shingu, P. H. (1990). *Z. Metallkde.* 81:380–385.
170. Zdujić, M., Poleti, D., Karanovié, Lj, Kobayashi, K. F., Shingu, P. H. (1994). *Mater. Sci. Eng.* A185:77–86.
171. Peng, Z., Suryanarayana, C., Froes, F. H. (1993). In: deBarbadillo, J. J., et al., eds. *Mechanical Alloying for Structural Applications.* Materials Park, OH: ASM International, pp. 335–341.
172. Hellstern, E., Schultz, L., Bormann, R., Lee, D. (1988). *Appl. Phys. Lett.* 53:1399–1401.
173. Kaneyoshi, T., Takahashi, T., Hayashi, Y., Motoyama, M. (1992). In: Capus, J. M., German, R. M., eds. *Advances in Powder Metallurgy and Particulate Materials.* Vol. 7. Princeton, NJ: Metal Powder Industries Federation, pp. 421–429.
174. Kim, D. K., Okazaki, K. (1992). *Mater. Sci. For.* 88–90:553–560.
175. Kawanishi, S., Ionishi, K., Okazaki, K. (1993). *Mater. Trans. Jpn. Inst. Metals* 34:43–48.
176. Peng, Z., Suryanarayana, C., Froes, F. H. (1996). *Metall. Mater. Trans.* 27A:41–48.

177. Ivanov, E., Golubkova, G. V., Grigorieva, T. F. (1990). *Reactiv. Solids* 8:73–76.
178. Ivanov, E., Makhlouf, S. A., Sumiyama, K., Yamauchi, H., Suzuki, K., Golubkova, G. V. (1992). *J. Alloys Compounds* 185:25–34.
179. Benameur, T., Inoue, A., Masumoto, T. (1994). *Nanostructured Mater.* 4:303–322.
180. Schultz, L., Hellstern, E., Zorn, Z. (1988). *Z. Phys. Chem.* 157:203–210.
181. Liu, Z. G., Guo, J. T., Hu, Z. Q. (1995). *Mater. Sci. Eng.* A192/193:577–582.
182. Itsukaichi, T., Norimatsu, T., Umemoto, M., Okane, I., Wu, B.-Y. (1992). In: Tamura, I., ed. *Heat and Surface'92.* Tokyo, Japan: Japan Tech. Info. Center, pp. 305–308.
183. Xu, Y., Makhlouf, S. A., Ivanov, E., Wakoh, K., Sumiyama, K., Suzuki, K. (1994). *Nanostructured Mater.* 4:437–444.
184. Uenishi, K., Kobayashi, K. F., Ishihara, K. N., Shingu, P. H. (1992). *Mater. Sci. For.* 88–90:453–458.
185. Sá Lisboa, R. D., Perdigão, M. N. R. V., Kiminami, C. S., Botta F, W. J. (2002). *Mater. Sci. For.* 386–388:59–64.
186. Lerf, R., Morris, D. G. (1990). *Mater. Sci. Eng.* A128:119–127.
187. Zhang, L. P., Shi, J., Zin, J. Z. (1994). *Mater. Lett.* 21:303–306.
188. Mukhopadhyay, D. K., Suryanarayana, C., Froes, F. H., Yolton, C. F. (1993). In: deBarbadillo, J. J., et al., eds. *Mechanical Alloying for Structural Applications.* Materials Park, OH: ASM International, pp. 131–138.
189. Chen, G. H., Suryanarayana, C., Froes, F. H. (1993). In: deBarbadillo, J. J., et al., eds. *Mechanical Alloying for Structural Applications.* Materials Park, OH: ASM International, pp. 367–375.
190. Ahn, J. H., Lee, K. Y. (1995). *Mater. Trans. Jpn. Inst. Metals* 36:297–304.
191. Avvakumov, E. G., Karakchiev, L. G., Gusev, A. A., Vinokurova, O. B. (2002). *Mater. Sci. For.* 386–388:245–250.
192. Welham, N. J. (1998). *J. Mater. Res.* 13:1607–1613.
193. Chou, T. C., Nieh, T. G., Wadsworth, J. (1992). *Scripta Metall. Mater.* 27:881–886.
194. García-Pacheco, G., Cabañas-Moreno, J. G., Cruz-Gandarilla, F., Yee-Madeira, H., Umemoto, M. (2002). *Mater. Sci. For.* 386–388:281–286.
195. López Hirata, V. M., Juárez Martinez, U., Cabañas-Moreno, J. G. (1995). *Mater. Sci. For.* 179–181:261–266.
196. Sherif El-Eskandarany, M., Aoki, K., Sumiyama, K., Suzuki, K. (1997). *Scripta Mater.* 36:1001–1009.
197. Morris, M. A., Morris, D. G. (1990). In: Yavari, A. R., Desré, P., eds. *Multilayer Amorphisation by Solid-State Reaction and Mechanical Alloying.* Vol. 51. Colloq. C4., Suppl. 14. Les Ulis Cedex, France: Les editions de Physique, pp. 211–217.
198. Xi, S., Zhou, J., Zhang, D., Wang, X. (1996). *Mater. Lett.* 26:245–248.
199. Kim, K. J., Sherif El-Eskandarany, M., Sumiyama, K., Suzuki, K. (1993). *J. Non-Cryst. Solids* 155:165–170.
200. Kim, K. J., Sumiyama, K., Suzuki, K. (1994). *J. Non-Cryst. Solids* 168:232–240.
201. Kis-Varga, M., Beke, D. L. (1996). *Mater. Sci. For.* 225–227:465–470.
202. Ivanov, E. Y., Grigorieva, T. F. (1997). *Solid State Ionics* 101–103:235–241.
203. Mizutani, U., Imaeda, C., Murasaki, S., Fukunaga, T. (1992). *Mater. Sci. For.* 88–90:415–422.
204. Cardellini, F., Contini, V., Mazzone, G., Vittori, M. (1993). *Scripta Metall. Mater.* 28:1035–1038.
205. Fair, G. H., Wood, J. V. (1994). *J. Mater. Sci.* 29:1935–1939.
206. Bonetti, E., Scipione, G., Valdre, G., Cocco, G., Frattini, R., Macri, P. P. (1993). *J. Appl. Phys.* 74:2053–2057.
207. Fair, G. H., Wood, J. V. (1993). *Powder Metall.* 36:123–128.
208. Wolski, K., Le Caër, G., Delcroix, P., Fillit, R., Thévenot, F., Le Coze, J. (1996). *Mater. Sci. Eng.* A207:97–104.

209. Hashii, M. (1999). *Mater. Sci. For.* 312–314:139–144.
210. Liu, Z. G., Guo, J. T., He, L. L., Hu, Z. Q. (1994). *Nanostructured Mater.* 4:787–794.
211. Oleszak, D., Matyja, H. (2000). *Mater. Sci. For.* 343–346:320–325.
212. Elkalkouli, R., Grosbras, M., Dinhut, J. F. (1995). *Nanostructured Mater.* 5:733–743.
213. Gonzalez, G., Sagarzazu, A., Villalba, R., Ochoa, J., D'Onofrio, L. (2001). *Mater. Sci. For.* 360–362:355–360.
214. Goya, G. F. (2002). *Mater. Sci. For.* 386–388:491–496.
215. Baláž, P., Bastl, Z., Havlík, T., Lipka, J., Toth, I. (1997). *Mater. Sci. For.* 235–238:217–222.
216. Pan, C. W., Hung, M. P., Chang, Y. H. (1994). *Mater. Sci. Eng.* A185:147–152.
217. Yelsukov, E. P., Dorofeev, G. A., Barinov, V. A., Grigorieva, T. F., Boldyrev, V. V. (1998). *Mater. Sci. For.* 269–272:151–156.
218. Ruuskanen, P. (1998). *Mater. Sci. For.* 269–272:139–144.
219. Trudeau, M. L., Schulz, R., Zaluski, L., Hosatte, S., Ryan, D. H., Doner, C. B., Tessier, P., Strom-Olsen, J. O., Van Neste, A. (1992). *Mater. Sci. For.* 88–90:537–544.
220. Oleszak, D., Jachimowicz, M., Matyja, H. (1995). *Mater. Sci. For.* 179–181:215–218.
221. Burgio, N., Iasonna, A., Magini, M., Martelli, S., Padella, F. (1991). *Il Nuovo Cimento* 13D:459–476.
222. Alonso, T., Liu, Y., McCormick, P. G. (1992). *J. Mater. Sci. Lett.* 11:164–166.
223. Arcos, D., Rangavittal, N., Vazquez, M., Vallet-Regí, M. (1998). *Mater. Sci. For.* 269–272:87–92.
224. Šepelák, V., Rogachev, A.Yu, Steinike, U., Uecker, D-Chr, Krumeich, F., Wissmann, S., Becker, K. D. (1997). *Mater. Sci. For.* 235–238:139–144.
225. Al-Hajry, A., Al-Assiri, M., Enzo, S., Hefner, J., Al-Heniti, S., Cowlam, N., Al-Shahrani, A. (2001). *Mater. Sci. For.* 360–362:343–348.
226. Thompson, J. R., Politis, C., Kim, Y. C. (1988). *Mater. Sci. Eng.* 97:31–34.
227. Ivanov, E., Grigorieva, T. F. (1998). *Mater. Sci. For.* 269–272:241–246.
228. Takai, S., Moriyama, M., Esaka, T. (1998). *Mater. Sci. For.* 269–272:93–98.
229. Chen, Y., Williams, J. S. (1996). *Mater. Sci. For.* 225–227:881–888.
230. Gennari, F. C., Castro, F. J., Urretavizcaya, G. (2001). *J. Alloys Compounds* 321: 46–53.
231. Bobet, J.-L., Chevalier, B., Song, M. Y., Darriet, B., Etourneau, J. (2002). *J. Alloys Compounds* 336:292–296.
232. Hwang, S., Nishimura, C. (2002). *Mater. Sci. For.* 386–388:615–620.
233. Chen, J., Takeshita, H. T., Chartouni, D., Kuriyama, N., Sakai, T. (2001). *J. Mater. Sci.* 36:5829–5834.
234. Ruggeri, S., Lenain, C., Roué, L., Liang, G., Huot, J., Schulz, R. (2002). *J. Alloys Compounds* 339:195–201.
235. Chen, J., Dou, S. X., Liu, H. K. (1996). *J. Alloys Compounds* 244:184–189.
236. Ahn, J. H., Kim, Y. J., Choi, C. J. (2002). *Mater. Sci. For.* 386–388:609–614.
237. Crew, D. C., McCormick, P. G., Street, R. (1995). *Scripta Metall. Mater.* 32:315–318.
238. Wee, L., McCormick, P. G., Street, R. (1999). *Scripta Mater.* 40:1205–1208.
239. Chernenko, V. A., Wee, L., McCormick, P. G., Street, R. (2000). *J. Mater. Sci.* 35: 613–616.
240. Larson, J. M., Luhman, T. S., Merrick, H. F. (1977). In: Meyerhoff, R. W., ed. *Manufacture of Superconducting Materials.* Materials Park, OH: ASM International, pp. 155–163.
241. Tracy, M. J., Groza, J. R. (1992). *Nanostructured Mater.* 1:369–378.
242. Groza, J. R., Tracy, M. J. H. (1993). In: deBarbadillo, J. J., et al., eds. *Mechanical Alloying for Structural Applications.* Materials Park, OH: ASM International, pp. 327–334.
243. Portnoy, V. K., Tretjakov, K. V., Streletskii, A. N., Berestetskaya, I. V. (2000). *Mater. Sci. For.* 343–346:453–458.

244. Pabi, S. K., Murty, B. S. (1996). *Mater. Sci. Eng.* A214:146–152.
245. Tomasi, R., Pallone, E. M. J. A., Botta F., W. J. (1999). *Mater. Sci. For.* 312–314: 333–338.
246. Portnoy, V. K., Blinov, A. M., Tomilin, I. A., Kuznetsov, V. N., Kulik, T. (2002). *J. Alloys Compounds* 336:196–201.
247. Esaki, H., Tokizane, M. (1992). *Mater. Sci. For.* 88–90:625–630.
248. Ochiai, S., Shirokura, T., Doi, Y., Kojima, Y. (1991). *ISIJ Int.* 31:1106–1112.
249. Lu, L., Lai, M. O., Zhang, S. (1994). *Mater. Design* 15:79–86.
250. Lu, L., Lai, M. O., Zhang, S. (1995). *J. Mater. Proc. Technol.* 48:683–690.
251. Schropf, H., Kuhrt, C., Arzt, E., Schultz, L. (1994). *Scripta Metall. Mater.* 30:1569–1574.
252. Huang, B. L., Perez, R. J., Lavernia, E. J., Luton, M. J. (1996). *Nanostructured Mater.* 7:67–79.
253. Itsukaichi, T., Shiga, S., Masuyama, K., Umemoto, M., Okane, I. (1992). *Mater. Sci. For.* 88–90:631–638.
254. Murty, B. S., Joardar, J., Pabi, S. K. (1996). *Nanostructured Mater.* 7:691–697.
255. Ivanov, E., Grigorieva, T. F., Golubkova, G. V., Boldyrev, V. V., Fasman, A. B., Mikhailenko, S. D., Kalinina, O. T. (1988). *Mater. Lett.* 7:51–54.
256. Mikhailenko, S. D., Kalinina, O. T., Dyunusov, A. K., Fasman, A. B., Ivanov, E., Golubkova, G. V. (1991). *Siberian J. Chem. No.* 5:93–104.
257. Schaffer, G. B., Heron, A. J. (1993). In: deBarbadillo, J. J., et al., eds. *Mechanical Alloying for Structural Applications.* Materials Park, OH: ASM International, pp. 197–203.
258. Liu, Z. G., Guo, J. T., Hu, Z. Q. (1996). *J. Alloys Compounds 1996* 234:106–110.
259. Oleszak, D., Portnoy, V. K., Matyja, H. (1999). *Mater. Sci. For.* 312–314:345–350.
260. Baláž, P., Havlík, T., Briančin, J., Kammel, R. (1995). *Scripta Metall. Mater.* 32:1357–1362.
261. Wang, K. Y., Shen, T. D., Wang, J. T., Quan, M. X. (1991). *Scripta Metall. Mater.* 25:2227–2231.
262. Wang, K. Y., Shen, T. D., Wang, J. T., Quan, M. X. (1993). *J. Mater. Sci.* 28:6474–6478.
263. Oh, T. S., Choi, J. S., Hyun, D. B. (1995). *Scripta Metall. Mater.* 32:595–600.
264. Kong, L. B., Ma, J., Zhu, W., Tan, O. K. (2001). *J. Mater. Sci. Lett.* 20:1241–1243.
265. Kong, L. B., Ma, J., Zhu, W., Tan, O. K. (2002). *J. Alloys Compounds* 336:242–246.
266. Aning, A. O., Hong, C., Desu, S. B. (1995). *Mater. Sci. For.* 179–181:207–213.
267. Calka, A., Radlinski, A. P. (1989). *Scripta Metall.* 23:1497–1501.
268. Liu, K. W., Mücklich, F., Birringer, R. (2001). *Intermetallics* 9:81–88.
269. Liu, K. W., Mücklich, F. (2002). *Mater. Sci. Eng.* A329-331:112–117.
270. Liu, K. W., Mücklich, F., Pitschke, W., Birringer, R., Wetzig, K. (2001). *Z. Metallkde.* 92:924–930.
271. Kis-Varga, M., Beke, D. L., Mészáros, S., Vad, K., Kerekes, Gy, Daróczi, L. (1998). *Mater. Sci. For.* 269–272:961–968.
272. Izard, V., Record, M. C., Tedenac, J. C. (2002). *J. Alloys Compounds* 345:257–264.
273. Tan, M. (1994). *J. Mater. Sci.* 29:1306–1309.
274. Ding, J., McCormick, P. G., Street, R. (1992). *J. Alloys Compounds* 189:83–86.
275. Guo, W., Martelli, S., Magini, M. (1994). *Acta Metall. Mater.* 42:411–417.
276. Burgio, N., Guo, W., Martelli, S., Magini, M., Padella, F., Soletta, I. (1990). In: Froes, F. H., deBarbadillo, J. J., eds. *Structural Applications of Mechanical Alloying* pp. 175–183. Materials Park, OH: ASM International.
277. Frefer, A., Suryanarayana, C., Froes, F. H. (1993). In: Froes, F. H., Caplan, I. L., eds. *Titanium '92: Science and Technology.* Warrendale, PA: TMS, pp. 933–940.
278. Oleszak, D., Burzynska-Szyszko, M., Matyja, H. (1993). *J. Mater. Sci. Lett.* 12:3–5.
279. Guo, W., Martelli, S., Burgio, N., Magini, M., Padella, F., Paradiso, E., Soletta, I. (1991). *J. Mater. Sci.* 26:6190–6196.
280. Ahn, J. H., Lee, K. R., Cho, H. K. (1995). *Mater. Sci. For.* 179–181:153–158.

281. Ahn, J. H., Chung, H. (1994). In: Ravi, V. A., Srivatsan, T. S., Moore, J. J., eds. *Processing and Fabrication of Advanced Materials III.* Warrendale, PA: TMS, pp. 227–237.
282. Ameyama, K., Okada, O., Hirai, K., Nakabo, N. (1995). *Mater. Trans. Jpn. Inst. Metals* 36:269–275.
283. Suryanarayana, C., Chen, G. H., Frefer, A., Froes, F. H. (1992). *Mater. Sci. Eng.* A158:93–101.
284. Chen, G. H., Suryanarayana, C., Froes, F. H. (1991). *Scripta Metall. Mater.* 25: 2537–2540.
285. Suryanarayana, C., Chen, G. H., Froes, F. H. (1992). In: Froes, F. H., et al., eds. *Advancements in Synthesis and Processing.* Covina, CA: SAMPE, pp. M671–M683.
286. Chen, G. H., Suryanarayana, C., Froes, F. H. (1992). In: Capus, J. M., German, R. M., eds. *Advances in Powder Metallurgy and Particulate Materials.* Vol. 7. Princeton, NJ: Metal Powder Industries Federation, pp. 183–194.
287. Jurczyk, M., Jankowska, E., Nowak, M., Jakubowicz, J. (2002). *J. Alloys Compounds* 336:265–269.
288. Skakov, YuA., Edneral, N. V., Frolov, E. V., Povolozki, J. A. (1995). *Mater. Sci. For.* 179–181:33–38.
289. Blouin, M., Guay, D., Huot, J., Schulz, R. (1997). *J. Mater. Res.* 12:1492–1500.
290. Aoki, K., Memezawa, A., Masumoto, T. (1994). *J. Mater. Res.* 9:39–46.
291. Suryanarayana, C., Sundaresan, R., Froes, F. H. (1992). *Mater. Sci. Eng.* A150:117–121.
292. Abe, O., Suzuki, Y. (1996). *Mater. Sci. Forum* 225–227:563–568.
293. Schiffini, L., Mulas, G., Daturi, M., Ferretti, M. (1993). In: deBarbadillo, J. J., et al., eds. *Mechanical Alloying for Structural Applications.* Materials Park, OH: ASM International, pp. 457–461.
294. López-Hirata, V. M., Zhu, Y. H., Rodriguez-Hernández, J. C., Saucedo Muñoz, M. L., Arce-Estrada, E. M. (1999). *Mater. Sci. For.* 312–314:133–138.
295. Hong, L. B., Bansal, C., Fultz, B. (1994). *Nanostructured Mater.* 4:949–956.
296. Froes, F. H., Suryanarayana, C., Eliezer, D. (1992). *J. Mater. Sci.* 27:5113–5140.
297. Froes F. H., Suryanarayana C. (1996). In: Stoloff, N. S., Sikka, V. K., eds. *Physical Metallurgy and Processing of Intermetallic Compounds.* New York: Chapman & Hall, pp. 297–350.
298. Morris, D. G. (1998). *Mechanical Behavior of Nanostructured Materials. Materials Science Foundations 2.* Zeurich, Switzerland: Trans Tech.
299. Koch, C. C., Morris, D. G., Lu, K., Inoue, A. (1999). *MRS Bull.* 24(2):54–58.
300. McFadden, S. X., Mishra, R. S., Valiev, R. Z., Zhilyaev, A. P., Mukherjee, A. K. (1999). *Nature* 398:684–686.
301. Suryanarayana, C., Koch, C. C. (2000). *Hyperfine Interact.* 130:5–44.
302. Lu, L., Wang, L. B., Ding, B. Z., Lu, K. (2000). *J. Mater. Res.* 15:270–273.
303. Lu, L., Sui, M. L., Lu, K. (2000). *Science* 287:1463–1465.
304. Wang, Y., Chen, M., Zhou, F., Ma, E. (2002). *Nature* 419:912–995.
305. Suryanarayana, C. (2002). *Int. J. Non-equilibrium Proc.* 11:325–345.
306. Frefer, A., Suryanarayana, C., Froes, F. H. (1993). In: Moore, J. J., et al., eds. *Advanced Synthesis of Engineered Materials.* Materials Park, OH: ASM International, pp. 213–219.
307. Skakov, Yu A. (2000). *Mater. Sci. For.* 343–346:597–602.
308. Zhang, D. L., Ying, D. Y. (2001). *Mater. Sci. For.* 360–362:323–328.
309. Schaffer, G. B. (1992). *Scripta Metall. Mater* 27:1–5.
310. Guerrero-Paz, J., Robles-Hernández, F. C., Martínez-Sánchez, R., Hernández-Silva, D., Jaramillo-Vigueras, D. (2001). *Mater. Sci. For.* 360–362:317–322.
311. Singh, D., Suryanarayana, C., Mertus, L., Chen, R.-H. (2003). *Intermetallics* 11:373–376.
312. Gauthier, V., Bernard, F., Gaffet, E., Munir, Z. A., Larpin, J. P. (2001). *Intermetallics* 9:571–580.

313. Gauthier, V., Bernard, F., Gaffet, E., Vrel, D., Gailhanoud, M., Larpin, J. P. (2002). *Intermetallics* 10:377–389.
314. McQuillan, A. D. (1950). *Proc. R. Soc. (London)* A204:309.
315. Froes, F. H., Eylon, D., Suryanarayana, C. (1990). *JOM* 42(3):26–29.
316. Mukhopadhyay, D. K., Suryanarayana, C., Froes, F. H. (1993). In: Froes, F. H., Caplan, I., eds. *Titanium'92: Science and Technology*. Warrendale, PA: TMS, pp. 829–835.
317. Bououdina, M., Luklinska, Z., Guo, Z. X. (2001). *Mater. Sci. For.* 360–362:421–426.
318. Yen, B. K., Aizawa, T., Kihara, J. (1996). *Mater. Sci. Eng.* A220:8–14.
319. Ma, E., Pagan, J., Cranford, G., Atzmon, M. (1993). *J. Mater. Res.* 8:1836–1844.
320. Lee, P. Y., Chen, T. R., Yang, J. L., Chin, T. S. (1995). *Mater. Sci. Eng.* A192–193: 556–562.
321. Patankar, S. N., Xiao, S. Q., Lewandowski, J. J., Heuer, A. H. (1993). *J. Mater. Res.* 8:1311–1316.
322. Huang, J. Y., Ye, L. L., Wu, Y. K., Ye, H. Q. (1995). *Metall. Mater. Trans.* 26A: 2755–2758.
323. Lou, T., Fan, G., Ding, B., Hu, Z. (1997). *J. Mater. Res.* 12:1172–1175.
324. Radev, D. D., Klissurski, D. (1994). *J. Alloys Compounds* 206:39–41.
325. Lee, W. H., Reucroft, P. J., Byun, C. S., Kim, D. K. (2001). *J. Mater. Sci. Lett.* 20: 1647–1649.
326. Pallone, E. M. J. A., Hanai, D. E., Tomasi, R., Botta F, W. J. (1998). *Mater. Sci. For.* 269–272:289–294.
327. Coreño-Alonso, O., Cabañas-Moreno, J. G., Cruz-Rivera, J. J., Florez-Díaz, G., De Ita, A., Quintana-Molina, S., Falcony, C. (2000). *Mater. Sci. For.* 343–346:290–295.
328. Maric, R., Ishihara, K. N., Shingu, P. H. (1996). *J. Mater. Sci. Lett.* 15:1180–1183.
329. Pabi, S. K., Joardar, J., Manna, I., Murty, B. S. (1997). *Nanostructured Mater.* 9:149–152.
330. Guo, F. Q., Lu, K. (1998). *Phil. Mag. Lett.* 77:181–186.
331. Stoloff, N. S., Davies, R. G. (1966). *Prog. Mater. Sci.* 13:1–84.
332. Suryanarayana, C. (2002). In: Westbrook, J. H., Fleischer, R. L., eds. *Intermetallic Compounds. Principles and Practice: Vol. 3, Progress*. Chichester, UK: Wiley, pp. 749–764.
333. Suryanarayana, C., Zhou, E., Peng, Z., Froes, F. H. (1994). *Scripta Metall. Mater.* 30:781–785.
334. Ahn, J.-H., Ha, K.-H., Ko, J.-W., Kim, H.-D., Chung, H. (1992). In: Hayakawa, H., Koshizuka, N., eds. *Advances in Superconductivity IV*. Tokyo: Springer-Verlag pp. 459–462.
335. Thompson, J. R., Politis, C., Kim, Y. C. (1992). *Mater. Sci. Forum* 88–90:545–552.
336. Simoneau, M., Lvallee, F., L'Esperance, G., Trudeau, M. L., Schulz, R. (1992). In: Yavari, A. R., ed. *Ordering and Disordering in Alloys*. London: Elsevier, pp. 385–393.
337. Gümbel, A., Eckert, J., Fuchs, G., Nenkov, K., Müller, K.-H., Schultz, L. (2002). *Appl. Phys. Lett.* 80:2725–2727.
338. Politis, C. (1985). *Physica* 135B:286–289.
339. Whang, S. H., Li, Z. X. (1989). *Modern Phys. Lett.* B3:665–672.
340. Batalla, E., Zwartz, E. G. (1990). *J. Mater. Res.* 5:1802–1805.
341. Luo, J. S., Lee, H. G., Sinha, S. N. (1994). *J. Mater. Res.* 9:297–304.
342. Gümbel, A., Ledig, L., Hough, D., Oerkel, C.-G., Skrotzki, W., Eckert, J., Schultz, L. (1999). *Mater. Sci. For.* 312–314:61–66.
343. Sun, J., Zhang, J. X., Fu, Y. Y., Hu, G. X. (2002). *Mater. Sci. Eng.* A329–331:703–707.
344. Carlsson, A. E., Meschter, P. J. (1989). *J. Mater. Res.* 4:1060.
345. Zedalis, M., Fine, M. E. (1983). *Scripta Metall.* 17:1247–1251.
346. Qi, M., Zhu, M., Li, G. B., Sui, H.-X., Yang, D. Z. (1993). *J. Mater. Sci. Lett.* 12:66–69.
347. Gauvin, R., Bernier, M. A., Joy, D. C., Schmidt, M. (1993). In: deBarbadillo, J. J., et al.,

eds. *Mechanical Alloying for Structural Applications*. Materials Park, OH: ASM International, pp. 93–100.

348. Sato, K., Ishizaki, K., Chen, G. H., Frefer, A., Suryanarayana, C., Froes, F. H. (1993). In: Moore, J. J., et al., eds. *Advanced Synthesis of Engineered Structural Materials*. Materials Park, OH: ASM International, pp. 221–225.
349. Radhakrishnan, R., Bhaduri, S., Hanegar, C. H. Jr. (1997). *JOM* 49(1):41–45.
350. Vasudevan, A. K., Petrovic, J. J. (1992). *Mater. Sci. Eng.* A155:1–17.
351. Corrias, A., Ennas, G., Licheri, G., Marongiu, G., Paschina, G. (1991). *Mater. Sci. Eng.* A145:123–125.
352. Corrias, A., Ennas, G., Licheri, G., Marongiu, G., Musinu, A., Paschina, G. (1993). In: deBarbadillo, J. J., et al., eds. *Mechanical Alloying for Structural Applications*. Materials Park, OH: ASM International, pp. 451–456.
353. Calka, A., Radlinski, A. P. (1991). *Appl. Phys. Lett.* 58:119–121.
354. Ruuskanen, P., Heczko, O. (1993). *Key Engng. Mater.* 81–83:159–168.
355. Calka, A., Wexler, D. (2002). *Mater. Sci. For.* 386–388:125–134.
356. Morris, M. A., Morris, D. G. (1991). *J. Mater. Sci.* 26:4687–4696.
357. Corrias, A., Ennas, G., Morangiu, G., Musinu, A., Paschina, G., Zedda, D. (1995). *Mater. Sci. Eng.* A204:211–216.
358. Calka, A., Radlinski, A. P. (1990). *J. Less Common Metals* 161:L23–L26.
359. Savyak, M., Uvarova, I., Timofeeva, I., Isayeva, L., Kirilenko, S. (2000). *Mater. Sci. For.* 343–346:411–416.
360. Millet, P., Hwang, T. (1996). *J. Mater. Sci.* 31:351–355.
361. Matteazzi, P., Basset, D., Miani, F., LeCaër, G. (1993). *Nanostructured Mater.* 2:217–229.
362. Krasnowski, M., Matyja, H. (2000). *Mater. Sci. For.* 343–346:302–307.
363. Tokumitsu, K. (1997). *Mater. Sci. For.* 235–238:127–132.
364. Ivanov, E. (1993). In: deBarbadillo, J. J., et al., eds. *Mechanical Alloying for Structural Applications*. Materials Park, OH: ASM International, pp. 171–176.
365. Ivanov, E., Golubkova, G. V. (1992). *J. Alloys Compounds* 190:L25–L26.
366. Huang, H., McCormick, P. G. (1997). *J. Alloys Compounds* 256:258–262.
367. Oleszak, D., Matyja, H. (2000). *Mater. Sci. For.* 343–346:320–325.
368. Tanaka, T., Nasu, S., Ishihara, K. N., Shingu, P. H. (1991). *J. Less Common Metals* 171:237–247.
369. Tokumitsu, K., Umemoto, M. (2002). *Mater. Sci. For.* 386–388:479–484.
370. Basset, D., Matteazzi, P., Miani, F. (1993). *Mater. Sci. Eng.* A168:149–152.
371. Liu, Z. G., Umemoto, M., Tsuchiya, K. (2002). *Metall. Mater. Trans.* 33A:2195–2203.
372. Umemoto, M., Liu, Z. G., Liu, D. Y., Takaoka, H., Tsuchiya, K. (2002). *Mater. Sci. For.* 386–388:199–204.
373. Tanaka, T., Nasu, S., Nakagawa, K., Ishihara, K. N., Shingu, P. H. (1992). *Mater. Sci. For.* 88–90:269–274.
374. Wang, G. M., Calka, A., Campbell, S. J., Kaczmarek, W. A. (1995). *Mater. Sci. For.* 179-181:201–205.
375. Tokumitsu, K., Umemoto, M. (2001). *Mater. Sci. For.* 360–362:183–188.
376. Campbell, S. J., Wang, G. M., Calka, A., Kaczmarek, W. A. (1997). *Mater. Sci. Eng.* A226-228:75–80.
377. Murphy, B. R., Courtney, T. H. (1994). *Nanostructured Mater.* 4:365–369.
378. Krivoroutchko, K., Kulik, T., Matyja, H., Fadeeva, V. I., Portnoy, V. K. (2001). *Mater. Sci. For.* 360-362:385–390.
379. Gaffet, E., Marco, P., Fedoroff, M., Rouchaud, J. C. (1992). *Mater. Sci. For.* 88–90: 383–390.
380. Sherif El-Eskandarany, M., Sumiyama, K., Suzuki, K. (1995). *J. Mater. Res.* 10:659–667.
381. Malchere, A., Gaffet, E. (1993). In: deBarbadillo, J. J., et al., eds. *Mechanical Alloying for Structural Applications*. Materials Park, OH: ASM International, pp. 297–305.

382. Tomasi, R., Pallone, E. M. J. A., Botta, F., W. J. (1999). *Mater. Sci. For.* 312–314:333–338.
383. Ye, L. L., Quan, M. X. (1995). *Nanostructured Mater.* 5:25–31.
384. Sherif El-Eskandarany, M., Konno, T. J., Sumiyama, K., Suzuki, K. (1996). *Mater. Sci. Eng.* A217/218:265–268.
385. Umemoto, M., Liu, Z. G., Masuyama, K., Tsuchiya, K. (1999). *Mater. Sci. For.* 312–314:93–102.
386. Wu, N. Q., Lin, S., Wu, J. M., Li, Z. Z. (1998). *Mater. Sci. Technol.* 14:287–291.
387. Wanikawa, S., Takeda, T. (1989). *J. Jpn. Soc. Powder Powder Metall.* 36:672–676.
388. Calka, A., Kaczmarek, W. A. (1992). *Scripta Metall. Mater.* 26:249–253.
389. Calka, A., Williams, J. S. (1992). *Scripta Metall. Mater.* 27:999–1004.
390. Calka, A., Nikolov, J. I., Ninham, B. W. (1993). In: deBarbadillo, J. J., et al., eds. *Mechanical Alloying for Structural Applications.* Materials Park, OH: ASM International, pp. 189–195.
391. Wang, G. M., Campbell, S. J., Calka, A., Kaczmarek, W. A. (1997). *J. Mater. Sci.* 32:1461–1467.
392. Xueming, M. A., Gang, J. I. (1996). *J. Alloys Compounds* 245:L30–L32.
393. Wang, G. M., Millet, P., Calka, A., Campbell, S. J. (1995). *Mater. Sci. For.* 179–181: 183–188.
394. Sherif El-Eskandarany, M., Omori, M., Ishikuro, M., Konno, T. J., Takada, K., Sumiyama, K., Suzuki, K. (1996). *Metall. Mater. Trans.* 27A:4210–4213.
395. Eckert, J., Jost, K., De Haas, O., Schultz, L. (1997). *Mater. Sci. For.* 235–238: 133–138.
396. Müller, B., Ledig, L., Hough, D., Oertel, C. -G., Skrotzki, W., Gümbel, A., Eckert, J. (2000). *Mater. Sci. For.* 343–346:689–694.
397. Gümbel, A., Eckert, J., Handstein, A., Skrotzki, W., Schultz, L. (2000). *Mater. Sci. For.* 343–346:924–930.
398. Calka, A., Williams, J. S. (1992). *Mater. Sci. For.* 88–90:787–794.
399. Chen, Y., Gerald, J. F., Williams, J. S., Willis, P. (1999). *Mater. Sci. For.* 312–314: 173–178.
400. Koyano, T., Lee, C. H., Fukunaga, T., Mizutani, U. (1992). *Mater. Sci. For.* 88–90: 809–816.
401. Sherif El-Eskandarany, M., Sumiyama, K., Aoki, K., Suzuki, K. (1992). *Mater. Sci. For.* 88–90:801–808.
402. Calka, A., Nikolov, J. J., Williams, J. S. (1996). *Mater. Sci. For.* 225–227:527–532.
403. Kaczmarek, W. A., Ninham, B. W., Onyszkiewicz, I. (1995). *J. Mater. Sci.* 30:5514–5521.
404. Millet, P., Calka, A., Williams, J. S., Vantenaar, G. J. H. (1993). *Appl. Phys. Lett.* 63:2505–2507.
405. Bab, M. A., Mendoza-Zélis, L., Damonte, L. C. (2001). *Acta Mater.* 49:4205–4213.
406. Mendoza-Zélis, L., Bab, M. A., Damonte, L. C., Sánchez, F. H. (1999). *Mater. Sci. For.* 312–314:179–184.
407. Sherif El-Eskandarany, M., Sumiyama, K., Aoki, K., Masumoto, T., Suzuki, K. (1994). *J. Mater. Res.* 9:2891–2898.
408. Bodart, M., Moret, F., Baccino, R. (1992). In: Capus, J. M., German, R. M., eds. *Advances in Powder Metallurgy and Particulate Materials.* Vol. 7. Princeton, NJ: Metal Powder Industries Federation, pp. 207–219.
409. Li, Z. L., Williams, J. S., Calka, A. (1997). *J. Appl. Phys.* 81:8029–8034.
410. Li, Z. L., Williams, J. S., Calka, A. (1998). *Mater. Sci. For.* 269–272:271–276.
411. Wexler, D., Calka, A., Colburn, S. J. (1998). *Mater. Sci. For.* 269–272:219–224.
412. Chen, Y., Ninham, B. W., Ogarev, V. (1995). *Scripta Metall. Mater.* 32:19–22.
413. Sherif El-Eskandarany, M. (1994). *J. Alloys Compounds* 203:117–126.
414. Suryanarayana, C., Chen, G. H., Froes, F. H. (1992). *Scripta Metall. Mater.* 26: 1727–1732.

415. Saji, S., Neishi, Y., Araki, H., Minamino, Y., Yamane, T. (1995). *Metall. Mater. Trans.* 26A:1305–1307.
416. Chin, Z. H., Perng, T. P. (1997). *Mater. Sci. For.* 235–238:73–78.
417. Chin, Z. H., Perng, T. P. (1997). *Appl. Phys. Lett.* 70:2380–2382.
418. Shen, T. D., Koch, C. C. (1995). *Nanostructured Mater.* 5:615–629.
419. Ogino, Y., Miki, M., Yamasaki, T., Inuma, T. (1992). *Mater. Sci. For.* 88–90:795–800.
420. Ogino, Y., Yamasaki, T., Miki, M., Atsumi, N., Yoshioka, K. (1993). *Scripta Metall. Mater.* 28:967–971.
421. Guo, W., Martelli, S., Padella, F., Magini, M., Burgio, N., Paradiso, E., Franzoni, U. (1992). *Mater. Sci. For.* 88–90:139–146.
422. Miki, M., Yamasaki, T., Ogino, Y. (1993). *Mater. Trans. Jpn. Inst. Metals* 34:952–959.
423. Aoki, K., Memezawa, A., Masumoto, T. (1992). *Appl. Phys. Lett.* 61:1037–1039.
424. Sakurai, K., Lee, C. H., Kuroda, N., Fukunaga, T., Mizutani, U. (1994). *J. Appl. Phys.* 75:7752–7755.
425. Calka, A. (1991). *Appl. Phys. Lett.* 59:1568–1569.
426. Kuhrt, C., Schropf, H., Schultz, L., Arzt, E. (1993). In: deBarbadillo, J. J., et al., eds. *Mechanical Alloying for Structural Applications.* Materials Park, OH: ASM International, pp. 269–273.
427. Bampton, C. C., Rhodes, C. G., Mitchell, M. R., Vassiliou, M. S., Graves, J. A. (1990). In: Yavari, A. R., Desré, P., eds. *Multilayer Amorphisation by Solid-State Reaction and Mechanical Alloying.* Vol. 51. Colloq. C4., Suppl. 14. Les Ulis Cedex, France: Les editions de Physique, pp. 275–282.
428. Fadeeva, V. I., Sviridov, I. A., Kochetov, G. A., Baldokhin, Yu. V. (2000). *Mater. Sci. For.* 343–346:296–301.
429. Gaffet, E., Malhouroux-Gaffet, N., Abdellaoui, M. (1993). *J. Alloys Compounds* 194: 339–360.
430. Malhouroux-Gaffet, N., Gaffet, E. (1993). *J. Alloys Compounds* 198:143–154.
431. Umemoto, M., Shiga, S., Okane, I. (1993). In: Henein, H., Oki, T., eds. *Processing Materials for Properties.* Warrendale, PA: TMS, pp. 679–682.
432. Umemoto, M., Shiga, S., Raviprasad, K., Okane, I. (1995). *Mater. Sci. For.* 179–181:165–170.
433. Yelsukov, E. P., Fomin, V. M., Nemtsova, O. M., Shklyaev, D. A. (2002). *Mater. Sci. For.* 386–388:181–186.
434. Garcia-Escoral, A., Adeva, P., Cristina, M. C., Martin, A., Carmona, F., Cebollada, F., Martin, V. E., Leonato, M., Gonzalez, J. M. (1991). *Mater. Sci. Eng.* A134:1394–1397.
435. Muñoz-Palos, J. M., Cristina, M. C., Adeva, P. (1996). *Mater. Trans. Jpn. Inst. Metals* 37:1602–1606.
436. Riffel, M., Schilz, J. (1998). *J. Mater. Sci.* 33:3427–3431.
437. Umemoto, M., Liu, Z. G., Omatsuzawa, R., Tsuchiya, K. (2000). *Mater. Sci. For.* 343–346:918–923.
438. Yen, B. K., Aizawa, T., Kihara, J. (1997). *Mater. Sci. For.* 235–238:157–162.
439. Iwamoto, N., Uesaka, S. (1992). *Mater. Sci. For.* 88–90:763–770.
440. Fei, G. T., Liu, L., Ding, X. Z., Zhang, L. D., Zheng, Q. Q. (1995). *J. Alloys Compounds* 229:280–282.
441. Ma, E., Pagan, J., Cranford, G., Atzmon, M. (1993). *J. Mater. Res.* 8:1836–1844.
442. Lee, P. Y., Chen, T. R., Yang, J. L., Chin, T. S. (1995). *Mater. Sci. Eng.* A192/193: 556–562.
443. Gaffet, E., Malhouroux-Gaffet, N. (1994). *J. Alloys Compounds* 205:27–34.
444. Perdigão, M. N. R. V., Jordão, J. A. R., Kiminami, C. S., Botta Filho, W. J. (1997). *Mater. Sci. For.* 235–238:151–156.
445. Kumar, K. S., Mannan, S. K. (1989). In: Liu, C. T., Taub, A. I., Stoloff, N. S., Koch, C.

C., eds. *High Temperature Ordered Intermetallic Alloys III*. Vol. 133. Pittsburgh, PA: Mater. Res. Soc., pp. 415–420.
446. Viswanadham, R. K., Mannan, S. K., Kumar, K. S. (1988). *Scripta Metall.* 22:1011–1014.
447. Omuro, K., Miura, H. (1991). *Jpn. J. Appl. Phys.* 30:L851–853.
448. Zhang, D. L., Massalski, T. B. (1994). *J. Mater. Res.* 9:53–60.
449. Oehring, M., Bormann, R. (1991). *Mater. Sci. Eng.* A134:1330–1333.
450. Zotov, N., Parlapanski, D. (1994). *J. Mater. Sci.* 29:2813–2820.
451. Bokhonov, B., Ivanov, E., Boldyrev, V. V. (1993). *J. Alloys Compounds* 199:125–128.
452. Lü, L., Lai, M. O., Su, Y., Teo, H. L., Feng, C. F. (2001). *Scripta Mater.* 45:1017–1023.
453. Brand, K., Suryanarayana, C., Kieback, B. F., Froes, F. H. (1996). In: Cadle, T. M., Narasimhan, K. S., eds. *Advances in Powder Metallurgy and Particulate Materials—1996*. Vol. 2. Princeton, NJ: Metal Powder Industries Federation, pp. 49–58.
454. Brand, K., Suryanarayana, C., Kieback, B. F., Froes, F. H. (1996). *Mater. Sci. For.* 225–227:471–476.

11

Disordering of Intermetallics

11.1 INTRODUCTION

It has long been known that materials containing partially ordered phases are stronger than those with wholly disordered or fully ordered phases. This is due to the fact that at a certain value of the long-range order parameter, S, superdislocations separate into unlinked singles. Thus, it is of interest to study the mechanical behavior of materials in various states of partial order. Disordering phenomena of ordered alloys have also been studied in an effort to understand the mechanism of disordering and to investigate the different methods by which the disordered phase could be retained at room temperature. Disordered intermetallics are of interest because they could have a higher ductility, and consequently higher formability, than their ordered counterparts.

It was mentioned in Chapter 10 that intermetallic compounds exist mostly in the ordered state and that they suffer from unacceptably low levels of ambient temperature ductility. This has been the major obstacle in preventing their widespread use. The room temperature ductility of intermetallics could be increased either by disordering them or by refining their grain sizes [1,2]. Several techniques have been employed to disorder the ordered intermetallics. These include irradiation [3,4], rapid solidification processing [5,6], laser quenching [7], vapor quenching [8], chemical vapor deposition on a cold finger [9], or heavy plastic deformation [10]. This chapter deals with the phenomenon of disordering of ordered intermetallic compounds by mechanical milling (MM), a process of heavy plastic deformation. It will be shown that the technique of MM could lead to grain refinement (producing even nanocrystalline materials), disordering of the ordered lattice, and synthesis of metastable phases. All of these effects, either singly or in combination, can lead to enhanced ductility of disordered intermetallics.

Large amounts of plastic deformation result in the generation of a variety of defect structures (dislocations, vacancies, stacking faults, grain boundaries, etc.), and these destabilize the ordered nature of the lattice leading to the formation of a disordered phase. The disordered phase could be either a crystalline phase (solid solution or an intermetallic with a different crystal structure) or, if the disordering is very severe, an amorphous phase could form. This chapter discusses only the formation of disordered crystalline phases from the ordered intermetallic phases. Formation of amorphous phases by MM of intermetallics is discussed in Chapter 12 and so will not be treated further here.

When an ordered intermetallic undergoes heavy deformation, one or more of the following transformations could occur:

1. The grain size could be refined down to nanometer dimensions;
2. Chemical long-range order (LRO) may decrease or completely disappear as indicated by a decrease in the intensity or complete absence of the superlattice reflections;
3. Chemical short-range order (SRO) may decrease and form a completely random solid solution;
4. A change in crystal structure could occur; or
5. An amorphous phase could form.

The first observation of disordering of an ordered compound by MM was reported to occur in $ZnFe_2O_4$ by Ermakov et al. in 1982 [11]. Disordering of Fe_3Si by mechanical grinding was reported later in 1983 [12]. Formation of amorphous phases by MM of ordered intermetallics was reported even earlier in 1981 [13]. A majority of the work in this area of disordering of intermetallics by MM has been carried out by Hans Bakker and his colleagues at the Van der Waals–Zeeman Laboratory at the University of Amsterdam in The Netherlands. They published a comprehensive review on this aspect of disordering of intermetallics a few years ago [14].

11.2 METHODOLOGY

A variety of ordered intermetallic compounds with the B2 (CoZr, NiAl, RuAl, etc.), $L1_2$ (Ni_3Si, Fe_3Ge, Zr_3Al, etc.), A15 (Nb_3Al, V_3Ga, Nb_3Au, etc.), B8 (Fe_3Ge_2, Mn_3Sn_2, etc.), and other crystal structures have been disordered by MM. The prealloyed intermetallics were subjected to MM in a vibratory ball mill under vacuum. The mill consisted of a stainless steel container with a hardened steel bottom, the central part of which consisted of a tungsten carbide disk. A single hardened steel ball was kept in motion by a vibrating frame on which the container was mounted. A few grams of the powder were milled at a time.

The progress of disordering has been monitored by several methods including X-ray diffraction (XRD) techniques to measure the lattice parameter and long-range order parameter, S, and by measurement of superconducting transition temperature and magnetic susceptibility (if the compound is superconducting in the initial state), Mössbauer measurements, differential scanning calorimetry and differential thermal analysis, and microhardness. In addition, AC magnetic susceptibility measurements and magnetization measurements in high fields and also as a function of temperature have been very useful in providing valuable information.

11.3 TYPES OF DEFECTS GENERATED DURING DISORDERING AND THERMODYNAMIC STABILITY

The intense plastic deformation associated with mechanical alloying (MA) and MM can introduce a variety of defects into the ordered lattice, which raise the free energy of the system. Furthermore, disordering of the lattice also can raise the free energy of the system. Thus, the difference in the free energy of the system before and after milling, ΔG (milling), can be represented as

$$\Delta G \text{ (milling)} = \Delta G \text{ (disorder)} + \Delta G \text{ (grain boundaries and other defects)} \tag{11.1}$$

where ΔG (disorder) is the increase in free energy due to the disordering of the lattice and ΔG (grain boundaries and other defects) is the increase in free energy due to the introduction of defects such as grain boundaries, stacking faults, and so on. Consequently, the type of phase produced (solid solution, disordered intermetallic, or amorphous phase) is determined by the relative position of the free energy of the product phase with respect to the several possible competing phases and the parent phase. Thus, as shown in Figure 11.1, if the free energy of the disordered intermetallic exceeds the free energy of the amorphous phase, then the amorphous phase becomes more stable. On the other hand, if the free energy of the disordered intermetallic is lower than that of the amorphous phase, then the disordered intermetallic is the more stable phase. If the solid solution phase has a lower free energy than either of the other two competing phases (disordered intermetallic and the amorphous phase), then the solid solution becomes stable. While a detailed analysis of the nature and density of

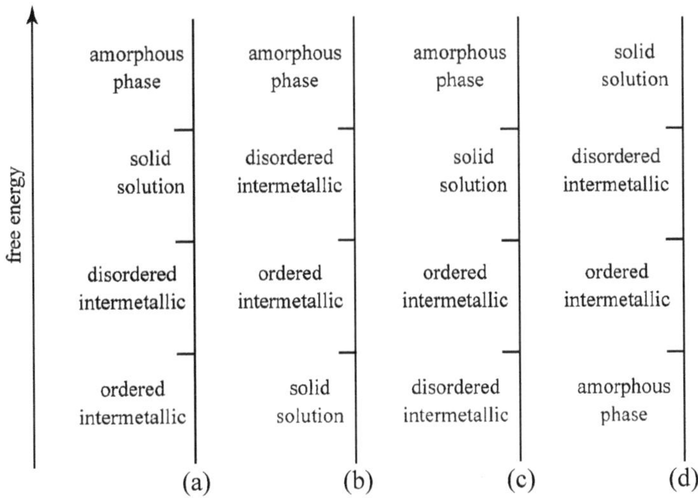

Figure 11.1 Relative positions of the free energies of competing phases in the system. Assuming that solid solution, ordered intermetallic, disordered intermetallic, and amorphous phase are the competing phases, then the stability of the phase is decided by its free energy relative to that of the others. For example, in (a), the ordered intermetallic is more stable, in (b) it is the solid solution, in (c) it is the disordered intermetallic, and in (d) it is the amorphous phase.

defects introduced into the system is beyond the scope of this book, one can briefly discuss how the disordering of the lattice could take place by the introduction of defects.

The atomic disorder in an intermetallic can be manifested in different ways [14,15]:

Antisite disorder
Triple-defect disorder
Quadruple-defect disorder, and
Redistribution of interstitials

11.3.1 Antisite Disorder

Let us consider the example of β-brass (CuZn) intermetallic as a typical example of materials crystallizing in the B2 structure (Figure 11.2). This structure can be considered to consist of two sublattices—α sublattice occupied by the Cu atoms and the β sublattice occupied by the Zn atoms, each forming a primitive cubic lattice. (Alternatively, the α sublattice can be occupied by the Zn atoms and the β sublattice by the Cu atoms. There is no difference between these two situations.) Thus, the B2 structure can be considered as made up of two interpenetrating primitive cubic sublattices.

At 0 K, the alloy is completely in the ordered state and the component atoms are expected to occupy the "right" lattice sites. However, occupancy of the lattice sites could be different at elevated temperatures. A majority of the atoms continue to occupy the right lattice sites at temperatures higher than 0 K, and almost up to the order–disorder transformation temperature, T_c. Above T_c, the alloy is completely disordered, i.e., the component Cu and Zn atoms are randomly distributed in equal proportions over both sublattices. Occupation of the "wrong" sublattices by the atomic species at temperatures below T_c, is referred to as *antisite disorder*. For example, in the above case of β-brass (CuZn), occupation of the β sublattice by Cu atoms and occupation of the α sublattice by Zn atoms is considered to be antisite disorder. Figure 11.3 shows a schematic of the antisite disorder in a B2 compound. Figure 11.3a shows a perfectly ordered arrangement of the component atoms, while

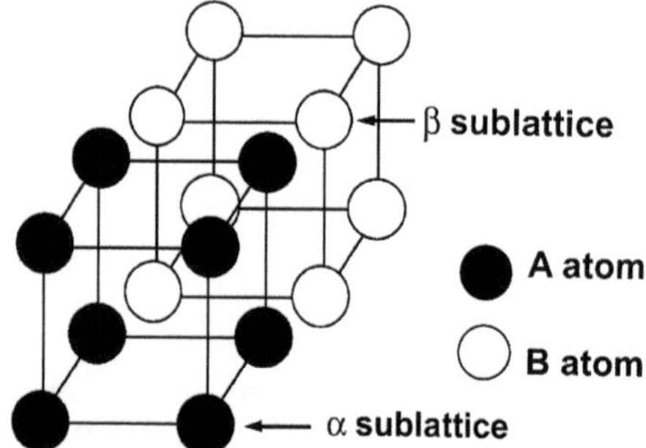

Figure 11.2 The B2 structure exhibited by ordered intermetallic compounds such as CuZn, NiAl, and CsCl. The structure can be visualized as consisting of two interpenetrating primitive cubic sublattices, one made up of Cu (or Ni or Cs) atoms and the other Zn (or Al or Cl).

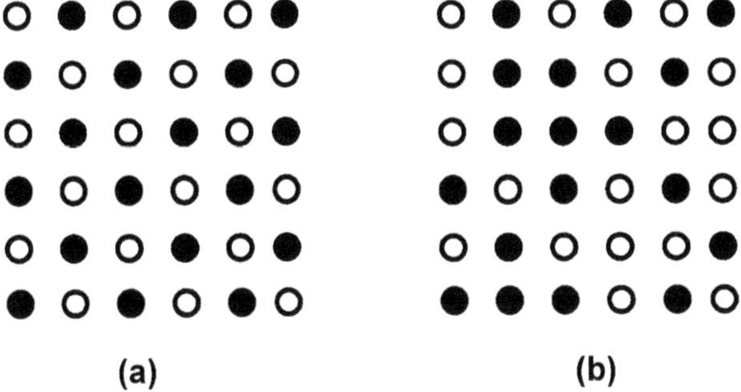

Figure 11.3 Schematic of antisite disorder in a B2-type compound. (a) represents a completely ordered structure and (b) shows the antisite order. Notice that some of the "black" atoms are occupying the "white" sublattice.

Figure 11.3b represents the situation when some of the constituent atoms occupy the wrong sites. Even though the lattice is not completely disordered, a significant part of the strain resides in the disordered lattice; this is because atoms on the wrong sublattice are to be accommodated on lattice sites, where they do not fit. This introduces strain into the lattice. The presence of antisite disorder can be monitored by following the changes in lattice parameter and/or degradation in the superconducting transition temperature, T_c. A consequence of antisite disorder in the lattice is an increase in the lattice parameter. This type of disorder is observed in almost every type of ordered alloy. Some examples are a number of mechanically milled compounds with the A15 structure (Nb_3Sn, V_3Ga, NiV_3), $L1_2$ structure (Ni_3Al, Ni_3Si, Fe_3Ge), and B2 structure (CoGa, CoAl, AlRu).

11.3.2 Triple-Defect Disorder

A different type of disorder has been observed in some ordered intermetallic compounds. In this case only one of the two components can be accommodated on the wrong sublattice, while compensating vacancies are formed on the other sublattice. In an equiatomic compound such as CoGa, for example, the transition metal Co atoms can substitute on the Ga sublattice, resulting in an antisite disorder. However, the Ga atoms stay on their own sublattice. The requirement of equal number of sublattice sites then dictates that for each wrong Co atom, two vacancies are formed on the Co sublattice. Thus, vacancies on the Co sublattice in combination with Co antisite atoms in a ratio of 2:1 constitute the *triple-defect disorder*. This type of defect is shown schematically in Figure 11.4. It has the important consequence that CoGa will not show a transformation to a different structure upon prolonged milling because Ga atoms, remaining on their own sublattice, "freeze" or "lock" the original B2 structure. On the other hand, intermetallic compounds that exhibit antisite disorder of both components usually transform by milling to a disordered structure characteristic of a random arrangement of atoms, i.e., either to a solid solution of one component in the lattice structure of another, or to the amorphous state.

A clear indication of the vacancy-type disorder is the decrease of the lattice parameter because of relaxation around the high concentration of vacancies, whereas the antisite disorder is accompanied by an increase in the lattice parameter.

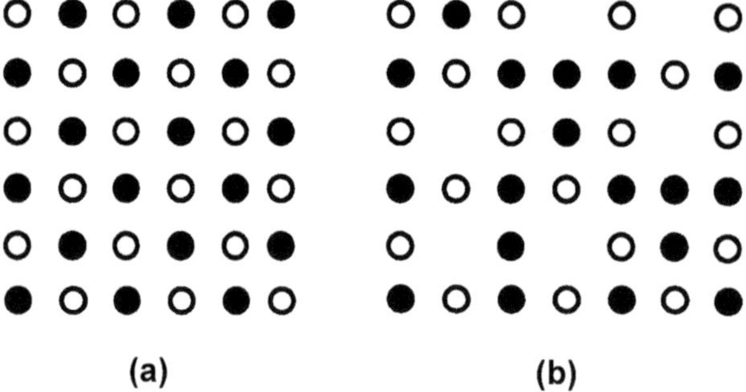

Figure 11.4 Schematic of triple-defect disorder. (a) A completely ordered structure and (b) a triple-defect disorder. Notice that the "black" atoms are on the "white" sublattice and that two vacancies are present on the "black" sublattice.

In addition to a change in the lattice parameter (either decrease or increase), formation of these defects can also affect the magnetic behavior of the compound. For example, it was reported that the B2 $Fe_{60}Al_{40}$ compound, which is paramagnetic at $T \geq 300$ K, becomes strongly ferromagnetic in the disordered bcc state as the Fe-Fe nearest-neighbor population increases [16]. On the other hand, heavy deformation of equiatomic FeRh, which also has the B2 structure, transformed to a metastable fcc structure, similar to the high-temperature γ-solid solution [17]. While the B2-FeRh is strongly ferromagnetic above room temperature, its metastable γ-fcc polymorph is paramagnetic, with spin-glass behavior at low temperatures [18].

11.3.3 Quadruple-Defect Disorder

The vacancy-type disorder related to that mentioned above can also occur in other intermetallic compounds. For example, in the C15 compounds with the AB_2 structure, one could visualize the occurrence of *quadruple-defect disorder*. This involves a combination of three vacancies on the β sublattice, the sublattice that is mainly occupied by the B atoms, and one antisite B atom on the α sublattice, the sublattice that is mainly occupied by A atoms [19,20].

In principle, one could also consider the presence of higher order vacancy-type defects. These vacancy-type defects (triple, quadruple, etc.) preserve the correct stoichiometry as well as the given ratio of the sublattice sites.

Miedema analysis [21] enables calculation of the enthalpy associated with the different type of defects produced in the intermetallics. Their relative values will then determine the actual type of defect created in the mechanically alloyed/milled material, noting that the enthalpy of the milled material is going to be slightly different in the annealed and milled condition for the same material. For example, Modder et al. [15] estimated the enthalpies for formation of antisite disorder and quadruple defects in AB_2-type compounds (Table 11.1). It may be noted that in all the compounds except $GdMg_2$, the enthalpy for quadruple-defect formation is lower than that for antisite disorder. MM of these compounds was conducted and the XRD technique was utilized to detect the nature of the phases present and calculation of the lattice parameter. The presence of antisite disorder was inferred from the change in the nature of the phase and the quadruple defect by the decrease in the lattice parameter.

Table 11.1 Estimated Enthalpies for the Formation of Antisite Disorder and Quadruple Defects in C15-type GdX_2 Intermetallic Compounds

Compound	ΔH for quadruple defects, kJ/mole (per defect)	ΔH for antisite disorder, kJ/mole
$GdAl_2$	55	230
$GdPt_2$	35	320
$GdIr_2$	23	310
$GdRh_2$	40	330
$GdMg_2$	70	48

Source: Data from Ref. 15.

In confirmation of the predictions made by the Miedema model, experimentally it was shown that the $GdAl_2$, $GdPt_2$, $GdIr_2$, and $GdRh_2$ compounds generate quadruple-defect disorder whereas $GdMg_2$ generates the antisite disorder. Even though the model only gives estimates of the enthalpy values, it provides a very good predictive capability regarding the type of defect produced in mechanically milled materials.

11.3.4 Redistribution of Interstitials

Assuming that atoms are spherical, their arrangement into different crystal structures produces interstices (holes) between the atoms, with their sizes smaller than those of the atoms. There are two types and sizes of interstices in the different crystal structures, and these are termed tetrahedral and octahedral interstices. In one type, four atoms surround an interstice in the shape of a tetrahedron and hence this is referred to as a tetrahedral interstice. In the other, six atoms surround an interstice in the shape of an octahedron; hence this is referred to as an octahedral interstice. Figure 11.5 shows these two types of interstices in a face-centered cubic (fcc) structure. Similar types of interstices exist in the hexagonal close-packed (hcp) and body-centered cubic (bcc) structures also. The only difference between the interstices in these structures is in their coordinates, number, and size. The properties of these interstices are summarized in Table 11.2.

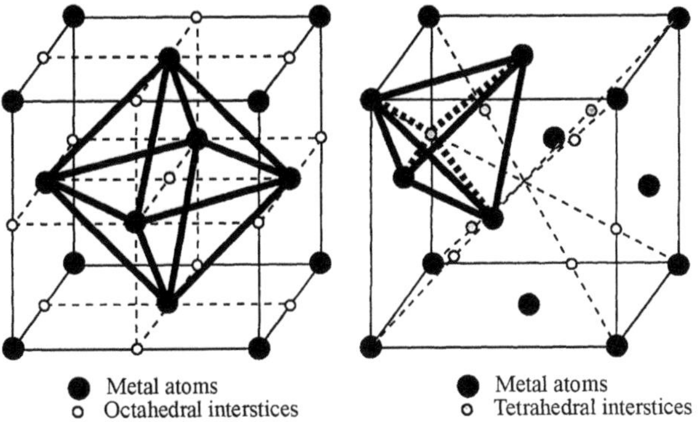

Figure 11.5 Schematics of the tetrahedral and octahedral interstices in a face-centered cubic lattice.

Table 11.2 Coordinates, Size, and Number of Tetrahedral and Octahedral Interstices in the bcc, fcc, and hcp Structures[a]

Property	BCC	FCC	HCP
Coordinates of the tetrahedral sites	½, ¼, 0, and equivalent sites	¼, ¼, ¼, and equivalent sites	0, 0, ⅜; 0, 0, ⅝; ⅓, ⅔, ⅛; ⅓, ⅔, ⅞
Coordinates of the octahedral sites	0, 0, ½; ½, ½, 0, and equivalent sites	0, 0, ½; ½, ½, ½, and equivalent sites	⅓, ⅔, ¼; ⅔, ⅓, ¾
Size of tetrahedral site (r_i/r_A)	0.291	0.225	0.225
Size of octahedral site (r_i/r_A)	0.155	0.414	0.414
Number of tetrahedral sites per host atom	6	2	2
Number of octahedral sites per host atom	3	1	1

[a] The coordinates are given with reference to the unit cell of the host lattice. The sizes of the interstitial sites are given in terms of the radius ratio (r_i/r_A) where r_i is the radius of the largest atom that can "fit" into the interstitial position and r_A is the radius of the host atoms. The number of the interstitial sites is given in terms of the number of sites per host atom.

New structures could be generated by placing atoms of a different nature at the interstitial sites. Depending on which of the interstices and how many of them are "filled," completely different structures could be produced. For example, the cubic ZnS structure can be produced by "filling" half of the tetrahedral interstitial sites in a hypothetical fcc lattice of Zn with S atoms (ions). A CaF_2-type structure can be produced by filling all the tetrahedral interstitial sites with F atoms (ions) in an fcc lattice of Ca atoms (ions). It has been reported that some of the atoms could change their positions in the mechanically milled materials and thus form new structures with different properties. It was reported that the magnetization behavior of the Fe_3Ge_2 and Mn_3Sn_2 structures is completely opposite for both the compounds, even though both of them have the same Ni_3Sn_2 structure in the ordered condition. Whereas the magnetization in ferromagnetic Fe_3Ge_2 increases on disordering, that for the ferrimagnetic Mn_3Sn_2 decreases. This has been explained based on the concept that the transition metal atoms are transferred during milling from the fully occupied octahedral sites to the empty tetrahedral sites [14]. Thus, redistribution of the interstitials could lead to some type of disordering.

Antisite disorder and triple-defect disorder have also been observed in compounds disordered by quenching from high temperatures [22] as well as by irradiation [23], mostly in the superconducting compounds. However, redistribution of interstitials appears to be quite unique to mechanically milled materials, suggesting that MM can introduce a very unique type of disordering in ordered lattices.

In addition to the above types of disorder created in the lattice, grain refinement (often resulting in the formation of nanostructured material) increases the grain boundary area and this also raises the free energy of the system. Furthermore, introduction of lattice strain into the crystal and other types of crystal defects also raise the free energy of the system. Thus, the sum of the energy of all these effects (disordering, creation of grain boundaries, and other defects) will be the total energy introduced into the material during milling.

11.4 THEORETICAL BACKGROUND

It has been reported that upon milling, the long-range order parameter (S) in the intermetallic is gradually reduced and, in many cases, the material may become totally disordered ($S = 0$), e.g., Ni_3Al [24,25]. Figure 11.6 shows the variation of S with milling time in Ni_3Al and Ni_3Al-B powders. Even though the kinetics of disordering are different, the S value reaches 0 in both the cases after milling for sufficient time, i.e., 5 h for Ni_3Al and 8 h for Ni_3Al-B powders. In other cases, S is reduced with milling time but does not reach 0, i.e., partial order and partial disorder coexist in compounds such as CuTi [26] and AlRu [27]. In other cases, the S value does not decrease at all and is maintained at $S = 1$; but, with continued milling the material becomes amorphous, e.g., in CoZr [28]. These three situations are schematically represented in Figure 11.7. Thus, upon milling, an ordered intermetallic can transform, with or without complete loss of long-range order, either into a disordered crystalline phase (solid solution) or an amorphous phase. If the product is a crystalline phase, the material has an extremely fine grain size, usually in the nanometer range. Figure 11.8 summarizes the types of changes that can occur on milling an ordered intermetallic compound [29,30].

Even if the intermetallic does not become completely disordered on milling under "normal" conditions, it could be made to fully disorder by changing the process parameters. It has been reported by Poecht et al. [31] that the FeAl intermetallic did not become completely disordered, i.e., $S \neq 0$, on milling in a Fritsch P0 vibratory ball mill at an intensity of 1000 ms^{-2} at room temperature. However, by increasing the milling intensity, the authors reduced the value of S to zero (Figure 11.9). The NiAl intermetallic also did not become fully disordered on milling. Changing the milling parameters also did not have an effect in decreasing the value of S to zero [32,33]. This may be related to the high value of ΔH_{ord}, the ordering enthalpy. It has been demonstrated that ternary alloying additions, e.g., Fe and Cr, to the extent of 18 at% or more could completely disorder the NiAl compound [34] (Fig. 11.10a). It has also been noted that the S value reached zero faster when Fe was added than when Cr was added [35]. This was explained by the fact that Fe could easily dissolve in NiAl due

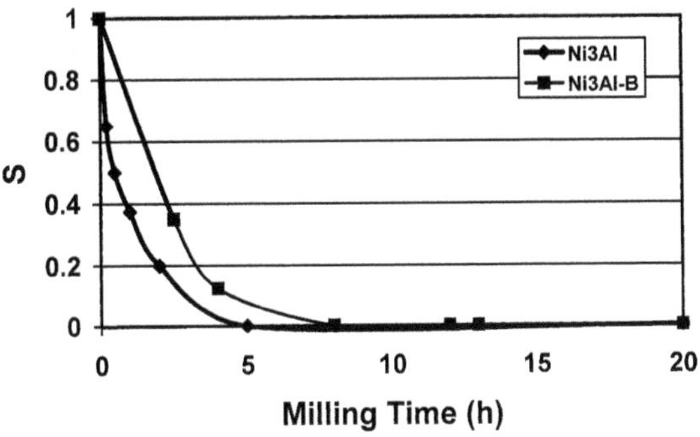

Figure 11.6 Variation of the long-range ordered parameter, S, as a function of milling time in Ni_3Al [24] and Ni_3Al-B [25] powders.

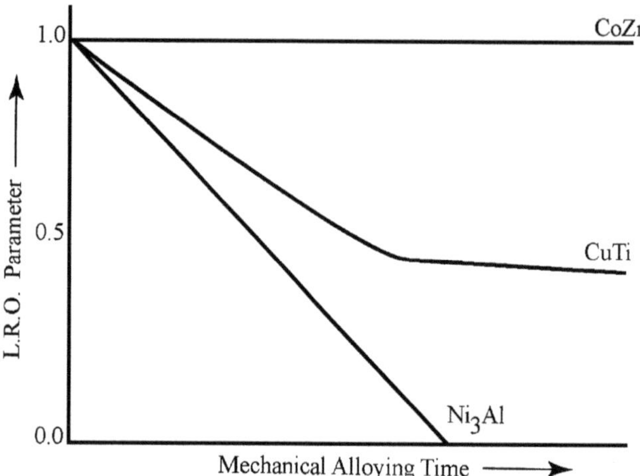

Figure 11.7 Variation of long-range order parameter, S, with milling time. Three different situations, i.e., complete loss of order (Ni_3Al), partial loss of order (CuTi), and no loss of order (CoZr) on milling, have been represented.

to the isomorphous nature of FeAl and NiAl, which is not the case with Cr (Fig. 11.10b).

Disordering of intermetallics has been studied primarily by XRD techniques [36]. The presence of superlattice reflections in the XRD pattern confirms the presence of ordering in the lattice. For example, in a B2-type of structure, the superlattice reflections are those for which $(h + k + l)$ is odd, whereas $(h + k + l)$ is even for the fundamental reflections. Here h, k, and l are the Miller indices of the planes. The intensity of the superlattice peaks decreases with increasing milling time due to an increase in the amount of disordering (decrease of long-range order parameter, S). Figure 11.11 shows a series of XRD patterns of the mechanically milled FeAl ordered compound [37]. One can notice a decrease in the intensity of the superlattice reflections with increasing milling time due to a decrease in S. In addition, both the fundamental and superlattice reflections show peak broadening due to a decrease in crystallite size and increase in lattice strain. After sufficient milling, the superlattice reflections completely disappear, indicating that the compound has become completely disordered. Therefore, the degree of disordering can be estimated by measuring the value of S at different stages of milling, which in turn can be calculated by measuring the integrated intensity of the superlattice reflections relative to those of the fundamental reflections. The value of the long-range order parameter, S, is then calculated using the relationship:

$$S^2 = \frac{I_{s(dis)}/I_{f(dis)}}{I_{s(ord)}/I_{f(ord)}} \tag{11.2}$$

where I_s and I_f represent the integrated intensities of the superlattice and fundamental reflections, respectively, and the subscripts (dis) and (ord) refer to the partially disordered and fully ordered states, respectively [36].

It has also been reported that the lattice parameter of the alloy increases slightly (0.3–0.8%) in the disordered state, if disordering occurs by antisite disorder [38] (Fig. 11.12), whereas the lattice parameter decreases if disordering occurs by the vacancy-

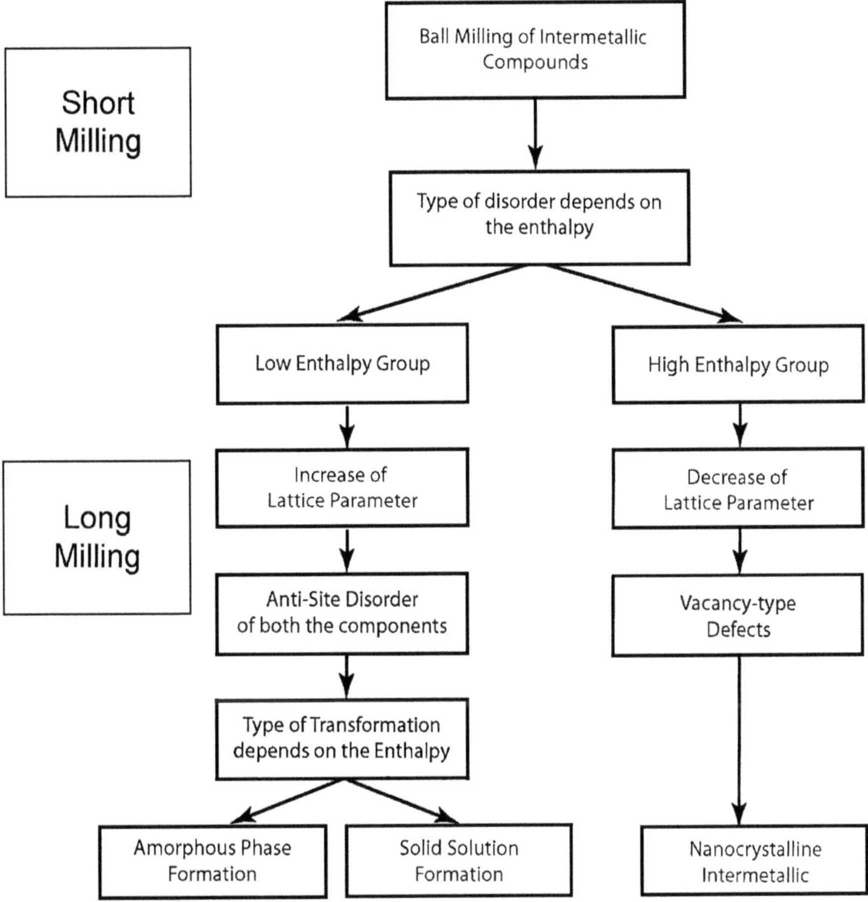

Figure 11.8 Schematic of the expected situation on mechanically milling an intermetallic compound.

type defects [14]. Lattice expansion is a natural consequence of disordering, which destroys the best packing of atoms realized in the completely ordered alloy. The magnitude of increase or decrease of lattice parameter depends on the relative difference in the atomic sizes of the two component atoms. For example, the increase in the lattice parameter for Nb_3Sn is much larger than in Nb_3Au due to the fact that the Nb and Au atoms are more similar in size than the Nb and Sn atoms. The lattice parameter continues to increase even after the compound has been completely disordered and formed a solid solution as indicated for the mechanically milled Ni_3Al-B powder (Fig. 11.13) [39]. One of the explanations offered for this behavior is the progressive reduction in short-range order that is still present when the long-range order has been completely destroyed in the alloy [25]. A similar situation was found earlier in the Cu_3Au alloy quenched from different temperatures above the order–disorder transition temperature [40]. As will be discussed later (Chapter 15), contamination of the milled powder by impurities from the milling atmosphere and/or milling tools can also increase the lattice parameter of the alloy. A confirmation that

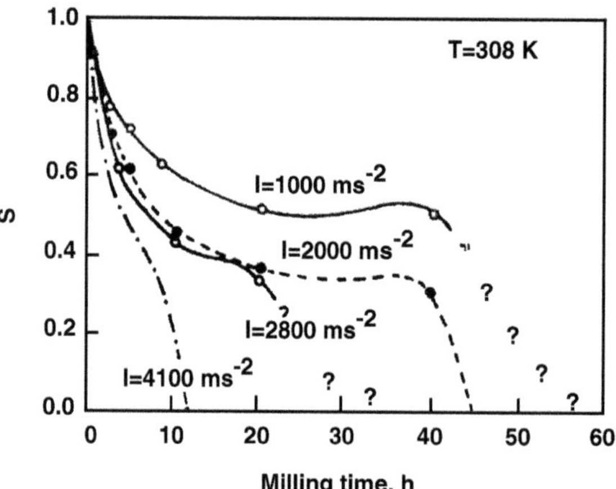

Figure 11.9 Effect of milling intensity on the long-range order parameter, S, of mechanically milled FeAl [31]. Note that while the S value did not reach zero at a milling intensity of 1000 ms^{-2}, it reached 0 when the milling intensity was raised to 4100 ms^{-2}.

this is due to a progressive reduction in the short-range order and not a contamination effect will be when the increased lattice parameter is completely recovered by annealing the samples. An inverse linear relationship has been shown to exist between the lattice expansion and the relative long-range order parameter (Fig. 11.14).

Yavari et al. [41] made an interesting observation during a study of the disordering of FeAl by MM. They noted that while there was gradual loss of long-range ordering in a coarse-grained alloy, fully ordered and fully disordered regions were found to coexist in the nanocrystalline FeAl intermetallic. This was inferred by deconvoluting the 110 peak of the $Fe_{60}Al_{40}$ compound milled for 30 min (Fig. 11.15). The peak appears to consist of two peaks, one positioned at the diffraction angle corresponding to that of the fully ordered B2 phase and the other corresponding to the fully disordered bcc solid solution.

The explanation provided for this observation is as follows. When a coarse-grained intermetallic is milled, disordering and other transformations start at the deformation bands as dislocations move, pile up, and eventually lead to grain refinement. Under these circumstances, the measured properties, such as lattice parameters, defect densities, and magnetizations, reflect only the average values since the intermetallic contains regions, which are partially ordered and partially disordered. In this situation, the properties show a continuous variation between those of the completely ordered and completely disordered states. However, in those cases when one starts with a nanocrystalline ordered alloy, the disordered regions that form during milling are of the order of the grain size. Therefore, it is possible to have some grains that are completely disordered and some other grains that are completely ordered. Thus, the lattice parameter of the ordered state changes discontinuously to that of the disordered state with the destruction of long-range order of the B2 structure.

It may also be pointed out in this context that during MM both loss of long-range order and grain refinement take place simultaneously. The difference is that the

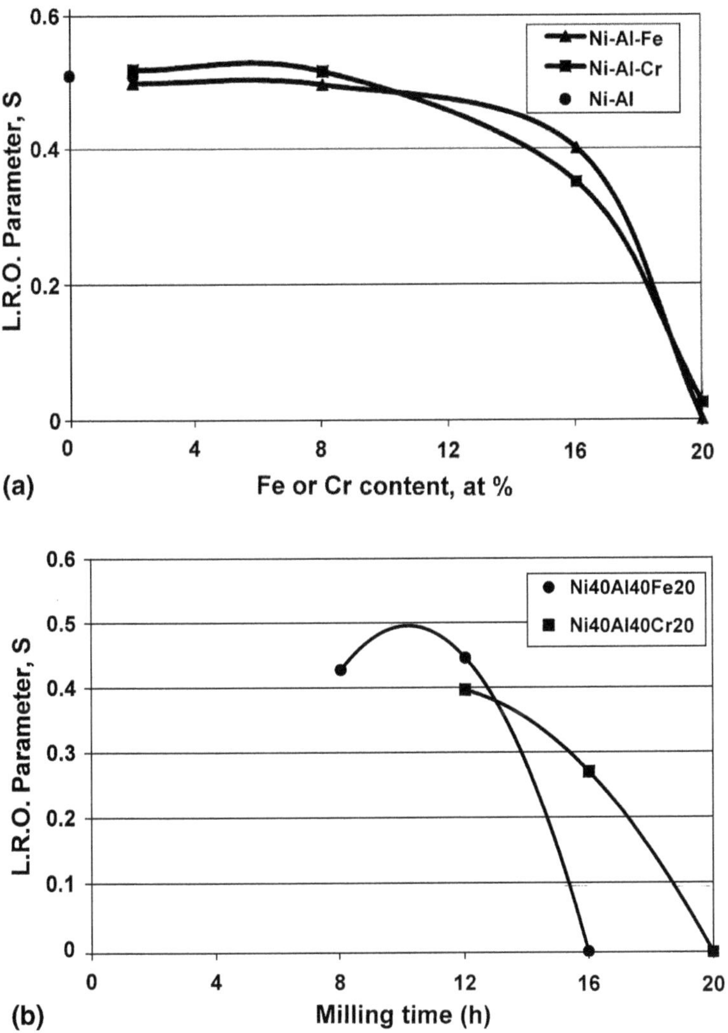

Figure 11.10 (a) Effect of ternary additions of Fe and Cr on long-range order parameter, S, of mechanically milled NiAl [34]. (b) Change in the long-range order parameter, S, of mechanically milled NiAl(Cr) and NiAl(Fe) as a function of milling time [35].

long-range order parameter is less than 1 in the disordered state. Thus, during disordering of nanocrystalline FeAl intermetallic, the lattice parameter seems to increase suddenly for a particular time of milling, and this increased lattice parameter corresponds to that of the disordered FeAl (Fig. 11.16), explaining the coexistence of fully ordered ($S = 1$) and fully disordered ($S = 0$) regions in the milled material.

Measurement of change in lattice parameter is not possible if there is a change in the crystal structure due to disordering. Mössbauer spectroscopy techniques and measurement of superconducting transition temperature and/or magnetic susceptibility have also been employed to study the disordering phenomenon. Recently, Zhou

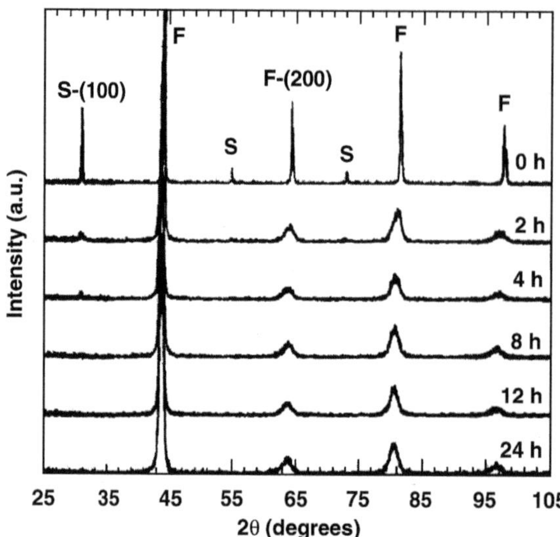

Figure 11.11 X-ray diffraction patterns of the ordered FeAl compound as a function of milling time. Notice that the intensity of the superlattice reflections (marked S) decreases with increasing milling time and that they eventually disappear. The reflections marked F are fundamental [37].

and Bakker [42] demonstrated that magnetic measurements are very powerful in determining the nature of disordering if one of the atoms involved has magnetic moments.

Cho and Koch [28] and Zhou and Bakker [42] studied the disordering behavior of the CoZr compound having the B2 structure by MM. Even though, with milling time, the intensities of both the fundamental and superlattice XRD peaks decreased

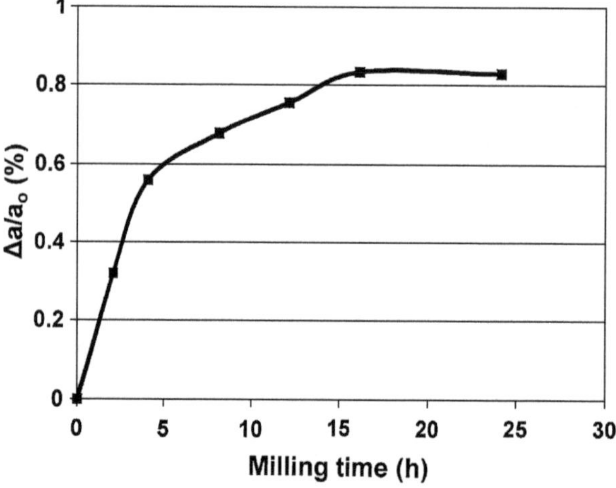

Figure 11.12 Increase in lattice parameter with milling time for the FeAl powder. The increase is attributed to antisite disorder, wherein the larger aluminum atoms are accommodated on the smaller iron atom sites [38].

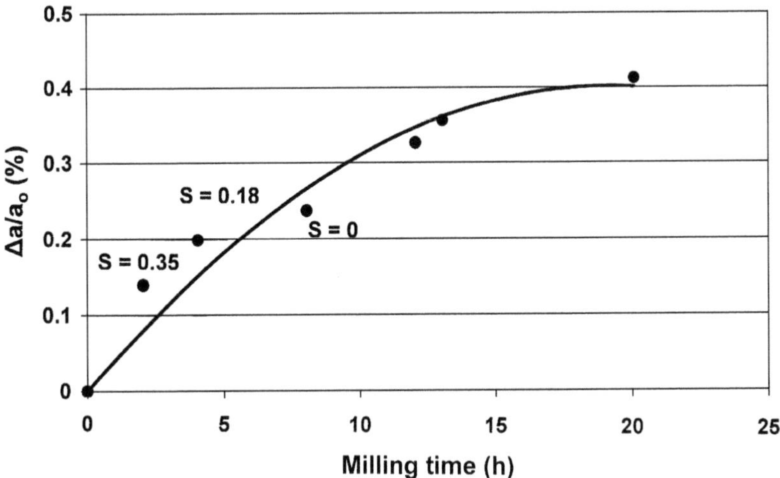

Figure 11.13 Lattice expansion in mechanically alloyed $Ni_3Al + B$ powder as a function of milling time. The long-range order parameter, *S*, is also indicated in the figure. Note that the lattice parameter continues to increase even after the lattice is completely disordered, i.e., $S = 0$ [39].

and their peaks broadened, due to lattice strain and crystallite refinement, the relative intensities of the superlattice peaks did not show any significant change with reference to the fundamental peaks. With continued milling the alloy became amorphous. Based on these observations it was concluded that the grain boundary energy in this system was high enough to drive the crystal-to-amorphous transition, and not destruction of long-range order [28]. However, in the early stages of milling, Zhou and Bakker [42]

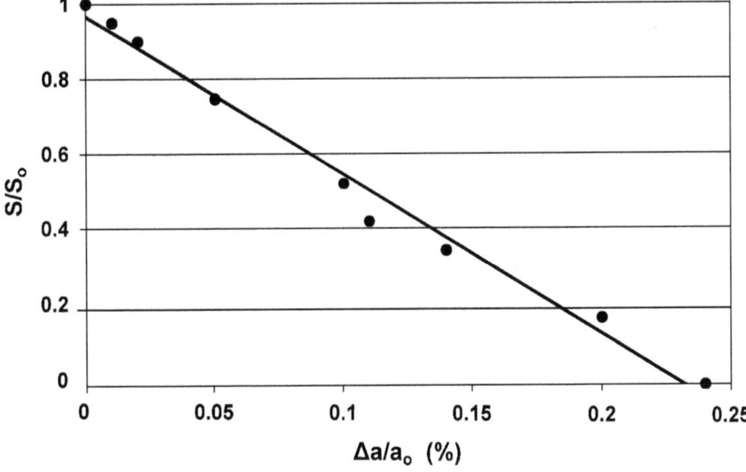

Figure 11.14 Relationship between lattice expansion and relative long-range order parameter (S/S_0) in mechanically milled Ni_3Al-B powder. Note that the long-range order parameter, *S*, decreases and that lattice expansion ($\Delta a/a_0$) increases with milling time [25].

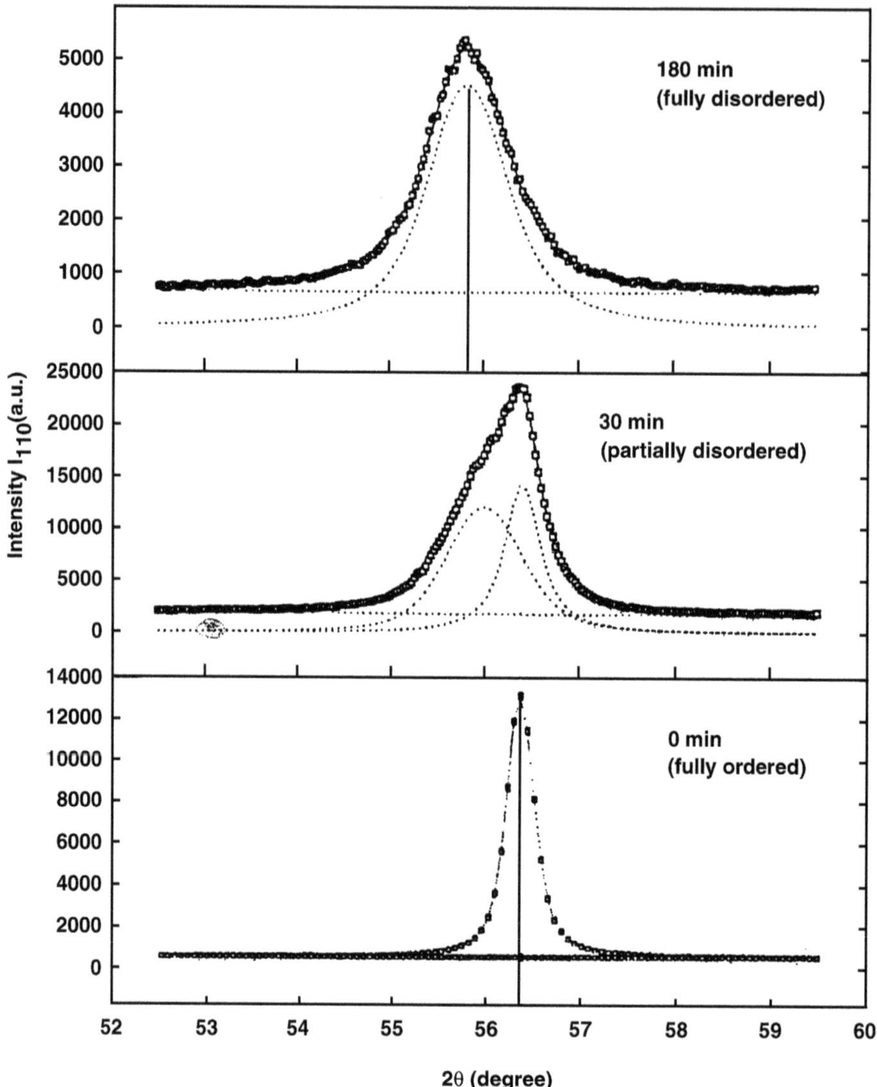

Figure 11.15 (110) Bragg peaks of the $Fe_{60}Al_{40}$ ordered nanocrystalline powder as a function of milling time. Note that at a milling time of 30 min, the peak appears to be a convolution of the ordered B2 and disordered bcc peak intensities [41].

observed that magnetization in the mechanically milled CoZr alloy, measured at 4.2 K, increased with milling time up to about 40 h. They also did not observe any reduction in the relative intensity of the superlattice reflections. The lattice parameter of the disordered nanocrystalline phase also increased with milling time. The increase in magnetization has been explained as follows: In the completely ordered alloy, the ferromagnetic Co atoms are surrounded by Zr atoms, which makes the magnetization low. However, in the partially disordered alloy, when the Co atoms substitute on the Zr sublattice, these antisite Co atoms have Co atoms on the Co sublattice as nearest

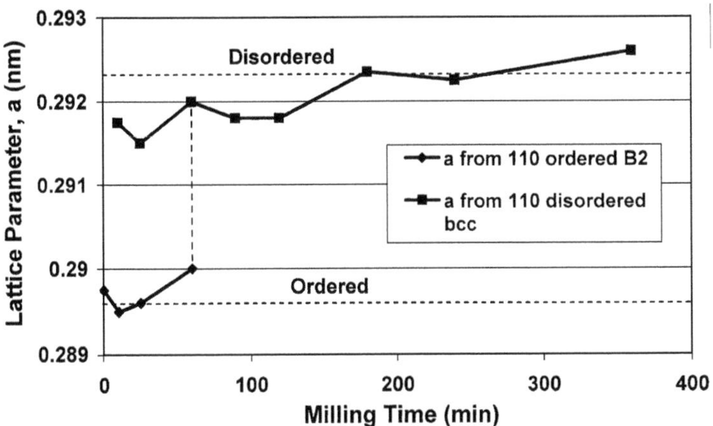

Figure 11.16 Variation of lattice parameter with milling time for the nanocrystalline FeAl intermetallic. Note the sudden increase in lattice parameter at 1 h of milling when the alloy becomes completely disordered [41].

neighbors, leading to an increase in the magnetization. From these two observations (increase in the magnetization and the lattice parameter), Zhou and Bakker [42] concluded that disordering of CoZr occurs by atomic disorder and grain refinement (i.e., creation of additional grain boundaries). Thus, it appears that the magnetic properties are more sensitive to, and a better probe for, studying disordering (in alloys that contain atoms with magnetic moments) than relative intensities of low-intensity and broadened XRD peaks. A combination of techniques can always shed additional evidence and provide an accurate account of the mechanism of disordering.

11.5 PHASE SELECTION

Mechanical milling of ordered intermetallics has been shown to result in one of the three following transformations [14,30,43]:

1. Formation of a solid solution of one component in the other, i.e., formation of the terminal solid solution based on the major component. This has been observed in compounds such as Nb_3Al, V_3Ga, Ni_3Al, Fe_3Ge, Ni_2V, and $NbAu_2$,
2. Formation of an amorphous phase, observed, for example, in Nb_3Sn, NiZr, NiV_2, and CoZr, or
3. Formation of a different phase with a complex crystal structure, noted in Ni_3Sn_2 and $TiSi_2$.

Thus, the sequence of transformations of an ordered intermetallic on MM can be represented as:

$$\text{Ordered intermetallic} \rightarrow \text{disordered intermetallic (or solid solution)} \rightarrow \text{amorphous phase} \tag{11.3}$$

Whether an ordered intermetallic transforms to a solid solution or an amorphous phase on milling is determined by the relative free energy values of the amorphous and crystalline phases with respect to the energy stored in the intermetallic by MM (see Fig. 11.1). For example, the intermetallic will transform to the solid solution if the enthalpy of the amorphous state as estimated by the Miedema analysis [21,44] is higher than that of the solid solution. Thus, if ΔG (milling) $> \Delta G^{a\text{-}c}$, where $\Delta G^{a\text{-}c}$ represents the difference in free energy between the ordered crystalline and amorphous phases, complete amorphization occurs. On the other hand, if ΔG (milling) $< \Delta G^{a\text{-}c}$, no amorphization occurs; instead, a solid solution forms.

Conclusions similar to the above can also be reached by observing the nature of phase diagrams. If the phase diagram shows that the ordered intermetallic transforms to a (disordered) solid solution before it melts, i.e., it is a reversibly ordered intermetallic, then milling of the intermetallic will result in the formation of a (disordered) solid solution. Continued milling may then produce the amorphous phase. But if the ordered intermetallic melts congruently (irreversibly ordered or permanently ordered intermetallic), then milling of the intermetallic will produce an amorphous phase directly. This has been shown to be true in a number of cases [14]. Thus, introduction of mechanical energy into the system is equivalent to heating the alloy to higher temperatures (Fig. 11.17), i.e., an alloy can become disordered either at higher temperatures or on MM. Depending on the nature of the phase, either melting or formation of an amorphous phase can occur in some cases. It has been noted that among about 700 binary intermetallic compounds only very few compounds such as CuZn and Cu_3Au are reversibly ordered compounds, i.e., they exhibit the order–disorder transformation before melting. Table 11.3 lists the compounds that have been disordered by MM. In the table, the third column gives the type of disorder: a question mark means that the type of disorder has not been investigated; "antisite" means antisite disorder, "triple defect" means triple-defect disorder, "quadruple" means quadruple-defect disorder, and "red. int." means redistribution of interstitials. The fourth column gives the type of transformation: "Amorphous" means that the compound has become amorphous after milling and "SS" means that the compound has transformed to a solid solution phase; the crystal structure of the solid solution is

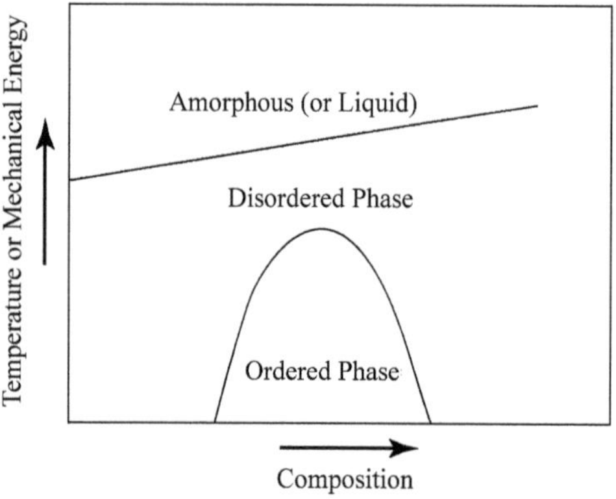

Figure 11.17 Equivalence of mechanical energy and temperature in disordering of ordered intermetallic compounds.

Table 11.3 Disordering of Intermetallics by Mechanical Milling

Compound	Crystal structure	Type of Disorder	Transformation	Phase diagram	Ref.
Nb_3Al	A15	?	SS (bcc)	?	44, 45
Nb_3Au	A15	Antisite	SS (bcc)	SS > T_c	46, 47
Nb_3Sn	A15	Antisite	Amorphous	Liq.	48, 49
NiV_3	A15	?	SS (bcc)	?	50
V_3Ga	A15	?	SS (bcc)	SS > T_c	51
CoAl	B2	Triple defect	None	Irrelevant	52
CoGa	B2	Triple defect	None	Irrelevant	53
CoZr	B2	Antisite	Amorphous	Liq.	28, 42
FeAl	B2	Antisite	SS (bcc)		32, 37
NiAl	B2	Triple defect	None	Irrelevant	32, 54
RuAl	B2	Triple defect?	None	Irrelevant?	27
Fe_3Ge_2	$B8_2$	Red. int.	None	Irrelevant	55
Mn_3Sn_2	$B8_2$	Red. int.	None	Irrelevant	55
LaAu	B27	?	Amorphous	LTP → HTP → liq.	56
NiZr	B_r	?	Amorphous	Liq.	57, 58
Fe_3Ge	$L1_2$	Antisite	SS (bcc)	?	14
Ni_3Al	$L1_2$	Antisite	SS (fcc) + amorphous	?	24, 38, 59
Ni_3Ge	$L1_2$	?	SS (fcc) + amorphous	?	57, 60
Ni_3Si	$L1_2$	Antisite	SS (fcc)	?	28, 61, 62
Zr_3Al	$L1_2$	?	Amorphous	?	63
$Co_{41}Cr_{59}$	σ phase	?	None	σ → δ → liq.	64
CoV	σ phase	Disorder	Amorphous	Liq.	65
$Cr_{53}Fe_{47}$	σ phase	Disorder	SS (bcc)	SS > T_c	66
FeV	σ phase	?	SS (bcc)	SS > T_c	57, 64
NiV_2	σ phase	?	Amorphous	σ → σ′ → liq.	50
Ni_4V_6	σ phase	?	Amorphous	σ → σ′ → liq.	57
Co_2Ge	Co_2Si	Antisite	Amorphous	LTP → HTP → liq.	67
Co_2Ge	Ni_2In	Antisite	Amorphous	HTP → liq.	67
Co_3Sn_2	Orthorhombic	Red. int.	LTP → HTP	LTP → HTP	68
Cu_2MnAl		Disorder			69
Fe_2Sc	C14	?	Amorphous	?	70
Fe_2Y	C15	?	Amorphous	?	70
$GdAl_2$	C15	Quadruple	None	Liq.	71
$GdIr_2$	C15	Quadruple	None	Liq.	15
$GdMg_2$	C15	Antisite	SS		15
$GdPt_2$	C15	Quadruple	None	Liq.	15
$GdRh_2$	C15	Quadruple	None	Liq.	15
$Nb_{60}Au_{40}$	A15 + Tetra.	?	SS (bcc)	SS > T_c	46
$Nb_{58}Au_{42}$	AlB_2 + Tetra.	?	SS (bcc/fcc)	SS > T_c	46
$Nb_{50}Au_{50}$	AlB_2 + Tetra.	?	SS (fcc)	SS > T_c	46
$NbAu_2$	AlB_2	?	SS (fcc)	SS > T_c	46
$Ni_{45}Nb_{55}$	W_6Fe_7	?	Amorphous	?	72

Table 11.3 Continued

Compound	Crystal structure	Type of Disorder	Transformation	Phase diagram	Ref.
Ni_3Sn	DO_{19}	?	Amorphous	LTP → HTP → liq.	73
Ni_3Sn_2	Orthorhombic	Red. int.	LTP → HTP	LTP → HTP	74
$NiTi_2$	E93	?	Amorphous	?	75
Ni_3V	DO_{22}	?	SS (fcc)	SS > T_c	50
Ni_2V	$MoPt_2$	?	SS (fcc)	SS > T_c	50
Pd_3Zr			SS (fcc)		76
RuAl	B2		A2 (bcc SS)		77
$TiAl_3$	DO_{22}	Antisite	SS (fcc)	?	78
Ti_5Si_3	$D8_3$	?	Amorphous	Liq.	78
$TiSi_2$	C54	?	C54 → C49	Irrelevant	78
YCo_2	Cu_2Mg	?	Amorphous	?	13
YCo_3	$CeNi_3$	?	Amorphous	?	13
Y_2Co_7	Co_7Er_2	?	Amorphous	?	13
YCo_5	Co_5Y	?	Amorphous	?	13
Y_2Co_{17}	$Ni_{17}Th_2$	?	Amorphous	?	13

Source: Adapted from Ref. 14.

mentioned within parentheses. "None" means that no transformation was observed, and LTP → HTP means a transformation from the low-temperature phase to a high-temperature phase in the phase diagram. Information on the type of transformation from the phase diagram is given in column 5. A question mark means that either the information is not available or not clear; "liq" means that the compound melts directly from its crystal structure, suggesting amorphization; SS > T_c means the formation of a solid solution above a critical temperature, which could point in the direction of formation of a solid solution.

It should also be realized that contamination could play a significant role in determining the nature of the final phase formed after milling. For example, it was shown that MM of the Nb_3Al intermetallic in a SPEX mill produced the expected bcc solid solution phase after 3 h. But when milling was conducted in an improperly sealed container, an amorphous phase was produced after 10 h, which on heat treatment transformed to the crystalline NbN phase. This suggests that the amorphous phase in this system had formed because of nitrogen contamination of the powder during milling [45]. Similarly, contamination of the milled near-equiatomic TiAl powders resulted in the formation of an fcc phase, which was misinterpreted as the formation of a disordered form of the γ-TiAl phase [79].

The crystal structure of the phase alone does not appear to decide whether the intermetallic transforms to a solid solution or an amorphous phase on MM. The three intermetallics—Ni_3Al, Ni_3Ge, and Zr_3Al—have the $L1_2$-type cubic structures. On MM, both Ni_3Al and Ni_3Ge form a disordered nanocrystalline fcc solid solution and do not show any tendency for amorphization. On the other hand, Zr_3Al amorphizes completely after a short milling time. These differences in the milling behavior have been attributed partly to the differences in the lattice stability in terms of their respective disordered fcc phases [80].

It was mentioned earlier that materials in a partially ordered state are stronger than those that are completely disordered or fully ordered. It was shown that the microhardness of mechanically milled Ni_3Al powders exhibited a pronounced sharp maximum corresponding to $S = 0.5$ [24]. This work suggests that the greater is the ordering energy, the steeper will be this maximum of strength (or hardness). It may also be mentioned in passing that all ordered states (from $S = 0$ to $S = 1$) cannot be accessed by traditional methods of disordering. For example, in equilibrium, the order parameter for Cu_3Au jumps discontinuously from 0.8 to zero at the critical temperature. However, MM can be used to obtain different degrees of order so that the effect of order parameter on structure and mechanical properties of the alloys can be investigated.

11.6 REORDERING KINETICS

There have also been several studies on reordering kinetics of disordered phases obtained by MM (see, for example, Refs. 81–85). It was reported that on heating the milled powder in a differential scanning calorimeter, exothermic transformation occurs over a wide range of temperatures from about 100°C to about 600°C (Fig. 11.18). Because of the complex shape of the signal it can be concluded that the transformation takes place in different stages. It was further noted that the transformations that occurred were irreversible since on repeating the thermal cycle on the same specimen no sign of transformation was detected.

The different processes that contribute to each component of the complex differential scanning colorimetric peak have been analyzed using isothermal measurements and XRD methods. The low- temperature low-intensity peak, occurring around 170°C, has been attributed to the reestablishment of short-range order. This was confirmed by the results obtained from extended electron energy loss fine structure (EXELFS) studies. No other changes could be detected, after long-term annealing at 150°C, than the localized creation of the right-ordered atomic environment around the constituent atoms. On heating the powders to 200°C, a temperature just above the low-temperature transformation peak, XRD analysis indicated a slight lattice contraction and an increase of the average crystallite size. Lattice contraction is expected to occur in the ordered state due to the efficient packing of atoms. The absence of such a peak during reheating of the mechanically milled elemental powders also indirectly confirms this hypothesis since chemical reordering cannot occur in pure metal powders. The effective activation energy of this process is about 100 kJ/mol for the Ni_3Al system and does not appear to depend sensibly on the initial degree of order.

The second and more intense major peak has its maximum at about 370°C. The peak temperature was found to decrease with increasing milling time but reached a constant value in the completely disordered samples. This peak is associated with the simultaneous evolution of ordering and grain growth [82,85]. The apparent activation energy for the transformation has been found to be lower than expected during ordering of a conventional alloy. The explanation could be that the high density of point defects such as vacancies and antisite defects, generated during milling, assist the diffusive processes and help in achieving reordering easily.

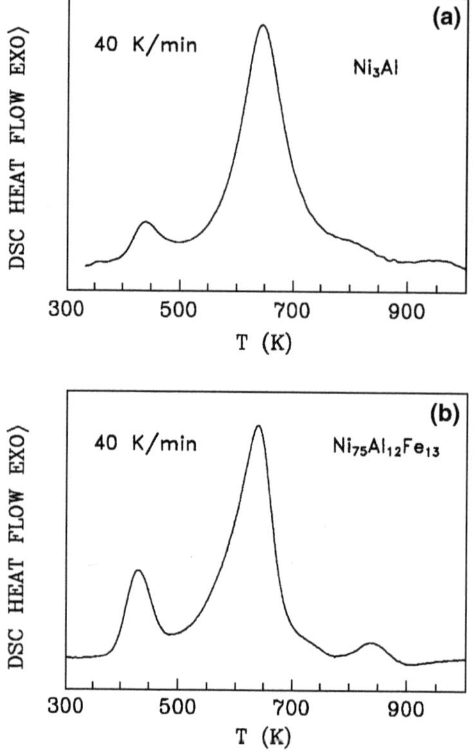

Figure 11.18 DSC curves from (a) Ni_3Al milled for 4 h and (b) $Ni_{75}Al_{12}Fe_{13}$ milled for 8 h. Two distinct exothermic peaks can be seen in (a) and an additional low-intensity peak at higher temperatures in (b) [82].

Most of the mechanically milled disordered intermetallics exhibit these two peaks in their differential scanning colorimetric plots, with the difference that the actual temperatures and peak shapes and heights could be different depending on the nature and composition of the alloy and milling time. In ternary intermetallics, however, a third low-temperature peak, often overlapping the second major peak, is observed (Fig. 11.18a). Although a clear explanation is not available for the occurrence of this peak, it is thought that this could be due to annealing out of dislocations, usually not mobile at temperatures as low as those of the first two peaks.

REFERENCES

1. Stoloff, N. S., Sikka, V. K., eds. (1996). *Physical Metallurgy and Processing of Intermetallic Compounds*. New York: Chapman & Hall.
2. Westbrook, J. H., Fleischer, R. L., eds. *Intermetallic Compounds: Principles and Practice, Vol. 3, Progress*. Chichester, UK: John Wiley & Sons Ltd.
3. Russel, K. C. (1985). *Prog. Mater. Sci.* 28:229–434.
4. Liu, B. X. (1999). In: Suryanarayana, C., ed. *Non-equilibrium Processing of Materials*. Oxford: Pergamon, pp. 197–224.

5. Koch, C. C. (1988). *Int. Mater. Rev.* 33:201–219.
6. Boettinger, W., Aziz, M. J. (1989). *Acta Met.* 37:3379–3391.
7. West, J. A., Manos, J. T., Aziz, M. J. (1991). In: Johnson, L. A., Pope, D. P., Stiegler, J. O., eds. *High Temperature Ordered Intermetallics IV*. Pittsburgh, PA: Materials Research Society, pp. 859–864.
8. Harris, S. R., Pearson, D. H., Garland, C. M., Fultz, B. (1991). *J. Mater. Res.* 6:2019–2021.
9. Haubold, T., Bohn, R., Birringer, R., Gleiter, H. (1992). *Mater. Sci. Eng.* A153:679–683.
10. Stoloff, N. S., Davies, R. G. (1966). *Prog. Mater. Sci.* 13:1–84.
11. Ermakov, A. E., Yurchikov, E. E., Elsukov, E. P. (1982). *Fiz Tverd Tela No* 4:1947–1952.
12. Elsukov, E. P., Barinov, V. A., Galakhov, V. R., Yurchikov, E. E., Ermakov, A. E. (1983). *Phys. Met. Metallogr.* 55(2):119–123.
13. Ermakov, A. E., Yurchikov, E. E., Barinov, V. A. (1981). *Phys. Met. Metallogr.* 52(6):50–58.
14. Bakker, H., Zhou, G. F., Yang, H. (1995). *Prog. Mater. Sci.* 39:159–241.
15. Modder, I. W., Kuin, M. J., Bakker, H. (1998). *Mater. Sci. For.* 269–272:619–624.
16. Gialanella, S., Amils, X., Baró, M. D., Delcroix, P., Le Caër, G., Lutterotti, L., Suriñach, S. (1998). *Acta Mater.* 46:3305–3316.
17. Navarro, E. PhD thesis, University of Complutense, Madrid, Spain, May 1998.
18. Navarro, E., Multigner, E. M., Yavari, A. R., Hernando, A. (1996). *Europhys. Lett.* 35:307.
19. Modder, I. W., Bakker, H. (1997). *Phys. Stat. Sol.* (b)199:369–378.
20. Zhou, G. F., Bakker, H. (1995). *Phys. Rev.* B52:9437–9445.
21. Bakker, H. (1998). Enthalpies in alloys: Miedema's semi-empirical model. *Mater. Sci. Foundations*. Vol. 1.
22. Lo Cascio, D. M. R., Bakker, H. (1991). *J. Phys. C: Condens. Matter.* 3:5227–5229.
23. Fähnle, M. (1982). *J. Low Temp. Phys.* 46:3–25.
24. Jang, J. S. C., Koch, C. C. (1990). *J. Mater. Res.* 5:498–510.
25. Gialanella, S., Cahn, R. W., Malagelada, J., Suriñach, S., Baró, M. D., Yavari, A. R. (1992). In: Chen, H., Vasudevan, V. K., eds. *Kinetics of Ordering Transformations in Metals*. Warrendale, PA: TMS, pp. 161–175.
26. Seki, Y., Johnson, W. L. (1990). In: Clauer, A. H., deBarbadillo, J. J., eds. *Solid State Powder Processing*. Warrendale, PA: TMS, pp. 287–297.
27. Hellstern, E., Fecht, H. J., Fu, Z., Johnson, W. L. (1989). *J. Appl. Phys.* 65:305–310.
28. Cho, Y. S., Koch, C. C. (1993). *J. Alloys & Compounds* 194:287–294.
29. Bakker, H., Modder, I. W., Zhou, G. F., Yang, H. (1997). *Mater. Sci. For.* 235–238: 477–486.
30. Suryanarayana, C. (2002). In: Westbrook, J. H., Fleischer, R. L., eds. *Intermetallic Compounds. Principles and Practice: Vol. 3, Progress*. Chichester, UK: John Wiley & Sons Ltd, pp. 749–764.
31. Poecht, P., Tominez, E., Chaffron, L., Martin, G. (1995). *Phys. Rev.* B52:4006–4016.
32. Schröpf, H., Kuhrt, C., Arzt, E., Schultz, L. (1994). *Scripta Metall. Mater.* 30:1569–1574.
33. Murty, B. S., Joardar, J., Pabi, S. K. (1996). *J. Mater. Sci. Lett.* 15:2171–2172.
34. Murty, B. S., Joardar, J., Pabi, S. K. (1996). *Nanostructured Mater.* 7:691–697.
35. Murty, B. S., Joardar, J., Pabi, S. K. (1997). *Mater. Sci. Eng.* A226:41–44.
36. Suryanarayana, C., Norton, M. G. (1998). *X-Ray Diffraction: A Practical Approach*. New York, NY: Plenum Publishing Corp.
37. Gialanella, S. (1995). *Intermetallics* 3:73–76.
38. Gialanella, S., Newcomb, S. B., Cahn, R. W. (1992). In: Yavari, A. R., ed. *Ordering and Disordering in Alloys*. London: Elsevier Appl. Sci, pp. 67–78.
39. Cahn, R. W. (1992). In: Liu, C. T., Cahn, R. W., Sauthoff, G., eds. *Ordered Intermetallics–Physical Metallurgy and Mechanical Behaviour*. The Netherlands: Kluwer Acad Pub Dordrecht, pp. 511–524.

40. Betteridge, W. (1949). *J. Inst. Metals* 75:559–570.
41. Yavari, A. R., Negri, D., Navarro, E., Deriu, A., Hernando, A., Botta, W. J. (1999). *Mater. Sci. For.* 312–314:229–236.
42. Zhou, G. F., Bakker, H. (1996). *Scripta Mater.* 34:29–35.
43. Bakker, H., Zhou, G. F., Yang, H. (1995). *Mater. Sci. For.* 179–181:47–52.
44. Oehring, M., Bormann, R. (1990). *J. Phys. Colloq.* 51:C4-169–C4-174.
45. Mukhopadhyay, D. K., Suryanarayana, C., Froes, F. H. (1995). In: Phillips, M., Porter, J., eds. *Advances in Powder Metallurgy & Particulate Materials–1995*. Vol. 1. Princeton, NJ: Metal Powder Industries Federation, pp. 123–133.
46. Di, L. M., Bakker, H., Bárczy, P., Gácsi, Z. (1993). *Acta Metall. Mater.* 41:2923–2932.
47. Di, L. M., Bakker, H. (1992). *J. Appl. Phys.* 71:5650–5653.
48. Di, L. M., Loeff, P. I., Bakker, H. (1991). *J. Less-Common Metals* 168:183–193.
49. Koch, C. C., Cho, Y. S. (1992). *Nanostructured Mater.* 1:207–212.
50. Yang, H., Bakker, H. (1993). In: deBarbadillo, J. J., et al. eds. *Mechanical Alloying for Structural Applications*. Materials Park, OH: ASM International, pp. 401–408.
51. Di, L. M., Bakker, H. (1991). *J. Phys. C: Condens. Matter.* 3:3427–3432.
52. Di, L. M., Bakker, H., de Boer, F. R. (1992). *Physica* B182:91–98.
53. Di, L. M., Bakker, H., Tamminga, Y., de Boer, F. R. (1991). *Phys. Rev.* B44:2444–2451.
54. Gialanella, S., Guella, M., Baró, M. D., Malagelada, J. M., Suriñach, S. (1993). In: deBarbadillo, J. J., et al., eds. *Mechanical Alloying for Structural Applications*. Materials Park, OH: ASM International, pp. 321–325.
55. Zhou, G. F., Bakker, H. (1994). *Phys. Rev.* B49:12507–12518.
56. Loeff, P. I., Bakker, H. (1988). *Scripta Metall.* 22:401–404.
57. Suzuki, K., Fukunaga, T. (1993). *J. Alloys Compounds* 194:303–309.
58. Weeber, A. W., Bakker, H. (1988). *J. Less-Common Metals* 141:93–102.
59. Koch, C. C., Jang, J. S. C., Lee, P. Y. (1989). In: Arzt, E., Schultz, L., eds. *New Materials by Mechanical Alloying Techniques*. Oberursel, Germany: DGM Informationgesellschaft, pp. 101–110.
60. Wernick, J. H. (1967). In: Westbrook, J. H., ed. *Intermetallic Compounds*. New York, NY: John Wiley, p. 197.
61. Zhou, G. F., Bakker, H. (1994). *Acta Metall. Mater.* 42:3009–3017.
62. Jang, J. S. C., Tsau, C. H. (1993). *J. Mater. Sci.* 28:982–988.
63. Gialanella, S., Yavari, A. R., Cahn, R. W. (1992). *Scripta Metall. Mater.* 26:1233–1238.
64. Yang, H., Bakker, H. (1994). *Mater. Sci. For.* 150–151:109.
65. Zhou, G. F., Bakker, H. (1994). *Intermetallics* 2:103–110.
66. Abe, O., Suzuki, Y. (1996). *Mater. Sci. For.* 225–227:563–568.
67. Zhou, G. F., Bakker, H. (1993). *Phys. Rev.* B48:13383–13398.
68. Di, L. M., Zhou, G. F., Bakker, H. (1993). *Phys. Rev.* B47:4890–4895.
69. Robinson, J. S., McCormick, P. G., Street, R. (1995). *J. Phys. C: Condens. Matter.* 7:4259–4269.
70. Xia, S. K., Larica, C., Rodriguez, V. A., Rizzo Assunção, F. C., Baggio-Saitovitch, E. (1996). *Mater. Sci. For.* 225–227:389–394.
71. Zhou, G. F., Bakker, H. (1994). *Phys. Rev. Lett.* 73:344–347.
72. Schwarz, R. B., Koch, C. C. (1996). *Appl. Phys. Lett.* 49:146–148.
73. Di, L. M., Loeff, P. I., Bakker, H. (1989). *Phys. Stat. Sol.* (a)117:K99–K101.
74. Zhou, G. F., Di, L. M., Bakker, H. (1993). *J. Appl. Phys.* 73:1521–1527.
75. Schwarz, R. B., Petrich, R. R. (1988). *J. Less Common Metals* 140:171–184.
76. Katona, G. L., Kis-Varga, M., Beke, D. L. (2002). *Mater. Sci. For.* 386–388:193–198.
77. Liu, K. W., Mücklich, F., Birringer, R. (2001). *Intermetallics* 9:81–88.
78. Oehring, M., Yan, Z. H., Klassen, T., Bormann, R. (1992). *Phys. Stat. Sol.* (a)131:671–689.
79. Suryanarayana, C. (1995). *Intermetallics* 3:153–160.
80. Benameur, T., Yavari, A. R. (1992). *J. Mater. Res.* 7:2971–2977.

81. Yavari, A. R. (1993). *Acta Metall.* 41:1391–1396.
82. Baró, M. D., Suriñach, S., Malagelada, J. (1993). In: deBarbadillo, J. J., et al., eds. *Mechanical Alloying for Structural Applications*. Materials Park, OH: ASM International, pp. 343–348.
83. Cardellini, F., Contini, V., Mazzone, G. (1995). *Scripta Metall. Mater.* 32:641–646.
84. Gialanella, S., Delorenzo, R., Marino, F., Guella, M. (1995). *Intermetallics* 3:1–8.
85. Baró, M. D., Malagelada, J., Suriñach, S., Clavaguera, N., Clavaguera-Mora, M. T. (1992). In: Yavari, A. R., ed. *Ordering and Disordering in Alloys*. London: Elsevier, pp. 55–66.

12

Solid-State Amorphization

12.1 INTRODUCTION

A solid alloy with a liquid-like (noncrystalline) atomic arrangement is called a metallic glass or an amorphous (metallic) alloy. A glass is obtained when a liquid is cooled to a state of rigidity without crystallizing; such an alloy will exhibit a glass transition temperature. On the other hand, if a material with a similar structure is made by some processes other than cooling from the liquid state, then it is called an amorphous alloy.

Amorphous alloys were first synthesized by vapor deposition in the form of thin films by Buckel and Hilsch [1]. However, the synthesis of a noncrystalline (glassy) phase by rapid solidification from the liquid state by Pol Duwez and his associates in 1960 [2] paved the way for an enormous amount of activity during the last four decades. These metallic glasses have an unusual combination of properties such as high strength, good bend ductility, high fracture toughness, good corrosion resistance, and desirable soft magnetic properties. Because of this excellent combination of properties, metallic glasses have found a variety of industrial applications, the most important of which is for core laminations in distribution transformers. Information on the synthesis, properties, and applications of metallic glasses produced by the technique of rapid solidification processing (RSP) from the melt may be found in several reviews and books [3–10].

Noncrystalline alloys (or amorphous alloys or glasses) can now be formed in alloy systems by different nonequilibrium (far from equilibrium) processing techniques such as rapid solidification from the liquid state, vapor deposition, plasma processing, and laser processing [3,4,9,10]. In these methods there is a change in the state of matter, i.e., a solid phase is formed from either the liquid or the vapor phase, and the high effective "quenching" rate and the associated undercooling have been shown to be responsible for formation of the glassy phase. There are also methods of amorphizing a solid without passing through the liquid state at any stage. Known as solid-state amorphization reactions (SSARs), these include irradiation [11], hydrogen-assisted amorphization [12,13], interdiffusion of elemental metals [14], pressure-induced vitrification [15], and heavy mechanical deformation (cold rolling followed by low-temperature annealing). An amorphous phase was reported to form when the intermetallic Zr_3Rh was annealed in a hydrogen atmosphere at a relatively low temperature of 200°C [12]; this was attributed to the fast diffusivity of hydrogen into the intermetallic and formation of an amorphous hydride phase. Schwarz and Johnson [14] have reported amorphization on annealing vapor-deposited multilayer films of La and Au; again attributed to the fast diffusivity of Au in La. Cold rolling of thin foils of Ni and Zr followed by a low-temperature anneal also produced an amorphous phase [16]. Apart from all these methods, which are being practiced to a limited extent, mechanical alloying (MA) and mechanical milling (MM) have been the most popular methods for producing amorphous alloys in the solid state [17]. This chapter reviews the different aspects of amorphous phases produced in binary and higher order alloy systems by MA/MM. The topics discussed here include the thermodynamic and kinetic aspects of formation, mechanisms and models of phase formation, and theoretical models for predicting the amorphous phase–forming range. The chapter concludes with a comparison of the amorphous phases produced by MA and RSP, two popular nonequilibrium processing techniques.

12.2 AMORPHOUS PHASES BY MA/MM

The synthesis of an amorphous phase in the Ni-Nb system by MA starting from blended elemental powders of Ni and Nb was first reported in 1983 by Koch et al. [18]. Since the Ni-40 at% Nb alloy was known to be a good glass former by RSP, they investigated the effect of MA on the powder blend. They observed that the X-ray diffraction (XRD) pattern of the mechanically alloyed powder blend, milled for 14 h in a SPEX shaker mill, was typical of an amorphous phase and that the pattern was similar to the one obtained in the RSP alloy. Thus, they concluded that MA of powder blends could produce amorphous phases. This gave rise to research activity in the solid-state synthesis of amorphous alloy phases, and since then there have been numerous reports on formation of amorphous phases by MA in several binary and ternary (and a few higher order) alloy systems. In fact, a majority of the recent literature on MA is concerned with the synthesis and characterization of amorphous alloys obtained by MA and MM [17]. This has been mostly due to the earlier huge success on the production and commercialization of metallic glasses by RSP [19], and to investigate whether MA/MM could offer any advantages. Amorphization of intermetallic compounds (e.g., YCo_3, Y_2Co_7, and YCo_5) by MM was reported earlier by Ermakov et al. [20,21]. Apart from these two combinations, mixtures of intermetallics and intermetallics/elemental metal powders were also amorphized.

Thus, it is now known that amorphous alloy phases could be synthesized by MA or MM starting from

1. Blended elemental powder mixtures
2. Prealloyed powders and/or intermetallics

3. Mixtures of intermetallics, or
4. Mixtures of intermetallics and elemental powders

One of the early reviews, comprehensive at the time of writing and the only one devoted exclusively to the topic of amorphous phases obtained by milling of powders, was published in 1988 by Weeber and Bakker [22].

An exhaustive listing of the alloy systems and the composition ranges in which amorphous phases have been found to form by MA and MM is presented in Tables 12.1 and 12.2, respectively. Since amorphous phase formation is known to be critically dependent on the milling conditions, details about the mill, milling container, grinding medium, ball-to-powder weight ratio (BPR), and other parameters have also been included in these tables.

12.3 DIFFERENCE BETWEEN AMORPHOUS PHASE FORMATION BY MA AND MM

A point to be considered is whether there exists a difference in the way an amorphous phase forms by MA from a mixture of powders (A and B) and by MM from the intermetallic compound (A_mB_n).

Starting from blended elemental powders of metals A and B, an amorphous phase could form by MA via different reaction routes. It can form either directly or through the formation of an intermetallic or a solid solution phase. These different possibilities may be represented as:

$$mA + nB \rightarrow (A_mB_n)_{amorphous} \tag{12.1}$$

$$mA + nB \rightarrow (A_mB_n)_{crystalline} \rightarrow (A_mB_n)_{amorphous} \tag{12.2}$$

$$mA + nB \rightarrow A_m(B_n) \rightarrow (A_mB_n)_{amorphous} \tag{12.3}$$

$$mA + nB \rightarrow A_m(B_n) \rightarrow (A_mB_n)_{crystalline} \rightarrow (A_mB_n)_{amorphous} \tag{12.4}$$

where $A_m(B_n)$ represents a solid solution of B in A, $(A_mB_n)_{crystalline}$ represents a crystalline intermetallic phase, and $(A_mB_n)_{amorphous}$ represents the amorphous phase with the composition A_mB_n. Even though an amorphous phase has eventually formed in all cases, in every case, except the first one, it is preceded by the formation of a solid solution and/or an intermetallic. In some cases, a (supersaturated) solid solution was shown to directly transform to an amorphous phase. Occurrence of amorphization by all these routes has been reported in the literature.

The basic difference between MA and MM is that no alloying is required during MM when prealloyed powders or intermetallics are used. However, during MA, when one starts with blended elemental powders, alloying is a prerequisite before amorphization can occur and therefore additional time is required for alloying to occur. On the other hand, when the starting material is an intermetallic, i.e., during MM, no alloying is needed but only amorphization. Thus, intuitively one would expect that the times required for amorphization are shorter during MM than during MA, provided that all the milling conditions remain the same. Weeber et al. [342] reported that the prealloyed Ni-Zr powder could be amorphized much faster than the elemental blend. This was explained as being due to the reduction in energy requirement for amorphization by an amount equal to the energy required for alloying. Since prealloyed powders are less prone to oxidation, crystallization temperature of the amorphous MM powder was

Table 12.1 Composition Ranges and Details of Milling Parameters for Formation of Amorphous Phases in Blended Elemental Powder Mixtures by Mechanical Alloying[a]

System	Mill	Vial material	Grinding medium material	BPR	Speed (rpm)	MA time (h)	Glass-forming range (at%)	Ref.
Ag-Pd	SPEX 8000			10.4:1	—	65	50 Pd	23
Al-18.2Co	Planetary ball mill QM-1F	SS	Hardened steel	20:1	200	80		24
Al + $Al_{13}Co_4$ (Al-15Co)	Fritsch P5	SS	SS	10:1	150	100		25
Al-15Cr	Ball mill	SS	SS	120:1		1000	Partly Am.	26
Al-20Cr	Ball mill	SS	SS	120:1		[b]	Partly Am.	26
Al-20Cu-15Fe	Planetary ball mill QM-1SP	Steel	Hardened steel	10:1	200–300	300		27
Al-20Cu-15Mn	Fritsch P5			15:1	Int. 5	510		28
Al-15Fe	Planetary ball mill	Hardened steel	Hardened steel	20:1	—	300	—	29
Al-20Fe	Ball mill					170		30
Al-20Fe	Fritsch P7	Hardened steel	Hardened steel	6:1	—	40		31
Al-20 Fe	Planetary ball mill	Hardened steel	Cr-steel	20:1		240		32
Al-24.4Fe	Ball mill	SS	SS	90:1				33
Al-1 to 25Fe	SPEX mill	Steel	521000 steel	10:1	—	50	25 Fe	34
Al-25Fe	Horizontal ball mill	SS	SS		90	600		35
Al-33Fe	Ball mill		SS	90:1		454		36
Al-40Fe	Horizontal ball mill	SS	SS	50:1	90			35
Al-Fe	Conventional ball mill	SS	SS	90:1	—	454	17–33 Fe	37
Al-Fe	Planetary ball mill	Steel	Steel		230	180	20–50 Fe	38
Al-27.4Fe-28.7C	SPEX8000	Hardened steel	SS	2.3:1	—	48		39
Al-5Fe-5Nd	High-energy shaker mill					240	Partial; coexists with fcc-Al	40
Al-7Fe-3Zr	SPEX 8000	Hardened steel	Hardened steel	10:1	—	30	Partly Am.	41
Al-25Ge-10Fe	Vibratory ball mill	SS	SS	5:1	—	80		42
Al-50Hf								43
Al-14Mn	Planetary ball mill			60:1	720	300	[c]	44
Al-6Mn-2Ce	Fritsch P5	—	—	13:1	180	400		45
Al-20Mn-30Si	Fritsch			20:1	450–650	85		46
Al-50Nb	Rod mill					400		47
Al-50Nb	Rod mill	SS	SS	30:1	—	40		48

Al-Nb	SPEX 8000	Steel	52100 Steel	10:1	—			49
Al-40Ni	Low-energy vibrating mill							50
$(Al_{88}Ni_8Co_4)_{100-x}Zr_x$	Fritsch			10:1	120		$x = 4$	51
$(Al_{88}Ni_8Co_4)_{100-x}Zr_x$	Fritsch planetary mill	SS		10:1		280		52
$Al_{80}Ni_8Fe_4Gd_8$								53
Al-33 Ni-33Ti	Ball mill	SS	SS	100:1		1000		54
Al-30Si-15Fe-5Ni	SPEX 8000	Hardened steel	Hardened steel	15:1	—	200	—	55
Al-50Ta	Ball mill	SS	SS	30:1		20		56
Al-50Ta		SS	SS	36:1, 108:1	—	400		57
Al-Ta	Rod mill	SS	SS	30:1		400	33–67 Ta	58, 59
Al-6Ti	Attritor	SS	Hardened steel	220:1		1300		60
Al-12.4Ti	Attritor	SS	Hardened steel	220:1	175	400		60
Al-25Ti	Fritsch P7	Hardened steel	Hardened steel	10:1	645	28		61
Al-33.3Ti	Rod mill	SS	SS			400		62
Al-35 to 48Ti	Planetary ball mill	Steel		30:1		90		63
Al-45Ti	Planetary ball mill			60:1		40		64
Al-23Ti-23C	Planetary ball mill	Steel		30:1		90		63
Al-12Ti-7Fe	Planetary ball mill	SS	SS			100		65
$Al_{85}Y_8Ni_5Co_2$	Planetary ball mill Retsch PM 4000	Hardened steel	Hardened steel	15:1		100	Only a small amount of Am.	66
$Al_{85}Y_8Ni_5Co_2$	Fritsch P5	—	—	13:1	180	280		45
$Al_{88}Y_2Ni_6Fe_4$	Planetary ball mill Retsch PM 4000	Hardened steel	Hardened steel	15:1		100	Only a small amount of Am.	66
Al-50 Zr	Rod mill	SS	SS	30:1		400		67
Al_2O_3-38 mol% ZrO_2	Planetary ball mill G7	Steel	Steel	7:1	500/700	72	Am + trace Al_2O_3	68, 69
Al_2O_3-38 mol% ZrO_2-3 mol% Y_2O_3	Planetary ball mill G7	Steel	Steel	7:1	500/700	72	Am + trace Al_2O_3	68
Al_2O_3-40 mol% $ZrSiO_4$	Planetary ball mill G7	Steel	Steel	7:1	500/700	240		69
Au-45La	Fritsch P0	WC-lined hardened steel	WC-lined hardened steel			130		70
$BiFeO_3$-25 mol% $ZnFe_2O_4$		SS	SS		80–110	250		71
Co-20B	Fritsch P5	Tempered steel		10:1	320	45		72

Table 12.1 Continued

System	Mill	Vial material	Grinding medium material	BPR	Speed (rpm)	MA time (h)	Glass-forming range (at%)	Ref.
Co-33B	Fritsch P5	Tempered steel		10:1	320	13		72, 73
Co-33B	Fritsch P5	Tempered steel		5.5:1		70	Partly Am.	74
Co-B	Fritsch P5	Tempered steel	SS	10:1	250	18–34	33–50 Co	75
Co-Nb	Attritor, Mitui Miike MA 1D	SS	SS		300	60	15–20 Nb	76
Co-25Si	Planetary ball mill	Hardened tool steel	Hardened tool steel	11.3:1		220		77
Co-66.7Si	Lab ball mill	SS	SS	64:1	360	25		78
Co-Si	Fritsch P5	SS	SS	9.6:1		200	33 Si	79
Co-Si	Lab ball mill	SS	SS	64:1	360	30	30–70 Si	80
Co-10 and 20Sn		SS		150:1	110	300	Partly Am.	81
Co-25Ti	Fritsch P5	SS	SS	17:1	250	3		82, 83
Co-25Ti	Fritsch P5	SS	SS	10:1		48		82, 83
Co-50Ti	Fritsch P5	SS	SS	17:1	250	6		84
Co-Ti	Attritor, Mitui Miike MA 1D	SS	SS		300	50	15–20 Ti	76
Co-Ti	Fritsch P5	Tempered Cr steel	Tempered Cr steel	14:1	2.1 s^{-1}	24	25–67	85
Co-V						60	33–60 V	86
Co-15Zr	High-energy ball mill				300	8.5		87
Co-45Zr	Planetary ball mill			13:1		30		88
Co-Zr	Planetary ball mill					60	8–73 Zr	86
Cr-30Cu	Vibrational mill		SS	45:1		100	d,e	89
Cr-15 to 65Fe	SPEX 8000	Hardened steel	Hardened steel	7:1	—		28 Fe	90
Cr-33.3Nb	Fritsch P7			4:1		60	Partly Am.	91
Cr-Nb	Planetary mill						30–68 Nb	92
Cr-Nb-Si	Fritsch P7	Hardened steel	Hardened steel			24		93
Cr-25Si	Planetary ball mill	Hardened tool steel	Hardened tool steel	11.3:1		220		77
Cu-42Nb-14Ge	Lab mill	WC-Co	WC-Co	50:1		24		94
Cu-42Nb-14Si	Lab mill	WC-Co	WC-Co	50:1		24		94
Cu-35Nb-20Sn	Lab mill	WC-Co	WC-Co	50:1		24		95

Cu-42Nb-14Sn	Lab mill	WC-Co	WC-Co	50:1		24		94, 95
Cu-50Nb-5Sn	Lab mill	WC-Co	WC-Co	50:1		24		95
Cu-11Ni-18P	Horizontal ball mill	SS	SS	30:1	80	300		96
Cu-20P	Horizontal ball mill	SS	SS	30:1	80	800	Partly Am.	97
Cu-50 and 70Sb	Fritsch P0	SS	Hardened steel	220:1			Sb becomes Am.	98
Cu-10 and 20Sn		SS		150:1	110	300	Partly Am.	81
$Cu_{86-x}Sn_xP_{14}$	High-energy planetary mill	Steel	Hardened steel	10:1 to 15:1	—	28, $x = 4$; 20, $x = 5$; 12, $x = 8$; 32, $x = 10$	4–10 Sn	99
Cu-40Ti	Planetary ball mill	Steel	Hardened steel	10:1 to 15:1	—	16		100
Cu-42Ti	Fritsch P5	SS	SS	4:1	436	43		101
Cu-50Ti	SPEX 8000					16	Am + TiH_2	102
Cu-50Ti	SPEX 8000	Hardened steel	SS	6:1	—	16		103
Cu-60Ti	SPEX 8000				—	42		104
Cu-Ti	SPEX 8000	Hardened steel		8.3:1	—	16	25–50 Ti 25–70 Ti with 4 to 10 at.% H_2	105
Cu-Ti		Tool steel or WC	WC	3:1 to 10:1		6	13–90 Ti	106
Cu-55[(1-x)Ti + $x$$TiH_2$]	SPEX 8000	Hardened steel or Cu-Be			—		$x = 0$ and 0.02	107
Cu-20Ti-20Zr	Planetary ball mill	Steel	Hardened steel	10:1 to 15:1		16		100
Cu-50V	Fritsch P5	Cu-Be	Cu-Be		573	120		108
Cu-W	Fritsch P5	Tempered steel	Steel				30–90 W	109
Cu-40Zr	Fritsch	Steel		13:1		20		110
Cu-40Zr	Planetary ball mill	Steel		13:1		30		88
Cu-40 to 60Zr	SPEX 8000	Hardened tool steel	440C steel	10:1	—	13		111
Cu-42Zr	SPEX 8001	Hardened steel				4		112
Cu-50Zr	SPEX 8000	Hardened steel	SS	6:1	—	24		113
Cu-Zr	Fritsch	Steel		13:1		30	40–60 Zr	110, 111
$Cu_{90-x}Zr_xTi_{10}$ ($x = 20$–80)	SPEX 80000	High-speed steel	Cr steel	5:1	—	5–10	—	114
$Cu_{80-x}Zr_xTi_{20}$ ($x = 30$–50)	SPEX 80000	High-speed steel	Cr steel	5:1	—	5–10	—	114
$Cu_{70-x}Zr_xTi_{30}$ ($x = 30$–40)	SPEX 80000	High-speed steel	Cr steel	5:1	—	5–10	—	114
$Cu_{40}Zr_{20}Ti_{40}$	SPEX 80000	High-speed steel	Cr steel	5:1	—	5–10	—	114

Table 12.1 Continued

System	Mill	Vial material	Grinding medium material	BPR	Speed (rpm)	MA time (h)	Glass-forming range (at%)	Ref.
Fe + FeB (Fe-20B)	Fritsch P5			5:1		350		115
Fe-20 B	Planetary ball mill					100		116
Fe-50B	Fritsch		WC	50:1		30		117
Fe-50B	Ball mill	SS	Hardened steel	10:1		250		118
Fe-60B	Ball mill	SS	Hardened steel	10:1		300		118
Fe-B	Planetary ball mill	SS		15:1	200		35–40 B	119
Fe-15B-10Si	Fritsch P5			5:1		350		115
Fe-25C	Ball mill	SS	Hardened steel			700	Mostly Am. (+ α-Fe)	120
Fe-70C	Ball mill	SS	SS	100:1		1000		121
Fe-C	Planetary ball mill	SS		15:1	200	72	20–25C	119
Fe-C	Ball mill	SS	SS	100:1		200	17–60C	122
Fe-32C-14Si	Ball mill	SS	SS	100:1		500		121
Fe-16C-16Si	Ball mill	SS	SS	100:1		500		121
Fe-30C-25Si	Ball mill	SS	SS	100:1		500		121
Fe-15C-29Si	Ball mill	SS	SS	100:1		500		121
$(Fe_{0.5}Co_{0.5})_{60}Cu_2V_8B_{30}$	Planetary ballmill	SS		15:1	240	100	—	123
Fe-50Cr	Fritsch P5	SS	SS		430	200	Milled in N_2 atmosphere	124
$Fe_{83-x}Cr_xC_{17}$ x = 10 to 60	Planetary ball mill			11.3:1		200		125
$Fe_{80-x}Cr_xN_{20}$ x = 0 to 24	Planetary ball mill			11.3:1		200		125
$(Fe_{0.5}Cu_{0.5})_{83-87}Zr_{13-17}$	Planetary ball mill	Hardened steel	Hardened steel	15:1		85		126
$(Fe_{0.5}Cu_{0.5})_{85}Zr_{15}$	Planetary ball mill	Hardened steel	Hardened steel	15:1		85		127
$Fe_{83-x}Mo_xC_{17}$ x = 5–60	Planetary ball mill			11.3:1		200		125
Fe-50Nb	SPEX 8000	High-speed steel	Cr steel	5:1	—	5		128
Fe + Ni + Fe-B ($Fe_{40}Ni_{40}B_{20}$)	Fritsch P5	SS	SS	9.6:1	573	280		101
Fe + Ni + Fe-B + Fe-P ($Fe_{40}Ni_{40}P_{14}B_6$)	Fritsch P5	SS	SS	9.6:1	573	24		101

$Fe_{40}Ni_{40}P_{20-x}Si_x$ (x = 6, 10, 14)	Fritsch P7	SS	SS	5:1	—	—	Partly Am.	129
Fe-38Ni-12Si-10B	SPEX 8000	Hardened tool steel		8:1	—	30	Partly Am.	130
Fe-P	Planetary ball mill	SS		15:1	200		≈ 25P	131
Fe-10 and 20Si	Planetary ball mill					300		116
Fe-15Si-15B	Planetary ball mill					300		116
Fe-Si	Lab ball mill	SS	SS	64:1	360	250	10–40 Si	80
Fe-Si								132
Fe-67Sn							Partly Am.	133, 134
Fe-33Ti	SPEX 8000	WC	SS		—	16		135
Fe-Ti	SPEX 8000	WC-coated	SS		—			136
Fe-36Tb	High-energy vibration ball mill	WC	WC	10:1				137
Fe-V	SPEX 8000	WC	WC	6:1	—	24–48	Partly Am.	138
$Fe_{75-x}V_xC_{25}$ (x = 0–30)	Planetary mill			11.3:1		200		125
Fe-30W	Fritsch P5	Hardened steel	Hardened steel	10:1	153	1700		139
Fe-50W	Fritsch P5	Hardened steel	Hardened steel	10:1	153	1700		139
Fe-16 and 20Zr	Fritsch P5	Steel	Steel			60		140
Fe-33Zr	Fritsch P5	Tempered steel		30:1 to 90:1				141
Fe-40Zr	Planetary ball mill			13:1		30		88
Fe-60Zr	Fritsch	Steel		13:1		20		110
Fe-Zr	Fritsch P5	Steel	Steel			60	20 Zr	140
Fe-Zr							22–70 Zr	110, 142
Fe-Zr-B							21–25 Zr and 0–15 B	143
Ge-S	Planetary ball mill	Hard bearing steel		40:1		60	61–72 S	144
Ge-10, 30, 40, 50 Se	Fritsch	SS	SS	—	300, 600	—	50 Se	145
Ge-25 to 75 w/o Si	Fritsch P7/P5	Tempered steel						146
Hf-50 Co	SPEX 8001	Hardened steel	—	—	—	3.75	—	147
Hf-Cu		Hardened steel	Hardened steel or WC			17	20–90 Cu	148
Hf-Cu	SPEX 8000	WC	WC	5:1	—	12	30–70 Cu	149, 150
Hf-Ni	SPEX 8000	WC	WC	5:1	—	12	15–65 Ni	150
Hf-Ni		Hardened steel	Hardened steel or WC			17	10–85 Ni	148

Table 12.1 Continued

System	Mill	Vial material	Grinding medium material	BPR	Speed (rpm)	MA time (h)	Glass-forming range (at%)	Ref.
Hf-37Pd	SPEX 8001	Hardened steel	—	—	—	6	—	151, 152
La-Al-Ni	Planetary ball mill	WC-Co	WC-Co	20:1	180		30–70 La 10–50 Ni	153
Mg-Ni	Planetary ball mill	—	—	10:1	—	—	50–80 Ni	154
Mg-50Ni	SPEX 8000	SS	SS	10:1		10		155
$Mg_2Ni + Ni$	Planetary ball mill	—	—	10:1	—	—	50 Ni	154
$MgNi_2 + Mg$	Planetary ball mill	—	—	10:1	—	—	50 Ni	154
Mg-Ni	Planetary ball mill	—	—	15:1	—	120	40–70 Ni	156
Mg-57Al-5Ca	SPEX 8000		52100 steel		—	1		157
Mg-15Y-25Cu	Planetary ball mill Retsch PM 4000	Hardened steel	Hardened steel	15:1		170	Partly Am.	66
Mg-10Y-30Cu	Planetary ball mill Retsch PM 4000	Hardened steel	Hardened steel	15:1		170	Partly Am.	66
Mg-15Y-30Cu	Planetary ball mill Retsch PM 4000	Hardened steel	Hardened steel	15:1		170	Partly Am.	66
Mg-10Y-35Cu	Planetary ball mill Retsch PM 4000	Hardened steel	Hardened steel	15:1		170	Partly Am.	66
Mg-15Y-30Cu + 30 vol% Y_2O_3	Planetary ball mill Retsch PM 4000	Hardened steel	Hardened steel	15:1	180		Partly Am.	158
$(Mg\text{-}15Y\text{-}30Cu)_{100-x}$ + $(C, Si, Ce, Ca)_x$ x = 0.5–10 at%	Planetary ball mill Retsch PM 4000	Hardened steel	Hardened steel	15:1	180		Partly Am.	158
Mn-25Si	Planetary ball mill	Hardened tool steel	Hardened tool steel	11.3:1	—	220		77
Mn-66.7Si	Lab ball mill	SS	SS	64:1		25		78
Mn-40Zr	Planetary ball mill			13:1		30		88
Mo-47 Ni	SPEX 8000D Simoloyer	SS	SS	5:1	—	36	Partly Am.	159
Mo-50Ni	SPEX 8000	Hardened tool steel		10:1	—	28	Partly Am.	160
Mo-25Si	Planetary ball mill			20:1	600 m/s^2	10 min		161
Mo-25Si	Planetary ball mill	Hardened tool steel	Hardened tool steel	11.3:1	—	220		77

Mo-67 Si	DSP-1P Planetrary ball mill		Refractory alloy balls	30:1	—	35		162
Nb-25 and 33.3Al	Vibration mill	SS	SS	5:1		40–20		163
Nb-Al	SPEX 8000	Steel	52100 steel	10:1	—		25–85 Al	49, 164, 165
Nb-Fe	Planetary ball mill QM-4H	Tool steel	SS	10:1	—	—	30–70 Fe	166
Nb-25Ge	SPEX 8000				—	5	12–34 Ge	167
Nb-Ge	High-energy ball mill	WC or hardened tool steel	WC or hardened tool steel			10–16	18–27 Ge	168
Nb-Ge	SPEX 8000	Hardened tool steel	440C steel	10:1	—	10–15	12–30 Ge	169
Nb-Ge							25–27	168
$Nb_{75}Ge_{25-x}Al_x$							x = 6,12, 19	168
Nb-(x) Ge-(25-x) Si	High-energy ball mill	WC or hardened tool steel	WC or hardened tool steel			8	x = 6–19	168
Nb-Mn							35–60 Mn	92
Nb-16Si	SPEX 8000		SS	15:1	—	4		170
Nb-25Si	High-energy ball mill	WC or hardened tool steel	WC or hardened tool steel			8		168
Nb-27Si	SPEX 8000	WC-lined steel	52100 steel	10:1	—	12		171
Nb-50Si	Planetary ball mill QF-1	Steel	Steel	34:1		150		172
Nb-66.7Si	Fritsch P7			4:1		60	Partly Am.	91
Nb-Sn							25	94, 173
$(Nb_{0.7}Ta_{0.3})_5Si_3$	SPEX 8000	WC-lined steel	52100 steel	10:1	—			171
$(Nb_{0.8}Ta_{0.2})_5Si_3$	SPEX 8000	WC-lined steel	52100 steel	10:1	—			171
Nd-77Fe-8B	High-energy ball mill	WC	WC	40:1	750	100		170
Nd-74Fe-15Mo	Lab ball mill	SS	SS	100:1	95	100	Am + α-Fe + Mo	174, 175
Nd-82Fe-7Ti	Lab ball mill	SS	SS	100:1	95	100	Am + α-Fe	175
Nd-74Fe-15Ti	Lab ball mill	SS	SS	100:1	95	100	Am + α-Fe	174
Nd-74Fe-15V	Lab ball mill	SS	SS	100:1	95	100	Am + α-Fe	174, 175
Nd-81Fe-4Ti- 4Mo	Lab ball mill	SS	SS	100:1	95	100	Am + α-Fe + Mo	174
Ni-Al	Lab ball mill	Hardened steel	Hardened steel	10:1			27–35 Al	176
Ni-25Al-25Ti	SPEX 8000	Hardened steel		10:1	—	30	Partly Am.	177
Ni-Mg	Planetary ball mill			10:1		180	20–50 Mg	154
Ni-50Mo	SPEX 8000	Hardened tool steel		10:1	—	10–32	Partly Am.	178, 179
Ni-20Nb	Fritsch P6			10:1		80		180
Ni-40Nb	Fritsch P6			10:1		85		180

Table 12.1 Continued

System	Mill	Vial material	Grinding medium material	BPR	Speed (rpm)	MA time (h)	Glass-forming range (at%)	Ref.
Ni-40Nb	SPEX 8000	Hardened tool steel	52100 steel	3:1	—	9–11		18, 181
Ni-40Nb	Lab ball mill	SS	Hardened steel			200		182
Ni-40Nb	Fritsch P6	—	—	10:1	—	85		183
Ni-55Nb	Ball mill	SS	SS			11		184
Ni-60Nb	Fritsch P6			10:1		95		180
Ni-Nb	Planetary ball mill		Cr-steel	4:1			20–80 Nb	185
Ni-Nb	SPEX 8000	Hardened tool steel	440C steel	10:1	—		21–80 Nb	186
Ni-Nb								187
Ni-18P	Horizontal rotating ball mill	SS	SS	30:1	80	800	Partly Am.	97
Ni-30Si	Planetary ball mill	Hardened tool steel	Hardened tool steel	11.3:1		167		77
Ni-33Si	Planetary ball mill	Hardened tool steel	Hardened tool steel	11.3:1		220		77
Ni-33.3Si	Fritsch P5	SS	SS	9.6:1		200		79
Ni-Sn							20–40 Sn	186
Ni-25Ta	Planetary ball mill	WC-Co	WC-Co			20	Partly Am.	188
Ni-50Ta	Planetary ball mill	WC-Co	WC-Co			20	Partly Am.	188
Ni-Ta	Fritsch P7			5:1			40–80 Ta	189
Ni-40Ti	Planetary ball mill					200		190
Ni-40Ti	Planetary ball mill	Hardened steel	Hardened steel	30:1	720	14 in N_2 20 in Ar 32 in O_2		191, 192
Ni-50Ti	Ball mill	SS	WC			20		193
Ni-50Ti	SPEX 8000	Hardened steel	SS	6:1	—	24		113
Ni-50Ti	SPEX 8000	Tool steel				25–35		194
Ni-66.7Ti	Ball mill	SS	SS			17		184
Ni-Ti		Hardened steel	440C steel				30–67 Ti	195
Ni-V	Conventional ball mill	SS	Steel		110	400	40—65 V	196
Ni-31Zr	Attritor, Mitui Miike MA 1D	SS	SS		300	16		197

Ni-32Zr	Fritsch P5	Steel	Steel	13:1		20		88
Am Ni-32Zr + Zr	Fritsch P7	Steel	Steel			20	35–72 Zr	198
Ni-35Zr	Fritsch P5	Steel	Steel					199
Ni-38Zr	Uni ball mill					60 HE mode 240 LE mode	Partly Am.	200
Ni-50Zr	SPEX 8000	Hardened tool steel	440C steel	10:1	—	18		201
Ni-50Zr	SPEX 8000	Hardened steel	SS	6:1	—	24		113
Ni-50Zr	Super Misuni NEV-MA8	SS	SS	7:1	—	15	−180°C No Am. 25°C Partly Am. 200°C Am	202
Ni-60Zr	SPEX 8000	Hardened tool steel	440C steel	10:1	—	12		201
Ni + $NiZr_2$	SPEX 8000	Hardened tool steel	440C steel	10:1	—	15	20–60 Zr	203
$Ni_{11}Zr_9 + NiZr_2$	SPEX 8000	Hardened tool steel	440C steel	10:1	—		15–76 Zr	201, 203
$Ni_{10}Zr_7$ + $Ni_{21}Zr_8$			Hardened steel			21		204
Am-Ni-32Zr + Zr						20	35–72 Zr	198
Ni-Zr							15–73 Zr	110, 142, 198, 205–207
Ni-Zr	Fritsch P5	SS	SS		710	15–20	20–80 Zr	208
Ni-Zr	Planetary ball mill	Steel	Hardened steel				35–45 Zr	209
Ni-Zr	Vibratory ball mill		WC or 440C steel	54:1		32	30–80 Zr	210
Ni-Zr	High-energy ball mill	Steel		4:1		120	30–80 Zr	206
Ni-Zr							17–70 Zr	211
Pd-17Si	Lab ball mill	SS	Hardened steel			200		212
Pd-20Si	Fritsch P5	Tempered steel		5:1		10–15		213
Pd-20Si		WC-Co	WC-Co			13	Am + Pd_3Si	214
Pd-Si	SPEX 8000	Hardened steel	SS	5:1	—	8	19 Si	215
Pd-Si	Fritsch P5						13.5–25 Si	216
Pd-Ti		WC-Co	WC-Co				42–85 Ti	214
PTFE-Cu	Vibration high-energy mill	SS	SS	47:1				216
PTFE-Ni	Vibration high-energy mill	SS	SS	47:1				216

Table 12.1 Continued

System	Mill	Vial material	Grinding medium material	BPR	Speed (rpm)	MA time (h)	Glass-forming range (at%)	Ref.
Se-Te	Fritsch P5	SS	SS	—	—	150	0–20 Te	217
Si-30Fe								218
Si-10 to 20Sn	Fritsch P7/P5	Tempered steel		1.4:1				219
Sm-87.5Fe	Planetary ball mill Retsch PM 4000						Am (Sm-Fe) + α-Fe	220
Sm-Fe	SPEX 8000	Hardened steel	Hardened steel	10:1	—	48	Am (Sm-Fe) + α-Fe	221
Ta-30Al	Rod mill	SS	SS	36:1	85	400		222
Ta-33Al	Rod mill	SS	SS	30:1	85	400		223
Ta-Al	Ball mill	SS	SS	108:1	85	300	10–90 Al	224
Ta-20 to 50Cu	Fritsch P5	SS	SS	4:1	Int. 7 or 5	15	20 Cu	225
Ta-20 to 50Cu	Fritsch P5	Cu-Be	Cu-Be	4:1	Int. 7 or 5	75	30 Cu	225
Ta-30Cu	Super Misuni NEV-MA8	Cu-Be	Cu-Be			60 (200 °C)	Partly Am.	226
Ta-30Cu	Fritsch P5	Cu-Be	Cu-Be	7:1	Int. 5	120		227
Ta-30Cu	Planetary ball mill	Steel	Steel	15:1		100		228
Ta-Cu	Planetary ball mill	Steel	Steel	15:1			30–50 Cu	229
Ta-5Fe	Fritsch P5	SS	SS	4:1		30		225
Ta-Ni	Fritsch P7	High speed steel	Cr-steel	5:1		120	20–60 Ni	189
Ta-Ni	SPEX 8000	High speed steel	Cr-steel	2:1	—	15–20	10–80 Ni	230
Ta-37.5Si	SPEX 8000	WC-lined SS	52100 steel	5:1	—	12		171, 231
Ti-24 Al	SPEX 8000	Steel	52100 steel	10:1				232
Ti-25Al	SPEX 8000	Hardened tool steel			—	9–16		181
Ti-25Al	SPEX 8000	WC		8:1	—	20–24		233
Ti-25Al	Ball mill	SS	SS	100:1		50		234
Ti-25Al	Vibration mill	SS	SS	5:1		25	Am + Microcrystalline	163
Ti-50Al	Ball mill	SS	SS	100:1		500		234
Ti-50Al	Fritsch P5	SS	SS	18:1		20		235
Ti-50Al	Szegvari attritor					60		236
Ti-50Al	Attritor	SS	SS	15:1	150	15		237

Ti-50Al	Planetary ball mill			60:1	720	30 in N_2 100 in Ar		238
Ti-50Al	Planetary ball mill			20:1		40		239
Ti-50Al		SS	Hardened steel	70:1 to 100:1		75		240
Ti-50Al	Horizontal ball mill	SS	440C steel	40:1		500		241
Ti-50Al	Horizontal ball mill	SS	SS	30:1 to 60:1	150	80		242
Ti-50Al	Vibration mill	SS	SS	5:1				163
Ti-50Al	Vibratory ball mill	SS	SS	15:1		80		237
Ti-50Al	Planetary ball mill Retsch PM 4000	SS	SS	15:1	150			237, 242
Ti-50Al		SS	SS	100:1	—	800		243
Ti-60Al	Vibration mill	SS	SS	5:1		25		163
Ti-60Al	Fritsch P5	Hardened steel	Hardened steel	10:1		25–30	20—60 Al	244
Ti-Al	SPEX 8000	Steel	52100 steel	10:1				245
Ti-Al							45–65 Al	143
Ti-Al	Lab mill	SS	SS	100:1	95		50–75 Al	246
Ti-Al	Fritsch	Hardened steel	Hardened steel	10:1			20–50 Al	247
Ti-Al	Fritsch P7	Hardened steel	Hardened steel	10:1	490	40	10–60 Al	61
Ti-Al	Fritsch P5	Hardened steel	Hardened steel	10:1		15–75	20–50 Al	248
Ti-Al	Fritsch P5	Steel	Steel	10:1			10–75 Al	61
Ti-Al	Rod mill	SS	SS	30:1	85	200	25–67 Al	249
Ti-48Al-2Cr	Horizontal ball mill		SS	30:1	100	500	Partly Am.	250
Ti-48Al-2Mn-2Nb	FritschP5	SS	SS	20:1	200	20		251
TiH_2-22Al-23Nb	Fritsch P5	SS	SS	10:1	—	40	Partial (?)	252
Ti-40Al-10Ni	SPEX 8000	Steel	Steel	8:1	—	40		253
Ti-25Al-25Ni	SPEX 8000	Steel	Steel	8:1	—	20		253
Ti-25Al-25Ni	Ball mill	SS	SS	100:1		100		54
Ti-20Al-20Ni	SPEX 8000	Steel	Steel	8:1	—	40		253
Ti-15Al-15Ni	SPEX 8000	Steel	Steel	8:1	—	20		253
Ti-10Al-30Ni	Ball mill	SS	SS	100:1		100		54
Ti-Al-Ni	Ball mill	SS	SS	100:1		500	0–50 Al, 0–50 Ni, 30–90 Ti	54
Ti-50Co	SPEX 8000			15:1				236
Ti-50Cu		SS	Hardened steel	70:1 to 100:1		75		240
Ti-50Cu		WC-Co	WC-Co					214
Ti-Cu		Hardened tool steel or WC-Co	WC-Co	3:1 to 10:1			10–87 Cu	106
Ti-Cu		Hardened steel	Hardened steel or WC			17	10–90 Cu	148
Ti-Cu	Fritsch P7	WC	WC	10:1	490	20	10–50 Cu	254

Table 12.1 Continued

System	Mill	Vial material	Grinding medium material	BPR	Speed (rpm)	MA time (h)	Glass-forming range (at%)	Ref.
Ti-20Cu-24Ni-4Si-2B	SPEX 8000	Hardened steel	Hardened steel	5:1	—	32		255
Ti-40 to 60Mn	Fritsch P6	Ti	Ti	5:1 to 7:1	460	55–40		256
Ti-24.3 Ni	Ball mill				110	400		257
Ti-33Ni	Ball mill	SS	SS			17		184
Ti-33Ni	SPEX 8000	Hardened steel	Hardened steel					258
Ti-35Ni	SPEX 8000	Hardened steel	Hardened steel			11		181
Ti-40Ni	Planetary ball mill			30:1	720	15 min		259
Ti-50Ni	Ball mill	SS	SS	90:1		300		260
Ti-Ni	Fritsch P7	WC	WC	10:1	Int. 6	20 h	10–70 Ni	254
Ti-Ni	SPEX 8000	Hardened steel					35–50 Ni	261
$Ti_{50}Ni_xAl_{50-x}$							x = 10, 25	253
$Ti_{60}Ni_xAl_{40-x}$							x = 15, 20	253
$Ti_{60}Ni_{1-x}Cu_x$ (x = 10, 20, and 30)	Fritsch P7	WC	WC	10:1		14–40	x = 30 partly Am.	262
Ti-(40-x)Ni-xCu	Fritsch P7	WC	WC	10:1	490	18	x = 10–20	254
Ti-(50-x)Ni-xCu	Fritsch P7	WC	WC	10:1	490	18	x = 10–30	254
Ti-(60-x)Ni-xCu	Fritsch P7	WC	WC	10:1	490	18	x = 10–40	254
Ti-18Ni-15Cu	SPEX 8000	Steel	52100 steel	10:1	—	16		263
Ti-Ni-Cu	Fritsch P7	WC	WC	10:1		14–40	10–30 Ni 10–30 Cu	262
Ti-18Ni-10Fe-16Si	SPEX 8000			6:1	—	30		264
Ti-Pd	SPEX 8000	WC-Co or hardened steel	WC-Co or hardened steel	5:1		17	15–58 Pd	214, 265
Ti-30Pd-20Cu		WC-Co	WC-Co					214
Ti-35Si	Ball mill					180		266
Ti-37.5Si	Fritsch P5	Cr steel	Cr steel	10:1	153		Am + Ti_5Si_3	267
Ti-37.5Si	Vibration mill	SS	SS	5:1		60		163
Ti-37.5Si	Horizontal ball mill	SS	440C steel	40:1		500		241
Ti-37.5Si	SPEX 8000	SS	Hardened steel	5:1	—	24		268
Ti-66.7Si	Lab ball mill	SS	SS	64:1		25		269
Ti-Si	Vibro mill	Hardened steel	WC-Co			37	16–63 Si; Partly Am. 63–7 Si	270

Ti-Si	Ball mill	SS	SS	50:1		300	20–50 Si	271
Ti-50V	Fritsch P5		SS	20:1		7		272
Ti-50Zr	Fritsch P5	SS	SS	5:1		24		273
Ti-50Zr	Fritsch P5	SS	SS	20:1		285		269
Ti-38Zr-17Ni	Fritsch P7	SS	SS	8:1	3,200	7		274
$TiH_{1.924}$-25Al	SPEX 8000	Steel	52100 steel	10:1	—	30		275
$TiH_{1.924}$-50Al	SPEX 8000	Steel	52100 steel	10:1		30		275
Ti-22Al-23Nb	Fritsch P5	—	—	—	335	40	Partly Am.	276
TiH_2-22Al-23Nb	Fritsch P5	—	—	—	335	40	Partly Am.	276
V-40Ni	Ball mill	SS	SS		110	800		270
V-25Si	Fritsch P5	SS	SS	10:1	280	30		277
V-37.5Si	SPEX 8000	WC-lined SS	52100 steel	10:1	—	12		231
V-Si							25–50 Si	278
W-50Fe	Ball mill	SS	SS	30:1	85	300		279
W-50Fe	Ball mill	SS	SS	30:1	250	400		280
W-Fe	Fritsch P5	Hardened steel	Hardened steel	20:1			30–70 Fe	281
Y-Ni-B-C	Super Misuni NEV MA-8	Hardened steel	Hardened steel	10:1	—	300		282
				15:1		64		
Y-Ni-B-C + 15 wt% Fe	Super Misuni NEV MA-8	Hardened steel	Hardened steel	10:1	—	64		282
Y_2O_3-BaO_2-CuO	SPEX 8000	WC	WC	6:1	—	25		283
Zn-Ti	Planetary ball mill	Steel	Hardened steel	6:1 to 10:1	—	26	40 Ti	284
Zr-30Al	Attritor		ZrO_2 balls	10:1		45		285
Zr-50Al	SPEX 8000	Steel	52100 steel	10:1	—	7		286
Zr-Al	SPEX 8000	SS	SS	4:1	—	24	17.5–30 Al	287
Zr-Al	SPEX 8000	SS	SS	4:1	—	24	17.5–40 Al	288
Zr-Al	SPEX 8000	Hardened steel	WC	4:1	—	12	17.5–40 Al	289
$Zr_{65}Al_{10}Cu_{25-x}Ni_x$ $x = 5–20$	Planetary ball mill Retsch PM 4000	Hardened steel	Hardened steel	15:1		100		290
$Zr_{65}Al_{7.5}Cu_{17.5}Ni_{10}$	Planetary ball mill Retsch PM 4000	Hardened steel	Hardened steel	15:1		60		66
$Zr_{65}Al_{7.5}Cu_{17.5}Ni_{10}$ + 10 vol% AlN, TiN, or Si_3N_4	NEV-MA8 vibrating mill	SS	SS	15:1	—	40		291
$Zr_{55}Cu_{30}Al_{10}Ni_5$	Planetary ball mill Retsch PM 4000	Hardened steel	Hardened steel	14:1	150	100–200		292
$Zr_{50}Al_{13.8}Cu_{26.2}Ni_{10}$	Planetary ball mill Retsch PM 4000	Hardened steel	Hardened steel	15:1		60		66

Table 12.1 Continued

System	Mill	Vial material	Grinding medium material	BPR	Speed (rpm)	MA time (h)	Glass-forming range (at%)	Ref.
$Zr_{41.2}Al_{13.8}Cu_{26.2}Ni_{18.8}$	Planetary ball mill Retsch PM 4000	Hardened steel	Hardened steel	15:1		60		66
$(Zr_{65}Al_{7.5}Cu_{17.5}Ni_{10})_{100-x}Fe_x$ ($x = 0$–20)	Planetary ball mill Retsch PM 4000	Hardened steel	Hardened steel	15:1		100		293, 294
$Zr_{60}Al_{10}Cu_{18}Ni_9Co_3$				6:1		36		295, 296
$Zr_{60}Al_{10}Cu_{18}Ni_9Co_3$	SPEX 8000			6:1			Partly Am.	297
$Zr_{55}Cu_{30}Al_{10}Ni_5$ + 5 to 15 vol% CaO	Planetary ball mill Retsch PM 4000	Hardened steel	Hardened steel	14:1	150	100–200	Am + CaO	292
$Zr_{55}Cu_{30}Al_{10}Ni_5$ + 5 to 30 vol% ZrC	Planetary ball mill Retsch PM 4000	Hardened steel	Hardened steel	14:1	150	100–200	Am. + ZrC	292
Zr-Co	Fritsch P5	Steel	Steel	15:1		60	27–92 Co	298
Zr-50Cu		SS	SS	40:1		200		299, 300
Zr-50Cu	Horizontal ball mill	SS	SS	20:1	80 240	400	Am No Am phase	301
Zr-40Fe	Fritsch			13:1		20		110
Zr-Fe	Fritsch P5	Steel	Steel	15:1		60	30–78 Fe	298
Zr-Fe		Hardened steel	Hardened steel or WC			17	20–50 Fe	148
Zr-Mn	Fritsch P5	Steel	Steel	15:1		60	20–85 Mn	298
Zr-Ni	Fritsch P5	Steel	Steel	15:1		60	27–83 Ni	298
Zr-25Ni-20Al	Planetary ball mill Retsch PM 4000	Steel	Steel	23:1	200	44		302
$Zr-NiZr_2$	SPEX 8000	Hardened tool steel	440C steel	10:1	—	15	24–27 Ni	203
Zr-Pd							45–60 Pd	303
Zr-20Pd-10Ni	Planetary ball mill	Tempered Cr steel	Tempered Cr steel	14:1	250	48		304
Zr-25Rh	SPEX 8001	Hardened steel	—	—	—		Partly Am.	305
Zr-50Rh	SPEX 8001	Hardened steel	—	—	—	6		305

$Zr_{55}Ti_{2.5}Cu_{27.5}Ni_{15}$	Planetary ball mill Retsch PM 4000	Hardened steel	Hardened steel	15:1		100		290
$Zr_{65}Ti_{7.5}Cu_{17.5}Ni_{10}$	Planetary ball mill Retsch PM 4000	Hardened steel	Hardened steel	15:1		60		66
$Zr_{50}Ti_{13.8}Cu_{26.2}Ni_{10}$	Planetary ball mill Retsch PM 4000	Hardened steel	Hardened steel	15:1		60		66
$Zr_{55}Ti_{13.8}Cu_{26.2}Ni_{18.8}$	Planetary ball mill Retsch PM 4000	Hardened steel	Hardened steel	15:1		60		66
$Zr_{48.9}Ti_{9.5}Cu_{29.6}Ni_{12}$	Planetary ball mill Retsch PM 4000	Hardened steel	Hardened steel	15:1		100		290
$Zr_{22.5}Ti_{30}Cu_{30}Ni_{17.5}$	Planetary ball mill Retsch PM 4000	Hardened steel	Hardened steel	15:1		100		290
$Zr_{11.9}Ti_{33.4}Cu_{42.7}Ni_{10}$	Planetary ball mill Retsch PM 4000	Hardened steel	Hardened steel	15:1		100		290
$Zr_5Ti_{45}Cu_{30}Ni_{20}$	Planetary ball mill Retsch PM 4000	Hardened steel	Hardened steel	15:1		100		290
Zr-29V		WC-lined hardened steel	Hardened steel			140		299
ZrO_2-20 mol% MgO	Planetary G7	Steel	Steel	7:1	500–700	188	Am + cubic ZrO_2	68
ZrO_2-TiO_2	Lab planetary ball mill AGO-2	Steel	Steel		700	15 min	Partial Am ZrO_2	306
$ZrO_2.nH_2O$-TiO_2 nH_2O	Lab planetary ball mill AGO-2	Steel	Steel		700	15 min		306

SS, Stainless steel.

[a] All the compositions are expressed in atomic percent, unless otherwise specified. The subscripts in column 1 (unless it is the chemical formula of the compound) represent the atomic percentage of the element in the powder mixture.

[b] MA/1000 h + anneal at 740 K + quench.

[c] In hydrogen atmosphere.

[d] Argon-nitrogen mixture.

[e] Argon-air mixture.

Table 12.2 Amorphous Phases Formed in Intermetallics/Prealloyed Powders by Mechanical Milling

Compound	Mill	Vial material	Grinding medium material	BPR	Speed (rpm)	MA time (h)	Glass-forming range (at%)	Ref.
AgPd	SPEX 8000	—	—	10.4:1	—	65		23
$Al_{13}Co_4$	Fritsch P5	SS	SS	10:1	150	200		25
Al-15 Cr (RS)	Ball mill	SS	SS	120:1		1000	Partly Am.	307
Al-20Cr (RS)	Ball mill	SS	SS	120:1				307
$Al_{67}Mn_8Ti_{24}Nb_1$	Planetary mill	SS	WC	20:1	—	60		308
AlTa	Rod mill					40		309
$BiFeO_3$						65		310
C (graphite)	SPEX 8000	WC	WC	4:1	—	206		311, 312
Graphite (C)	Ball Mill					1000		313
α-Co_2Ge	Fritsch P0	WC bottom-lined steel	Hardened steel			180		314
β-Co_2Ge	Fritsch P0	WC bottom-lined steel	Hardened steel			36		314
Co_5Y		SS	SS		80–110	300		71
CoZr	SPEX 8000	Hardened steel	440 martensitic steel	10:1	—	4–6 h at −85 and −190°C		315
CoZr	Vibrating mill	Hardened steel	440C SS	10:1	—	80		316
Cr	Vibrational mill		SS	45:1		100	Ar-N_2 mixture	89
Cr	Fritsch P5	SS	SS	—	—	640	In a nitrogen atm. (19.1 at% N)	317
Cr-28Fe		Hardened steel	Hardened steel			2000		190
CuZr	SPEX 8000	Hardened tool steel	440C steel	10:1	—			111
Cu_3Zr_2	SPEX 8000	Hardened tool steel	440C steel	10:1	—	13		111
Ge	Fritsch P7/P5	Tempered steel					Partly Am.	318
$GeSe_2$	SPEX 8000	YSZ	YSZ	—	—	0.5		319
Ge_3Se_4	SPEX 8000	YSZ	YSZ	—	—	1		319
Ge_4Se_5	SPEX 8000	YSZ	YSZ	—	—	51.5		319
$HfFe_2$	Retsch MM2 vibratory mill	Steel	Steel					320
La-25Al-20Ni	Planetary ball mill	WC-Co	WC-Co	20:1	180	12		321
$Mg_{70}Zn_{30}$	Planetary ball mill				200	58		322
Nb-20 Ge	SPEX 8000	Hardened tool steel	440C steel	10:1	—	10–15		169
$Nb_{55}Ni_{45}$	Planetary ball Mill							323
$Nd_2Fe_{14}B$	SPEX 8000	Hardened steel	Hardened steel	10:1	—	24	Am + α-Fe	324

Nd-77Fe-8B	High-energy ball mill		SS	18:1		24		325
Ni_3Al	Invicta vibrator mill BX920/2	Hardened tool steel	440C SS	10:1		50	Partly Am.	326
$Ni_{60}Nb_{40}$	Fritsch P6			10:1		85		183
$Ni_{45}Nb_{55}$	Ball mill	SS	SS			14		184
$Ni_{45}Nb_7V_{48}$	Planetary ball mill						Partly Am.	323
$Ni_{35-45}Nb_{45-50}V_{5-20}$	Planetary ball mill						Partly Am.	323
$Ni_{35-45}Nb_{32}V_{23-33}$	Planetary ball mill						Partly Am.	323
Ni_5Si_2	Fritsch P5	SS	SS	9.6:1		200		79
NiTi	SPEX 8000	Hardened steel	440C steel	10:1		18 at 220°C, 13 at 60°C, 2 at −190°C		315
$NiTi_2$	Ball mill	SS	SS			14		184
$Ni_{45}V_{55}$	Planetary ball mill						Partly Am.	323
$Ni_{10}Zr_7$	Fritsch P0	WC	Hardened steel			95		327
NiZr	SPEX 8000	Hardened tool steel	440C steel	10:1	—	18		201
NiZr	Fritsch P5	SS	SS	30:1		48		328
Ni-60 Zr	SPEX 8000	Hardened tool steel	440C steel	10:1	—	8		201
$NiZr_2$	Super Misuni NEV-MA 8	SS	SS	7:1				202
Pd-32Ni-20P	Planetary ball mill	WC-Co	WC-Co	20:1	180	24		321
Sb-12Ga	Ball mill	Agate				16		329
Se	Planetary ball mill						At <50 °C	330
Se	Super Misuni Nisshin Giken			8:1		10		331, 332
Se	High-Energy Ball Mill	SS	SS	10:1		4 h 10 min		333
Si	Fritsch P7/P5	Tempered steel	Steel	1.4:1	Int. 5 or 10	70–95		219
Si	SPEX 8000	Hardened tool steel	440C steel	8:1	—	38	Partly Am.	334
Si	Vibratory mill	Hardened Steel	Hardened Steel	15:1				126
SiC (α)	WL- Planetary ball mill	Hardened Steel	Hardened Steel	40:1		60	Partly Am.	335
Talc	Vibration mill	Steel	Steel					336
Ti-33Ni	Ball mill	SS	SS			14		184
Ti-33Ni	SPEX 8000	Hardened steel	Hardened steel					258
TiNi	SPEX 8000	Tool steel	440C steel	10:1	—	13		337
Ti-18Ni-15Cu (RS)	SPEX 8000					16		263
Ti_5Si_3	Fritsch P5	Cr steel	Cr steel	10:1	228		Partly Am.	267
TiV	Fritsch P5		SS	20:1		100^b		272

Table 12.2 Continued

Compound	Mill	Vial material	Grinding medium material	BPR	Speed (rpm)	MA time (h)	Glass-forming range (at%)	Ref.
V	Fritsch P5	SS	SS		520		In a N_2 atm (24.3 at% N)	317
YCo_2	Ball mill	Steel	Steel		110	300		20, 113, 338
YCo_3								20
YCo_5	Ball mill	Steel	Steel		110	300		113, 338
YNi_2B_2C	Planetary ball mill Retsch PM 4000	WC	WC	56:1		50		339
$YPd_5B_3C_{0.3}$	Planetary ball mill Retsch PM 4000	WC	WC	56:1		50		339
Zr_2Co	Planetary ball mill	Hardened steel			300 m/s^2			340
ZrCo	Planetary ball mill	Hardened steel			600 m/s^2			340
$ZrFe_2$	Retsch MM2 vibratory mill	Steel	Steel					320
Zr_2Ni								202
$Zr_{70}Pd_{30}$	Fritsch P5	Cr-Steel	Cr-Steel	14:1	2.1 s^{-1}	72		341
$ZrSiO_4$	Planetary G7	Steel	Steel	7:1	500–700	240		68

SS, Stainless steel.

higher. However, in some cases it has been reported that the times for the formation of a fully amorphous phase are approximately the same in both MA and MM cases (reported in Ref. 184).

Formation of a crystalline intermetallic phase has been reported to occur before amorphization in Cu-Ti [105], Fe-Sn [134], Nb-Ge [169], Nb-Si [172], Nb-Sn [173], Ni-Zr [209], and Ti-Ni-Fe-Si [264], among others. In some instances, it has also been reported that a solid solution forms first which on continued milling becomes amorphous. This has been reported to be true in case of Ta-Al [223], Ti-Al [232, 245], Zr-Al [289,343], and several other alloy systems. In some systems, it was reported that MA of the blended elemental powder mixture results, with increasing milling time, in the sequential formation of phases of solid solution, followed by an intermetallic and finally an amorphous phase [163,344]. Whether an intermetallic or a solid solution phase forms before amorphization depends on the relative free energies of the solid solution and the intermetallic phases; the phase with the lower free energy forms first. On the other hand, if the amorphous phase has the lowest free energy of all these competing phases, it forms directly without any other phase forming prior to that. (See Chapter 11 for a brief thermodynamic explanation for this feature).

In general, the end product is the same irrespective of whether the starting material is a mixture of the elements, a single intermetallic phase, or a mixture of two or more such phases. This has been shown to be true in the Nb-Al system, where an amorphous phase was obtained whether the starting material was an intermetallic or blended elemental powder mixture. However, in the Mg-Zn system, it has been reported that an amorphous phase could be produced only by MM starting from the intermetallic $Mg_{70}Zn_{30}$ and not by MA starting from the blended elemental powder mixture of the Mg-30Zn composition [322]. On the other hand, an Fe-33Ti powder mixture could be amorphized by MA whereas the intermetallic Fe_2Ti could not be amorphized by MM [135]. Reasons for differences in these behaviors will be discussed later.

Amorphization in ordered alloys seems to follow the sequence [326]:

$$\begin{aligned}&\text{Ordered phase} \rightarrow \text{disordered phase (loss of long-range order)} \rightarrow \\ &\quad \text{fine grained (nanocrystalline) phase} \rightarrow \text{amorphous phase}\end{aligned} \tag{12.5}$$

The formation of an amorphous phase can occur, at least in some cases, without the loss of long-range order (see Chapter 11 for examples that illustrate this behavior).

12.4 EFFECT OF PROCESS VARIABLES

Even though it may appear that any alloy can be made amorphous under appropriate conditions of milling, this has not been possible in all the cases. It is possible that the right conditions have not been obtained in these cases. This is because a number of variables, the more important of them being milling energy, milling temperature, and impurity contamination, control the constitution of the final product. This observation is similar to what was reported in the case of amorphization of alloys by the technique of RSP. It was reported [3–10] that any alloy could be amorphized under appropriate solidification rates, but these rates may be so high that it may not possible to achieve them in practice.

The effect of process variables during MA on the amorphization behavior has been studied in several alloy systems. Among these, the most important variables studied are milling energy (including higher BPR) and milling temperature. Since powder contamination also has been shown to be a key factor in the amorphization process, this also has been discussed. The effect of other process variables was discussed in Chapter 5.

12.4.1 Milling Energy

The kinetic energy of the grinding medium (balls) ($E = \frac{1}{2}mv^2$), where m is the mass of the balls and v is the relative velocity with which they move) can be increased by increasing either the mass or the relative velocity of the balls. Increased milling energy (also achieved by use of higher BPR, increased speed of rotation, and so forth) is normally expected to introduce more strain and consequently increase the defect concentration in the powder, leading to easier and faster amorphization. However, higher milling energies also produce more heat (and higher temperatures), and this can result in the crystallization of the amorphous phase. Therefore, a balance between these two effects determines the nature of the final product phase. If the defect concentration predominates, then an amorphous phase forms. On the other hand, if the temperature rise is higher so that recovery and recrystallization occur, the amorphous phase, even if it had formed, crystallizes during the milling process.

It was reported in the Ni-Zr system that milling in a planetary ball mill at an intensity of 3 did not produce any amorphous phase [211]; this was due to the insufficient energy available at low intensities. However, when the intensity was increased to 5, amorphous phase formation was observed in a wide composition range of 17–70 at% Zr. At an intensity of 7, amorphous phase formation was observed, but only between 25 and 34 at% Zr. These observations suggest that with increasing milling energy, the heat generated is high enough to raise the temperature of the powder above the crystallization temperature of the amorphous phase. Consequently, crystallization of the amorphous phase takes place. Similar results were also reported in other systems [196,322,345,346]. Thus, it appears that the maximal amorphization range is observed at intermediate values of milling intensity—too low an intensity does not provide enough energy to amorphize, whereas at very high intensities the amorphous phase formed crystallizes.

It has also been reported that increasing the BPR has a similar effect. It was shown in the Al-Ta system that a fully amorphous phase was obtained only at a BPR of 36:1 or 108:1. When the BPR used was 12:1 a crystalline phase was obtained, whereas at a BPR of 324:1 a mixture of amorphous and crystalline phases was obtained [57].

It has been known for some time now that formation of metastable phases in general, and amorphous phases in particular, is observed only when the milling conditions during MA/MM are not very severe. For example, Gerasimov et al. [340] reported that an amorphous phase could be obtained only under soft milling conditions. "Soft" milling conditions means that the milling energy/intensity is not high. In this study, the authors had used small-diameter grinding medium. On the other hand, when "hard" milling conditions were used (by increasing the diameter of the grinding medium), only a crystalline phase was produced. Many such examples are available in the literature [213,244,278,301]. Therefore, it may be safely concluded that metastable phases (including amorphous phases) could be produced only under soft milling conditions. Hard milling conditions produce less "metastable" (where the departure from equilibrium is less than in the other case) phases, including the equilibrium phases.

In some cases an amorphous phase may form at both low and high milling intensities, provided that the temperature rise at the high milling intensity is not sufficient to cause crystallization of the amorphous phase produced. However, the difference may be in the time required for amorphization. It has been reported that amorphization in a Ni-38 at% Zr powder mixture requires 60 h in a high-energy mode in a Uni-Ball mill whereas 240 h is required at low and intermediate energies; even then the product is only partly amorphous [200].

12.4.2 Milling Temperature

There have been conflicting reports on the effect of milling temperature on the nature of the phase formed. Koch et al. [315] summarized the results of varying the temperature of the mill on the kinetics of amorphization in intermetallics. They concluded that generally a lower milling temperature led to faster formation of the amorphous phase. Since a nanostructured material can be easily produced at lower temperatures of milling, the increased grain boundary area drives the crystal-to-amorphous transformation. For example, the time required for amorphization in NiTi was 2 h at −190°C, 13 h at 60°C, and 18 h at 220°C [315]. Similar results were also reported by some others [327].

Figure 5.4a showed the grain size obtained at different milling temperatures for the CoZr intermetallic. It may be seen that the grain size is smaller at lower milling temperatures; therefore it is easier to form the amorphous phase at lower milling temperatures. Note also the band representing the critical grain size for amorphization. The milling time required for obtaining an amorphous phase as a function of the normalized milling temperature (ratio of the milling temperature to the melting temperature) is plotted for NiTi, CoZr, and $NiZr_2$ intermetallics in Fig. 5.4b. This figure confirms that the times are shorter at lower homologous temperatures [347]. While a majority of the limited number of studies follow the above behavior, there is at least one example in which the reverse behavior was reported. Lee et al. [202] observed that the amorphization kinetics were faster at higher temperatures of milling. They indicated that while no amorphization occurred in the NiZr and $NiZr_2$ intermetallics after milling for 15 h at −180°C, full amorphization was observed for the same milling time at 200°C. A partially amorphous phase formed at an intermediate temperature of 25°C. This was explained on the basis of the increased interdiffusion rates at higher temperatures. Later experiments by Koch [347] confirm that, in accordance with their earlier results, rapid amorphization was achieved in the $NiZr_2$ compound at lower temperatures.

Room temperature MA of Zr-Al blended elemental powders is known to produce an hcp solid solution of Al in Zr up to 10 at% Al. Amorphization was found to occur beyond this concentration range, i.e., the crystal-to-amorphous transition in the Zr-Al system seems to proceed under a polymorphic constraint. In other words, there is no observable two-phase (solid solution + amorphous phases) coexistence region as would be expected from a common tangent rule if a metastable two-phase equilibrium was established. In contrast to this, milling at higher temperatures, say up to 573 K, clearly revealed a two-phase solid solution + amorphous phase region. The authors calculated the Gibbs free energy values for the solid solution and amorphous phases and noted that the crossover point occurs at 10 at% Al at room temperature. On increasing the temperature, the crossover point shifted to higher Al contents and also increased the width of the two-phase region. In fact, as Figure 12.1 shows, the width of the two-phase region increased from almost zero at room temperature to about 17 at% Al at 573 K [343]. This was explained using the concept of γ, the parameter that represents the ratio of the frequencies of atomic jumps caused by external forcing and that due to thermal diffusion (see Chapter 9 for details of the discussion of the γ parameter). However, such studies have not been reported in other alloy systems.

12.4.3 Powder Contamination

Mechanical alloying/milling introduces contamination into the milled powder and this can substantially alter the constitution and stability of the powder product (see Chapter 15 for full details of this aspect). Generally, the presence of additional elements favors amorphization since the increased number of species makes it difficult to allow arrangement of the constituent atoms into a well-crystallized structure. (This is very similar to the confusion principle enunciated for the formation of amorphous phases by RSP of liquid alloys and so frequently obeyed [9,10].) For example, it has been reported that an amorphous phase was obtained in ball-milled Fe-Cr alloy powders only in the presence of oxygen; an intermetallic formed in the absence of oxygen [124]. A similar situation was obtained in MM of Nb_3Al powders where a disordered solid solution was produced when milled in a clean atmosphere whereas an amorphous phase formed in the presence of nitrogen [348]. Al-Ti [60] and Ni-Nb [187] systems also exhibited similar behavior. Substantial pickup of nitrogen and oxygen

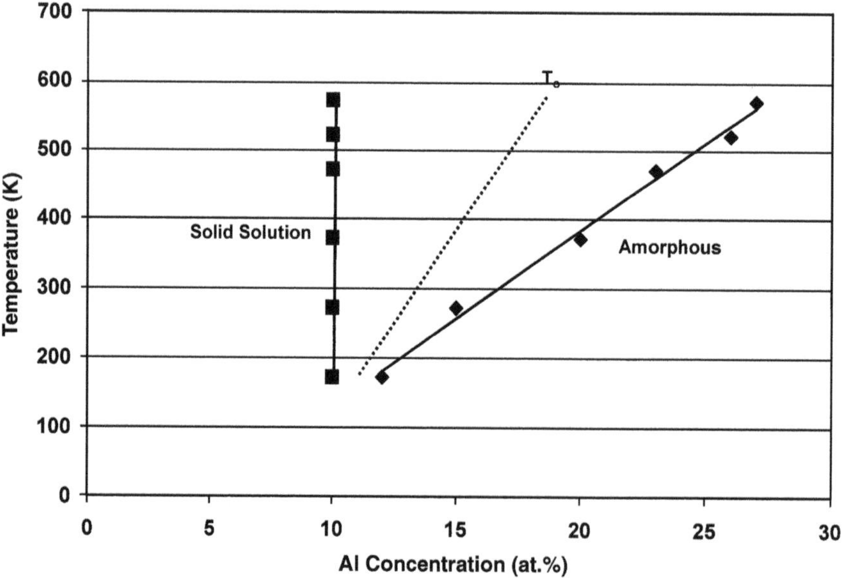

Figure 12.1 Increased two-phase (solid solution + amorphous) region at higher temperatures of milling in the Zr-Al system [343].

was found to be the reason for amorphization observed in Ti-50 at% Al powders. Improved atmospheric control prevented amorphization even after milling for 100 h [349].

While the above results suggest that nitrogen contamination has increased the stability of the amorphous phase, there are also reports where the stability of the amorphous phase was known to decrease due to contamination. That is, the amorphous phase is no longer stable, but a crystalline phase forms. (There are also some reports where continued milling of an amorphous phase results in the formation of equilibrium crystalline phases, referred to as mechanical crystallization. This aspect will be discussed later.) Formation of a crystalline phase has been reported on continued milling of Ti-Al powders after the formation of an amorphous phase [350]. Chemical analysis of the powder, fine-structure analysis by X-ray photoelectron spectroscopy techniques, and lattice parameter measurements strongly suggested that this fcc crystalline phase is due to increased nitrogen contamination of the milled powder and the crystalline phase has been identified as TiN [351]. Similar results were reported on milling of Zr-Al powder [289].

Sometimes, iron contamination from the grinding medium and milling container can be so high as to substantially alter the powder composition. For example, Sá Lisboa et al. [55] milled an Al-30 at% Si-15 at% Fe-5 at% Ni powder blend in a SPEX 8000 mill for 200 h to produce an amorphous phase. They noted that an amorphous phase was formed. However, chemical analysis of the powder showed that excessive contamination by Fe occurred and that the final composition of the powder was Al-18 at% Si-48 at% Fe-3 at%Ni. Therefore, it was concluded that formation of the amorphous phase was due to increased Fe concentration in the powder.

12.5 THERMODYNAMICS AND KINETICS OF AMORPHOUS PHASE FORMATION

Metallic glasses have been obtained by rapidly solidifying the melts of appropriate compositions in a number of binary, ternary, and higher order alloy systems [3–10,352]. It has been well established that, among other conditions, the likelihood of obtaining a glassy phase is very high near deep eutectics. This was explained on the basis of the higher reduced glass transition temperature, T_{rg} (defined as the ratio of the glass transition temperature, T_g, to the melting point of the alloy, T_m, i.e., $T_{rg} = T_g/T_m$). Since the glass transition temperature does not change very much with composition, but the melting point of the alloy is drastically reduced at the eutectic composition, the T_{rg} value is quite large. And the larger the T_{gr} value, the easier it is to obtain metallic glasses by quenching from the liquid state. This was one of the most important criteria in selecting the composition of an alloy to investigate whether it would form a glassy phase or not by rapid solidification from the liquid state [10]. In fact, it is the very large T_{rg} values of alloys that are responsible for obtaining bulk metallic glasses in millimeter section thickness in multicomponent alloys [353].

The situation in mechanically alloyed materials appears to be quite different. For example, most commonly amorphous phases are reported to form by MA around equiatomic compositions, irrespective of the nature of the phase diagram [354]. This observation can be rationalized with reference to Figure 12.2. A hypothetical binary phase diagram featuring some solid solubility on either end and also exhibiting the presence of an intermetallic phase is shown in Figure 12.2a. α is the terminal solid

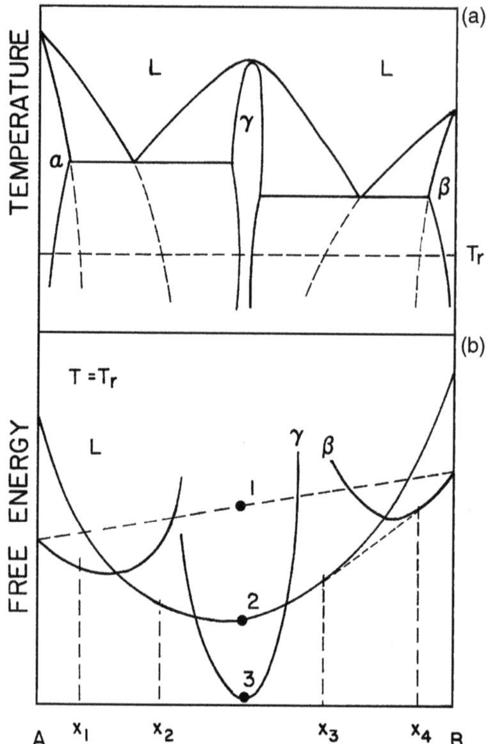

Figure 12.2 (a) Hypothetical binary phase diagram featuring limited solid solubility on both ends and also showing the presence of an intermetallic phase. (b) Gibbs free energy vs. composition at a temperature corresponding to T_r in the above diagram. Note that points 1, 2, and 3 represent the free energy values of the mixture of A and B, amorphous phase, and the intermetallic, respectively.

solution of B in A, β is the terminal solid solution of A in B, γ is the intermetallic, and L represents the liquid phase. The Gibbs free energy of the different phases at a temperature T_r is shown as a function of composition in Figure 12.2b, assuming that the free energy of the undercooled liquid fairly accurately represents the free energy of the amorphous phase. At this temperature, α, β, γ, and mixtures of these phases are thermodynamically stable, whereas L is metastable.

The equiatomic blended elemental powder mixture has a free energy, G_c, corresponding to point 1 in Figure 12.2b, half way along the straight line joining the free energies of the two pure metals A and B. If the two component elements A and B are allowed to interdiffuse freely, then the lowest free energy state corresponding to the formation of the γ phase, G_i (point 3), will be obtained. Even though thermodynamically this is the lowest free energy state, this can be kinetically prevented from occurring when the free energy G_a, corresponding to the formation of an amorphous phase (point 2), is reached.

Preventing the reaction c (crystalline elemental mixture) → i (intermetallic) from occurring and favoring the c → a (amorphous) to occur is possible by a proper choice of metals A and B, the reaction temperature T_r, and the reaction time t_r. Schwarz and

Johnson [14] proposed that two conditions must be satisfied for an amorphous phase to form from a thin-film couple (or a blended elemental powder mixture). The two components must have a large negative heat of mixing and asymmetrical diffusion coefficients. That is, one of the components must be capable of diffusing much faster than the other. The condition of a large negative heat of mixing in the liquid (amorphous) state provides the thermodynamic driving force for the reaction to occur. Other investigators [355,356] have also confirmed the importance of the large negative heat of mixing. Hellstern and Schultz [355] demonstrated this by studying the glass-forming ability of several titanium-containing transition metal alloys. Systems such as Cu-Ti, Ni-Ti, and Co-Ti with a large negative heat of mixing could be completely amorphized, whereas Fe-Ti, Mn-Ti, and Cr-Ti could be only partially amorphized due to their smaller heat of mixing. The isomorphous V-Ti system did not amorphize at all because a solid solution always formed on milling the powder mixture. Alloy systems with a positive heat of mixing have also been amorphized by MA in recent years. Thus, a large negative heat of mixing appears to be an essential condition for glass formation by RSP and not necessary for MA/MM. The second criterion of largely different diffusion rates into each other and in the amorphous alloy phase favors kinetics of the reaction path c $\rightarrow$ a in Figure 12.2b over that of reaction path c $\rightarrow$ i. This asymmetry in the diffusivities has been correlated with the difference in atomic sizes for several elements [354].

The reaction temperature is another important parameter. If the temperature at which milling is conducted is high, then formation of the amorphous phase would not be generally expected. This is because the temperature rise during milling and the high temperature of milling together may result in a situation wherein the temperature attained by the powder is higher than the crystallization temperature of the amorphous phase. Consequently, even if the amorphous phase were to form, it would crystallize. This has been shown to be true when higher milling intensities or higher BPR values were employed during milling, i.e., "hard" milling conditions were employed. In this respect, it is better that milling be carried out at a relatively low temperature if formation of an amorphous phase is desired. Furthermore, diffusion of one element into the other is slow at low temperatures; consequently, nucleation and growth of the intermetallic would not take place.

In some cases, it has been reported that an additional heat treatment, after MA of the powder blend, facilitates formation or completion of formation of the amorphous phase. In the case of an Al-15 at% Cr alloy powder mixture, only partial amorphous phase formation was observed after milling for 1000 h in a low-energy ball mill. However, when this milled powder was annealed at 740 K and quenched, a fully amorphous phase was obtained [26]. A similar observation was made in the Ta-Cu system [357]. In general, a lower temperature of milling favors amorphization. It was also reported that on milling a Co-Ti powder mixture, only an intimate mixture of the Co and Ti phases was observed at short milling time, say 1 h. But, when this fine-grained mixture was heated to a higher temperature (externally), amorphous phase formation was noted [85]. The authors designated this process as thermally enhanced glass formation reaction (TEGFR). It is important to realize that alloying is always facilitated when the components are distributed evenly and on a fine scale. That is why it was suggested that deformation-induced layer refinement during MA is an important component of the amorphization reaction in addition to the large negative heat of mixing [356].

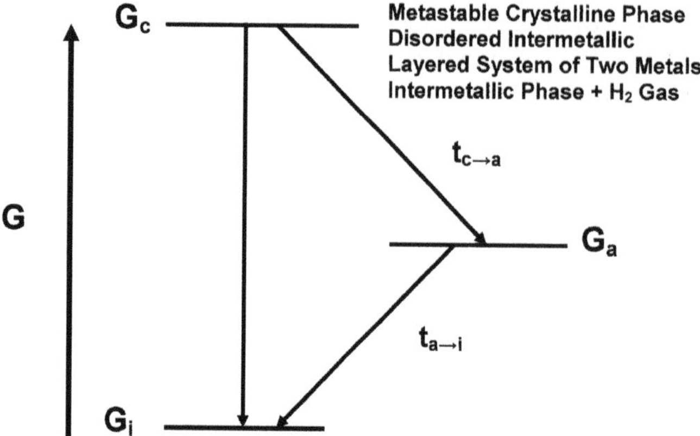

Figure 12.3 Schematic free energy diagram indicating the time scale criteria to be met for solid state amorphization reaction. G_i represents the free energy of the stable equilibrium crystalline phase, G_c the free energy of the metastable crystalline phase with higher free energy, and G_a the free energy of the amorphous phase. G_c can be achieved either by disordering the intermetallic, or mixing two layers of the metals, or by introducing hydrogen gas into the system.

The reaction time scales also are important because not only should one allow the formation of the amorphous phase, but one should also maintain it without transforming to the equilibrium phases. These concepts can be explained with reference to Figure 12.3. As in Figure 12.2, we assume that the free energies of the blended elemental mixture is G_c, that of the amorphous phase G_a, and that of the intermetallic is G_i, as shown schematically in Figure 12.3. To prevent the stable intermetallic from forming, the time scale for the formation of the amorphous phase from the blended elemental mixture, i.e., $t_{c \to a}$, should be much shorter than $t_{c \to i}$, the time scale for the formation of the intermetallic from the blended elemental powder mixture. Furthermore, since the amorphous phase is metastable, $t_{a \to i}$ should be much longer than $t_{c \to a}$ so that the stability of the amorphous phase can be increased. Thus, the kinetic conditions for the formation of an amorphous phase by solid-state reactions can be summarized as

$$t_{c \to a} \ll t_{c \to i} \qquad \text{and} \qquad t_{a \to i} \gg t_{c \to a} \tag{12.6}$$

12.6 MECHANISMS AND MODELS FOR AMORPHIZATION

The mechanism of amorphization by MA and/or MM is not clearly understood. Early investigators [20,21] assumed that the powder particles melted because of the very high temperatures experienced instantaneously by the powder particles due to the high rate of plastic deformation and the large amount of energy transferred to the powder particles by the milling medium. Subsequent quenching of the liquid by heat conduction into the less deformed, and hence cooler, interior regions of the particles resulted in the formation of the amorphous phase (as in RSP). However, energy input calculations and temperature estimates and measurements suggest that the temperature rise is not large enough for the powder particles to melt (see also Chapter 8). Therefore, it is unlikely that this is the mechanism by which amorphous phases are obtained in mechanically alloyed powders. Furthermore, were this mechanism true, the glass-forming composition ranges in mechanically alloyed and rapidly solidified alloys should be the same; and this is also not true, as will be shown later.

Researchers now believe that amorphization during MA is not purely a mechanical process and that a solid-state reaction similar to that observed in thin films [14] occurs during MA also. During MM, however, destabilization of the crystalline phase is thought to occur by the increase in free energy through accumulation of structural defects such as vacancies, dislocations, grain boundaries, and antiphase boundaries. The continuous decrease in grain size (and consequent increase in grain boundary area) and a lattice expansion would also contribute to the increase in free energy of the system [358]. It has also been reported that amorphization occurs when the strain in the slowly diffusing species reaches a maximum [61]. In the case of ordered alloys, amorphization was reported to occur when the long-range order parameter is less than 0.6 with a corresponding volume of expansion of about 2% [359]. It has been reported that the stored energy during MA can be about 50% of the enthalpy of fusion, whereas by cold rolling or wire drawing it is only a small fraction [360]. This explains why amorphization is not possible by simple cold working operations but is possible by MA/MM methods. These lattice defects introduced into the material during MA raise the free energy of the intermetallic system to a level higher than that of the amorphous phase; consequently, it becomes possible for the amorphous phase to form.

Liu and Chang [361] have recently undertaken a detailed study of the amorphization behavior of $Fe_{70-x-y}Co_xNi_yZr_{10}B_{20}$ ($x = 0, 7, 21$ and $y = 14, 21, 28$) alloy powders by MA and MM. They noted that both the blended elemental and prealloyed powders first transformed to a supersaturated solid solution of bcc α-Fe and an amorphous phase was obtained later in Co-free alloys and Co-containing alloys with high Ni/Co ratios of 1 and 3. No amorphous phase was obtained in Co-containing alloys with a lower Ni/Co ratio of 0.33. Thus, depending on the composition of the powders, three different routes have been noted:

1. Fully amorphous phase for all Co-free alloys and Co-containing alloy of $Fe_{42}Co_7Ni_{21}Zr_{10}B_{20}$
2. A mixture of amorphous phase, nanocrystalline bcc α-Fe solid solution and residual $Zr_{70}Ni_{30}$ crystalline phases for $Fe_{56}Co_7Ni_7Zr_{10}B_{20}$, and
3. A mixture of single supersaturated nanocrystalline bcc α-Fe solid solution and residual $Zr_{70}Ni_{30}$ crystalline phases for the $Fe_{42}Co_{21}Ni_7Zr_{10}B_{20}$

The formation of the bcc α-Fe solid solution as a transition phase was explained on the basis of the lower heat of mixing between the different alloying elements and the enhanced diffusion process. It was also reported that the dissolution of Ni, Co, and Zr into the Fe lattice increases the elastic energy for the formation of the solid solution due to atomic size mismatch effect. This results in larger lattice parameter for the bcc α-Fe solid solution and leads to destabilization of the crystalline phase and formation of the glassy phase. The increase in lattice strain could also be due to an increase in grain boundary volume fraction (due to formation of nanocrystalline structure) and mechanical deformation in addition to the size mismatch.

Irradiation of crystalline materials by energetic particles and electrons has been known to cause amorphization when the following criteria are obeyed [11]:

The intermetallic compound has a narrow or zero homogeneity range.

The order-disorder transition temperature of the intermetallic, T_c is higher than the melting temperature, T_m, i.e., the intermetallic is of the permanently (or irreversibly) ordered variety.

The two components (elements) are separated by more than two groups in the periodic table.

The intermetallic has a complex crystal structure.

The fraction of the solute atoms, $f_B \geq 1/3$.

Intermetallics have also been amorphized by MM when the above criteria were generally followed. However, there have been several exceptions to the above empirical rules (too many to be ignored). For example, compounds with reasonably wide homogeneity ranges have also been amorphized. Furthermore, a number of compounds containing only 25 at% of the solute atoms (fraction of solute atoms = ¼) have been made amorphous. In view of these observations, it should be realized that the above criteria may only be used as guidelines and *not* that if they are not obeyed amorphization will not be observed.

Beke et al. [362,363] estimated the elastic mismatch energy stored in an ordered solid solution when its long-range chemical order is destroyed. They showed that if $T_c > T_m$ (see the second condition above) and the ratio of the elastic mismatch energy to the ordering energy is high enough, amorphization is expected to occur. This has been proved to be a valid criterion in many cases of amorphization observed by irradiation.

Recently, attempts have been made to predict the ability of an alloy to become amorphous under MA/MM conditions. For this purpose, Miedema's coordinate scheme, which is a two-dimensional representation of the electronic properties of the metals, is employed. In these diagrams, $|\Delta\phi| = |\phi_A - \phi_B|$ and $\Delta(1/n_{WS}^{1/3}) = (1/n_{WS}^{1/3})_A - (1/n_{WS}^{1/3})_B$ are plotted as the ordinate and abscissa, respectively, to study the regularity of the formation of binary amorphous alloys. ϕ and n_{WS} are the Miedema coordinates. ϕ represents the chemical potential (close to the work function), which is also a measure of the electronegativity, and n_{WS} is the electron density at the boundary of the Wigner-Seitz atomic cell [364]. Zhang [365] studied 51 binary TM-TM (TM = transition metal) alloy systems synthesized by MA. Of these, 37 formed the amorphous phase and 14 did not form the amorphous phase. When the results of the Miedema coordinates are plotted, it was possible to separate regions indicating formation and absence of the amorphous phase by an empirical straight line:

$$|\Delta\phi| = 3.825\left|\Delta\left(\frac{1}{n_{WS}^{1/3}}\right)\right| \tag{12.7}$$

Of the 37 amorphous phases formed in the mechanically alloyed powders, 6 were located in the non–amorphous phase forming region, and of the 14 non–amorphous phase formers, 5 were located in the amorphous phase–forming region. Obviously, this is not a very satisfactory model.

Both the above Miedema coordinates are related to the electronic behavior and correspond to electron factors. However, it is well recognized that size factor (relative sizes of the constituent atoms) also plays an important role in determining the formation of an amorphous phase. Accordingly, Zhang and Zhang [366] constructed two-dimensional plots with the coordinates:

$$X = |(R_A - R_B)/R_A| \tag{12.8}$$

$$Y = |\Delta\phi/\Delta n^{1/3}| \tag{12.9}$$

where R_A and R_B are the atomic radii of the elements A and B, respectively, $\Delta\phi = \phi_A - \phi_B$ and $\Delta n = n_A - n_B$, and ϕ (electronegativity) and n (electron density) are Miedema's coordinates. X and Y are the size and electron factors. The values of R were taken as half of the interatomic distance of the element calculated from the room temperature crystal structure. This is reasonable because amorphization by MA occurs at room temperature. The physical significance of this plot is that the greater the values of X and Y, the greater are the short-range forces between unlike atoms and the smaller is the tendency for crystallization during amorphous phase formation. The equation dividing the regions where amorphous phase formation occurs and does not occur is given by:

$$Y = 2.52X^{-1/4} \tag{12.10}$$

Considering the same 51 systems mentioned above, of the 37 systems that formed the amorphous phase experimentally, 33 were located in the amorphous phase–forming region, and of the 14 systems that did not form the amorphous phase, 11 were located in the amorphous phase nonforming region. Therefore, it may be considered that the accuracy of correctly predicting the formation of the amorphous phase is 86.3%, up from the value of 78.4% when only the electron factors were considered.

Since both atomic sizes and heats of formation are important in the formation of amorphous phases, Zhang [367] plotted $(R_A - R_B)/R_B$ (where R_A and R_B are the atomic radii of the components A and B, respectively) against ΔH_f^a, heat of formation with short-range chemical order at the equiatomic composition of the amorphous phase. He observed that the regions where an amorphous phase formed or did not form could be separated by a straight line given by the equation:

$$\frac{R_A - R_B}{R_B} = 0.068\Delta H_f^a + 0.716 \tag{12.11}$$

Figure 12.4 shows a plot of known systems of binary transition metals processed by MA. It separates the regions for the formation and absence of amorphous phases using the above criterion. The proportion of correct predictions for the glass-forming alloys was 89.2% whereas that for the non-glass-forming alloys was only 71.4%.

Taking Cu-Sn-P as an example, Zhang et al. [99] noted that the nearest neighbor distances are smaller for P-P pairs and Cu-Cu pairs than Sn-Sn pairs due to the smaller size of Cu and P atoms (atomic radii are r_{Cu} = 0.128 nm, r_P = 0.109 nm, and r_{Sn} = 0.155 nm). They also noted that the theoretical glass-forming ability for the constituent binary alloys is different. For example, according to the above criterion, Cu-Sn is in the non–amorphous phase–forming region, Cu-P is in the amorphous region, and Sn-P is on the borderline. Thus, the glass forming ability increases in the sequence Cu-Sn, Sn-P, and Cu-P. Experimentally, it was observed that Cu-P alloys could not be amorphized by RSP techniques [368]. MA produced only partial amorphization in this system at 20 at% P [97]. This could be due to the large electronegativity difference

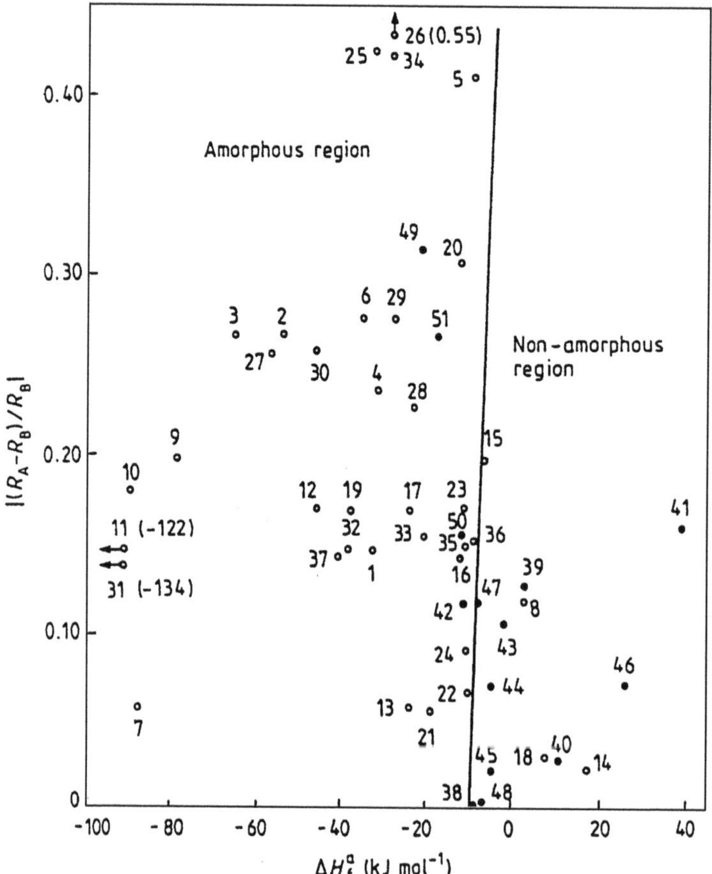

Figure 12.4 Plot of $\Delta R/R$ vs. ΔH showing the regions in which the amorphous phases form and do not form [367]. Description of thermochemical coordinates for the amorphous systems by mechanical alloying. O, Formed amorphous alloys: 1, CoNb [14]; 2, CoZr [15]; 3, NiZr [15]; 4, CuZr [15]; 5, MnZr [15]; 6, FeZr [15]; 7, PdTi [2]; 8, CuTa [16]; 9, RuZr [17]; 10, RhZr [17]; 11, PdZr [18]; 12, NiTi [19]; 13, NiV [20]; 14, CuCr [21]; 15, VZr [18]; 16, CuTi [23]; 17, FeTi [23]; 18, CuV [24]; 19, CoTi [23]; 20, MnTi [23]; 21, CoV [25]; 22, FeV [26]; 23, CrTi [23]; 24, NiMo [27]; 25, CoY [28]; 26, CuEu [17]; 27, NiHf [3]; 28, CuHf [3]; 29, FeHf [17]; 30, CoHf [17]; 31, PtZr [29]; 32, NiTa [30]; 33, Feta [31]; 34, CoGd [32]; 35, NbCr [33]; 36, TaZr [33]; 37, NiNb [1]. ●, Non-formed amorphous systems: 1, NiCr [35]; 2, CuAg [34]; 3, CuRu [36]; 4, FeAg [37]; 5, NiMn [35]; 6, VTi [19]; 7, TiMo [38]; 8, CuNi [9]; 9, CuW [40]; 10, CoMn [25]; 11, CoCr [25]; 12, FeNb [41]; 13, WZr [42]; 14, CrZr [2].

between Cu and P resulting in the easy formation of the Cu_3P compound. However, addition of Sn to Cu-P alloy increases the relative size difference and reduces the electronegativity difference. Both of these factors favor amorphization. Accordingly, the authors were able to produce an amorphous phase in $Cu_{86-x}Sn_xP_{14}$ for $x = 4, 5, 8,$ and 10 [99]. These empirical observations clearly demonstrate the importance of atomic size and electronegativity differences between the components in deciding the amorphous phase formation in alloy systems.

A theoretical model was also proposed in which amorphization was assumed to be realized through interstitial impurity formation during MA [369]. It was assumed that amorphization occurred when impurity atoms penetrated into interstitial sites and distorted the lattice locally. When the local distortions reached some critical value, the long-range order of the lattice was destroyed (destabilization of the crystalline phase) and an amorphous phase formed. It was also shown that the minimal concentration of solute atoms needed to amorphize a binary alloy system by MA was strongly related to the atomic size ratio of the constituents as observed in RSP investigations [370,371]. For example, in fcc-based metal systems, an amorphous phase was formed by MA when the atomic size ratio was between 0.68 and 1.83, whereas the size ratio was between 0.66 and 1.31 for bcc-based metal systems [369].

Fecht and Johnson [372] explained the phenomenon of solid-state amorphization based on polymorphous phase diagrams. Figure 12.5 shows a temperature–composition plot (phase diagram) where temperatures corresponding to equal entropy, T_i^s, and equal free energy, T_o, for the liquid and solid phases are plotted for different compositions. T_{g_o} is the ideal glass transition temperature [373]. The way to interpret this phase diagram is that if the composition of the alloy is above c^* and if the temperature is below T^* (or T_g), the alloy will transform catastrophically into the amorphous phase. It has been known that one of the elements in alloys forming amorphous phases by MA is an anomalously fast diffuser, which is further enhanced by the increased defect density due to heavy plastic deformation. Thus, when the

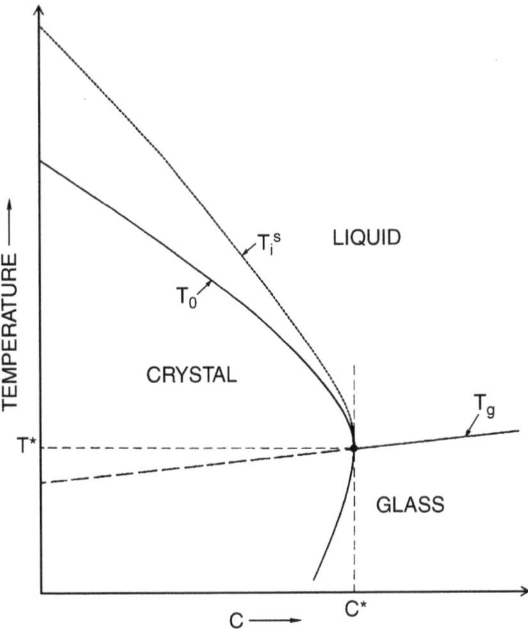

Figure 12.5 Schematic polymorphous phase diagram showing the possibility of crystal-to-glass transition at relatively low temperatures. T_i^s and T_o represent temperatures corresponding to equal entropy and equal free energy of the solid and liquid phases for a particular composition of the alloy. T^* and c^* are the critical temperature and composition, respectively [372].

concentration of the fast-diffusing component in the relatively immobile component reaches the critical concentration of c^*, amorphization occurs catastrophically. It will be instructive to determine if this critical concentration, c^*, and the minimal solute concentration proposed by Egami and Waseda [370] for amorphization are equal or if they have some relationship to each other.

It was recently shown that large negative values of the interaction parameter ω_C^A promote amorphization [125]. The interaction parameter ω_C^A is given by the equation:

$$\omega_C^A = RT\varepsilon_C^A \tag{12.12}$$

where R is the gas constant, T is the absolute temperature, and ε_C^A is the interaction coefficient defined as $\varepsilon_C^A = \mathrm{d}\ln \gamma_C/\mathrm{d}X_A$ where γ_C is the activity coefficient and X_A is the atom concentration of component A.

12.7 CRYSTALLIZATION BEHAVIOR OF AMORPHOUS ALLOYS

Amorphous alloys are metastable in nature and therefore they tend to crystallize when heated above the crystallization temperature of the amorphous phase. Crystallization temperatures of amorphous alloys are determined by continuous heating of the amorphous samples in a differential thermal analyzer or a differential scanning calorimeter. By heating the amorphous alloy at different heating rates and noting the change in the crystallization temperature, one could calculate the activation energy for crystallization using the Kissinger method (see Sec. 7.7).

A number of research investigations were conducted on the crystallization behavior of rapidly solidified glassy alloys. It was shown that crystallization of metallic glasses could occur in three different ways—polymorphic, primary, and eutectic (Figure 12.6). In the polymorphic type of crystallization, the amorphous alloy transforms into a crystalline phase without a change in the chemical composition. In the case of primary crystallization, which seems to be the most common variety, the amorphous phase first transforms to a crystalline solid solution phase based on the solvent metal. The remaining amorphous phase can then transform either in a polymorphic mode into a crystalline phase or in a eutectic mode to a mixture of two crystalline phases. In the eutectic mode, the amorphous phase transforms simultaneously into two crystalline phases. (Even though it may appear that a solid phase is transforming into a mixture of two solid phases and so this should be called a eutectoid reaction, it is referred to as a eutectic reaction because the amorphous phase is considered as a supercooled liquid.) Some excellent reviews (even though dated) are available explaining the different aspects of crystallization of amorphous alloys obtained by RSP techniques [374–376].

As Tables 12.1 and 12.2 clearly indicate, a number of amorphous phases have been synthesized by MA/MM techniques. Some of the systems are identical or at least similar to those investigated by RSP and some are entirely different.

A number of researchers have reported differential thermal analysis/differential scanning calorimetry (DTA/DSC) investigations to determine the transformation (glass transition and crystallization) temperatures and measure the activation energies for crystallization. A few investigators have even conducted transmission electron microscopy (TEM) studies to confirm the formation of the amorphous phase in the mechanically alloyed/milled powders. But there are hardly any detailed investigations

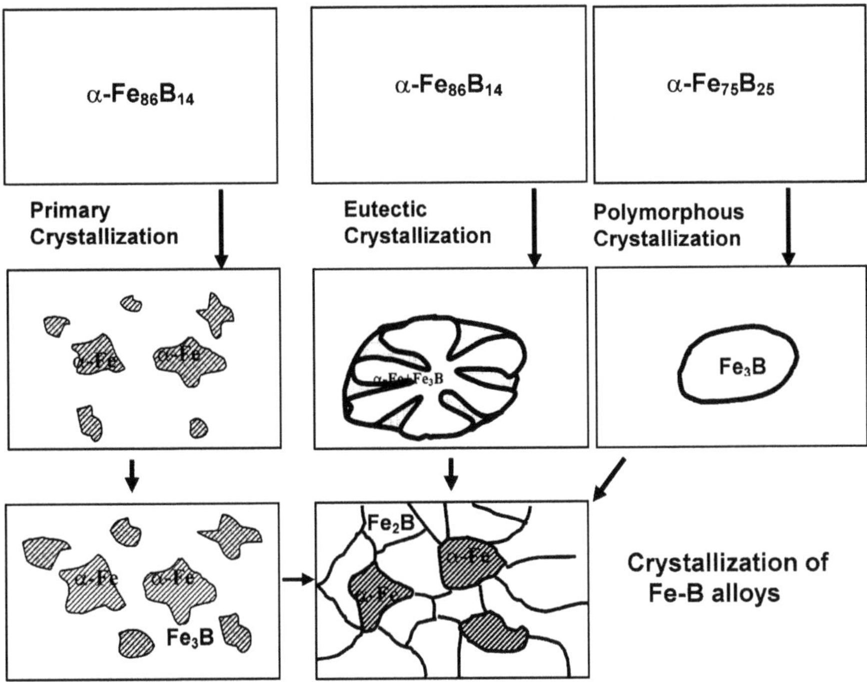

Figure 12.6 Schematic diagrams of typical crystallization reactions in amorphous Fe-B alloys.

of the crystallization behavior of mechanically alloyed amorphous phases by TEM methods. This is in stark contrast to rapidly solidified alloys, where a combination of both DTA/DSC and TEM investigations was the rule rather than the exception. This difference in the two cases may be related to the ease of specimen preparation in the case of rapidly solidified ribbons (some ribbons may even be electron transparent in the as-solidified condition) and the associated specimen preparation difficulties for the mechanically alloyed powders.

It has been noted that the radial distribution functions and structure factors of amorphous alloys of the same composition processed by RSP and MA/MM techniques are similar. This is true in a number of cases [112,152,377].

However, differences have been noticed in the crystallization temperatures of amorphous alloys of the same composition processed by RSP and MA/MM techniques. Cardoso et al. [40] reported that the first T_x for the Al-5 at% Fe-5 at% Nd processed by MA was 405.5°C, much higher than the value of 218°C for the glassy phase processed by RSP. On the other hand, Suñol et al. [129] reported that the T_x for the Fe-40 at% Ni-(20-x) at% P-x at% Si (x = 6, 10, and 14) amorphous alloys was less than 407°C for the melt-spun ribbons, whereas it was only 277°C for MA powders. Furthermore, it has been reported that the number of stages in which the amorphous phase transforms into the crystalline phases is also generally different. It is noticed that the amorphous alloy processed by MA/MM crystallizes in fewer stages than that processed by RSP.

Cardoso et al. [40] studied the crystallization behavior of rapidly solidified and mechanically alloyed Al-5 at% Fe-5 at% Nd amorphous alloys. They reported that the mechanically alloyed powder underwent primary crystallization on annealing forming fcc-Al crystals in an amorphous matrix. Annealing at a higher temperature resulted in complete crystallization of the remaining amorphous phase to a mixture of the equilibrium crystalline phases, i.e., $Al_{11}Nd_3$ and $Al_{10}Fe_2Nd$. The rapidly solidified ribbon also underwent primary crystallization in the first stage. However, at higher temperature of annealing a metastable ternary Al-Fe-Nd phase formed. At still higher temperatures, of course, the equilibrium phases had formed. Thus, while the amorphous phase in the MA powder crystallized in two stages, the RSP ribbons transformed in three stages. A similar situation was also reported for amorphous Co-Ti alloys processed by RSP [378] and those processed by MA [85]. The RSP alloy crystallized in two stages, whereas the mechanically alloyed powder transformed in one single step.

A completely different situation was reported in some other alloy systems; the number of stages was more for crystallization of the MA powders than the RSP ribbons. Zhou et al. [291] noted that crystallization of MA powders went through two stages for the crystallization of a Zr-7.5 at% Al-17.5 at% Cu-10 at% Ni alloy, while it went through only one step for the RSP metallic glass. Seidel et al. [379] also reported a similar behavior. This difference was suggested to be due to increased oxygen content (more than 0.2 at%) in the mechanically alloyed powder [380].

A low-temperature relaxation peak was observed in mechanically alloyed powders but not in the rapidly solidified amorphous ribbons [129]. Thus, there appear to be very noticeable differences in the crystallization behavior of amorphous phases obtained by MA and RSP techniques. Other major differences in the glass forming abilities and composition ranges of formation of amorphous phases will be discussed later.

12.8 MECHANICAL CRYSTALLIZATION

As mentioned earlier blended elemental mixtures or intermetallics or combinations of them can be amorphized by MA or MM. In some cases a crystalline phase forms after the formation of an amorphous phase in certain alloy systems, especially those based on reactive metals like titanium or zirconium. But it was convincingly shown that this was a contaminant phase and therefore to be discounted as a product of crystallization of the amorphous phase. However, in some alloy systems a true crystallization event of the amorphous phase takes place on milling the powder containing the amorphous phase. This phenomenon is referred to as mechanical crystallization, since crystallization occurs on the application of mechanical energy. This was first reported by Trudeau et al. in 1999 [345].

Trudeau et al. [345] chose two commercially available metallic glasses manufactured by Allied Signal—Metglas 2605S-2 ($Fe_{78}Si_9B_{13}$) and Metglas 2605CO ($Fe_{66}Co_{18}Si_1B_{15}$)—and subjected them to MM in a SPEX mill. They noted that crystalline phases were produced in both cases on MM with the difference that the Co-containing glass crystallized at shorter milling times. As mentioned earlier, the temperature of the powder rises during milling (see Chapter 8 for the details) and therefore one could explain this crystallization of the glassy alloy based on the temperature rise during milling. The early crystallization of the Co-containing glass was attributed to its lower crystallization temperature of 441°C, in comparison to the higher value of 553°C for the Metglas 2605CO alloy. With a view to change the crystallization temperature, these investigators have added Co or Ni to the $Fe_{78}Si_9B_{13}$ Metglas sample and investigated the crystallization behavior. It was observed that Co addition resulted in rapid crystallization whereas the Ni addition was found to stabilize the amorphous phase, even though the crystallization temperature of the glassy sample is known to be decreased by Ni additions. Therefore, it does not appear that the crystallization behavior is in any way directly related to the crystallization temperature of the glassy/amorphous alloy. Crystallization of the Metglas 2605S-2 was also observed by other investigators [381,382].

Similar observations of formation of a crystalline phase after the formation of the amorphous phase have been reported in recent years. For example, a mechanically induced solid-state devitrification (MISSD) phenomenon was reported in a Zr-20 at% Pd-10 at% Ni powder mixture on milling in a planetary ball mill [304]. They noted that a fully amorphous phase was obtained on milling the powder for 48 h, but on continued milling for 130 h the amorphous phase crystallized into a nanocrystalline phase, which on milling for 150 h transformed into a "big cube" structure (closely related to the Zr_2Ni- and Zr_2Cu-type phases). The authors have explained that this crystallization of the amorphous phase occurred due to the temperature increase during milling. A similar observation was made during milling of pure metal Se [331].

Even more interestingly, some of the investigators have reported that cyclic transformations occur during milling of some systems, e.g., Co-Ti [82–85]. That is, when the blended elemental powders were milled, an intermetallic phase was reported to form first. Continued milling resulted in amorphization of the alloy, which on further milling again transformed to the crystalline phase. It was reported that this process of crystalline → amorphous → crystalline phase changes could be repeatedly observed by changing the milling time. Reasons for this transformation were given in Section 10.6.

The number of instances where blended elemental powders and/or intermetallics have been amorphized are many (see Tables 12.1 and 12.2). But the reverse behavior of crystallization of amorphous alloys on milling was observed only in a few cases, and these reports are also from select groups. It is a well-known fact that a crystalline phase is more stable than an amorphous phase. However, the crystalline phase could be amorphized under nonequilibrium conditions by storing additional energy into the system, thus destabilizing the crystalline phase with respect to the amorphous phase. Therefore, it is difficult to understand that a metastable phase transforms into a stable phase on storing extra energy. Thus, it appears that this phenomenon of mechanical crystallization is not well understood and that more work is needed to ensure that this also is a general phenomenon.

12.9 BULK AMORPHOUS ALLOYS

During the 1970s it was realized that by a proper choice of the alloy system and the chemistry, it was possible to synthesize amorphous phases in some selected alloy systems even at solidification rates as low as 10^3 K/s. This enabled the thickness of the rapidly solidified ribbon to be substantially increased. Thus, millimeter-diameter rods of amorphous Pd-6 at% Cu-18 at% Si [383] and Pd-40 at% Ni-20 at% P [384] alloys could be produced. Fluxing to minimize the formation of crystalline nuclei has further increased the section thickness of the glassy alloys. It was also later realized that it is much easier to produce the amorphous state in multicomponent complex alloys. In fact, it was frequently observed that the more the number of components in the system, the easier it is to produce the glassy alloys. This was attributed to the principle of confusion or frustration because in a multicomponent alloy system the constituent atoms in the alloy find it "confusing" to go to the right lattice site and form the appropriate crystalline phase.

While investigating the fabrication of amorphous aluminum alloys during the late 1980s, Inoue and coworkers in Japan [385] and Johnson and coworkers in the United States [386] observed that the rare-earth-rich alloys, e.g., La-Al-Ni and La-Al-Cu, and Zr-Ti-Cu-Ni-Be alloys exhibited exceptional glass forming ability. By studying similar quaternary and higher order alloy systems, these authors were able to develop alloy compositions that formed the amorphous phase at cooling rates of less than 100 K/s with the critical casting thickness ranging upward to tens of millimeters. These are now referred to as bulk metallic glasses. Subsequently, a number of alloy systems based on La, Zr, Mg, Pd, and Fe have been prepared in the amorphous state [385,387–393].

There has been lot of activity in recent years on the synthesis/processing and characterization of these bulk amorphous alloys or bulk metallic glasses [353,385,387–393]. Traditionally, metallic glasses have been synthesized in micrometer section thicknesses by rapidly solidifying melts of appropriate compositions [3–11]. The new types of amorphous alloys are called bulk metallic glasses because they can be produced in section thicknesses of a few millimeters. This is possible because the critical solidification rate to achieve glass formation in these types of alloys is very low, i.e., typically 10^{-1} to 10^{1} K/s. This range should be contrasted with the critical solidification rates of approximately 10^{6} K/s needed for traditional or "thin" metallic glasses. Another important difference between thin and bulk metallic glasses is that the former contain usually two or three components (e.g., Fe-B, Pd-Si, Ni-Si-B), whereas the bulk metallic glasses contain usually at least four or five elements. The maximal section thickness achieved so far for bulk amorphous alloys by solidification methods is 72 mm in a $Pd_{40}Cu_{30}Ni_{10}P_{20}$ alloy [394].

Bulk metallic glasses have all the attributes of conventional metallic glasses obtained from the liquid state at high cooling rates (10^{5}–10^{6} K/s) and also are superior in properties and performance to those of corresponding crystalline alloys. However, they have additional advantages, which include the following:

1. The critical cooling rate (R_c) required for the formation of bulk metallic glasses is very low and reaches from about 10^{3} K/s down to about 0.1 K/s (Fig. 12.7).
2. The sample thickness (t_{max}) is increased to several millimeters (Fig. 12.7). The lowest cooling rate and highest sample thickness are almost comparable to those for the ordinary oxide and fluoride glasses.
3. Bulk metallic glasses have large reduced glass transition temperature, T_{rg}, generally above 0.6. The highest reported value is 0.73.
4. The new multicomponent bulk metallic glasses have a much wider supercooled liquid region, defined as the difference between crystallization temperature (T_x) and T_g, i.e., $\Delta T_x = T_x - T_g$ (Fig. 12.8), currently reaching a high value of 135 K.

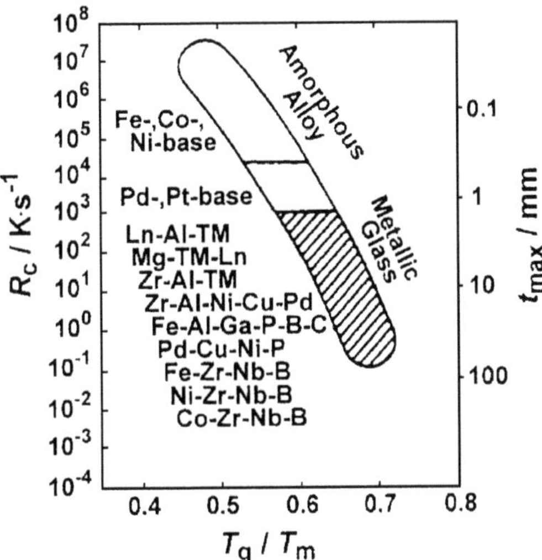

Figure 12.7 Relation between the maximal critical cooling rate for glass formation (R_c), maximal sample thickness (t_{max}), and the reduced glass transition temperature ($T_{rg} = T_g/T_m$) for multicomponent bulk amorphous alloy systems.

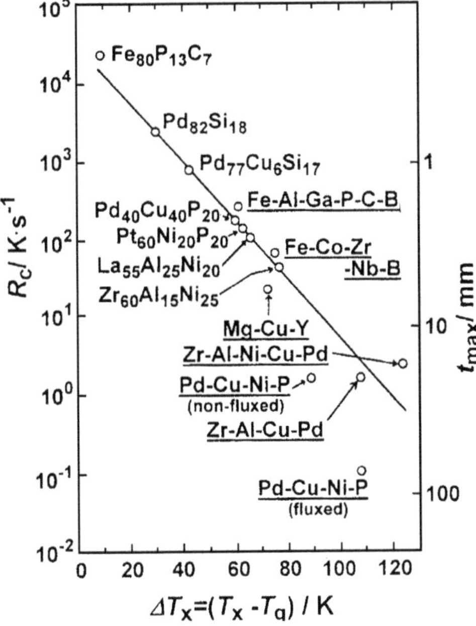

Figure 12.8 Correlation between the critical cooling rate for glass formation (R_c), maximal sample thickness (t_{max}), and the supercooled liquid region ($\Delta T_x = T_x - T_g$) for bulk amorphous alloys.

From the strong correlation among R_c, t_{max}, T_{rg}, and ΔT_x, it may be concluded that bulk metallic glasses are obtained when the alloy system has a high T_{rg} and exhibits a large ΔT_x. Based on the extensive investigations that have been conducted so far, three empirical rules have been formulated to achieve large glass forming ability in alloy systems [353]. These are as follows:

1. The alloy system should contain at least three elements. The larger the number of components, the easier it is to form the glassy phase at lower cooling rates and consequently with larger thickness.
2. A significant difference in atomic sizes (more than 12%) should exist among the main constituent elements.
3. The constituent elements should exhibit negative heats of mixing among them.

The reasons for the glass forming ability of alloys following the three empirical rules is related to (1) higher degrees of dense randomly packed atomic configurations, (2) new local atomic configurations that are different from those of the corresponding crystalline phases, and (3) a homogeneous atomic configuration of the different constituent atoms on a long-range scale. The small differences in the densities of 0.30–0.54% between the as-cast bulk metallic glasses and fully crystallized states have been attributed to the higher dense randomly packed atomic configurations of bulk amorphous alloys in comparison to the "conventional" metallic glasses synthesized at higher solidification rates.

The limitation on section thickness for "thin" metallic glasses is imposed by the required solidification rates to achieve glass formation in these alloys. It was also mentioned earlier (Sec. 12.5) that glass formation becomes easier when the reduced glass temperature, T_{rg}, is high. That is why alloys near eutectic compositions have been easy glass formers. If T_{rg} could be increased by adjusting the composition of the alloy, then the critical cooling rate for glass formation could be reduced and consequently the section thickness could be increased. This has been possible by choosing alloy compositions in such a way that the component elements form among themselves binary, ternary, and higher order eutectics. In contrast to thin metallic glasses, the bulk metallic glasses are characterized by a wide supercooled liquid region, i.e., a large temperature difference exists between the glass transition and crystallization temperatures for these glasses.

Based on an atomistic mechanism and extending the topological theory of glass formation to multicomponent systems, it was suggested that increasing the number of elements involved, increasing the atomic size ratio, increasing the attractive force between small and large atoms, and introducing a repulsive potential between small atoms will help formation of bulk metallic glasses [395].

Miracle and Senkov [396] also proposed a topological model for metallic glass formation through destabilization of the host crystalline lattice by susbstitutional and/or interstitial solute atoms. The model calculates the frequency of occupation of the solute atoms either in the substitutional or interstitial sites as a function of the strain energy associated with each site. The strain energy, in turn, depends on the solute and solvent elastic properties and relative sizes, and on temperature. The crystalline lattice is destabilized leading to amorphization when solute elements produce a critical internal strain required to change local coordination numbers. The critical concentration of a solute element required to destabilize the crystalline lattice of a binary alloy was calculated as a function of the radius ratio $R = R_B/R_A$ of the solute and solvent atoms. In the range of $0.5 < R < 1$, the critical concentration decreases, reaches a minimum at R about 0.8, and then increases as the size of the solute element decreases relative to the solvent.

Bulk amorphous alloys (we will use the term "amorphous alloys" instead of metallic glasses because they are synthesized by solid-state processes and not from the liquid state) have also been synthesized by MA in a number of multicomponent Zr-Ti-Cu-Ni, Zr-Al-Cu-Ni, Mg-Y-Cu, and other systems [66,397]. In comparison to RSP or casting techniques used to produce bulk metallic glasses, MA is an inexpensive and simpler technique to process the material into an amorphous state. Furthermore, it is easier to produce the amorphous phase in a wider composition range by MA than by any other technique. Phase diagram restrictions do not directly apply to the phases produced by MA, since the whole processing is carried out in the solid state. But the number of parameters that could be varied during MA is large and as such optimization of the process could be tricky. However, the complexity of these variables could be used to our advantage since we have at our disposal a number of variables that could be varied to produce the required phase in the desired alloy composition. Furthermore, changing the parameters, e.g., temperature of milling and/or the composition of the alloy, could easily produce a composite of amorphous and crystalline phases. Alternatively, the amorphous phase can be heat treated at sufficiently low temperatures and for short times to produce a uniform dispersion/precipitation of a crystalline phase in an amorphous matrix. Such combinations of a uniform dispersion of fine-grained (or nanocrystalline) phases in an amorphous matrix have been found to exhibit improved properties (e.g., improved strength and ductility) over alloys containing a monolithic amorphous phase.

The thermal stability of bulk amorphous alloys processed by MA appears to be affected by impurities such as oxygen. However, amorphization by MA can be achieved even at relatively high impurity levels, where bulk glass formation by slow cooling from the melt is not possible. The ability to reheat and deform glassy powders in the supercooled liquid state without crystallization allows consolidation and shaping of bulk glassy components using the easy flow of the supercooled liquid. Glass-matrix composites containing crystalline reinforcements appear to possess improved mechanical properties than the monolithic glassy alloys.

The mechanical behavior of bulk metallic glasses (obtained by die casting) and nanocomposites has been investigated in recent times [398–402]. It has been noted that while the fully amorphous alloys exhibit excellent tensile and compressive properties at room temperature, the presence of crystalline precipitates decreases these values and the fracture mode changes from ductile to brittle.

Bulk metallic glasses exhibit an excellent combination of properties such as high tensile strength, high elastic energy, good ductility, high fracture toughness, high impact fracture energy, high bending flexural strength, high bending fatigue strength, high corrosion resistance, good soft magnetic properties, fine and precise viscous deformability, good workability, good castability, and high consolidation tendency into bulk forms. Based on these properties several potential applications have been envisaged. For example, 3-mm-thick faceplate inserts made of Zr-based bulk metallic glasses have been used for high-end golf club heads, and these have been commercialized [403,404]. This application is based on the increased maximal stored elastic energy, which is roughly four times that of conventional crystalline materials due to doubling of the elastic strain limit achieved by forming a composite of an amorphous matrix containing a nanocrystalline phase. Another application for the bulk metallic glasses has been in developing "self-sharpening" kinetic energy penetrators [405].

12.10 THEORETICAL PREDICTIONS OF AMORPHOUS-PHASE-FORMING RANGE

It would be useful if one could predict if an alloy of a given composition will form an amorphous phase by MA/MM or not. The models discussed earlier (Sec. 12.6) were mostly empirical in the sense that they were proposed to fit the available experimental data to a model and draw conclusions. However, the predictive capability will help in designing new compositions to achieve the desired properties. Attempts have been made to predict whether metallic glasses form (glass forming ability) and to determine the glass-forming composition range (GFR) for metallic glasses formed by rapid solidification from the liquid state. Since no liquid → solid transformation is involved in the MA/MM process, these models could not be directly used to predict the amorphous phase–forming composition range in milled powders. Therefore, attempts were made to theoretically predict the composition ranges in which amorphous alloys are formed using the heats of formation of alloys.

Miedema and coworkers [364,406] developed methods to calculate the formation enthalpies of alloys. By comparing the enthalpies of formation of different competing phases in an alloy system, e.g., solid solution and an amorphous phase, one could calculate the composition range over which an amorphous phase is stable. These methods have been summarized by Bakker [407].

Miedema et al. [364,406] proposed that for a solid solution the formation enthalpy consists of three terms:

$$\Delta H\ (\text{solid solution}) = \Delta H^{\mathrm{c}} + \Delta H^{\mathrm{e}} + \Delta H^{\mathrm{s}} \tag{12.13}$$

where the superscripts c, e, and s represent the chemical, elastic, and structural contributions to the enthalpy of the solid solution. The chemical interaction contribution to the enthalpy of mixing of the solid solution consists of a negative contribution from the electronegativity difference between the two elements and a positive contribution from the difference in the electron densities of the two elements. ΔH^{c} can be calculated from the equation:

$$\Delta H^{c}_{AB} = c_A c_B (c_A \Delta H^{o}_{B\ in\ A} + c_B \Delta H^{o}_{A\ in\ B}) \tag{12.14}$$

where c_A and c_B are the atom fractions of the A and B components in the alloy. $\Delta H^{o}_{A\ \mathrm{in}\ B}$ is the enthalpy of solution of element A in the other element B at infinite dilution for a binary alloy system. Values for $\Delta H^{o}_{A\ in\ B}$ have been calculated by de Boer et al. for transition metal–transition metal and transition metal–nontransition metal alloy systems [408].

The elastic contribution to the enthalpy is due to the atomic size mismatch (difference in atomic volumes between the solvent and the solute) and can be calculated as:

$$\Delta H^{e}_{AB} = c_A c_B (c_A \Delta H^{e}_{B\ in\ A} + c_B \Delta H^{e}_{A\ in\ B}) \tag{12.15}$$

The values of $\Delta H^{e}_{A\ in\ B}$, the size mismatch contributions, can be calculated from the atomic radii and the bulk and shear modulus values of the component metals.

The structural contribution arises due to the differences between the valences and crystal structure of the solvent and solute elements, but it is usually ignored because its contribution to the enthalpy of formation of the solid solution is expected to be small. This term will be present only when alloying between two transition metals occurs.

In an amorphous alloy, the atoms arrange themselves in such a way that elastic mismatch is avoided but, in addition to a chemical term, there is a topological term due to the amorphous nature of the material. The total formation enthalpy for an amorphous alloy is then estimated as:

$$\Delta H\ (\text{amorphous}) = \Delta H^{c} + 3.5(c_A T_{m,A} + c_B T_{m,B}) \tag{12.16}$$

where $T_{m,A}$ and $T_{m,B}$ are the melting points, and c_A and c_B are the atomic fractions, of the elements A and B, respectively. Using the above relations, one can calculate the enthalpies of formation for the solid solution and amorphous phases. When these enthalpies of formation are plotted against composition, one can obtain the relative stabilities of the competing phases or phase mixtures by using the common tangent construction. Figure 12.9 shows the enthalpy vs. composition diagram for the Al-Fe system [34]. The common tangent construction for this diagram suggests that a fully amorphous phase should form between 25 and 60 at% Fe. At compositions less than 15 at% Fe and more than 90 at% Fe, only a solid solution is expected to be stable. In the intermediate ranges 15–25 at% Fe and 60–90 at% Fe, both the amorphous phase and solid solution are predicted to coexist. Similar calculations have been performed for many other alloy systems. Composition ranges for the formation of amorphous phases, calculated using this approach, are listed in Table 12.3.

Even though this model has been applied with considerable success to predict the formation and stability range of amorphous phases, a few words of caution are in order. First, we should realize that it is the free energy and not the enthalpy alone that decides the stability of a phase. In the Miedema calculations, the entropy term is

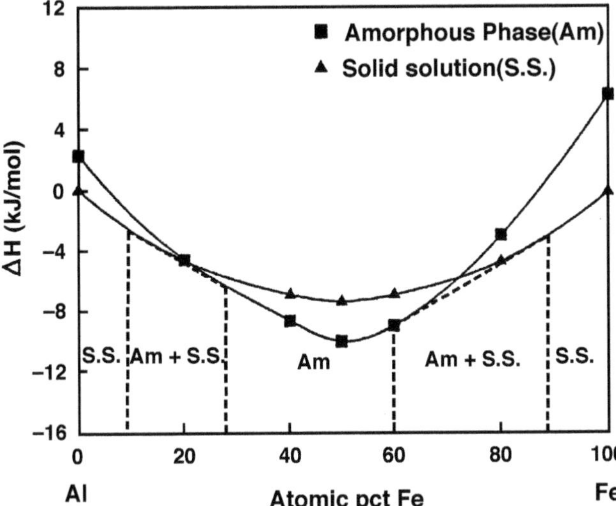

Figure 12.9 Enthalpy vs. composition diagrams for the solid solution and amorphous phases in the Al-Fe system based on Miedema's calculations.

Table 12.3 Comparison of the Predicted and Observed Glass-Forming Ranges in Mechanically Alloyed Powder Mixtures

	Amorphous phase-forming range (at% solute)				
System	Miedema model	Ref.	CALPHAD	Observed	Ref.
Al-Fe	25–60 Fe	34		25 Fe	34
Co-V				33–60 V	86
Co-Zr	22–61 Zr	88		27–92 Zr	298
Cu-Hf	21–71 Hf	22		30–70 Hf	149
Cu-Sn-P	3–62 Sn	99		2–3.4 Sn	99
Cu-Ti	25–72 Ti	22		13–90 Ti	106
Cu-Zr	26–51 Zr	88			
Fe-B				50–60 B	118
Fe-Zr	21–73 Zr	22		22–70 Zr	409
Fe-Zr	33–54 Zr	88			
Nb-Fe	30–70 Fe			30–70 Fe	166
Nb-Ge	18–27 Ge	168		12–30 Ge	169
Ni-Hf	16–75 Hf	22		35–85 Hf	149
Ni-Nb	20–69 Nb	22		20–79 Nb	187
Ni-Ti	23–76 Ti	22		28–72 Ti	195
Ni-V				45–60 V	410
Ni-Zr				20–80 Zr	22, 208
Ni-Zr	17–76 Zr	22		17–73 Zr	409
Ni-Zr	22–63 Zr	88	17–67 Zr	24–85 Zr	203
Pd-Ti	46–80 Ti	22		42–85 Ti	265
Pd-Zr				40–55 Zr	411
Ti-Al	15–88 Al	61	10–96 Al	10–75 Al	61
	35–75 Al	253			
Ti-Cu	46–65 Cu	254			
Ti-Ni	25–83 Ni	253, 254			

ignored. This has been justified because the Miedema model is strictly valid only at 4.2 K and therefore the enthalpy effects are far more important than the entropy effects at this low temperature. Furthermore, since we are dealing here with situations far from equilibrium, the kinetic path should be taken into account. The thermodynamic parameters (and the common tangent constructions) may then be used only as guidelines. Second, significant discrepancies have been observed between the calculated and experimentally observed heats of mixing. Third, the enthalpy calculations do not consider the occurrence of intermetallics in the phase diagram; only the terminal solid solutions and an amorphous phase are expected to be present under metastable conditions. Last, the prediction of glass-forming range using this model has been applied mostly to studies in the field of MA. Extremely few reports are available of application of this model to nonequilibrium phases in materials processed by other nonequilibrium processing techniques [365,366,412]. In spite of these apparent limitations, several researchers have calculated the glass-forming composition range using the Miedema model, and these values are compared with the experimentally observed values in Table 12.3.

Another common approach of thermodynamic evaluation of binary systems is by CALPHAD (calculation of phase diagrams) [413], which uses analytical expressions to describe the free energies of different competing phases. These expressions are derived from models with parameters obtained from numerical fits to measured equilibrium data. Extrapolations are then used to predict metastable equilibria. The CALPHAD method is usually preferred to the Miedema model because the former is based on more explicit thermodynamic data such as the equilibrium phase diagram. However, these calculations are also sometimes inadequate due to problems associated with insufficient experimental input, fitting errors, and invalid extrapolation to large undercooling.

Generally, the amorphous phase–forming range is experimentally determined by looking for the presence of broad peaks in XRD patterns of the processed alloy. As mentioned earlier, this may not always prove that the material produced is truly amorphous. It is much more reliable to measure the properties of the samples such as crystallization temperature, enthalpy of crystallization, saturation magnetization, or superconducting transition temperatures. When these properties are plotted against composition, a sudden change in the slope indicates the solute concentration up to which the amorphous phase is homogeneous [143].

It should be remembered that a good match is not always found between the observed and predicted amorphous phase–forming composition range. This is essentially because MA is a "complex" process and, further, the mechanically alloyed powders could pick up lot of impurities and get contaminated. The presence of these substitutional and interstitial impurities modifies the stability of the amorphous phase.

12.11 COMPARISON BETWEEN MA AND RSP

As mentioned earlier, the radial distribution functions of amorphous alloys synthesized by MA/MM or RSP techniques are similar. In fact, the structure factors for irradiated, rapidly solidified, and mechanically alloyed amorphous phases of different alloy systems are very similar [377]. The structural and magnetic properties of mechanically milled and sputtered YCo_2 and YCo_5 intermetallics are also very similar [113]. The crystallization temperatures of the amorphous alloys produced by MA have been found to be higher in some cases and lower in the others than those processed by the RSP techniques. The activation energy for crystallization of RSP alloys was much higher than for MA alloys [63,201].

Even though amorphization has been observed by both MA and MM, there are no quantitative data available to define the "critical" energy input (or other parameters) that could be equated with the critical cooling rate that should be exceeded to form the glassy phase in RSP studies. Modeling of MA/MM (currently in progress in different laboratories) may provide additional information in this direction.

The features of phase diagrams normally provide clues to the possibility of producing amorphous phases. Amorphous alloys are easily produced by RSP techniques in the vicinity of deep eutectics. This is because the reduced glass transition temperature T_{rg} is the highest at the eutectic point. The ease of forming an amorphous phase is higher the higher the T_{rg} value. But in the case of mechanically alloyed materials the amorphous phase is mostly obtained around the equiatomic composition. Furthermore, the composition ranges of the amorphous phases are much wider in alloys produced by MA than in those obtained by RSP. Note, for example, that Figure 12.10 shows the glass formation composition ranges for Zr-Fe, Zr-Co, and Zr-Ni alloy systems by both RSP and MA methods [414]. Whereas glass formation is observed in the vicinity of the eutectic compositions by RSP methods, amorphization by MA occurs in the central part of the phase diagram. Accordingly, the amorphous phase formation composition ranges for these Zr-based alloys are 30–78 at% Fe, 27–92 at% Co, and 27–85 at% Ni. Clearly, these composition ranges are much wider than observed for the rapidly solidified alloys.

Major differences were also noticed between the glass formability of alloys by the two techniques of MA/MM and RSP. Alloys near deep eutectics were the easiest to amorphize by RSP. It was very difficult to achieve amorphization at compositions corresponding to melting point maxima, alloys with a cascade of peritectic reactions, shallow eutectics, and in alloys with a positive heat of mixing. However, amorphization was achieved in a number of these cases by MA/MM. MA allowed easy amorphization of alloys corresponding to melting point maxima. For example, the intermetallic $Hf_{50}Ni_{50}$, with an orthorhombic structure, exists as a line compound and has a melting point of 1800 K; this was amorphized by MM [149,150]. Similarly, amorphization was reported in Cu-Hf alloys in the composition range of 60–70 at% Hf, close to and on either side of the Hf_2Cu compound having the highest melting point of the compounds in the system [149]. The intermetallic Ge_4Sb_5 could not be produced in the amorphous state by RSP techniques; but it was possible to do so by MM [319]. It has

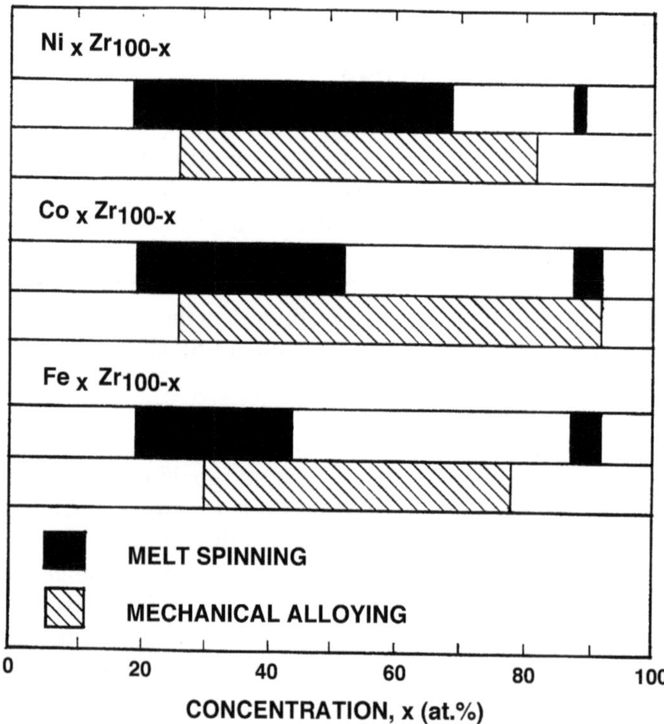

Figure 12.10 Comparison of glass formation composition ranges by rapid solidification processing and mechanical alloying methods in Zr-Fe, Zr-Co, and Zr-Ni alloy systems [414].

Table 12.4 Comparison of Amorphous-Phase–Forming Ranges (at%) by Mechanical Alloying and Rapid Solidification Processing Techniques

System	By RSP	By MA	Ref.
Ag-Pd	Not reported	50 Pd	23
Al-Fe	Not reported	20–50 Fe	38
		17–33 Fe	37
Al-Ta		33–67 Ta	58, 59
Co-B		33–50 B	75
Co-Ge	Not reported		415
Co-Nb		15–20 Nb	76
Co-Si		30–70 Si	80
Co-Ti		15–20 Ti	76
		25–67 Ti	85
Co-V		33–60 V	86
Co-Zr		8–73 Zr	86
Cr-Fe		28 Fe	90
Cr-Nb		30–68 Nb	92
Cu-P	Not reported	20 P	97
Cu-W		30–90 W	109
Fe-B		35–40 B	119
Fe-C		20–25 C	119
		17–60 C	122
$Fe_{83-x}Cr_xC_{17}$	24–50 Cr	10–60 Cr	125
$Fe_{83-x}Mo_xC_{17}$	5–26 Mo	5–60 Mo	125
Fe-P		25 P	131
Fe-Si		10–40 Si	80
Fe-Zr		22–70 Zr	110, 142
Ge-S		61–72 S	144
Hf-Cu		20–90 Cu	148
		30–70 Cu	149, 150
Hf-Ni		15–65 Ni	150
		10–85 Ni	148
La-Al-Ni		30–70 La	153
		10–50 Ni	
Mg-Ni		50–80 Ni	154
		40–70 Ni	156
Nb-Al		25–85 Al	164
Nb-Co		35–85 Co	92
Nb-Cr		32–70 Cr	92
Nb-Fe		35–75 Fe	92
Nb-Ge		18–27 Ge	168
		12–30 Ge	169
Nb-Mn		35–60 Mn	92
Nb-Ni	40–60 Ni	20–80 Ni	92, 185, 186
Ni-Mg		20–50 Mg	154
Ni-Mo	Not reported	50 Mo	178
Ni-Sn		20–40 Sn	186
Ni-Ta	35–65 Ta	40–80 Ta	189
	30–60 Ta	20–90 Ta	230
Ni-V		40–65 V	196
Ni-Zr		20–80 Zr	208
		30–80 Zr	210

Table 12.4 Continued

System	By RSP	By MA	Ref.
Pd-Si	15–23 Si	13.5–25 Si	216
Pd-Ti	Not reported	42–85 Ti	214, 265
Ta-Al	Not reported	10–90 Al	224
Ta-Cu		20–50 Cu	229
Ti-Al	Not reported	10–60 Al	61
		10–75 Al	61
		20–50 Al	232, 247
		20–60 Al	244
		25–67 Al	249
		45–65 Al	143
		50–75 Al	246
Ti-Al-Ni		0–50 Al, 0–50 Ni, 30–90 Ni	54
Ti-Cu	30–70 Cu	10–87 Cu	106
		10–90 Cu	148
Ti-Ni		10–70 Ni	254
		35–50 Ni	261
$Ti_{50}Ni_xAl_{50-x}$		x = 10, 25	253
$Ti_{60}Ni_xAl_{40-x}$		x = 15, 20	253
Ti-(40-x)Ni-xCu		x = 10–20	254
Ti-(50-x)Ni-xCu		x = 10–30	254
Ti-(60-x)Ni-xCu		x = 10–40	254
Ti-Ni-Cu		10–30 Ni, 10–30 Cu	262
Ti-Si		16–63 Si	270
Zn-Ti		40 Ti	284
W-Fe		30–70 Fe	281
Zr-Al		17.5–30 Al	287
		17.5–40 Al	288, 289
Zr-Co	20–57 Co	27–92 Co	298
Zr-Cu	25–90 Cu	40–60 Cu	110, 111
Zr-Fe	8–17 Fe	20–50 Fe	148
	60–80 Fe	30–78 Fe	140, 298, 416
Zr-Mn		20–50 Mn	298
Zr-Ni	20–40 Ni and 58–70 Ni	27–83 Ni	298
Zr-Pd	20–35 Pd	45–60 Pd	412

not been possible to amorphize alloy systems with a positive heat of mixing by RSP, but MA could produce them in the amorphous state. In general, it has been noted that it is easier to produce an amorphous phase, and in a wider composition range, in an alloy system by MA/MM than by RSP. There are some systems in which it is possible to produce an amorphous phase by MA but not by RSP techniques. Among those the role of contamination on the formation of amorphous phases by MA, at least in some cases, cannot be completely ruled out. Table 12.4 compares the amorphous phase–

forming composition ranges in some alloy systems produced by these two techniques. Amorphous phase formation has not been reported in all the listed alloy systems by both the techniques. Information is included where available.

Several investigations were conducted on the crystallization behavior of rapidly solidified amorphous alloys and also their transformation behavior on application of pressure (see the relevant chapters in Refs. 4–10). However, only a few reports are available on such transformations of amorphous phases produced by MA/MM. An important difference observed in amorphous phases synthesized by MA and RSP techniques is the degree of thermal relaxation. Since MA is conducted at or near room temperature, the amorphous alloy samples are in a very unrelaxed state [199]. However, it has also been reported that a low-temperature relaxation peak was observed in mechanically alloyed powders but not in RSP ribbons [129]. Such differences may prove to be very useful in probing the microscopic origin of the relaxation processes. Another important difference is in the nature of the transformation. For example, the amorphous phase at the Zr-24 at% Fe composition synthesized by MA directly transforms to the equilibrium Zr_3Fe phase. But, the first crystallization products of rapidly solidified (melt spun) alloy are metastable Zr_2Fe, ω-Zr, and α-Zr, which were not observed during the decomposition of the mechanically alloyed powder [417].

Amorphous phases are also synthesized by many other methods, such as laser processing, sputtering, and ion mixing. Differences have been observed in the amorphous phase–forming composition range between MA and these methods. Figure 12.11 compares the amorphous phase–forming composition ranges for the Ni-Nb system processed into the amorphous state by several different techniques. It may be noted that the homogeneity range for the formation of the amorphous phase is

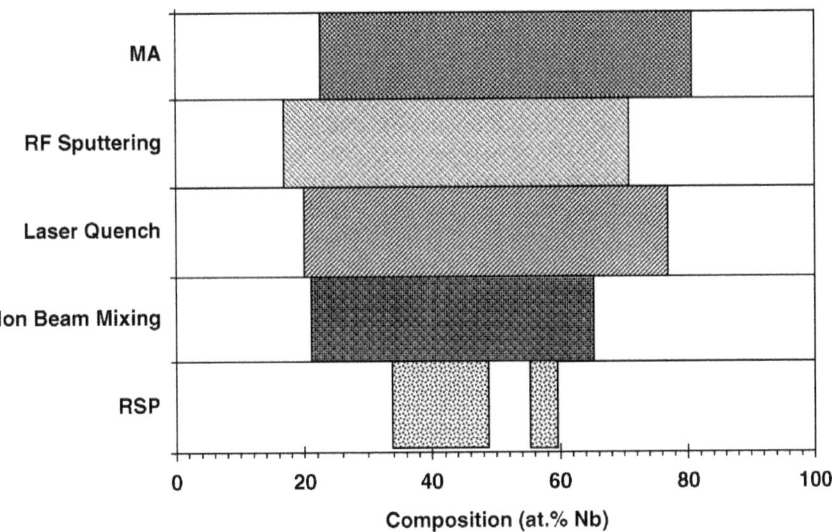

Figure 12.11 Composition of amorphous phase forming range achieved in the Ni-Nb system by different nonequilibrium processing routes.

quite wide for the mechanically alloyed powder. Alloys processed by laser quenching only come close to this range.

REFERENCES

1. Buckel, W., Hilsch, R. (1952). *Z. Phys.* 131:420.
2. Klement, W., Willens, R. H., Duwez, P. (1960). *Nature* 187:869–871.
3. Suryanarayana, C., ed. (1999). *Non-equilibrium Processing of Materials.* Oxford, UK: Pergamon.
4. Liebermann, H. H., ed. (1993). *Rapidly Solidified Alloys: Processes, Structures, Properties, Applications.* New York: Marcel Dekker.
5. Anantharaman, T. R., Suryanarayana, C. (1987). *Rapidly Solidified Metals: A Technological Overview.* Aedermannsdorf, Switzerland: TransTech.
6. Luborsky, F. E., ed. (1983). *Amorphous Metallic Alloys.* London, UK: Butterworths.
7. Anantharaman, T. R., ed. (1984). *Metallic Glasses: Production, Properties, and Applications.* Aedermannsdorf, Switzerland: TransTech.
8. Beck, H., Güntherodt, H. J., eds. *Glassy Metals.* Berlin: Springer-Verlag, Vol. 1, 1981; Vol. 2, 1983; Vol. 3, 1994.
9. Cahn, R. W. (1991). In: Zarzycki, J., ed. *Glasses and Amorphous Materials. Vol. 9 of Materials Science and Technology—A Comprehensive Treatment.* Weinheim: VCH, pp. 493–548.
10. Cahn, R. W., Greer, A. L. In: Cahn, R. W., Haasen, P., eds. *Physical Metallurgy.* Vol. 2. Amsterdam: Elsevier, pp. 1723–1830.
11. Russell, K. C. (1985). *Prog. Mater. Sci.* 28:229–434.
12. Yeh, X. L., Samwer, K., Johnson, W. L. (1983). *Appl. Phys. Lett.* 42:242–244.
13. Aoki, K. (2001). *Mater. Sci. Eng.* A304–306:45–53.
14. Schwarz, R. B., Johnson, W. L. (1983). *Phys. Rev. Lett.* 51:415–418.
15. Sharma, S. M., Sikka, S. K. (1996). *Prog. Mater. Sci.* 40:1–77.
16. Atzmon, M., Verhoven, J. D., Gibson, E. D., Johnson, W. L. (1984). *Appl. Phys. Lett.* 45:1052–1053.
17. Suryanarayana, C. (1995). *Bibliography on Mechanical Alloying and Milling.* Cambridge, UK: Cambridge International Science Publishing.
18. Koch, C. C., Cavin, O. B., McKamey, C. G., Scarbrough, J. O. (1983). *Appl. Phys. Lett.* 43:1017–1019.
19. See the articles starting on pp. 617 (Carl H Smith), 665 (Guthrie, M. S.), 691 (Rabinkin, A.; Liebermann, H. H.), and 737 (Suryanarayana, C.; Froes, F. H.) In: Liebermann, H. H., ed. *Rapidly Solidified Alloys: Processes, Structures, Properties, Applications.* New York: Marcel Dekker, 1993.
20. Ermakov, A. E., Yurchikov, E. E., Barinov, V. A. (1981). *Phys. Met. Metallogr.* 52(6):50–58.
21. Ermakov, A. E., Barinov, V. A., Yurchikov, E. E. (1982). *Phys. Met. Metallogr.* 54(5):90–96.
22. Weeber, A. W., Bakker, H. (1988). *Physica B* 153:93–135.
23. Aymard, L., Dehahaye-Vidal, A., Portemer, F., Disma, F. (1996). *J. Alloys Compounds* 238:116–127.
24. Sui, H. X., Zhu, M., Qi, M., Li, G. B., Yang, D. Z. (1992). *J. Appl. Phys.* 71:2945–2949.
25. Tsurui, T., Tsai, A. P., Inoue, A., Masumoto, T. (1995). *J. Alloys Compounds* 218:L7–L10.
26. Kobayashi, K. F., Tachibana, N., Shingu, P. H. (1990). *J. Mater. Sci.* 25:3149–3154.
27. Zhang, F. X., Wang, W. K. (1996). *J. Alloys Compounds* 240:256–260.

28. Eckert, J., Schultz, L., Urban, K. (1990). *Z. Metallkde.* 81:862–868.
29. Gu, X. J., Ye, F., Zhou, F., Lu, K. (2000). *Mater. Sci. Eng.* A278:61–65.
30. Wang, G., Zhang, D., Chen, H., Lin, B., Wang, W., Dong, Y. (1991). *Phys. Lett.* A155:57–61.
31. Morris, M. A., Morris, D. G. (1992). *Mater. Sci. For.* 88–90:529–535.
32. Zhou, F., Lück, R., Scheffer, M., Lang, D., Lu, K. (1999). *J. Non-Cryst. Solids* 250–252:704–708.
33. Nasu, S., Imaoka, S., Morimoto, S., Tanimoto, H., Huang, B., Tanaka, K., Kuyama, J., Ishihara, K. N., Shingu, P. H. (1992). *Mater. Sci. For.* 88–90:569–576.
34. Mukhopadhyay, D. K., Suryanarayana, C., Froes, F. H. (1995). *Metall. Mater. Trans.* 26A:1939–1946.
35. Oleszak, D., Shingu, P. H. (1997). *Mater. Sci. For.* 235–238:91–96.
36. Shingu, P. H., Ishihara, K. N., Uenishi, K., Kuyama, J., Huang, B., Nasu, S. (1990). In: Clauer, A. H., deBarbadillo, J. J., eds. *Solid State Powder Processing*. Warrendale, PA: TMS, pp. 21–34.
37. Huang, B., Ishihara, K. N., Shingu, P. H. (1997). *Mater. Sci. Eng.* A231:72–79.
38. Dong, Y. D., Wang, W. H., Liu, L., Xiao, K. Q., Tong, S. H., He, Y. Z. (1991). *Mater. Sci. Eng.* A134:867–871.
39. Zhang, D. L., Adam, G., Ammundsen, B. (2002). *J. Alloys Compounds* 340:226–230.
40. Cardoso, K. R., Garcia-Escorial, A., Botta, W. J., Lieblich, M. (2001). *Mater. Sci. For.* 360–362:161–166.
41. Rodrigues, C. A. D., Leiva, D. R., Cardoso, K. R., Kiminami, C. S., Botta, W. J. F. (2002). *Mater. Sci. For.* 386–388:33–38.
42. Burzynska-Szyske, M., Fadeeva, V. I., Matyja, H. (1997). *Mater. Sci. For.* 235–238:97–102.
43. Schwarz, R. B., Hannigan, J. W., Sheinberg, H., Tiainen, T. (1988). In: Gummeson, P. U. Gustafson, D. A., eds. *Modern Developments in Powder Metallurgy*. Vol. 21. Princeton, NJ: Metal Powder Industries Federation, pp. 415–427.
44. Chen, G., Wang, K., Wang, J., Jiang, H., Quan, M. (1993). In: deBarbadillo, J. J., et al., eds. *Mechanical Alloying for Structural Applications*. Materials Park, OH: ASM International, pp. 183–187.
45. Schurack, F., Börner, I., Eckert, J., Schultz, L. (1999). *Mater. Sci. For.* 312–314:49–54.
46. Roy, M., Singh, K., Bahadur, D. (1992). *J. Mater. Sci. Lett.* 11:858–861.
47. Sherif El-Eskandarany, M., Aoki, K., Suzuki, K. (1991). *Scripta Metall. Mater.* 25:1695–1700.
48. Sherif El-Eskandarany, M., Aoki, K., Suzuki, K. (1992). *J. Appl. Phys.* 71:2924–2930.
49. Peng, Z., Suryanarayana, C., Froes, F. H. (1996). *Metall. Mater. Trans.* 27A:41–48.
50. Makhlouf, S. A., Sumiyama, K., Suzuki, K. (1993). *J. Alloys Compounds* 199:119–124.
51. Benameur, T., Inoue, A., Masumoto, T. (1994). *Nanostructured Mater.* 4:303–322.
52. Benameur, T., Inoue, A. (1998). *Mater. Sci. For.* 269–272:163–168.
53. Dougherty, G. M., Shiflet, G. J., Poon, S. J. (1994). *Acta Metall. Mater.* 42:2275–2283.
54. Itsukaichi, T., Norimatsu, T., Umemoto, M., Okane, I., Wu, B. -Y. (1992). In: Tamura, I., ed. *Heat and Surface '92*. Tokyo: Japan Tech Info Center, pp. 305–308.
55. Sá Lisboa, R. D., Perdigão, M. N. R. V., Kiminami, C. S., Botta, W. J. F. (2002). *Mater. Sci. For.* 386–388:59–64.
56. Sherif El-Eskandarany, M., Aoki, K., Suzuki, K. (1992). *Appl. Phys. Lett.* 60:1562–1563.
57. Sherif El-Eskandarany, M., Aoki, K., Itoh, H., Suzuki, K. (1991). *J. Less Common Metals* 169:235–244.
58. Sherif El-Eskandarany, M., Aoki, K., Suzuki, K. (1992). *J. Alloys Compounds* 186:15–31.
59. Sherif El-Eskandarany, M., Aoki, K., Suzuki, K. (1992). *J. Non-Cryst. Solids* 150:472–477.

60. Saji, S., Abe, S., Matsumoto, K. (1992). *Mater. Sci. For.* 88–90:367–374.
61. Murty, B. S., Naik, M. D., Mohan Rao, M., Ranganathan, S. (1992). *Mater. For.* 16:19–26.
62. Kojima, Y., Senna, M., Shinohara, T., Ono, S., Sumiyama, K., Suzuki, K. (1995). *J. Alloys Compounds* 227:97–101.
63. Wu, N. Q., Wu, J. M., Li, Z. Z., Wang, G. X. (1997). *Mater. Trans. Jpn. Inst. Metals* 38:255–259.
64. Fan, G. J., Quan, M. X., Hu, Z. Q. (1995). *J. Mater. Sci.* 30:4847–4851.
65. Saji, S., Neishi, Y., Araki, H., Minamino, Y., Yamane, T. (1995). *Metall. Mater. Trans.* 26A:1305–1307.
66. Eckert, J., Seidel, M., Schultz, L. (1996). *Mater. Sci. For.* 225–227:113–118.
67. Sherif El-Eskandarany, M., Aoki, K., Suzuki, K. (1992). *Metall. Trans.* 23A:2131–2140.
68. Gaffet, E., Michael, D., Mazerolles, L., Berthet, P. (1997). *Mater. Sci. For.* 235–238:103–108.
69. Michel, D., Mazerolles, L., Berthet, P., Gaffet, E. (1995). *Eur. J. Solid State Inorg. Chem.* 32:673–682.
70. Loeff, P. I., Bakker, H. (1988). *Scripta Metall.* 22:401–404.
71. Suzuki, K. (1989). *J. Non-Cryst. Solids* 112:23–32.
72. Corrias, A., Ennas, G., Marongiu, G., Musinu, A., Paschina, G. (1993). *J. Mater. Res.* 8:1327–1333.
73. Atzeni, C., Pes, M., Sanna, U., Corrias, A., Paschina, G., Zedda, D. (1994). *Adv. Perform. Mater.* 1:243–253.
74. Corrias, A., Ennas, G., Marongiu, G., Paschina, G. (1991). *J. Mater. Sci.* 26:5081–5084.
75. Corrias, A., Ennas, G., Licheri, G., Marongiu, G., Musinu, A., Paschina, G. (1993). In: deBarbadillo, J. J., et al., eds. *Mechanical Alloying for Structural Applications.* Materials Park, OH: ASM International, pp. 451–456.
76. Kimura, H., Takada, F., Myung, W. N. (1988). *Mater. Sci. Eng.* 97:125–128.
77. Omuro, K., Miura, H. (1992). *Appl. Phys. Lett.* 60:1433–1435.
78. Okadome, K., Unno, K., Arakawa, T. (1995). *J. Mater. Sci.* 30:1807–1810.
79. Omuro, K., Miura, H. (1991). *Jpn. J. Appl. Phys.* 30:L851–853.
80. Sadano, H., Arakawa, T. (1996). *Mater. Trans. Jpn. Inst. Metals* 37:1099–1102.
81. López-Hirata, V. M., Juárez Martínez, U., Cabañas-Moreno, J. G. (1995). *Mater. Sci. For.* 189–191:261–266.
82. Sherif El-Eskandarany, M., Aoki, K., Sumiyama, K., Suzuki, K. (2002). *Acta Mater.* 50:1113–1123.
83. Sherif El-Eskandarany, M., Aoki, K., Sumiyama, K., Suzuki, K. (1997). *Appl. Phys. Lett.* 70:1679–1681.
84. Sherif El-Eskandarany, M., Aoki, K., Sumiyama, K., Suzuki, K. (1997). *Scripta Mater.* 36:1001–1009.
85. Sherif El-Eskandarany, M., Zhang, W., Inoue, A. (2003). *J. Alloys Compounds* 350:232–245.
86. Eckert, J., Schultz, L., Urban, K. (1990). *J. Less Common Metals* 166:293–302.
87. Kimura, H., Kimura, M., Ban, T. (1988). In: Lee, P. W., Moll, J. H., eds. *Rapidly Solidified Materials: Properties and Processing.* Materials Park, OH: ASM International, pp. 171–176.
88. Hellstern, E., Schultz, L. (1987). *Phil. Mag.* B56:443–448.
89. Ogino, Y., Maruyama, S., Yamasaki, T. (1991). *J. Less Common Metals* 168:221–235.
90. Xia, S. K., Rizzo Assunção, F. C., Baggio-Saitovich, E. (1996). *Mater. Sci. For.* 225–227:459–464.
91. Morris, M. A., Morris, D. G. (1991). *J. Mater. Sci.* 26:4687–4696.
92. Schanzer, M., Mehrer, H. (1990). *J. de Physique* 51(Colloq. 4, Suppl. 14):C4-87–C4-93.
93. Morris, M. A. (1992). *Mater. Sci. For.* 88–90:671–678.

94. Matsuki, K., Inoue, A., Kimura, H. M., Masumoto, T. (1988). *Mater. Sci. Eng.* 97:47–51.
95. Inoue, A., Kimura, H. M., Matsuki, K., Masumoto, T. (1987). *J. Mater. Sci. Lett.* 6:979–981.
96. Kim, K. J., Sherif El-Eskandarany, M., Sumiyama, K., Suzuki, K. (1993). *J. Non-Cryst. Solids* 155:165–170.
97. Kim, K. J., Sumiyama, K., Suzuki, K. (1994). *J. Non-Cryst. Solids* 168:232–240.
98. Kis-Varga, M., Beke, D. L. (1996). *Mater. Sci. For.* 225–227:465–470.
99. Zhang, B. W., Xie, H., Liao, S. (1999). *J. Mater. Proc. Technol.* 89–90:378–384.
100. Zhang, H., Naugle, D. G. (1992). *Appl. Phys. Lett.* 60:2738–2740.
101. Miura, H., Isa, S., Omuro, K. (1990). *Jpn. J. Appl. Phys.* 29:L339–342.
102. Battezzati, L., Baricco, M., Enzo, S., Schiffini, L., Soletta, I., Cocco, G. (1992). *Mater. Sci. For.* 88–90:771–778.
103. Baricco, M., Battezzati, L., Enzo, S., Soletta, I., Cocco, G. (1993). *Spectrochim. Acta* A49:1331–1344.
104. Hunt, J. A., Soletta, I., Enzo, S., Meiya, L., Havill, R. L., Battezzati, L., Cocco, G., Cowlam, N. (1995). *Mater. Sci. For.* 179–181:255–260.
105. Ivison, P. K., Cowlam, N., Soletta, I., Cocco, G., Enzo, S., Battezzati, L. (1991). *Mater. Sci. Eng.* A134:859–862.
106. Politis, C., Johnson, W. L. (1986). *J. Appl. Phys.* 60:1147–1151.
107. Baricco, M., Battezzati, L., Cocco, G., Soletta, I., Enzo, S. (1993). *J. Non-Cryst. Solids* 156–158:527–531.
108. Fukunaga, T., Mori, M., Inou, K., Mizutani, U. (1991). *Mater. Sci. Eng.* A134:863–866.
109. Gaffet, E., Louison, G., Harmelin, M., Faudot, F. (1991). *Mater. Sci. Eng.* A134:1380–1384.
110. Hellstern, E., Schultz, L. (1986). *Appl. Phys. Lett.* 48:124–126.
111. Jang, J. S. C., Koch, C. C. (1989). *Scripta Metall.* 23:1805–1810.
112. Al-Hajry, A., Al-Assiri, M., Al-Heniti, S., Enzo, S., Hefne, J., Al-Shahrani, A., Al-Salami, A. E. (2002). *Mater. Sci. For.* 386–388:205–210.
113. Boldrick, M. S., Wagner, C. N. J. (1991). *Mater. Sci. Eng.* A134:872–875.
114. Lee, P. Y., Lin, C. K., Chen, G. S., Louh, R. F., Chen, K. C. (1999). *Mater. Sci. For.* 312–314:67–72.
115. Suriñach, S., Baró, M. D., Segura, J., Clavaguera-Mora, M. T., Clavaguera, N. (1991). *Mater. Sci. Eng.* A134:1368–1371.
116. Ruuskanen, P., Heczko, O. (1993). *Key Eng. Mater.* 81–83:159–168.
117. Barinov, V. A., Tsurin, V. A., Elsukov, E. P., Ovechkin, L. V., Dorofeev, G. A., Ermakov, A. E. (1992). *Phys. Metals Metallogr.* 74(4):412–415.
118. Calka, A., Radlinski, A. P. (1991). *Appl. Phys. Lett.* 58:119–121.
119. Osagawara, T., Inoue, A., Masumoto, T. (1992). *Mater. Sci. For.* 88–90:423–430.
120. Nasu, T., Koch, C. C., Nagaoka, K., Itoh, N., Sakurai, K., Suzuki, K. (1991). *Mater. Sci. Eng.* A134:1385–1388.
121. Tanaka, T., Nasu, S., Nakagawa, K., Ishihara, K. N., Shingu, P. H. (1992). *Mater. Sci. For.* 88–90:269–274.
122. Tanaka, T., Nasu, S., Ishihara, K. N., Shingu, P. H. (1991). *J. Less Common Metals* 171:237–247.
123. Ji, Y., Wang, G., Li, F., Wang, G., Zhao, J., Zhang, S. (2001). *J. Mater. Sci. Lett.* 20:1267–1269.
124. Koyano, T., Takizawa, T., Fukunaga, T., Mizutani, U., Kamizuru, S., Kita, E., Tasaki, A. (1993). *J. Appl. Phys.* 73:429–433.
125. Omuro, K., Miura, H. (1995). *Mater. Sci. For.* 179–181:273–278.
126. Streletskii, A. N., Leonov, A. V., Beresteskaya, I. V., Mudretsova, S. N., Majorova, A. F., Ju Butyagin, P. (2002). *Mater. Sci. For.* 386–388:187–192.

127. Crespo, P., Hernando, A., Garcia-Escorial, A., Castaño, F. J., Multigner, M. (2000). *Mater. Sci. For.* 343–346:793–799.
128. Lin, C. K., Lee, P. Y., Kao, S. W., Chen, G. S., Louh, R. F., Hwu, Y. (1999). *Mater. Sci. For.* 312–314:55–60.
129. Suñol, J. J., Pradell, T., Clavaguera, N., Clavaguera-Mora, M. T. (2001). *Mater. Sci. For.* 360–362:525–530.
130. Miao, W. F., Li, G. S., Li, S. L., Wang, J. T. (1992). *Acta Phys. (Sinica)* 41:924–928.
131. Osagawara, T., Inoue, A., Masumoto, T. (1991). *Mater. Sci. Eng.* A134:1338–1341.
132. Carmona, F., Gonzalez, J. M., Martin, A., Martin, V. E. (1991). *J. Mag. Mag. Mater.* 101:119.
133. Le Caër, G., Matteazzi, P., Fultz, B. (1992). *J. Mater. Res.* 7:1387–1395.
134. Pan, C. W., Hung, M. P., Chang, Y. H. (1994). *Mater. Sci. Eng.* A185:147–152.
135. Enzo, S., Macri, P., Rose, P., Cowlam, N. (1993). In: deBarbadillo, J. J., et al., eds. *Mechanical Alloying for Structural Applications*. Materials Park, OH: ASM International, pp. 101–108.
136. Trudeau, M. L., Schulz, R., Zaluski, L., Hosatte, S., Ryan, D. H., Doner, C. B., Tessier, P., Ström-Olsen, J. O., Van Neste, A. (1992). *Mater. Sci. For.* 88–90:537–544.
137. Ruuskanen, P. (1998). *Mater. Sci. For.* 269–272:139–144.
138. Fultz, B., Le Caër, G., Matteazzi, P. (1989). *J. Mater. Res.* 4:1450–1455.
139. Bai, H., Michaelsen, C., Sinkler, W., Bormann, R. (1997). *Mater. Sci. For.* 235–238:361–366.
140. Hellstern, E., Schultz, L. (1988). *Mater. Sci. Eng.* 97:39–42.
141. Burgio, N., Iasonna, A., Magini, M., Martelli, S., Padella, F. (1991). *Il Nuovo Cimento* 13D:459–476.
142. Hellstern, E., Schultz, L., Bormann, R., Lee, D. (1988). *Appl. Phys. Lett.* 53:1399–1401.
143. Schultz, L. (1988). *Mater. Sci. Eng.* 97:15–23.
144. Shen, T. D., Wang, K. Y., Quan, M. X., Hu, Z. Q. (1993). *Appl. Phys. Lett.* 63:1637–1639.
145. Shirakawa, Y., Matsuda, T., Tani, Y., Shimosaka, A., Hidaka, J. *J. Non-Cryst. Solids* 293–295:764–768.
146. Gaffet, E. (1991). *Mater. Sci. Eng.* A149:85–94.
147. Al-Hajry, A., Al-Assiri, M., Enzo, S., Hefne, J., Al-Heniti, S., Cowlam, N., Al-Shahrani, A. (2001). *Mater. Sci. For.* 360–362:343–348.
148. Krauss, W., Politis, C., Weimar, P. (1988). *Metal Powder Rep.* 43:231–238.
149. Thompson, J. R., Politis, C., Kim, Y. C. (1988). *Mater. Sci. Eng.* 97:31–34.
150. Thompson, J. R., Politis, C., Kim, Y. C. (1992). *Mater. Sci. For.* 88–90:545–552.
151. Al-Assiri, M., Al-Hajry, A., Hefne, J., Al-Shahrani, A., Abboudy, S., Al-Heniti, S., Cowlam, N. (2000). *Mater. Sci. For.* 343–346:732–737.
152. Al-Hajry, A., Al-Assiri, M., Enzo, S., Cowlam, N., Hefne, J., Jones, L., Delogu, F., Brequel, H. (2002). *Mater. Sci. For.* 386–388:211–216.
153. Matsuki, K., Abe, F., Inoue, A., Masumoto, T. (1992). *Mater. Sci. For.* 88–90:313–320.
154. Yang, Q. M., Lei, Y. Q., Wu, J., Qang, Q. D., Lu, G. L., Chen, L. S. (1993). *Key Eng. Mater.* 81–83:169–173.
155. Ruggeri, S., Lenain, C., Roué, L., Liang, G., Huot, J., Schulz, R. (2002). *J. Alloys Compounds* 339:195–201.
156. Liu, W., Wu, H., Lei, Y., Wang, Q., Wu, J. (1997). *J. Alloys Compounds* 252:234–237.
157. Hazelton, L. E., Nielsen, C. A., Deshmukh, U. V., Pierini, P. E. (1990). In: Froes, F. H., deBarbadillo, J. J., eds. *Structural Applications of Mechanical Alloying*. Materials Park, OH: ASM International, pp. 243–250.
158. Weiss, B., Eckert, J. (2000). *Mater. Sci. For.* 343–346:129–134.
159. Martinez-Sanchez, R., Estrad-Guel, I., Jaramillo-Vigueras, D., De la Torre, S. D., Gaona-Tiburcio, C., Guerrero-Paz, J. (2002). *Mater. Sci. For.* 386–388:135–140.
160. Cocco, G., Enzo, S., Barrett, N., Roberts, K. J. (1989). *J. Less Common Metals* 154:177–186.

161. Bokhonov, B. B., Konstanchuk, I. G., Boldyrev, V. V. (1995). *J. Alloys Compounds* 218:190–196.
162. Zhang, H., Liu, X. (2001). *Int. J. Ref. Metals Hard Mater.* 19:203–208.
163. Oleszak, D., Burzynska-Szyszko, M., Matyja, H. (1993). *J. Mater. Sci. Lett.* 12:3–5.
164. Peng, Z., Suryanarayana, C., Froes, F. H. (1993). In: deBarbadillo, J. J., et al., eds. *Mechanical Alloying for Structural Applications*. Materials Park, OH: ASM International, pp. 335–341.
165. Peng, Z. (1994). MS thesis, University of Idaho, Moscow, ID, USA.
166. Yang, J. Y., Zhang, T. J., Cui, K., Li, X. G., Zhang, J. (1996). *J. Alloys Compounds* 242:153–156.
167. Kenik, E. A., Bayuzick, R. J., Kim, M. S., Koch, C. C. (1987). *Scripta Metall.* 21:1137–1142.
168. Politis, C. (1985). *Physica B.* 135:286–289.
169. Cho, Y. S., Koch, C. C. (1993). *Mater. Sci. Eng.* A161:65–73.
170. Harada, T., Kuji, T. (1996). *J. Alloys Compounds* 232:238–243.
171. Kumar, K. S., Mannan, S. K. (1989). In: Liu, C. T., Taub, A. I., Stoloff, N. S., Koch C. C., eds. *High Temperature Ordered Intermetallic Alloys III*. Vol. 133. Pittsburgh, PA: Mater Res Soc, pp. 415–420.
172. Li, B., Liu, L., Ma, X. M., Dong, Y. D. (1993). *J. Alloys Compounds* 202:161–163.
173. Kim, M. S., Koch, C. C. (1987). *J. Appl. Phys.* 62:3450–3453.
174. Hirosawa, S., Makita, K., Ikegami, T., Umemoto, M. (1992). Proc. 6th Int. Conf. on Ferrite, Tokyo: Japan Soc. Powder Powder Metall. , pp. 1100–1103.
175. Itsukaichi, T., Umemoto, M., Okane, I., Horosawa, S. (1993). *J. Alloys Compounds* 193:262–265.
176. Ivanov, E., Grigorieva, T. F., Golubkova, G. V., Boldyrev, V. V., Fasman, A. B., Mikhailenko, S. D., Kalinina, O. T. (1988). *Mater. Lett.* 7:51–54.
177. Luo, Z. G., Guo, J. T., Hu, Z. Q. (1996). *J. Alloys Compounds* 234:106–110.
178. Cocco, G., Enzo, S., Barrett, N. T., Roberts, K. J. (1992). *Phys. Rev.* B45:7066–7076.
179. Li, M., Enzo, S., Soletta, I., Cowlam, N., Cocco, G. (1993). *J. Phys. C: Condens. Matter.* 5:5235–5244.
180. Enayati, M. H., Chang, I. T. H., Schumacher, P., Cantor, B. (1997). *Mater. Sci. For.* 235–238:85–90.
181. Enzo, S., Bonetti, E., Soletta, I., Cocco, G. (1991). *J. Phys. D: Appl. Phys.* 24:209–216.
182. Nasu, T., Nagaoka, K., Takahashi, S., Suganuma, E., Sekiuchi, T., Fukunaga, T., Suzuki, K. (1989). *Mater. Trans. Jpn. Inst. Metals* 30:620–623.
183. Schumacher, P., Enayati, M. H., Cantor, B. (1999). *Mater. Sci. For.* 312–314:351–356.
184. Schwarz, R. B., Koch, C. C. (1986). *Appl. Phys. Lett.* 49:146–148.
185. Petzoldt, F. (1988). *J. Less Common Metals* 140:85–92.
186. Tiainen, T. J., Schwarz, R. B. (1988). *J. Less Common Metals* 140:99–112.
187. Lee, P. Y., Koch, C. C. (1987). *J. Non-Cryst. Solids* 94:88–100.
188. Merk, N., Tanner, L. E. (1991). *Scripta Metall. Mater.* 25:309–313.
189. Lee, P. Y., Chen, T. R. (1994). *J. Mater. Sci. Lett.* 13:888–890.
190. Xia, S. K., Baggio-Saitovich, E., Rizzo Assunção, F. C., Peña Rodriguez, V. A. (1993). *J. Phys. C: Condens. Matter.* 5:2729–2738.
191. Wang, K. Y., Shen, T. D., Wang, J. T., Quan, M. X. (1991). *Scripta Metall. Mater.* 25:2227–2231.
192. Wang, K. Y., Shen, T. D., Wang, J. T., Quan, M. X. (1993). *J. Mater. Sci.* 28:6474–6478.
193. Sun, D. Z., Cheng, L. Z., Zhang, Y. M., Ho, K. Y. (1992). *J. Alloys Compounds* 186:33–35.
194. Cocco, G., Enzo, S., Schiffini, L., Battezzati, L. (1988). *Mater. Sci. Eng.* 97:43–46.
195. Schwarz, R. B., Petrich, R. R., Saw, C. K. (1985). *J. Non-Cryst. Solids* 76:281–302.
196. Fukunaga, T., Homma, Y., Suzuki, K., Misawa, M. (1991). *Mater. Sci. Eng.* A134:987–991.
197. Kimura, H., Takada, F. (1988). *Mater. Sci. Eng.* 97:53–57.

198. Weeber, A. W., Bakker, H. (1988). *J. Phys. F.: Metal Phys.* 18:1359–1369.
199. Brüning, R., Altounian, Z., Ström-Olsen, J. O., Shultz, L. (1988). *Mater. Sci. Eng.* 97: 317–320.
200. Calka, A., Radlinski, A. P. (1991). *Mater. Sci. Eng.* A134:1350–1353.
201. Lee, P. Y., Koch, C. C. (1987). *Appl. Phys. Lett.* 50:1578–1580.
202. Lee, C. H., Mori, H., Fukunaga, T., Mizutani, U. (1990). *Jpn. J. Appl. Phys.* 29:540–544.
203. Lee, P. Y., Koch, C. C. (1988). *J. Mater. Sci.* 23:2837–2845.
204. Weeber, A. W., Bakker, H., deBoer, F. R. (1986). *Europhys. Lett.* 2:445–448.
205. Schultz, L., Hellstern, E., Thoma, A. (1987). *Europhys. Lett.* 3:921–926.
206. Weeber, A. W., Bakker, H. (1988). *J. Less Common Metals* 141:93–102.
207. Petzoldt, F., Scholz, B., Kunze, H. D. (1988). *Mater. Sci. Eng.* 97:25–29.
208. Mizutani, U., Lee, C. H. (1990). *J. Mater. Sci.* 25:399–406.
209. Weeber, A. W., Wester, A. J. H., Haag, W. J., Bakker, H. (1987). *Physica* B145:349–352.
210. Haruyama, O., Asahi, N. (1992). *Mater. Sci. For.* 88–90:333–338.
211. Eckert, J., Schultz, L., Hellstern, E., Urban, K. (1988). *J. Appl. Phys.* 64:3224–3228.
212. Nasu, T., Nagaoka, K., Takahashi, S., Fukunaga, T., Suzuki, K. (1989). *Mater. Trans. Jpn. Inst. Metals* 30:146–149.
213. Padella, F., Paradiso, E., Burgio, N., Magini, M., Martelli, S., Guo, W., Iasonna, A. (1991). *J. Less Common Metals* 175:79–90.
214. Politis, C., Thompson, J. R. (1987). In: Tenhover, M., Johnson, W. L., Tannner, L. E., eds. *Science and Technology of Rapidly Quenched Alloys*. Vol. 80. Pittsburgh, PA: Mater. Res. Soc., pp. 91–96.
215. Zhang, D. L., Massalski, T. B. (1994). *J. Mater. Res.* 9:53–60.
216. Ishida, T., Tamaru, S. (1993). *J. Mater. Sci. Lett.* 12:1851–1853.
217. Itoh, K., Misawa, M., Fukunaga, T. (2001). *J. Non-Cryst. Solids* 293–295:575–579.
218. Malhouroux-Gaffet, N., Gaffet, E. (1993). *J. Alloys Compounds* 198:143–154.
219. Gaffet, E., Harmelin, M. (1990). In: Froes, F. H., deBarbadillo, J. J., eds. *Structural Applications of Mechanical Alloying*. Materials Park, OH: ASM International, pp. 257–264.
220. Schnitzke, K., Schultz, L., Wecker, J., Katter, M. (1990). *Appl. Phys. Lett.* 57:2853–2855.
221. Ding, J., McCormick, P. G., Street, R. (1992). *J. Alloys Compounds* 189:83–86.
222. Sherif El-Eskandarany, M., Aoki, K., Suzuki, K. (1990). *J. Less Common Metals* 167:113–118.
223. Sherif El-Eskandarany, M. (1996). *Metall. Mater. Trans.* 27A:3267–3278.
224. Sherif El-Eskandarany, M., Itoh, F., Aoki, K., Suzuki, K. (1990). *J. Non-Cryst. Solids* 117–118:729–732.
225. Fukunaga, T., Nakamura, K., Suzuki, K., Mizutani, U. (1990). *J. Non-Cryst. Solids* 117–118:700–703.
226. Lee, C. H., Fukunaga, T., Mizutani, U. (1991). *Mater. Sci. Eng.* A134:1334–1337.
227. Lee, C. H., Mori, M., Fukunaga, T., Sakurai, K., Mizutani, U. (1992). *Mater. Sci. For.* 88–90:399–406.
228. Liu, L., Chu, Z. Q., Dong, Y. D. (1992). *J. Alloys Compounds* 186:217–221.
229. Veltl, G., Scholz, B., Kunze, H. D. (1991). *Mater. Sci. Eng.* A134:1410–1413.
230. Lee, P. Y., Yang, J. L., Lin, H. M. (1998). *J. Mater. Sci.* 33:235–239.
231. Viswanadham, R. K., Mannan, S. K., Kumar, K. S. (1988). *Scripta Metall.* 22:1011–1014.
232. Suryanarayana, C., Chen, G. H., Frefer, A., Froes, F. H. (1992). *Mater. Sci. Eng.* A158:93–101.
233. Bonetti, E., Valdre, G., Enzo, S., Cocco, G. (1993). *J. Alloys Compounds* 194:331–338.
234. Itsukaichi, T., Shiga, S., Masuyama, K., Umemoto, M., Okane, I. (1992). *Mater. Sci. For.* 88–90:631–638.
235. Suzuki, T., Ino, T., Nagumo, M. (1992). *Mater. Sci. For.* 88–90:639–646.

236. Nash, P., Kim, H., Choo, H., Ardy, H., Hwang, S. J., Nash, A. S. (1992). *Mater. Sci. For.* 88–90:603–610.
237. Ahn, J. H., Choi, C. J., Chung, H. S. (1993). In: Ravi, V. A., Srivatsan, T. S., eds. *Processing and Fabrication of Advanced Materials for High Temperature Applications II.* Warrendale, PA: TMS, pp. 33–44.
238. Chen, G., Wang, K. (1993). In: deBarbadillo, J. J., et al., eds. *Mechanical Alloying for Structural Applications.* Materials Park, OH: ASM International, pp. 149–155.
239. Qi, M., Zhu, M., Li, G. B., Sui, H. X., Yang, D. Z. (1993). *J. Mater. Sci. Lett.* 12:66–69.
240. Park, Y. H., Hashimoto, H., Watanabe, R. (1992). *Mater. Sci. For.* 88–90:59–66.
241. Ahn, J. H., Chung, H. S., Watanabe, R., Park, Y. H. (1992). *Mater. Sci. For.* 88–90:347–354.
242. Ahn, J. H., Lee, K. R., Cho, H. K. (1995). *Mater. Sci. For.* 179–181:153–158.
243. Kambara, M., Uenishi, K., Kobayashi, K. F. (2000). *J. Mater. Sci.* 35:2897–2905.
244. Guo, W., Iasonna, A., Magnini, M., Martelli, S., Padella, F. (1994). *J. Mater. Sci.* 29: 2436–2444.
245. Frefer, A., Suryanarayana, C., Froes, F. H. (1993). In: Moore, J. J., et al., eds. *Advanced Synthesis of Engineered Structural Materials.* Materials Park, OH: ASM International, pp. 213–219.
246. Itsukaichi, T., Masuyama, K., Umemoto, M., Okane, I., Cabanãs-Moreno, J. G. (1993). *J. Mater. Res.* 8:1817–1828.
247. Burgio, N., Guo, W., Martelli, S., Magini, M., Padella, F., Soletta, I. (1990). In: Froes, F. H., deBarbadillo, J. J., eds. *Structural Applications of Mechanical Alloying.* Materials Park, OH: ASM International, pp. 175–183.
248. Guo, W., Martelli, S., Burgio, N., Magini, M., Padella, F., Paradiso, E., Soletta, I. (1991). *J. Mater. Sci.* 26:6190–6196.
249. Sherif El-Eskandarany, M. (1996). *J. Alloys Compounds* 234:67–82.
250. Ahn, J. H., Chung, H. (1994). In: Ravi, V. A., Srivatsan, T. S., Moore, J. J., eds. *Processing and Fabrication of Advanced Materials III.* Warrendale, PA: TMS, pp. 227–237.
251. Zhang, F., Lu, L., Lai, M. O., Liang, T. F. (1999). *Mater. Sci. For.* 312–314:109–114.
252. Bououdina, M., Guo, Z. X. (2002). *Mater. Sci. Eng.* A332:210–222.
253. Nagarajan, R., Ranganathan, S. (1994). *Mater. Sci. Eng.* A179/180:168–172.
254. Murty, B. S., Ranganathan, S., Mohan Rao, M. (1992). *Mater. Sci. Eng.* A149:231–240.
255. Zhang, L. C., Xu, J. (2002). *Mater. Sci. For.* 386–388:47–52.
256. Chu, B. L., Chen, C. C., Perng, T. P. (1992). *Metall. Trans.* 23A:2105–2110.
257. Fukunaga, T., Misawa, M., Suzuki, K., Mizutani, U. (1992). *Mater. Sci. For.* 88–90:325–332.
258. Schwarz, R. B., Petrich, R. R. (1988). *J. Less Common Metals* 140:171–184.
259. Wang, K. Y., He, A., Quan, M., Chen, G. (1993). In: deBarbadillo, J. J., et al., eds. *Mechanical Alloying for Structural Applications.* Materials Park, OH: ASM International, pp. 21–25.
260. Esaki, H., Tokizane, M. (1992). *Mater. Sci. For.* 88–90:625–630.
261. Battezzati, L., Enzo, S., Schiffini, L., Cocco, G. (1988). *J. Less Common Metals* 145:301–308.
262. Murty, B. S., Mohan Rao, M., Ranganathan, S. (1990). *Scripta Metall. Mater.* 24:1819–1824.
263. Sundaresan, R., Jackson, A. G., Krishnamurthy, S., Froes, F. H. (1988). *Mater. Sci. Eng.* 97:115–119.
264. Bahadur, D., Dunlap, R. A., Foldeaki, M. (1996). *J. Alloys Compounds* 240:278–284.
265. Thompson, J. R., Politis, C. (1987). *Europhys. Lett.* 3:199–205.
266. Radlinski, A. P., Calka, A. (1991). *Mater. Sci. Eng.* A134:1376–1379.
267. Oehring, M., Yan, Z. H., Klassen, T., Bormann, R. (1992). *Phys. Stat. Sol.* (a)131:671–689.

268. Counihan, P. J., Crawford, A., Thadhani, N. N. (1999). *Mater. Sci. Eng.* A267:26–35.
269. Parlapanski, D., Denev, S., Ruseva, R., Gatev, E. (1991). *J. Less Common Metals* 171:231–236.
270. Fukunaga, T., Homma, Y., Misawa, M., Suzuki, K. (1990). *J. Non-Cryst. Solids* 117–118:721–724.
271. Yamasaki, T., Ogino, Y., Morishita, K., Fukuoka, K., Atou, F., Syono, Y. (1994). *Mater. Sci. Eng.* A179/180:220–223.
272. Aoki, K., Memezawa, T., Masumoto, T. (1994). *J. Mater. Res.* 9:39–46.
273. Aoki, K., Memezawa, T., Masumoto, T. (1992). *Appl. Phys. Lett.* 61:1037–1039.
274. Takasaki, A., Han, C. H., Furuya, Y., Kelton, K. F. (2002). *Phil. Mag. Lett.* 82:353–361.
275. Cardelllini, F., Contini, V., Mazzone, G., Montone, A. (1997). *Phil. Mag.* B76:629–638.
276. Bououdina, M., Luklinska, Z., Guo, Z. X. (2001). *Mater. Sci. For.* 360–362:421–426.
277. Liu, L., Casadio, S., Magini, M., Nannetti, C. A., Qin, Y., Zheng, K. (1997). *Mater. Sci. For.* 235–238:163–168.
278. Kaloshkin, S. D. (2000). *Mater. Sci. For.* 342–346:591–596.
279. Sherif El-Eskandarany, M., Sumiyama, K., Suzuki, K. (1996). *Sci. Rep. RITU* A42:31–38.
280. Sherif El-Eskandarany, M., Sumiyama, K., Suzuki, K. (1997). *Acta Mater.* 45:1175–1187.
281. Herr, U., Samwer, K. (1992). *Nanostructured Mater.* 1:515–521.
282. Eckert, J., Jost, K., De Haas, O., Schultz, L. (1997). *Mater. Sci. For.* 235–238:133–138.
283. Schiffini, L., Mulas, G., Daturi, M., Ferretti, M. (1993). In: deBarbadillo, J. J., et al., eds. *Mechanical Alloying for Structural Applications.* Materials Park, OH: ASM International, pp. 457–461.
284. Zhang, H., Su, Y., Wang, L., Wu, L., Tan, Z., Zhang, B. (1994). *J. Alloys Compounds* 204:27–31.
285. Biswas, A., Dey, G. K., Haq, A. J., Bose, D. K., Banerjee, S. (1996). *J. Mater. Res.* 11:599–607.
286. Chen, G. H., Suryanarayana, C., Froes, F. H. (1993). In: deBarbadillo, J. J., et al., eds. *Mechanical Alloying for Structural Applications.* Materials Park, OH: ASM International, pp. 367–375.
287. Ma, E., Atzmon, M. (1991). *Phys. Rev. Lett.* 67:1126–1129.
288. Ma, E., Atzmon, M. (1993). *J. Alloys Compounds* 194:235–244.
289. Fecht, H. J., Han, G., Fu, Z., Johnson, W. L. (1990). *J. Appl. Phys.* 67:1744–1748.
290. Eckert, J., Seidel, M., Schlorke, N., Kübler, A., Schultz, L. (1997). *Mater. Sci. For.* 235–238:23–28.
291. Zhou, C. R., Lu, K., Xu, J. (2000). *Mater. Sci. For.* 343–346:116–122.
292. Deledda, S., Eckert, J., Schultz, L. (2002). *Mater. Sci. For.* 386–388:71–76.
293. Seidel, M., Eckert, J., Schultz, L. (1997). *Mater. Sci. For.* 235–238:29–34.
294. Lin, R., Seidel, M., Jiang, J. Z., Eckert, J. (1998). *Mater. Sci. For.* 269–272:461–466.
295. Sagel, A., Wunderlich, R. K., Fecht, H. J. (1997). *Mater. Sci. For.* 235–238:389–394.
296. Sagel, A., Wunderlich, R. K., Fecht, H. J. (1997). *Mater. Lett.* 33:123–127.
297. Sagel, A., Wunderlich, R. K., Fecht, H. J. (1998). *Mater. Sci. For.* 269–272:81–86.
298. Eckert, J., Schultz, L., Urban, K. (1988). *J. Less Common Metals* 145:283–291.
299. Weeber, A. W., Bakker, H. (1988). *Mater. Sci. Eng.* 97:133–135.
300. Ahn, J. H., Park, Y. K. (1999). *J. Mater. Sci. Lett.* 18:17–19.
301. Ahn, J. H., Zhu, M. (1998). *Mater. Sci. For.* 269–272:201–206.
302. Illeková, E., Jergel, M., Kuhnast, F.-A., Held, O. (1998). *Mater. Sci. For.* 269–272:583–588.
303. Bokhonov, B., Konstanchuk, I., Ivanov, E., Boldyrev, V. V. (1992). *J. Alloys Compounds* 187:207–214.
304. Sherif El-Eskandarany, M., Saida, J., Inoue, A. (2003). *Metall. Mater. Trans.* 34A:893–898.

305. Al-Assiri, M., Al-Hajry, A., Hefne, J., Enzo, S., Cowlam, N., Al-Shahrani, A., Al-Salami, A. E. (2001). *Mater. Sci. For.* 360–362:379–384.
306. Avvakumov, E. G., Karakchiev, L. G., Gusev, A. A., Vinokurova, O. B. (2002). *Mater. Sci. For.* 386–388:245–250.
307. Kobayashi, K. F., Tachibana, N., Shingu, P. H. (1990). *J. Mater. Sci.* 25:801–804.
308. Sun, J., Zhang, J. X., Fu, Y. Y., Hu, G. X. (2002). *Mater. Sci. Eng.* A329–331:703–707.
309. Sherif El-Eskandarany, M., Aoki, K., Suzuki, K. (1991). *J. Alloys Compounds* 177:229–244.
310. Sakurai, M., Fukunaga, T., Sumiyama, K., Suzuki, K. (1991). *J. Jpn. Soc. Powder Powder Metall.* 38:63–66.
311. Tang, J. (1996). *Mater. Sci. For.* 225–227:477–482.
312. Tang, J., Zhao, W., Li, L., Falster, A. U., Simmons, W. B. Jr, Zhou, W. L., Ikuhara, Y., Zhang, J. H. (1996). *J. Mater. Res.* 11:733–738.
313. Welham, N. J., Williams, J. S. (1998). *Carbon* 36:1309–1315.
314. Zhou, G. F., Bakker, H. (1993). *Phys. Rev.* B48:13383–13398.
315. Koch, C. C., Pathak, D., Yamada, K. (1993). In: deBarbadillo, J. J., et al., eds. *Mechanical Alloying for Structural Applications.* Materials Park, OH: ASM International, pp. 205–212.
316. Cho, Y. S., Koch, C. C. (1993). *J. Alloys Compounds* 194:287–294.
317. Fukunaga, T. (2001). *J. Non-Cryst. Solids* 293–295:187–192.
318. Gaffet, E. (1991). *Mater. Sci. Eng.* A136:161–169.
319. Nasu, T., Araki, F., Uemura, O., Usuki, T., Kameda, Y., Takahashi, S., Tokumitsu, K. (2001). *Mater. Sci. For.* 360–362:203–210.
320. Damonte, L. C., Mendoza-Zélis, L., Sánchez, F. H. (1998). *Mater. Sci. For.* 269–272:625–630.
321. Inoue, A., Matsuki, K., Masumoto, T. (1992). *Mater. Sci. For.* 88–90:305–312.
322. Calka, A., Radlinski, A. P. (1989). *Mater. Sci. Eng.* A118:131–135.
323. Skakov, YuA., Djakonova, N. P., Sviridova, T. A., Shelekhov, E. V. (1998). *Mater. Sci. For.* 269–272:595–600.
324. Alonso, T., Yang, H., Liu, Y., McCormick, P. G. (1992). *Appl. Phys. Lett.* 60:833–834.
325. Nakamura, T., Inoue, A., Matsuki, K., Masumoto, T. (1989). *J. Mater. Sci. Lett.* 8:13–16.
326. Jang, J. S. C., Koch, C. C. (1990). *J. Mater. Res.* 5:498–510.
327. Chen, Y., Le Hazif, R., Martin, G. (1992). *Mater. Sci. For.* 88–90:35–41.
328. Aoki, K., Memezawa, A., Masumoto, T. (1993). *J. Mater. Res.* 8:307–313.
329. Tonejc, A., Duževič, D., Tonejc, A. M. (1991). *Mater. Sci. Eng.* A134:1372–1375.
330. Takeuchi, T., Koyano, T., Utsumi, M., Fukunaga, T., Kaneko, K., Mizutani, U. (1994). *Mater. Sci. Eng.* A179/180:224–228.
331. Guo, F. Q., Lu, K. (1998). *Phil. Mag. Lett.* 77:181–186.
332. Lu, K., Guo, F. Q., Zhao, Y. H., Jin, Z. H. (1999). *Mater. Sci. For.* 312–314:43–48.
333. Zhao, Y. H., Jin, Z. H., Lu, K. (1999). *Phil. Mag. Lett.* 79:747–754.
334. Shen, T. D., Koch, C. C., McCormick, T. L., Nemanich, R. J., Huang, J. Y., Huang, J. G. (1995). *J. Mater. Res.* 10:139–148.
335. Yang, X. Y., Wu, Y. K., Ye, H. Q. (2001). *J. Mater. Sci. Lett.* 20:1517–1518.
336. Liao, J., Senna, M. (1992). *Mater. Sci. For.* 88–90:753–758.
337. Yamada, K., Koch, C. C. (1993). *J. Mater. Res.* 8:1317–1326.
338. Fukamichi, K., Goto, T., Fukunaga, T., Suzuki, K. (1991). *Mater. Sci. Eng.* A133:245–247.
339. Gümbel, A., Ledig, L., Hough, D., Oertel, C.-G., Skrotzki, W., Eckert, J., Schultz, L. (1999). *Mater. Sci. For.* 312–314:61–66.

340. Gerasimov, K. B., Gusev, A. A., Ivanov, E. Y., Boldyrev, V. V. (1991). *J. Mater. Sci.* 26:2495–2500.
341. Sherif El-Eskandarany, M., Saida, J., Inoue, A. (2002). *Acta Mater.* 50:2725–2736.
342. Weeber, A. W., Van der Meer, K., Bakker, H., de Boer, F. R., Thijse, B. J., Jungste, J. F. (1986). *J. Phys. F: Metal. Phys.* 16:1897–1900.
343. Sheng, H. W., Lu, K., Ma, E. (1999). *J. Appl. Phys.* 85:6400–6407.
344. Bonetti, E., Cocco, G., Enzo, S., Valdre, G. (1990). *Mater. Sci. Technol.* 6:1258–1262.
345. Trudeau, M. L., Schulz, R., Dussault, D., Van Neste, A. (1990). *Phys. Rev. Lett.* 64:99–102.
346. Trudeau, M. L. (1994). *Appl. Phys. Lett.* 64:3661–3663.
347. Koch, C. C. (1995). *Mater. Trans. Jpn. Inst. Metals* 36:85–95.
348. Mukhopadhyay, D. K., Suryanarayana, C., Froes, F. H. (1995). In: Phillips, M., Porter, J., eds. *Advances in Powder Metallurgy and Particulate Materials—1995*. Vol. 1. Princeton, NJ: Metal Powder Industries Federation, pp. 123–133.
349. Nash, P., Higgins, G. T., Dillinger, N., Hwang, S. J., Kim, H. (1989). In: Gasbarre, T. G., Jandeska, W. F., eds. *Advances in Powder Metallurgy—1989*. Vol. 2. Princeton, NJ: Metal Powder Industries Federation, pp. 473–479.
350. Suryanarayana, C. (1995). *Intermetallics* (3,153–160.
351. Sato, K., Ishizaki, K., Chen, G. H., Frefer, A., Suryanarayana, C., Froes, F. H. (1993). In: Moore, J. J., et al., eds. *Advanced Synthesis of Engineered Structural Materials*. Materials Park, OH: ASM International, pp. 221–225.
352. Anantharaman, T. R., Suryanarayana, C. (1971). *J. Mater. Sci.* 6:1111–1135.
353. Inoue, A. (2000). *Acta Mater.* 48:279–306.
354. Schwarz, R. B., Rubin, J. B. (1993). *J. Alloys Compounds* 194:189–197.
355. Hellstern, E., Schultz, L. (1987). *Mater. Sci. Eng.* 93:213–216.
356. Herbert, R. J., Perepezko, J. H. (2002). *Mater. Sci. For.* 386–388:21–26.
357. Lee, C. H., Fukunaga, T., Yamada, Y., Mizutani, U., Okamoto, H. (1993). *J. Phase Equilibria.* 14:167–171.
358. Gaffet, E., Faudot, F., Harmelin, M. (1992). *Mater. Sci. For.* 88–90:375–382.
359. Massobrio, C. (1990). *J. Physique* 51(Colloq. 4, Suppl. 14):C4-55–C4-61.
360. Eckert, J., Holzer, J. C., Krill, C. E. III, Johnson, W. L. (1992). *Mater. Sci. For.* 88–90:505–512.
361. Liu, Y. J., Chang, I. T. H. (2002). *Acta Mater.* 50:2747–2760.
362. Beke, D. L., Loeff, P. I., Bakker, H. (1991). *Acta Metall. Mater.* 39:1259–1266.
363. Beke, D. L., Bakker, H., Loeff, P. I. (1991). *Acta Metall. Mater.* 39:1267–1273.
364. Miedema, A. R. (1976). *Philips Tech. Rev.* 36:217.
365. Zhang, B. W. (1985). *Physica B.* 132:319.
366. Zhang, H., Zhang, B. W. (1995). *Physica* B205:263–268.
367. Zhang, H. (1993). *J. Phys. C: Condens. Matter.* 5:L337–L342.
368. Zhang, H., Zhang, B. W. (1993). *Physica* B192:247–252.
369. Chakk, Y., Berger, S., Weiss, B. Z., Brook-Levinson, E. (1994). *Acta Metall. Mater.* 42:3679–3685.
370. Egami, T., Waseda, Y. (1984). *J. Non-Cryst. Solids* 64:113–134.
371. Egami, T. (1996). *J. Non-Cryst. Solids* 205–207:575–582.
372. Fecht, H. J., Johnson, W. L. (1991). *Mater. Sci. Eng.* A133:427–430.
373. Kauzmann, W. (1948). *Chem. Rev.* 43:219–256.
374. Köster, U., Herold, U. (1981). In: Güntherodt, H. J., Beck, H., eds. *Glassy Metals I.* Berlin: Springer-Verlag, pp. 225–259.
375. Scott, M. G. (1983). In: Luborsky, F. E., ed. *Amorphous Metallic Alloys*. London: Butterworths, pp. 144–168.
376. Ranganathan, S., Suryanarayana, C. (1985). *Mater. Sci. For.* 3:173–185.

377. Wagner, C. N. J., Boldrick, M. S. (1993). *J. Alloys Compounds* 194:295–302.
378. Inoue, A., Kobayashi, K., Suryanarayana, C., Masumoto, T. (1980). *Scripta Metall.* 14:119–122.
379. Seidel, M., Eckert, J., Schultz, L. (1995). *J. Appl. Phys.* 77:5446–5448.
380. Eckert, J., Matten, N., Zinkevitch, M., Seidel, M. (1998). *Mater. Trans. Jpn. Inst. Metals* 39:623.
381. Bansal, C., Fultz, B., Johnson, W. L. (1994). *Nanostructured Mater.* 4:919–925.
382. Huang, B., Perez, R. J., Crawford, P. J., Sharif, A. A., Nutt, S. R., Lavernia, E. J. (1995). *Nanostructured Mater.* 5:545–553.
383. Chen, H. S. (1974). *Acta Metall.* 22:1505–1511.
384. Chen, H. S. (1976). *Mater. Sci. Eng.* 24:153–158.
385. Inoue, A. (2001). In: Inoue, A., Hashimoto, K., eds., *Amorphous and Nanocrystalline Materials: Preparation, Properties, and Applications.* Berlin: Springer-Verlag, pp. 1–51.
386. Peker, A., Johnson, W. L. (1993). *Appl. Phys. Lett.* 63:2342–2344.
387. Inoue, A. (1999). In: Suryanarayana, C., ed. *Non-equilibrium Processing of Materials.* Oxford: Pergamon, pp. 375–415.
388. Inoue, A. (1998). *Bulk Amorphous Alloys: Preparation and Fundamental Characteristics. Vol. 4 of Materials Science Foundations.* Zurich: TransTech.
389. Inoue, A. (1999). *Bulk Amorphous Alloys: Practical Characteristics and Applications. Vol. 6 of Materials Science Foundations.* Zurich: TransTech.
390. Johnson, W. L. (1999). *MRS Bull.* 24(10):42–56.
391. Johnson, W. L., Inoue, A., Liu, C. T., eds. (1999). *Bulk Metallic Glasses. Vol 554 of MRS Symp Proc.* Warrendale, PA: Mater. Res. Soc.
392. Inoue, A., Yavari, A. R., Johnson, W. L., Dauskardt, R. H., eds. (2001). *Supercooled Liquid, Bulk Glassy, and Nanocrystalline States of Alloys. Vol 644 of MRS Symp Proc.* Warrendale, PA: Mater. Res. Soc.
393. Schneider, S. (2001). *J. Phys.: Condens. Matter.* 13:7723–7736.
394. Inoue, A., Nishiyama, N., Kimura, H. M. (1997). *Mater. Trans. JIM* 38:179–183.
395. Egami, T. (2003). *J. Non-Cryst. Solids* 317:30–33.
396. Miracle, D. B., Senkov, O. N. (2003). *Mater. Sci. Eng.* A347:50–58.
397. Eckert, J. (1999). *Mater. Sci. For.* 312–314:3–12.
398. Leonhard, A., Xing, L. Q., Heilmaier, M., Gebert, A., Eckert, J., Schultz, L. (1998). *Nanostructured Mater.* 10:805–817.
399. Leonhard, A., Heilmaier, M., Eckert, J., Schultz, L. (1999). In: Johnson, W. L., Inoue, A., Liu, C. T., eds. *Bulk Metallic Glasses, vol. 554 of Symp. Proc.* Warrendale, PA: Mater. Res. Soc., pp. 137–142.
400. Roger-Leonhard, A., Heilmaier, M., Eckert, J. (2000). *Scripta Mater.* 43:459–464.
401. He, G., Bian, Z., Chen, G. L. (2001). *J. Mater. Sci. Lett.* 20:633–636.
402. Gu, X., Jiao, T., Kecskes, L. J., Woodman, R. H., Fan, C., Ramesh, K. T., Hufnagel, T. C. (2003). *J. Non-Cryst. Solids* 317:112–117.
403. Ashley, S. (1998). *Mech. Eng.* 120(6):72–74.
404. V.I.P. Vintage Model Amorphous Face Golf Club Catalog, Dunlop, Tokyo, 1998.
405. Conner, R. D., Dandliker, R. B., Johnson, W. L. (1998). *Acta Mater.* 46:6089–6102.
406. Miedema, A. R., de Chatel, P. F., deBoer, F. R. (1980). *Physica* B100:1.
407. Bakker, H. (1998). *Enthalpies in Alloys—Miedema's Semi-Empirical Model. Vol. 1 of Materials Science Foundations.* Zurich: TransTech.
408. deBoer, F. R., Boom, R., Mattens, W. C. M., Miedema, A. R., Niessen, A. K. (1988). In: deBoer, F. R., Pettifor, D. G., eds. *Cohesion in Metals.* Amsterdam: Elsevier.
409. Weeber, A. W., Loeff, P. I., Bakker, H. (1988). *J. Less Common Metals* 145:293–299.
410. Eckert, J., Schultz, L., Urban, K. (1989). In: Arzt, E., Schultz, L., eds. *New Materials by Mechanical Alloying Techniques.* Oberursel, Germany: DGM, pp. 85–90.

411. Loeff, P. I., Spit, F. H. M., Bakker, H. (1988). *J. Less Common Metals* 145:271–275.
412. Zhang, B. W., Tan, Z. S. (1988). *J. Mater. Sci. Lett.* 7:681–682.
413. Saunders, N., Miodownik, A. P. (1998). *CALPHAD*. Oxford, UK: Pergamon.
414. Hellstern, E., Schultz, L., Eckert, J. (1988). *J. Less Common Metals* 140:93–98.
415. Zhou, G. F., Bakker, H. (1994). *Phys. Rev. Lett.* 72:2290–2293.
416. Hellstern, E., Schultz, L. (1986). *Appl. Phys. Lett.* 49:1163–1165.
417. Biegel, W., Krebs, H. U., Michaelsen, C., Freyhardt, H. C., Hellstern, E. (1988). *Mater. Sci. Eng.* 97:59–62.

13

Nanostructured Materials

13.1 INTRODUCTION

Nanocrystalline materials are single- or multiphase polycrystalline solids with a grain size of the order of a few nanometers (1 nm = 10^{-9} m = 10 Å), typically 1–100 nm in at least one dimension. Since the grain sizes are so small, a significant volume of the microstructure in nanocrystalline materials is composed of interfaces, mainly grain boundaries. This means that a large volume fraction of the atoms resides in the grain boundaries. Consequently, nanocrystalline materials exhibit properties that are significantly different from, and often improved over, their conventional coarse-grained polycrystalline counterparts. Compared to the material with a more conventional grain size, i.e., larger than a few micrometers, nanocrystalline materials show increased strength, high hardness, extremely high diffusion rates, and consequently reduced sintering times for powder compaction. Several excellent reviews are available giving details on different aspects of processing, properties, and applications of these materials [1–7].

13.2 CLASSIFICATION AND CHARACTERISTICS OF NANOSTRUCTURED MATERIALS

Nanocrystalline materials can be classified into different categories depending on the number of dimensions in which the material has nanometer modulations. Thus, they can be classified into (1) layered or lamellar structures, (2) filamentary structures, and (3) equiaxed nanostructured materials. A layered or lamellar structure is a one-dimensional (1-D) nanostructure in which the magnitude of length and width are much greater than the thickness, which is only a few nanometers. One can also visualize a two-dimensional (2-D) rod-shaped nanostructure that can be termed filamentary and in this the length is substantially larger than width and diameter, which are of nanometer dimensions. The most common of the nanostructures, however, is basically equiaxed (all three dimensions are of nanometer size) and are termed nanostructured crystallites [three-dimensional (3-D) nanostructures] [8].

The nanostructured materials may contain crystalline, quasi-crystalline, or amorphous phases and can be metal, ceramic, polymer, or composite. If the grains are made up of crystals, the material is called nanocrystalline; and this is the most common variety of nanostructured materials. On the other hand, if the grains are made up of quasi-crystalline or amorphous (glassy) phases, they are termed nano-quasi-crystals and nanoglasses, respectively [2]. Gleiter [9] has further classified the nanostructured materials according to the composition, morphology, and distribution of the nanocrystalline component.

Table 13.1 shows this classification of the three types of nanostructures. Among these, maximal research is conducted on the synthesis, consolidation, and characterization of the 3-D nanostructured crystallites followed by the 1-D layered nanostructures. While the former are expected to find applications based on their high strength, improved formability, and good combination of soft magnetic properties, the latter are targeted for electronic applications. Relatively few investigations have been carried out on the 2-D filamentary nanostructures.

Figure 13.1 shows a schematic representation of a hard-sphere model of an equiaxed nanocrystalline metal. Two types of atoms can be distinguished: crystal atoms with nearest-neighbor configuration corresponding to the lattice and the boundary atoms with a variety of interatomic spacings, differing from boundary to boundary. A nanocrystalline metal contains typically a high number of interfaces (6×10^{25} m^{-3} for a 10-nm grain size) with random orientation relationships, and consequently, a substantial fraction of the atoms lies in the interfaces. The total intercrystalline region consists of grain boundaries, triple junctions, i.e., intersection lines of three or more adjoining crystals, and other interfaces.

One can consider the grains to have the regular 14-sided tetrakaidecahedron shapes, with the hexagonal faces representing the grain boundaries, and edges corresponding to triple junctions. Then, assuming that the maximal diameter of an inscribed sphere as the grain size d, and the intercrystalline component as an outer "skin" of the tetrakaidecahedron with a thickness of $\Delta/2$, where Δ is the average grain boundary thickness, the intercrystalline volume fraction has been calculated as:

$$V_{ic} = 1 - \left(\frac{d - \Delta}{d}\right)^3 \tag{13.1}$$

Table 13.1 Classification of Nanocrystalline Materials

Dimensionality	Designation	Typical method(s) of synthesis
One dimensional	Layered (lamellar)	Vapor deposition Electrodeposition
Two dimensional	Filamentary	Chemical vapor deposition
Three dimensional	Crystallites (equiaxed)	Gas condensation Mechanical alloying/milling

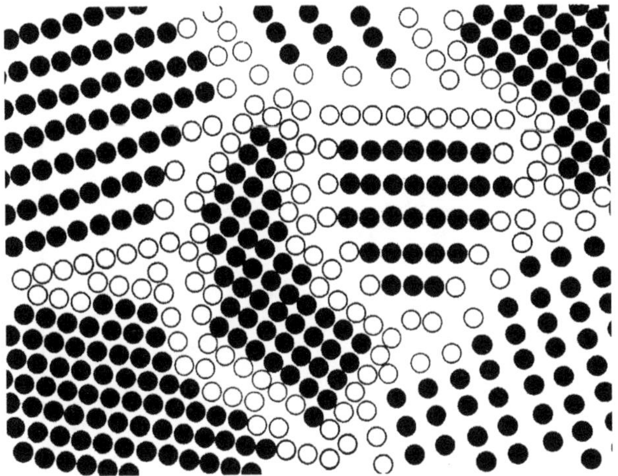

Figure 13.1 Schematic representation of an equiaxed nanocrystalline metal distinguishing between the atoms associated with the individual grains (filled circles) and those constituting the grain boundary network (open circles) [1].

The grain boundary component is given as:

$$V_{gb} = 3\Delta \frac{(d - \Delta)^2}{d^3} \tag{13.2}$$

and therefore, the volume fraction associated with triple junctions, V_{tj}, is given by:

$$V_{tj} = V_{ic} - V_{gb} \tag{13.3}$$

All of the above equations are valid for $d > \Delta$. Thus, it may be seen that the volume fraction of these components depends on the grain size of the metal. Figure 13.2 shows variation of the volume fraction of grain boundaries, triple junctions, and intercrystalline region as a function of the grain size of the metal. It may be noted that these values can be as much as 70–80% at very small grain sizes of 2–3 nm [10]. Because of the significant fraction of the interfacial component, nanocrystalline materials can be considered to be made up of two different components ("phases")—grains and the intercrystalline region. Thus, properties of nanocrystalline materials can be tailored by controlling the interfacial component via grain size control.

13.3 SYNTHESIS OF NANOSTRUCTURED MATERIALS

Nanocrystalline materials have been synthesized by a number of techniques starting from the vapor phase, liquid phase, or solid state. Two major approaches have been taken to synthesize nanocrystalline materials. One is the "top-down" approach in which a bulk material with conventional coarse grain sizes is reduced in its grain size until it reaches nanometer dimensions. Mechanical alloying/milling methods come under this category. The other one is termed the "bottom-up" approach, in which small clusters of material with nanometer dimensions are produced, and these are then consolidated to produce the "bulk" nanocrystalline material. A number of methods have been developed over the years using these two basic approaches. The

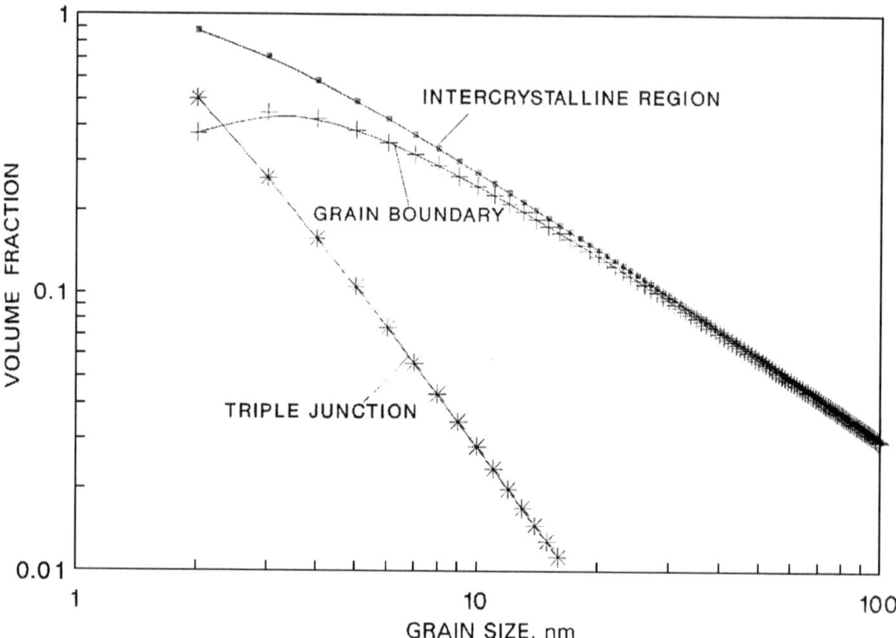

Figure 13.2 Effect of grain size on calculated volume fractions for intercrystalline regions, grain boundaries, and triple junctions, assuming a grain boundary width of 1 nm [10].

currently available methods of synthesis include inert gas condensation, mechanical alloying, spray conversion processing, electrodeposition, rapid solidification from the melt, physical vapor deposition, chemical vapor processing, coprecipitation, sol-gel processing, sliding wear, severe plastic deformation, spark erosion, plasma processing, autoignition, laser ablation, hydrothermal pyrolysis, thermophoretic forced flux system, quenching of the melt under high pressure, biological templating, sonochemical synthesis, and devitrification of amorphous phases. Research into newer methods is continuing so as to achieve a particular grain size or shape or other special features [11].

In practice any method capable of producing very-fine-grain materials can be used to synthesize nanocrystalline materials. The grain size, morphology, and texture can be varied by suitably modifying or controlling the process variables in these methods. Each of these methods has advantages and disadvantages and one should choose the appropriate method depending on the requirements. If a phase transformation is involved, e.g., liquid to solid or vapor to solid, then steps should be taken to increase the nucleation rate and decrease the growth rate during formation of the product phase. In fact, it is this strategy that is used during devitrification of metallic glasses to produce nanocrystalline materials [12].

Of all the methods indicated above, mechanical alloying (MA) has been used most extensively to synthesize nanocrystalline materials. This is essentially due to the fact that MA is a very simple room temperature technique that is capable of producing nanocrystalline structures in almost any type of material. The technique requires relatively simple and inexpensive equipment, and it is also possible to produce bulk quantities of material in the solid state. Scaling up of this technique to produce tonnage quantities on an industrial scale has already been proven.

The first report of formation of a nanostructured material synthesized by MA was issued by Thompson and Politis in 1987 [13], even though specific reference to formation of "nanometer order crystalline structures produced by mechanical alloying" was made by Shingu et al. in 1988 [14]. The field of synthesis of nanostructured materials by MA is growing very rapidly, but Koch [15] has summarized the results on the synthesis and structure of nanocrystalline structures produced by mechanical attrition; an updated version was published in 1997 [16].

13.4 MECHANISM OF FORMATION OF NANOSTRUCTURES

Grain sizes with nanometer dimensions have been observed in almost all mechanically alloyed pure metals, intermetallics, and alloys (if they continue to be crystalline). The minimal grain size achieved in mechanically alloyed materials has been reported to be a few nanometers, ranging typically from about 5 to 50 nm, but varying based on the material and processing conditions. Thus, it appears that synthesis of nanostructures by MA is a ubiquitous phenomenon and nanostructures could be produced by MA in every material. In spite of this, there have not been many detailed investigations to explain why and how nanometer-sized grains are obtained in these materials.

Hellstern et al. [17] studied the evolution of nanostructure formation in mechanically milled AlRu compound through detailed transmission electron microscopy (TEM) techniques. From high-resolution TEM observations, it was reported that deformation was localized within shear bands in the early stages of MA, due to the high deformation rates experienced during MA. These shear bands, which contain a high density of dislocations, have a typical width of approximately 0.5–1.0 μm. Small grains, with a diameter of 8–12 nm, were seen within the shear bands and electron diffraction patterns suggested significant preferred orientation. With continued milling, the average atomic level strain increased due to increasing dislocation density, and at a certain dislocation density within these heavily strained regions the crystal disintegrated into subgrains that are separated by low-angle grain boundaries. This resulted in a decrease of the lattice strain. The subgrains formed this way were of nanometer dimensions, often between 20 and 30 nm.

On further processing, deformation occurred in shear bands located in previously unstrained parts of the material. The grain size decreased steadily and the shear bands coalesced. The small-angle boundaries were replaced by higher angle grain boundaries, implying grain rotation, as reflected by the absence of texture in the electron diffraction patterns and random orientation of the grains observed from the lattice fringes in the high-resolution electron micrographs. Consequently, dislocation-free nanocrystalline grains were formed. This is the currently accepted mechanism of nanocrystal formation in mechanically alloyed/milled powders.

Li et al. [18] also proposed a model for the refinement of grain size during ball milling and noted that the grain size in the early stages of milling follows the relation:

$$d = Kt^{-2/3} \tag{13.4}$$

where d is the grain size, t is the time, and K is a constant.

Chung et al. [19] investigated the mechanisms of microstructure evolution during cryomilling of nickel containing hard AlN particles. They observed that, in the presence of 2 wt% AlN particles, the grain size of Ni was reduced to 35 nm after 8 h of cryomilling. In contrast, the grain size of Ni cryomilled under identical conditions, but without the hard particles, exceeded 100 nm. They also concluded that the rate of grain size refinement increased with increasing amount of hard particles. This was explained on the basis of interaction of dislocations with hard, nondeformable nitride particles and generation of thermal dislocations caused by the mismatch between the coefficients of thermal expansion between the Ni matrix and the hard AlN particles.

13.5 MINIMAL GRAIN SIZE

As noted earlier, the grain size of materials decreases with milling time and reaches a saturation level when a balance is established between the fracturing and cold-welding events. This minimal grain size is different depending on the material and milling conditions. For example, the grain size was smaller at lower milling temperatures. Some efforts have been made in recent years to rationalize the minimal grain size obtainable in different materials in terms of the material properties. The minimal

Table 13.2 Crystal Structure, Melting Point, and Minimal Grain Size (d_{min}) of Metals Obtained by Mechanical Milling

Metal	Crystal structure	Melting point (°C)	d_{min} (nm)	Ref.
Ag	FCC	960.8	20	22
Al	FCC	660.3	20	23
Cu	FCC	1084	14	24
Ge	FCC	937	10	25
Ir	FCC	2446	6	20
Ni	FCC	1455	7.5	26
Pd	FCC	1552	7	20
Rh	FCC	1962	7	20
Si	FCC	1414	3	27
Cr	BCC	1860	9	28
α-Fe	BCC	1536	3	29
Mo	BCC	2615	5	30, 31
Nb	BCC	2467	9	32
W	BCC	3410	9	33
Co	HCP	1495	10	34
Hf	HCP	2227	4	35
Mg	HCP	649	17	36
Ru	HCP	2310	11	17
Ti	HCP	1667	12	37
Zn	HCP	420	23	38
Zr	HCP	1852	12	24

grain size achievable by milling is determined by the competition between the plastic deformation via dislocation motion that tends to decrease the grain size, and the recovery and recrystallization behavior of the material that tends to increase the grain size [20,21]. This balance gives a lower bound for the grain size of pure metals and alloys.

The minimal grain size, d_{min}, obtained by MM was different for different metals and also varied with the crystal structure. Tables 13.2 and 13.3 give the d_{min} values reported for different metals and compounds, respectively. This list is not exhaustive since all the available data on solid solutions, and results of all the investigators for

Table 13.3 Crystal Structure, Melting Point, and Minimal Grain Size (d_{min}) of Compounds Obtained by Mechanical Milling

Compound	d_{min} (nm)	Ref.
Al_2O_3	13	32
CeO_2	10	39
Cr_3C_2	15	40
$CuFe_2O_4$	9	41
FeAl	12	42
$(Fe,Cr)_2O_3$	18	43
α-Fe_2O_3	8	44
FeSi	10	45
β-$FeSi_2$	15	46
FeTi	6	47
HfN	5	48
$La_{2/3}Ca_{1/3}MnO_3$	10	49
$LaNi_5$	50	47
Li_2O	20	50
Mg_2Ni	4	51
Mg_2Sn	30	52
$MnFe_2O_4$	8	53
$MoSi_2$	10	54
Nb_2Al	4	55
NbC	6	56
NiAl	8	55
NiO	10	57
Ni_3Si	15	58
NiSn	5	59
Pd_3Zr	8	60
RuAl	5	61
(Ru,Ir)Al	5	62
Sb_2Fe	12.5	63
SiC	8	32
$SmCo_5$	2.4	64
SnS_2	15	65
TiC	10	66
TiH_2	8	67
ZnO	10	39

each metal or compound have not been included. Reference has been made only to the source where the smallest grain size has been mentioned.

From the above data it may be noted that in most of the metals the minimal grain size attained was in the nanometer dimensions. Furthermore, metals with the bcc crystal structure reach much smaller values than metals with the other crystal structures. This is probably related to the difficulty of extensive plastic deformation and consequent enhanced fracturing tendency during MA/MM. Ceramics and compounds are much harder and usually more brittle than the metals on which they are based. Therefore, intuitively one expects that the minimal grain size of these compounds should be smaller than those of the pure metals. But this does not appear to be always true (Table 13.3).

13.5.1 Correlation with Material Properties

The minimal grain size, d_{min}, obtained by MM was plotted against the melting temperature of metals with different crystal structures in Figure 13.3a–c. In line with the suggestion of Eckert et al. [20], it may be noted from these figures that the minimal grain size achieved decreases with an increase in the melting temperature of the metal. This trend is amply clear in the case of metals with close-packed structures (fcc, Fig. 13.3a and hcp, Fig. 13.3c). Another point of interest is that the difference in grain size is much less among metals that have high melting temperatures; the minimal grain size is virtually constant. Thus, for the hcp metals Co, Ti, Zr, Hf, and Ru, the minimal grain size is almost the same even though the melting temperatures vary between 1495°C for Co and 2310°C for Ru.

An inverse relation as above is less obvious in the case of bcc metals. However, an important point with respect to the bcc metals is that majority of the bcc metals

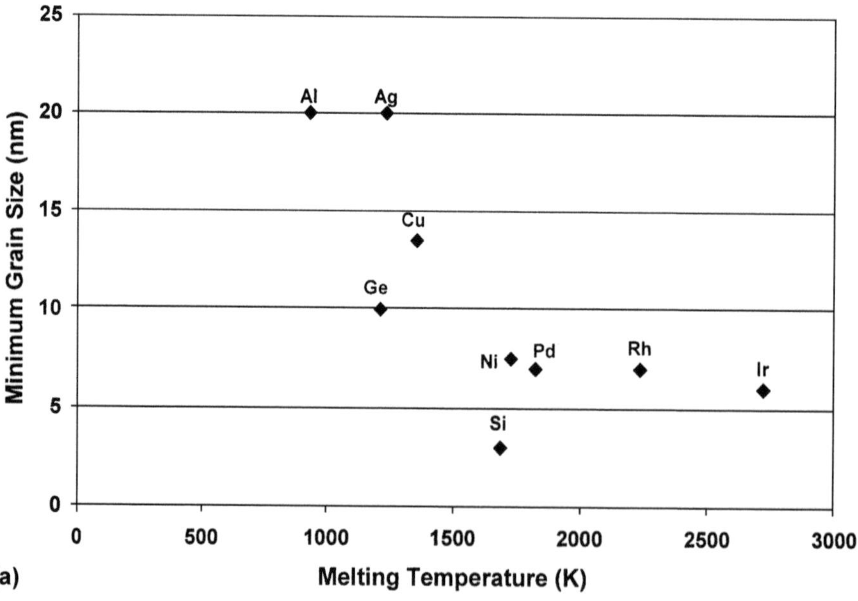

Figure 13.3 Minimal grain size obtained by mechanical milling of different pure metals with different crystal structures vs. melting temperature: (a) fcc, (b) bcc, and (c) hcp.

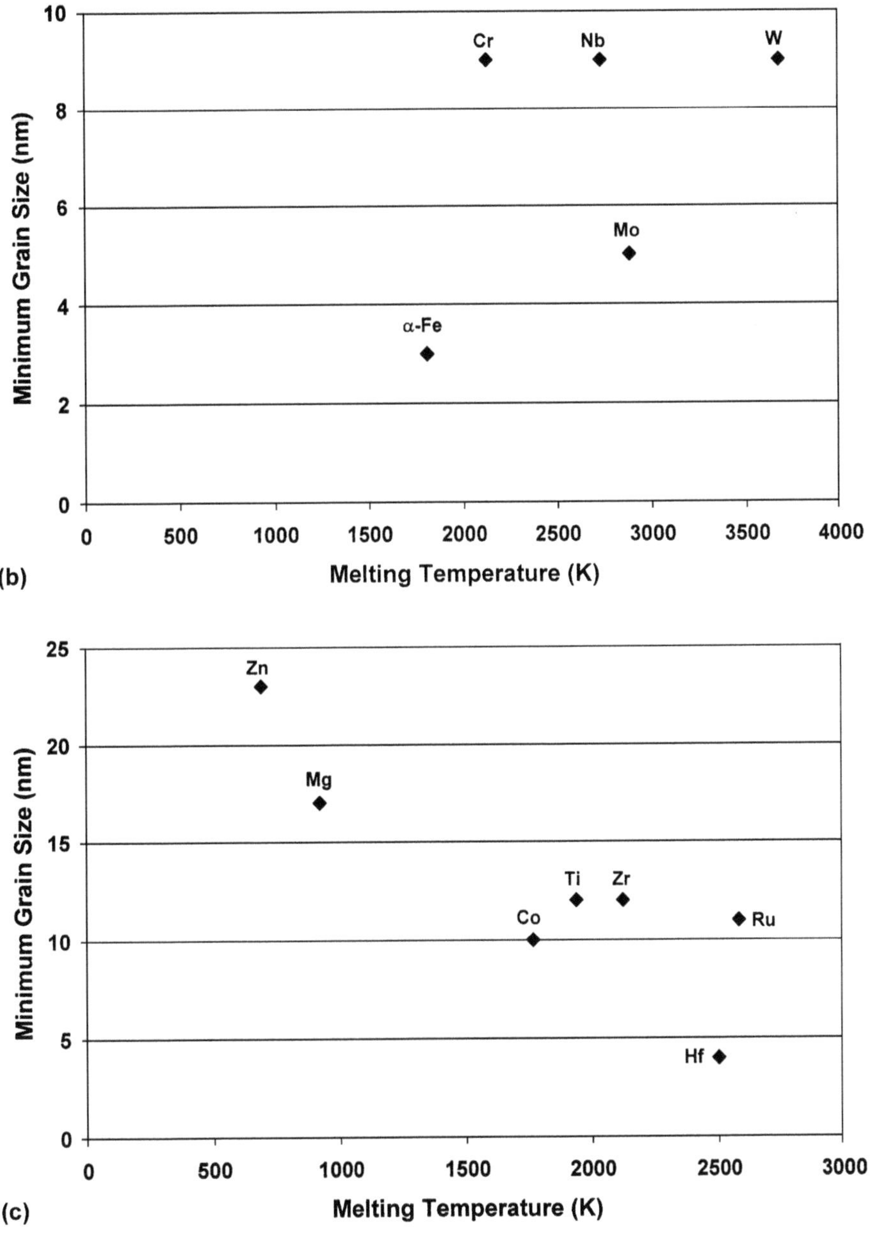

Figure 13.3 Continued.

have a high melting temperature. Consequently, the number of metals with a bcc structure and low melting temperatures are few. Therefore, based on the presently available data, it appears that there is no correlation between the minimal grain size and the melting temperature for the bcc metals. But, as noted above, the difference in the grain size is not significant, even for close-packed metals, which have high melting temperatures. Thus, the bcc metals also follow this trend. However, if milling is conducted on bcc metals with low melting temperatures, such as Na, K, Li, Cs, and Ba, it is possible that an inverse relationship is established in this case also. Most of these low-melting bcc metals are reactive, and so it may not be easy to conduct milling of these metals without the fear of contaminating them.

Mohamed and Xun [68] have investigated the minimal grain size obtainable by cryomilling in a two-phase Zn-22 wt% Al alloy. They obtained a minimal grain size of 33 nm for the Al phase and 41 nm for the Zn phase. These values appear to follow the expected trend but are smaller than those reported earlier (Table 13.2). The reason for the difference in d_{min} for Al was explained on the basis that Al in Zn-22 wt% Al alloy exists as a solid solution alloy of Zn in Al, and not as the pure element. But for the higher value of d_{min} for Zn, it was speculated that the purity of the starting material could be different.

Mohamed [69] has most recently analyzed the existing data on minimal grain size, d_{min}, of mechanically milled pure metals and correlated them with different properties of the metals. For example, it was shown that by plotting the normalized minimal grain size, $d_{min}/\boldsymbol{b}$, where $\boldsymbol{b}$ is the Burgers vector of the dislocations in that crystal structure, vs. melting temperature, T_m, the data could be fit to the equation:

$$d_{min}/\boldsymbol{b} = 112\ \exp(-0.0006T_m) \tag{13.5}$$

A similar equation could also be set up for the normalized minimal grain size vs. activation energy for self-diffusion, since the self-diffusion activation energy scales with the melting temperature.

Mohamed [69] has also shown that the minimal grain size of mechanically milled metals can be related to several other material parameters such as hardness, stacking fault energy, activation energy for self-diffusion, bulk modulus, and the equilibrium distance between two edge dislocations. It was shown that a straight line is obtained when log $(d_{min}/\boldsymbol{b})$ is plotted against log (H/G), where H is the hardness and G is the shear modulus, or log $(\gamma/G\boldsymbol{b})$, where γ is the stacking fault energy, or Q/R, where Q is the activation energy for self diffusion and R is the gas constant, or B, the bulk modulus. A linear relationship was also noted between d_{min} and log (L_c), where L_c is the critical equilibrium distance between two edge dislocations. Figure 13.4a–c shows the plots of $d_{min}/\boldsymbol{b}$ vs. melting temperature, T_m, activation energy for self-diffusion, Q, and normalized hardness, H/G. The correlation appears to be very good in all the three cases. Equations for these correlations may be summarized as follows:

$$d_{min}/\boldsymbol{b} = 112\ \exp(-0.0006T_m) \tag{13.5}$$

$$d_{min}/\boldsymbol{b} = 112\ \exp(-0.0037Q) \tag{13.6}$$

$$d_{min}/\boldsymbol{b} = 100\ \exp(-0.0046B) \tag{13.7}$$

13.5.2 Process Variables

Some information is available in the literature on the minimal grain size achieved under different processing conditions. Among these, the effect of milling energy, milling temperature, and alloying effects have been reported.

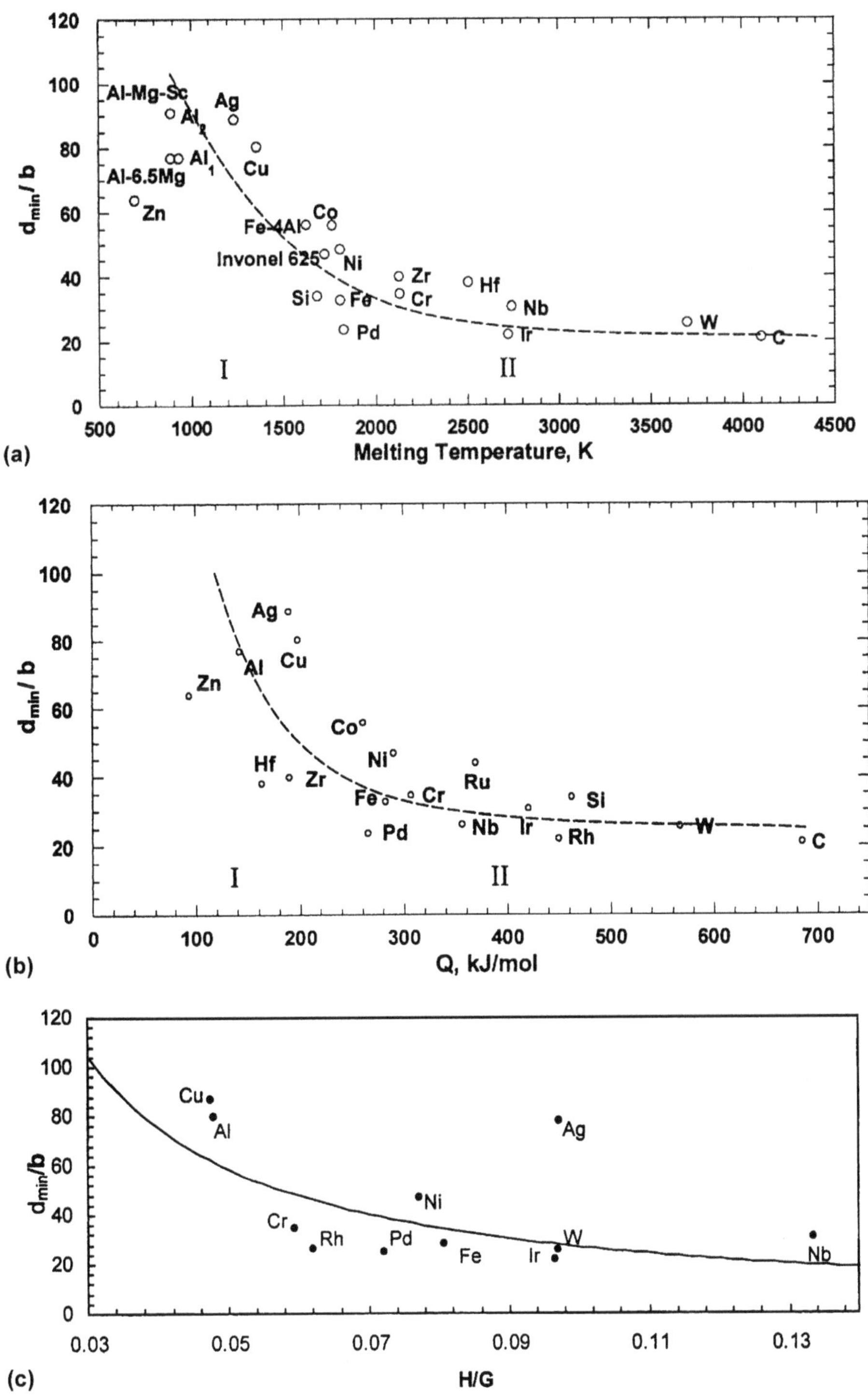

Figure 13.4 Normalized grain size ($d_{min}/\boldsymbol{b}$) obtained by milling vs. (a) melting temperature, T_m, (b) activation energy for self diffusion, Q, and (c) normalized hardness, H/G [68].

Milling Energy

Since the minimal grain size attained in a metal by mechanical milling is expected to depend on its mechanical properties, one suspects that neither the nature of the mill nor the milling energy will have any effect on the minimal grain size achieved. It was reported [70] that there was no significant effect of milling intensity on nanostructure formation in a Cu-Fe-Co powder blend, but results contrary to this expectation have been reported in the literature. It was shown that d_{min} was about 5 nm when the TiNi intermetallic powder was milled in a high-energy SPEX shaker mill, but only about 15 nm in the less energetic Invicta vibratory ball mill [71]. Similar trends were noted for variations in the ball-to-powder weight ratios (BPRs). The grain size of the niobium metal milled in the Invicta vibratory mill was about 26 $\pm$ 2 nm at a BPR of 5:1, but was only 18 $\pm$ 1 at a BPR of 10:1 [15]. Similar results were also reported for the pure metal Cu: 25 $\pm$ 5 nm for a BPR of 5:1 and 20 $\pm$ 1 for a BPR of 10:1. Furthermore, the kinetics of achieving this d_{min} could also depend on the milling energy, but no such studies have been reported so far.

It was also reported that during nanocrystal formation, the average crystal size increased and the internal lattice strain decreased at higher milling intensities due to the enhanced thermal effects [72]. In accordance with this argument, the grain size of Si milled at a high energy of 500 kJ/g was 25 nm, whereas that milled at a low energy of 20 kJ/g was only 4 nm [73].

Milling Temperature

It has been reported that the grain size of powders milled at low temperatures was smaller than that milled at higher temperatures. For example, the grain size of copper milled at room temperature was 26 $\pm$ 3 nm, while that milled at $-85\,^{\circ}$C was only 17 $\pm$ 2 nm [74]. Similar results were reported for other metals [75] as well as the CoZr intermetallic compound [16]. Milling of powders at higher temperatures also resulted in reduced root means square (rms) strain in addition to larger grain sizes [76].

Alloying Effects

Alloy composition appears to have a significant effect on the minimal grain size obtained after milling. Since solid solutions are harder and stronger than the pure metals on which they are based, it is expected that the minimal grain size is smaller for solid solutions than for pure metals. This can be understood on the basis of increased fragmentation tendency of the harder (and brittle) powders. Accordingly, Figure 13.5 shows that the minimal grain size in the mechanically alloyed Cu-rich Cu-Fe powders decreases with increasing Fe content. It may be noted that the grain size after milling for 24 h in a SPEX mill decreases from 20 nm for pure Cu to about 9 nm for Cu-55 at% Fe. This was attributed to solid solution strengthening of the Cu matrix by segregation of Fe atoms to stacking faults [77]. For the Fe-rich Fe-Cu powders also, the grain size was reported to decrease from 6 nm for pure Fe to 4.6 nm for Fe-15 at% Cu, again attributed to solid solution hardening [78].

A similar decrease in grain size with increasing Ni content was also reported for the B2-NiAl phase [21]. The grain size decreased from 10 nm for the Al-rich Al-46 at% Ni to about 5 nm for the Ni-rich Al-60 at% Ni. This observation again confirms that the alloy composition determines the final grain size. Since NiAl has a wide

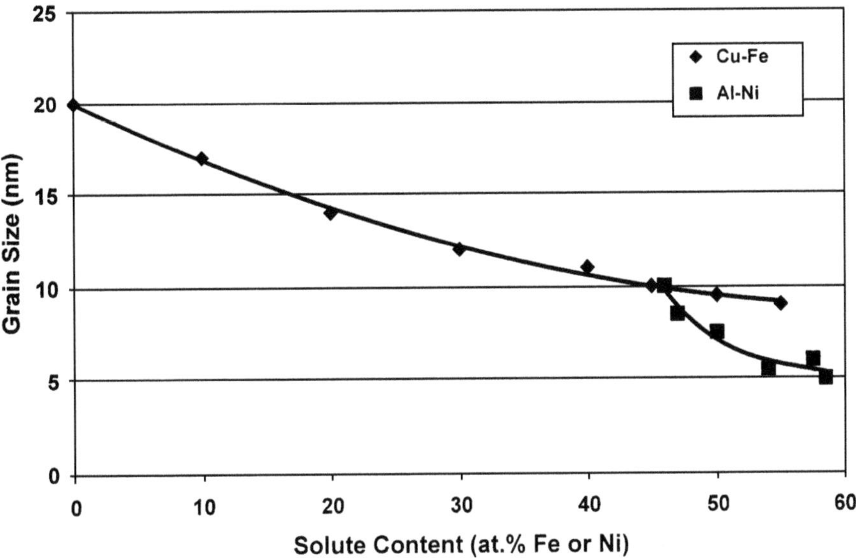

Figure 13.5 Minimal grain size obtained by milling of Cu-Fe and Al-Ni powders vs. solute content. The Fe-Cu powders were milled for 24 h in a SPEX 8000 mill and the Al-Ni powders for 100 h in a Fritsch Pulverisette P5 planetary ball mill.

range of homogeneity, this compound can accommodate constitutional vacancies in its lattice. It is known that for the Al-rich compositions, excess vacancies in the Ni sublattice lead to strong vacancy hardening and a large resistance against dislocation motion in coarse-grained polycrystalline material [79]. Furthermore, the large diffusivity enhances dynamic recovery for Al-rich compositions [80]. A combination of these two effects could prevent formation of very small grain sizes on milling the Al-rich Ni-Al alloys. The difficulty of cross-slip and low diffusivities of Ni-rich compositions hinder recovery. Thus, the transgranular fracture behavior coupled with little recovery is reported to be the reason for efficient mechanical attrition and consequent small grain sizes.

The above observations make it clear that the small grain sizes obtained in solid solution alloys are due to the effects of solid solution strengthening. Therefore, it is logical to expect that the grain size in the milled solid solution powder should be larger if the solid solution is softer than the pure metal and remain constant if there is no significant change in hardness. Results consistent with this observation were reported in Ni(Co) where the hardness remained the same in the solid solution as in the pure metal [74] and in Ni(Cu), Fe(Cu), and Cr(Cu), which exhibited solid solution softening [77].

Since a number of process variables (method to estimate the grain size, milling energy, milling temperature, alloying effects, contamination, and so forth) can influence the minimal grain size achieved, precautions should be taken and systematic investigations should be conducted in defining the minimal grain size achieved in different materials.

13.5.3 Theoretical Estimates

It has been suggested that the minimal grain size achievable by milling is determined by the balance between defect/dislocation structure introduced by the plastic deformation of milling and its recovery by thermal processes. This gives a lower bound for the grain size of pure metals and alloys and suggests that a small grain size itself provides a limit for further grain refinement by milling. Once the whole material has the nanocrystalline structure, further deformation can only be accomplished by grain boundary sliding, which does not lead to microstructural refinement anymore.

It was proposed that the limiting grain size is determined by the minimal grain size that can sustain a dislocation pile-up within a grain and by the rate of recovery [20]. Based on the dislocation pile-up model, the critical equilibrium distance between two edge dislocations in a pile-up, L_c (which could be assumed to be the crystallite or grain size in milled powders), was calculated using the equation [81]:

$$L_c = \frac{3G\boldsymbol{b}}{\pi(1-\nu)H} \tag{13.8}$$

where G is shear modulus, $\boldsymbol{b}$ is Burgers vector, ν is Poisson's ratio, and H is hardness of the material. According to the above equation, increased hardness results in smaller values of L_c (grain size), and an approximate linear relationship was observed between L_c and the minimal grain size obtained by milling of a number of metals [77].

It has been recently suggested that an inverse linear relationship exists between ΔH^{l-c} and the grain size [82]. ΔH^{l-c} is the enthalpy difference between the undercooled liquid (amorphous phase) and conventional polycrystalline metal. The difference between the enthalpy of crystallization, ΔH^{cryst}, and ΔH^{l-c} is the excess enthalpy, ΔH^{c-nc}, which can be expressed as

$$\Delta H^{c-nc} = \Delta H^{cryst} - \Delta H^{l-c} \tag{13.9}$$

ΔH^{c-nc} is actually the excess enthalpy of nanocrystalline metal with respect to the conventional polycrystalline metal. Assuming that the excess enthalpy of the nanocrystalline metal is concentrated in the grain boundaries, ΔH^{c-nc} can be expressed as

$$\Delta H^{c-nc} = 2\gamma g V_m/d \tag{13.10}$$

where γ is the grain boundary enthalpy, V_m is the molar volume, d is the average grain size, and g is the geometrical factor. Thus, a plot of ΔH^{c-nc} against $1/d$ would produce a straight line. On linearly extrapolating this straight line to $\Delta H^{c-nc} = \Delta H^{c-a}$, where ΔH^{c-a} is the difference in the enthalpy between the amorphous phase and the conventional polycrystalline metal, the enthalpy of the nanocrystallized product will be equal to that of the amorphous state. The corresponding grain size will be the thermodynamics-based grain size limit. A similar value for the grain size limit could also be obtained by plotting ΔH^{cryst} against $1/d$ and extrapolating this straight line to the value $\Delta H^{cryst} = 0$.

13.6 NANOCOMPOSITES

Nanocomposites are a new class of materials in which at least one of the phases (the matrix, the reinforcement, or both) is of nanometer dimensions. They have been generally classified based on the matrix (such as metal-matrix composites, MMCs; ceramic-matrix composites, CMCs; or polymer-matrix composites, PMCs). Another method of classification is based on the microstructure. According to Niihara [83], nanocomposites can be divided into four categories: intragranular, intergranular, hybrid, and nano/nano composites (Fig. 13.6). In the first three types, the reinforcement phase is in the nanometer level whereas the matrix is not. But in the last category, both the matrix and reinforcement are of nanometer dimensions and these two components are randomly distributed. In the intergranular type (Fig. 13.6a), the nanometer-sized reinforcement is distributed along the grain boundaries of the micrometer-sized matrix phase. In the intragranular-type (Fig. 13.6b), the reinforcement is inside the coarse matrix grains, whereas in the hybrid type (Fig. 13.6c), the reinforcement is both inside the grains and along the grain boundaries of the matrix.

Nanocomposites are expected to exhibit high strength, fracture toughness, stiffness, wear resistance, and high-temperature properties compared to their coarse-

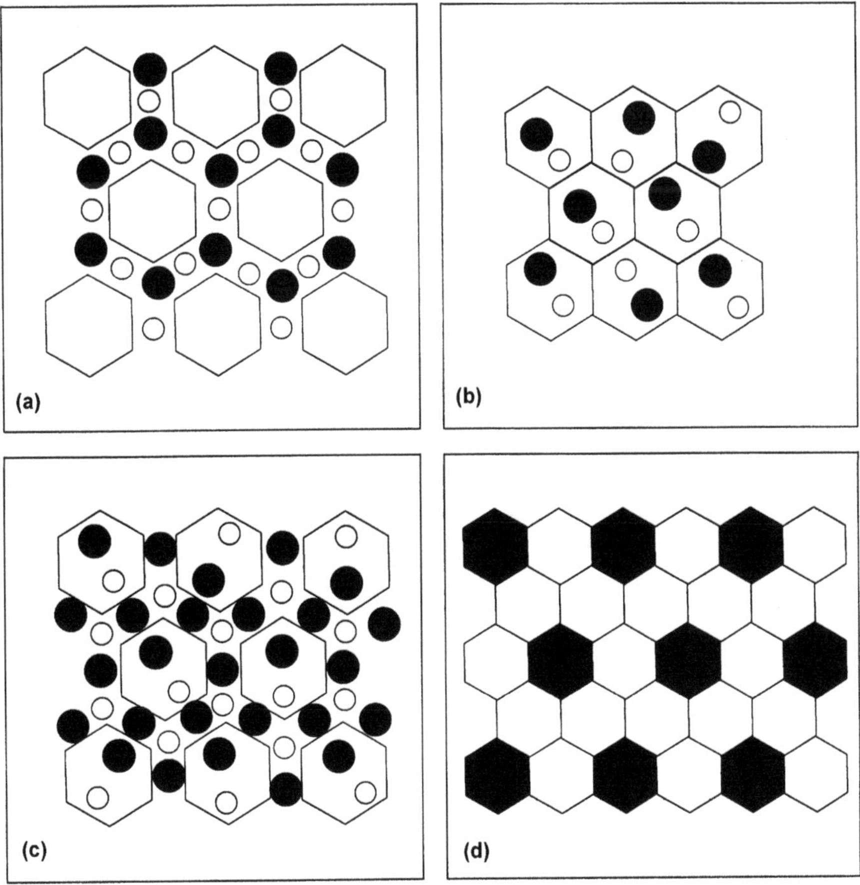

Figure 13.6 Classification of nanocomposites: (a) intergranular, (b) intragaranular, (c) hybrid, and (d) nano/nano. In the first three types, one of the phases, most probably the reinforcement, is in the nanometer level, whereas in the last category both phases are in nanometer level [83].

grained counterparts [84]. Hence they are potential candidate materials for structural applications in aerospace hardware and turbo engine parts and pumps. Magnetic nanocomposites containing small (single domain) magnetic particles, isolated electrically and magnetically by a nonmagnetic, nonmetallic component, also exhibit interesting magnetic properties. Therefore, there is considerable interest in developing such nanocomposites.

Natural nanocomposites like bone and nacre (the pearly, iridescent inner layer of the abalone shell) have been known for millions of years. For example, bone has impressive strength and toughness. It consists in large part of nanoscale, plate-like crystals of hydroxyapatite, $Ca_{10}(PO_4)_6(OH)_2$, dispersed in a matrix of collagen fibers. Hydroxyapatite and collagen by themselves are not particularly promising structural materials. But when these and other ingredients are assembled to form the complex microstructure of bone, the resulting nanocomposite offers properties that have proven extremely difficult to match with synthetic materials.

Nanocomposites are usually prepared by vacuum deposition or chemical methods. However, synthesis by mechanochemical methods offers unique advantages. Processing by these methods occurs at or near room temperature, and therefore process control is easy. The size of the matrix and/or reinforcement can be easily controlled by choosing proper milling conditions. The milling methods can be easily scaled up to industrial levels, using equipment developed to produce tons of oxide dispersion strengthened materials (see Chapter 4).

Several nanocomposites have been synthesized by MA. Nanocomposites have also been obtained when the amorphous phases obtained by MA/MM were crystallized at relatively low temperatures [85]. An important attribute of these nanocomposites is in preventing or minimizing grain growth up to very high temperatures. Reinforcement of Cu and Mg with Al_2O_3 was reported to prevent grain growth up to the melting point of the metals [86]. The grain size of a nanocrystalline mixture of intermetallics [$(Ni,Fe)_3Al$) and (Ni,Fe)Al] processed by MA was also shown to be more resistant to coarsening than that of a similar single-phase material [87]. Recently, there have also been efforts to develop intermetallic-ceramic nanocomposites such as γ-TiAl-Ti_5Si_3 by high-energy milling methods for elevated-temperature applications [88,89].

Nanocomposites by MA methods can be prepared in different ways. The two constituents can be put together and milled when only blending and attrition are expected to occur. Alternately, one could use mechanochemical methods (see Chapter 14) during which in situ reduction of oxides and simultaneous oxidation of the other metallic component could take place. For example, during milling of a mixture of iron oxide and a more reactive metal like Al, reduction of iron oxide to iron and oxidation of Al to Al_2O_3 occurs according to the reaction:

$$3Fe_3O_4 + 8Al \rightarrow 9Fe + 4Al_2O_3 \tag{13.11}$$

Reactions as mentioned above generate lot of heat, and so it is possible that they take place spontaneously in the self-propagating mode. In some cases, the heat generated is so large that the reaction takes place in a "run-away" mode and it is difficult to control the reaction. In such cases, addition of diluents slows down the reactions. The reader is advised to refer to Chapter 14 for details of the in situ reduction reactions. It may be noted that the grain size of the nanostructure formed during mechanochemical reactions is large due to the excess heat generated during the self-propagating high-temperature synthesis (SHS) reaction. If the SHS reaction does not take place and only the displacement reaction occurs, then the grain size can still be small.

Alloys prepared in the nanocrystalline state have been shown to have interesting properties that could lead to potential applications. Three such applications have been identified [90]. Details of these applications are discussed in Chapter 17.

13.7 PROPERTIES OF NANOCRYSTALLINE MATERIALS

Because of the very fine grain sizes and consequent high density of interfaces, nanocrystalline materials, including those prepared by MA methods, exhibit a variety of properties that are different from and often considerably improved over those of conventional coarse-grained materials. These include increased strength/hardness, enhanced diffusivity, improved ductility/toughness, reduced density, reduced elastic modulus, higher electrical resistivity, increased specific heat, higher coefficient of thermal expansion, lower thermal conductivity, and superior soft magnetic properties [1–7]. Among these, the most dramatic effect is seen in the mechanical properties [91].

13.7.1 Hardness and Strength

The mechanically alloyed nanocrystalline materials have a hardness that is 4–5 times higher than that of the conventional coarse-grained counterparts. This has been shown to be true in a number of cases [91,92]. The increase in hardness of mechanically alloyed/milled materials is due to a combination of several parameters. First and foremost is the effect of grain size, explained on the basis of the Hall-Petch relationship:

$$\sigma(\text{or } H) = \sigma_o(\text{or } H_o) + kd^{-1/2} \tag{13.12}$$

where σ is flow stress, H is hardness, d is grain size, and σ_o, H_o, and k are constants. This relationship has been used to explain the grain size dependence of strength and hardness of conventional coarse-grained polycrystalline materials. The strength (or hardness) increases with a decrease in grain size. The same equation, but with different constants, could be used to explain the grain size dependence on the strength of nanocrystalline materials also. Figure 13.7 shows the variation of hardness vs. $d^{-1/2}$, where d is the average grain size for mechanically alloyed/milled Fe [93], Nb_3Sn [93], Ti-24Al-11Nb (at%) [94], γ-TiAl [94], and Ni_3Al alloys containing oxide dispersoids [95]. It may be noted that the traditional Hall-Petch relationship is obeyed in these cases also.

The second factor contributing to the increased strength of milled powders is the lattice strain present in them. It was noted that the lattice strain increases with milling time, reaches a peak value, and may show a decrease in the last stages of milling (Fig. 13.8). Since the main contribution to lattice strain comes from the dislocations introduced during milling, the strain increases with milling time due to the increased dislocation density. However, beyond a particular milling time, nanometer-sized grains are produced, further grain refinement ceases due to the

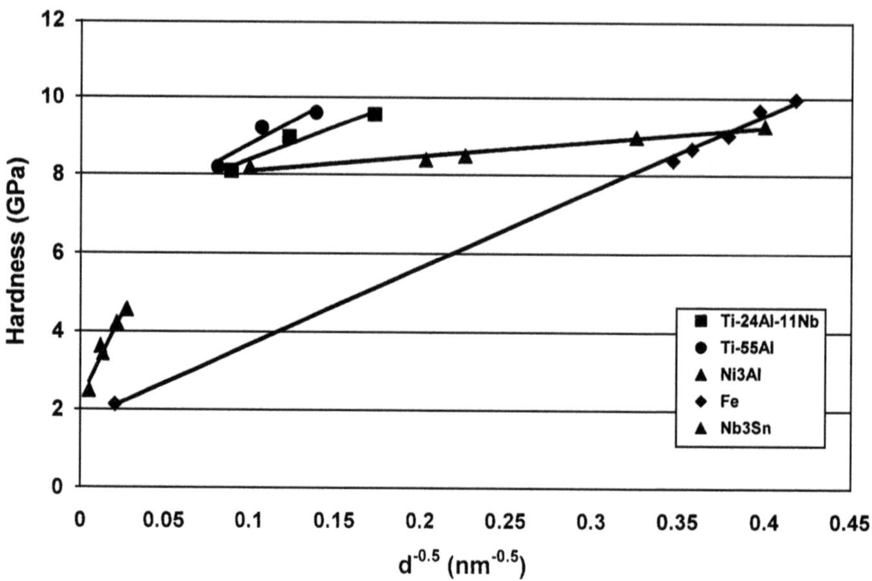

Figure 13.7 Variation of hardness vs. $d^{-1/2}$ for Fe, Nb_3Sn, Ti-24Al-11Nb, γ-TiAl, and Ni_3Al alloys. *d* is the average grain diameter. Note that the hardness increases with a decrease in grain size.

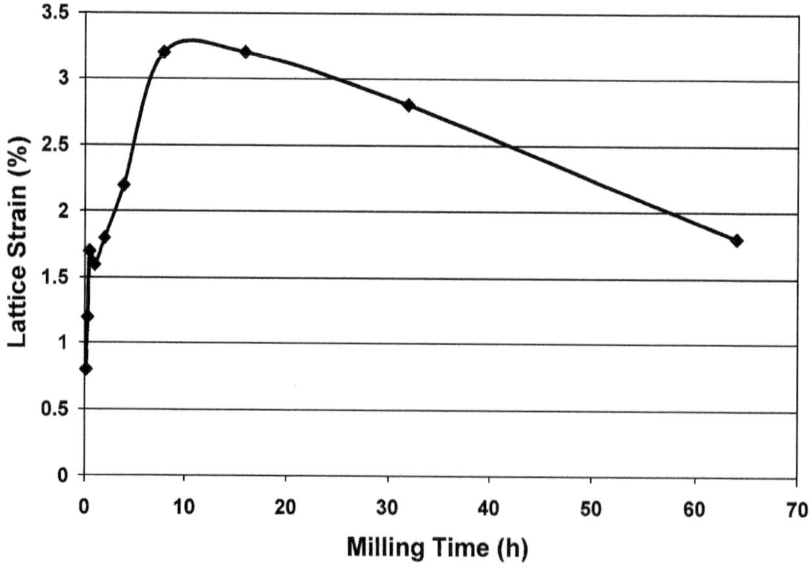

Figure 13.8 Average lattice strain in mechanically milled AlRu as a function of milling time. (After Ref. 17.)

inability of dislocation generation in these small grains; therefore, the strain remains constant. Since deformation in extremely small grains occurs not by formation and glide of dislocations but by grain boundary sliding accompanied by a reduction in overall dislocation density by annihilation at grain boundaries, the strain could decrease. It was shown that the hardness of nanocrystalline materials produced by MA/MM scales with the atomic level strain in the powder. It is also useful to remember that for a given grain size (not the smallest achieved), dislocation activity could still occur to a different extent depending on the milling conditions, and therefore the strain could be different. Thus, both grain size and lattice strain should be considered in explaining the hardening behavior of mechanically alloyed powders. In addition to the above two main factors, solid solution hardening and dispersion hardening also account for the increased hardness (and strength) of materials.

Even though nanocrystalline materials have been shown to follow the Hall-Petch behavior, they do so only up to a critical grain size. Below this critical size, there is an inverse relationship, i.e., hardness and strength decrease with a decrease in grain size. This critical grain size is dependent on the material properties but is about 10–20 nm in most cases. This suggests that the deformation mechanisms for nanocrystalline materials may be different from those of conventional polycrystalline materials. It has been shown in some cases that when the hardness of a material is plotted against the reciprocal of the square root of the grain size, a straight line is obtained over a wide a range of grain sizes. These grain sizes have been obtained by different methods, including conventional ingot solidification, rapid solidification, thermomechanical processing, and MA. Figure 13.9 shows the variation of hardness of γ-TiAl-type alloys over a very wide range of grain sizes obtained by different processing techniques, including MA. The inverse Hall-Petch behavior can be clearly noted in the figure [92]. Similar results have been reported in other cases also.

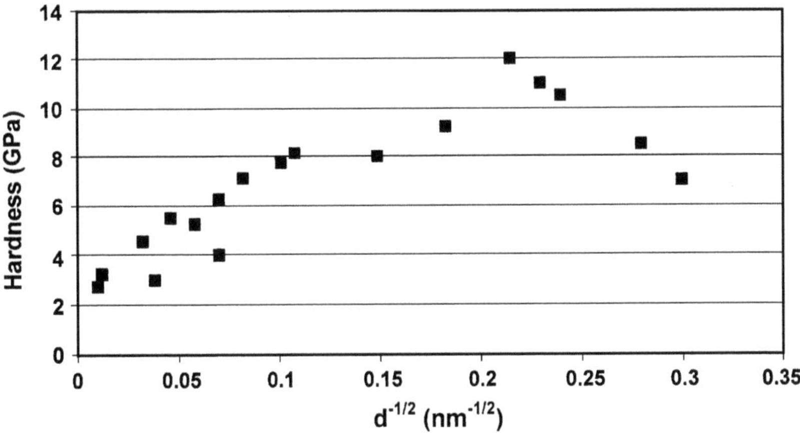

Figure 13.9 Variation of hardness vs. reciprocal of the square root of grain size for γ-TiAl alloys processed by different methods including ingot solidification, thermomechanical processing, rapid solidification, and MA. Note that the hardness increases with a decrease in grain size up to about 30 nm, following the normal Hall-Petch behavior. Below this value the hardness decreases with a decrease in grain size (inverse Hall-Petch behavior).

13.7.2 Ductility

The major driver for research in nanostructured materials has been the promised improvement in ductility of the inherently brittle intermetallics and ceramic compounds. It is well established that a decrease in grain size lowers the ductile-brittle transition temperature (DBTT) and therefore, nanocrystalline materials should also exhibit a lower DBTT than their coarse-grained counterparts. Further, the strain (creep) rate during Coble creep is given by:

$$\frac{d\varepsilon}{dt} = \frac{B\sigma\Omega\Delta D_b}{d^3kT} \tag{13.13}$$

where $d\varepsilon/dt$ is the creep rate, σ the applied stress, Ω the atomic volume, Δ the grain boundary width, D_b the grain boundary diffusivity, d the grain size, k the Boltzmann constant, T the temperature, and B a constant. Accordingly, if the grain size is decreased from 10 μm to 10 nm (by three orders of magnitude), the creep rate should increase by nine orders of magnitude. In addition, the grain boundary diffusivity is higher in nanocrystalline materials. The combination of these effects is expected to significantly increase the creep rate of nanocrystalline materials allowing the ceramics and intermetallics to be deformed plastically even at room temperature. The observation of plastic behavior in compression of the inherently brittle ceramics at low temperatures—CaF_2 at 80°C and TiO_2 at 180°C—gave the much needed fillip to activities in this area [96].

The results of ductility measurements on nanocrystalline materials are mixed and are sensitive to flaws and porosity, surface finish, and method of testing (e.g., tension vs. compression). Available results show that the elongation to failure is lower with a decrease in grain size. Furthermore, conventional materials that exhibit high ductility show reduced ductility in the nanocrystalline state. Improvements in ductility have been observed in the nanocrystalline state only when the coarse-grained counterparts show very limited ductility. This has been attributed to the different degrees of dislocation activity in the conventional materials; a significant amount of dislocation activity occurs in ductile materials, whereas it is limited in the ceramic materials. In addition, dislocation generation and movement are difficult in nanocrystalline materials.

However, recent results indicate that it has been possible to observe increased ductility in nanocrystalline high-strength materials (though not yet in mechanically milled materials). Lu et al. [97] recently prepared fully dense nanocrystalline Cu with a grain size of 27 nm by electrodeposition. They reported that this material had a tensile ductility of 30% even though the tensile strength was as high as 202 MPa. Coarse-grained Cu exhibits this large elongation only when the tensile strength is about 120 MPa. Large extensions of more than 5000% without any cracking were also reported when the copper specimen with a grain size of 20 nm was rolled at room temperature [98]. Large amount of ductilities are also achieved in a pure copper sample that has been subjected to 95% cold work and subsequently annealed at 200°C for 3 min to produce a bimodal microstructure. The microstructure consisted of micrometer-sized grains embedded inside a matrix of nanocrystalline and ultrafine (<300 nm) grains. The matrix grains impart high strength as expected from the Hall-Petch equation. The inhomogeneous microstructure induces strain hardening mechanisms that stabilize the tensile deformation, leading to a high tensile ductility. Thus, the yield strength in this condition is about 350 MPa and the elongation to failure is about 65% [99]. The elongation to failure compares favorably with that of annealed copper, but the yield strength is about 6–7 times higher than that of the annealed copper sample.

The promised room temperature superplasticity has not been observed in any nanocrystalline material so far. But the temperature at which superplasticity was observed is more than 400°C lower than that required for coarse-grained materials. For example, superplasticity has been reported in nanocrystalline Ni (20 nm) [100], Ni_3Al (50 nm) [100], and Ti_3Al (100 nm) at 350, 650, and 600°C, respectively, whereas the temperatures required for observance of superplasticity in the coarse-grained materials were 820, 1100, and 1100°C, respectively. The mechanical behavior of nanocrystalline materials is far from understood and lot more work needs to be done to elucidate this important aspect of nanocrystalline materials. A critical summary of the present status on this important aspect of ductility of nanostructured materials has been recently presented [101].

13.7.3 Enhanced Sinterability

Nanocrystalline materials have been shown to exhibit enhanced sintering behavior. This is reflected in the lowered sintering temperatures and increased sintering kinetics. Since nanocrystalline materials contain a very large fraction of atoms at the grain boundaries, the numerous interfaces provide a high density of short-circuit diffusion paths. Thus, the measured diffusivities in nanocrystalline materials are several orders of magnitude higher than lattice diffusion and even grain boundary diffusion. The most important consequence of increased diffusivity is that sintering of nanocrystalline powders can occur at temperatures much lower than those required for sintering coarse-grained polycrystalline powders. For example, TiO_2 with a grain size of 12 nm could be sintered at ambient pressures and at temperatures 400–600°C lower than those required for 1.3 μm powder, and without the need for any compacting or sintering aids [102]. Similarly, nanocrystalline titanium aluminides could be fully consolidated at temperatures about 400°C lower than those required for coarse-grained materials [94,103].

Another consequence of the fine grain size of nanocrystalline materials has been the increased solid solubility levels achieved in different alloy systems. For example, it was possible to have a solid solubility of 17 at% Hg in the immiscible Cu-Hg system [104]. Similarly, the solid solubility of Mg in Ti could be increased to high levels in the nanostructured condition, even though under equilibrium conditions the solid solubility is zero. Furthermore, no solid solubility increase was reported for coarse-grained milled powders [105]. Details of the mechanisms for increased solid solubility limits in nanocrystalline materials are discussed in Chapter 9.

13.7.4 Thermal Stability

Grain growth occurs in polycrystalline materials to reduce the interfacial energy and hence the total energy of the system. Since nanocrystalline materials have a highly disordered large interfacial component (and therefore are in a high-energy state), the driving force for grain growth is expected to be high. This driving force is proportional to the specific grain boundary area and therefore varies as the inverse of the grain size.

Knowledge of the thermal stability of nanocrystalline materials is important for both technological and scientific reasons. From a technological point of view, thermal stability is important for consolidation of nanocrystalline powders without coarsening the microstructure. From a scientific point of view, it would be instructive to check whether the grain growth behavior in nanocrystalline materials is similar to that in coarse-grained materials.

Contrary to expectations, experimental observations suggest that grain growth in nanocrystalline materials, prepared by any method, is very small (and almost negligible) up to a reasonably high temperature. It has been suggested that the resistance to grain growth observed in nanocrystalline materials results primarily from frustration, i.e., there is not sufficient driving force for grain growth [106]. This has been explained on the basis of structural factors such as narrow grain size distribution, equiaxed grain morphology, low-energy grain boundary structures, relatively flat grain boundary configurations, and porosity present in the samples. In some cases, however, abnormal grain growth has been reported [107]. Recently, Shaw et al. [108] investigated the thermal stability of nanostructured Al-Fe-Cr-Ti alloys prepared by MA and reported that the onset temperature grain growth coincided with the onset temperature for Al_6Fe precipitation. They also observed that inhibition to grain growth below 330°C was due to solute drag, while it was second-phase (Zener) drag above 450°C. Between 330°C and 450°C, it could be either solute drag, second phase drag, or both.

REFERENCES

1. Gleiter, H. (1989). *Prog. Mater. Sci.* 33:223–315.
2. Siegel, R. W. (1991). In: Cahn, R. W., ed. *Materials Science and Technology: A Comprehensive Treatment,* Vol. 15. *Processing of Metals and Alloys.* Weinheim: VCH, pp. 583–614.
3. Suryanarayana, C. (1995). *Int. Mater. Rev.* 40:41–64.
4. Suryanarayana, C., Koch, C. C. (1999). In: Suryanarayana, C., ed. *Non-equilibrium Processing of Materials.* Oxford: Pergamon, pp. 313–346.
5. Gleiter, H. (2000). *Acta Mater.* 48:1–29.
6. Suryanarayana, C., Koch, C. C. (2000). *Hyperfine Interactions* 130:5–44.
7. Koch, C. C. (2002). *Nanostructured Materials: Processing, Properties, and Applications.* Westwood, NJ: Noyes Publications and William Andrews Press.
8. Suryanarayana, C., Froes, F. H. (1992). *Metall. Trans.* 23A:1071–1081.
9. Gleiter, H. (1995). *Nanostructured Mater.* 6:3–14.
10. Suryanarayana, C., Mukhopadhyay, D., Patankar, S. N., Froes, F. H. (1992). *J. Mater. Res.* 7:2114–2119.
11. Nalwa, H. S. (2000). *Handbook of Nanostructured Materials and Nanotechnology.* Vol. 1. *Synthesis and Processing.* San Diego: Academic Press.
12. Lu, K. (1996). *Mater. Sci. Eng. Rep.* R16:161–221.
13. Thomson, J. R., Politis, C. (1987). *Europhys. Lett.* 3:199–205.
14. Shingu, P. H., Huang, B., Nishitani, S. R., Nasu, S. (1988). *Suppl. Trans. Jpn. Inst. Metals* 29:3–10.
15. Koch, C. C. (1993). *Nanostructured Mater.* 2:109–129.
16. Koch, C. C. (1997). *Nanostructured Mater.* 9:13–22.

17. Hellstern, E., Fecht, H. J., Garland, C., Johnson, W. L. (1989). In: McCandlish, L. E., Polk, D. E., Siegel, R. W., Kear, B. H., eds. *Multicomponent Ultrafine Microstructures*. Vol. 132. *Pittsburgh, PA: Mater. Res. Soc.* pp. 137–142.
18. Li, S., Wang, K., Sun, L., Wang, Z. (1992). *Scripta Metall. Mater.* 27:437–442.
19. Chung, K. H., He, J., Shin, D. H., Schoenung, J. M. (2003). *Mater. Sci. Eng.* A356: 23–31.
20. Eckert, J., Holzer, J. C., Krill, C. E. III, Johnson, W. L. (1992). *J. Mater. Res.* 7: 1751–1761.
21. Börner, I., Eckert, J. (1996). *Mater. Sci. For.* 225–227:377–382.
22. Xu, J., Yin, J. S., Ma, E. (1997). *Nanostructured Mater.* 8:91–100.
23. Bonetti, E., Pasquini, L., Sampaolesi, E. (1998). *Mater. Sci. For.* 269–272:999–1004.
24. Al-Hajry, A., Al-Assiri, M., Al-Heniti, S., Enzo, S., Hefne, J., Al-Shahrani, A., Al-Salami, A. E. (2002). *Mater. Sci. For.* 386–388:205–210.
25. Boolchand, P., Koch, C. C. (1992). *J. Mater. Res.* 7:2876–2882.
26. Schneider, M., Pischang, K., Worch, H., Fritsche, G., Klimanek, P. (2000). *Mater. Sci. For.* 343–346:873–879.
27. Shen, T. D., Koch, C. C., McCormick, T. L., Nemanich, R. J., Huang, J. Y., Huang, J. G. (1995). *J. Mater. Res.* 10:139–148.
28. Ogino, Y., Murayama, S., Yamasaki, T. (1991). *J. Less Common Metals* 168:221–235.
29. Caamaño, Z., Pérez, G., Zamora, L. E., Suriñach, S., Muñoz, J. S., Baró, M. D. (2001). *J. Non-Cryst Solids* 287:15–19.
30. Zdujic, M. V., Kobayashi, K. F., Shingu, P. H. (1991). *J. Mater. Sci.* 26:5502–5508.
31. Lucks, I., Lamparter, P., Mittemeijer, E. J. (2001). *Acta Mater.* 49:2419–2428.
32. Tomasi, R., Pallone, E. M. J. A., Botta, F. W. (1999). *Mater. Sci. For.* 312–314:333–338.
33. Fecht, H. J., Hellstern, E., Fu, Z., Johnson, W. L. (1989). In: Gasbarre, T. G., Jandeska, W. F. Jr., eds. *Advances in Powder Metallurgy*. Vol. 2. Princeton, NJ: Metal Powder Industries Federation, pp. 111–122.
34. Sort, J., Nogués, J., Amils, X., Suriñach, S., Muñoz, J. S., Baró, M. D. (2000). *Mater. Sci. For.* 343–346:812–919.
35. Bab, M. A., Mendoza-Zélis, L., Damonte, L. C. (2001). *Acta Mater.* 49:4205–4213.
36. Imamura, H., Sakasai, N., Kajii, Y. (1996). *J. Alloys Compounds* 232:218–223.
37. Wang, K. Y., Shen, T. D., Quan, M. X., Wei, W. D. (1993). *J. Mater. Sci. Lett.* 12: 1818–1820.
38. Zhang, X., Wang, H., Scattergood, R. O., Narayan, J., Koch, C. C. (2003). *Mater. Sci. Eng.* A344:175–181.
39. Tsuzuki, T., McCormick, P. G. (2000). *Mater. Sci. For.* 343–346:383–388.
40. He, J., Ice, M., Lavernia, E. J. (1999). *Mater. Sci. For.* 312–314:237–246.
41. Goya, G. F., Rechenberg, H. R., Jiang, J. Z. (1999). *Mater. Sci. For.* 312–314:545–550.
42. Amils, X., Nogués, J., Suriñach, S., Muñoz, J. S., Lutterotti, L., Gialanella, S., Baró, M. D. (1999). *Nanostructured Mater.* 11:689–695.
43. Michel, D., Mazerolles, L., Chichery, E. (1998). *Mater. Sci. For.* 269–272:99–104.
44. Jiang, J. Z., Lin, R., Mørup, S. (1998). *Mater. Sci. For.* 269–272:449–454.
45. Ahn, J.-H., Wang, G. X., Liu, H. K., Dou, S. X. (2001). *Mater. Sci. For.* 360–362: 595–602.
46. Tokumitsu, K., Wada, M. (2002). *Mater. Sci. For.* 386–388:473–478.
47. Zaluski, L., Zaluska, A., Tessier, P., Ström-Olsen, J. O., Schulz, R. (1996). *Mater. Sci. For.* 225–227:853–858.
48. Mendoza-Zélis, L., Bab, M. A., Damonte, L. C., Sánchez, F. H. (1999). *Mater. Sci. For.* 312–314:179–184.
49. Goya, G. F., Rechenberg, H. R., Ibarra, M. R. (2002). *Mater. Sci. For.* 386–388:433–438.

50. Indris, S., Heitjans, P. (2000). *Mater. Sci. For.* 343–346:417–422.
51. Singh, Arvind Kumar, Singh, Ajay Kumar, Srivastava, O. N. (1995). *J. Alloys Compounds* 227:63–68.
52. Urretavizcaya, G., Meyer, G. O. (2002). *J. Alloys Compounds* 339:211–215.
53. Arcos, D., Rangavittal, N., Vazquez, M., Vallet-Regí, M. (1998). *Mater. Sci. For.* 269–272:87–92.
54. Ahn, J. H., Kim, Y. (1999). *Mater. Sci. For.* 312–314:147–152.
55. Krivoroutchko, K., Kulik, T., Fadeeeva, V. I., Portnoy, V. K. (2002). *J. Alloys Compounds* 333:225–230.
56. Oleszak, D., Matyja, H. (2000). *Mater. Sci. For.* 343–346:320–325.
57. Jiang, J. Z., Gerward, L., Mørup, S. (1999). *Mater. Sci. For.* 312–314:115–120.
58. Jang, J. S. C., Tsau, C. H., Chen, W. D., Lee, P. Y. (1998). *J. Mater. Sci.* 33:265–270.
59. Ahn, J.-H., Wang, G. X., Liu, H. K., Dou, S. X. (2001). *Mater. Sci. For.* 360–362: 595–602.
60. Katona, G. L., Kis-Varga, M., Beke, D. L. (2002). *Mater. Sci. For.* 386–388:193–197.
61. Liu, K. W., Müklich, F., Birringer, R. (2001). *Intermetallics* 9:81–88.
62. Liu, K. W., Müklich, F., Pitschke, W., Birringer, R., Wetzig, K. (2001). *Z. Metallkde.* 92:924–930.
63. Kis-Varga, M., Beke, D. L., Mezzaros, S., Vad, K., Kerekes, Gy, Daróczi, L. (1998). *Mater. Sci. For.* 269–272:961–966.
64. Sort, J., Nogués, J., Suriñach, S., Muñoz, J. S., Chappel, E., Dupont, F., Chouteau, G., Baró, M. D. (2002). *Mater. Sci. For.* 386–388:465–472.
65. Baláz, P., Takacs, L., Ohtani, T., Mack, D. E., Boldizárová, E., Soika, V., Achimovicová, M. (2002). *J. Alloys Compounds* 337:76–82.
66. Baviera, P., Harel, S., Garem, H., Grosbras, M. (2000). *Mater. Sci. For.* 343–346:629–634.
67. Bhosle, V., Baburaj, E. G., Miranova, M., Salama, K. (2003). *Mater. Sci. Eng.* A356:190–199.
68. Mohamed, F. A., Xun, Y. (2003). *Mater. Sci. Eng.* A358:178–185.
69. Mohamed, F. A. (2003). *Acta Mater.* 51:4107–4119.
70. Galdeano, S., Chaffron, L., Mathon, M.-H., Vincent, E., De Novion, C.-H. (2001). *Mater. Sci. For.* 360–362:367–372.
71. Yamada, K., Koch, C. C. (1993). *J. Mater. Res.* 8:1317–1326.
72. Kuhrt, C., Schropf, H., Schultz, L., Arzt, E. (1993). In: deBarbadillo, J. J., et al., eds. *Mechanical Alloying for Structural Applications.* Materials Park, OH: ASM International, pp. 269–273.
73. Streleski, A. N., Leonov, A. V., Beresteskaya, I. V., Mudretsova, S. N., Majorova, A. F., Butyagin, P. Ju. (2002). *Mater. Sci. For.* 386–388:187–192.
74. Shen, T. D., Koch, C. C. (1995). *Mater. Sci. For.* 179–181:17–24.
75. Bonetti, E., Campari, E. G., Pasquini, L., Sampaolesi, E., Valdré, G. (1998). *Mater. Sci. For.* 269–272:1005–1010.
76. Hong, L. B., Bansal, C., Fultz, B. (1994). *Nanostructured Mater.* 4:949–956.
77. Eckert, J., Holzer, J. C., Krill, C. E. III, Johnson, W. L. (1992). *J. Mater. Res.* 7: 1980–1983.
78. Eckert, J., Holzer, J. C., Johnson, W. L. (1992). *Scripta Metall. Mater.* 27:1105–1110.
79. Noebe, R. D., Bowman, R. R., Nathal, M. V. (1993). *Int. Mater. Rev.* 38:193–232.
80. Kao, C. R., Chang, Y. A. (1993). *Intermetallics* 1:237–250.
81. Nieh, T. G., Wadsworth, J. (1991). *Scripta Metall. Mater.* 25:955–958.
82. Sun, N. X., Lu, K. (1999). *Phys. Rev.* B59:5987–5989.
83. Niihara, K. (1991). *J. Ceram. Soc. Jpn.* 99:974–982.
84. Bhaduri, S., Bhaduri, S. B. (1998). *JOM* 50(1):44–51.
85. Suryanarayana, C., Froes, F. H. (1993). *Nanostructured Mater.* 3:147–153.
86. Naser, J., Reinhemann, W., Ferkel, H. (1997). *Mater. Sci. Eng.* A234:467–469.

87. Scahffer, G. B. (1992). *Scripta Metall. Mater.* 27:1–5.
88. Bohn, R., Klassen, T., Bormann, R. (2001). *Intermetallics* 9:559–569.
89. Bohn, R., Klassen, T., Bormann, R. (2001). *Acta Mater.* 49:299–311.
90. Klassen, T., Bohn, R., Fanta, G., Oelerich, W., Eigen, N., Gärtner, F., Aust, E., Bormann, R., Kreye, H. (2003). *Z. Metallkde.* 94:610–614.
91. Morris, D. G. (1998). *Mechanical Behavior of Nanostructured Materials.* Zurich: TransTech.
92. Suryanarayana, C. (2002). *Int. J. Non-Eqm. Process.* 11:325–345.
93. Cho, Y. S., Koch, C. C. (1991). *Mater. Sci. Eng.* A141:139–148.
94. Suryanarayana, C., Korth, G. E., Froes, F. H. (1997). *Metall. Mater. Trans.* 28A:293–302.
95. Jang, J. S. C., Koch, C. C. (1988). *Scripta Metall.* 22:677–682.
96. Karch, J., Birringer, R., Gleiter, H. (1987). *Nature* 330:556–558.
97. Lu, L., Wang, L. B., Ding, B. Z., Lu, K. (2000). *J. Mater. Res.* 15:270–273.
98. Lu, L., Sui, M. L., Lu, K. (2000). *Science* 287:1463–1466.
99. Wang, Y., Chen, M., Zhou, F., Ma, E. (2002). *Nature* 419:912–915.
100. McFadden, S. X., Mishra, R. S., Valiev, R. Z., Zhilyaev, A. P., Mukherjee, A. K. (1999). *Nature* 398:684–686.
101. Koch, C. C., Morris, D. G., Lu, K., Inoue, A. (1999). *MRS Bull.* 24(2):54–58.
102. Siegel, R. W., Ramasamy, S., Hahn, H., Li, Z., Lu, T., Gronsky, R. (1988). *J. Mater. Res.* 3:1367–1372.
103. Suryanarayana, C., Korth, G. E. (1999). *Metals Mater. (Korea)* 5:121–128.
104. Ivanov, E. (1992). *Mater. Sci. For.* 88–90:475–480.
105. Suryanarayana, C., Froes, F. H. (1990). *J. Mater. Res.* 5:1880–1886.
106. Siegel, R. W. (1992). In: Wolf, D., Yip, S., eds. *Materials Interfaces: Atomic Level Structure and Properties.* London: Chapman & Hall, pp. 431–460.
107. Gertsman, V. Y., Birringer, R. (1994). *Scripta Metall. Mater.* 30:577–581.
108. Shaw, L., Luo, H., Villegas, J., Miracle, D. (2003). *Acta Mater.* 51:2647–2663.

14

Mechanochemical Processing

14.1 INTRODUCTION

Mechanochemical processing (MCP) is the term applied to the powder process in which chemical reactions and phase transformations take place due to application of mechanical energy. The technique of MCP had a long history with the first publication dating back to 1892 [1]. It was shown that the order of decomposition and sublimation of mercury and silver halides was different upon heating and trituration in a mortar. This study clearly established that chemical changes could be brought about not only by heating but also by mechanical action. However, the use of mechanically activated processes dates back to the early history of mankind, when fires were initiated by rubbing of flints one against the other. Ostwald coined the term "mechanochemical" in 1911 [2]. A simple way of differentiating between mechanically activated processes and MCP methods is that no chemical reactions or phase transformations take place during mechanical activation, whereas these do occur in MCP. However, some investigators [3] use the term MCP to include mechanical alloying (MA), mechanical milling (MM), and reaction milling (RM), the last one involving chemical reactions induced by mechanical activation. In this chapter we will use the term MCP to denote the process of mechanical activation to achieve chemical reactions. While the scientific basis underlying MCP was investigated from the very beginning, applications for the mechanochemical products were slow to come about mostly because of the limitations on the productivity of MCP reactors, the purity of the products, and the economics of the process. The general phenomenon of MCP has been a popular research topic in Germany [4,5], former USSR and Eastern Europe [6]. These researchers concentrated their efforts on both the fundamental principles of MCP and the potential applications for materials produced by MCP [7–13].

MCP involves conversion of mechanical energy into chemical energy to bring about chemical reactions. This has also been referred to in the literature as mechanosynthesis or mechanochemical synthesis. The effects of MCP include synthesis of novel materials, reduction/oxidation reactions, exchange reactions, decomposition of compounds, and phase transformations in both inorganic and organic solids [13]. The category of exchange reactions in MCP has received a lot of attention in recent years [3,14]. This emanated from the reports in 1989 that MA could be used to induce a wide variety of solid-solid and even liquid-solid chemical reactions [15,16]. For example, it was shown that CuO could be reduced to pure metal Cu by ball milling CuO at room temperature with a more reactive metal like Ca. Milling together of CuO and ZnO with Ca has resulted in the direct formation of β'-brass [16]. McCormick [3,14] presented an overview of the work from his group a few years ago. Gaffet et al. [17] presented a brief survey of the developments in the field of mechanical activation and mechanosynthesis. In fact, most of the work in this subarea of MA has been carried out by McCormick [3,14], Takacs [18], Matteazzi [19], and Welham. Takacs [18] presented an exhaustive review of the self-sustaining reactions induced by ball milling very recently. A few books are also available on the principles of MCP and the applications of their products [5,7,13,20,21].

Mechanochemical synthesis has generally been performed in high-energy mills such as SPEX mills, and rarely in low-energy mills, unless the heat of reaction is very high. Accordingly, while almost all the investigations of displacement reactions have been carried out in a SPEX mill, Xi et al. [22] studied the highly exothermic reaction between CuO and Al in a low-energy stirred ball mill. Similarly, several mixtures of Cr_2O_3, Fe_2O_3, and NiO were reduced with Al in a Fritsch P5 planetary ball mill [23].

14.2 THERMODYNAMIC ASPECTS

Most of the mechanosynthesis reactions studied in recent years have been displacement reactions of the type:

$$MO + R \rightarrow M + RO \tag{14.1}$$

where a metal oxide (MO) is reduced by a more reactive metal (reductant, R) to the pure metal M. In addition to the oxides, metal chlorides and sulfides have also been reduced to pure metals this way, and the results available to date are summarized in Table 14.1.

All solid-state reactions involve the formation of a product phase at the interface of the component phases. Thus, in the above example, the metal M forms at the interface between the oxide MO and the reductant R and physically separates the reactants (Fig. 14.1). Further growth of the product phase involves diffusion of atoms of the reactant phases through the product phase, which constitutes a barrier layer preventing further reaction from occurring. In other words, the reaction interface, defined as the nominal boundary surface between the reactants, continuously decreases during the course of the reaction. Consequently, kinetics of the reaction are slow. Since the reaction rates are determined by the diffusion rates of the reactants through the product phase, which acts as a barrier, elevated temperatures are required to achieve reasonable reaction rates.

The reactions listed in Table 14.1 are characterized by a large negative free energy change and are therefore thermodynamically feasible at room temperature.

The occurrence of these reactions at ambient temperatures is thus limited by kinetic considerations alone.

Mechanical alloying can provide the means to substantially increase the reaction kinetics of the reduction reactions. This is because the repeated cold welding and fracturing of powder particles increases the area of contact between the reactant powder particles by repeatedly bringing fresh surfaces into contact due to a reduction in particle size during milling. This allows the reaction to proceed without the necessity for diffusion through the product layer. As a consequence, reactions that normally require high temperatures occur at lower temperatures during MA or even without any externally applied heat. In addition, the high defect densities induced by MA accelerate diffusion processes. Alternatively, the particle refinement and consequent reduction in diffusion distances (due to microstructural refinement) can at least reduce the reaction temperatures significantly, even if they do not occur at room temperature.

Depending on the milling conditions, two entirely different reaction kinetics are possible [40,58,114]:

1. The reaction may extend to a very small volume during each collision, resulting in a *gradual* transformation, or
2. If the reaction enthalpy is sufficiently high, a *self-propagating high-temperature synthesis* (SHS) (also known as combustion reaction) can be initiated.

The latter type of reaction requires a critical milling time for combustion to be initiated. If the temperature of the vial is recorded during the milling process, the temperature initially increases slowly with time. After some time, the temperature increases abruptly, suggesting that ignition has occurred, and this is followed by a relatively slow decrease. This time at which the sudden increase in temperature occurs is referred to as the ignition time, t_{ig}, and these values are listed in Table 14.1. Figure 14.2a shows the variation of temperature with time during the milling operation. Note that the temperature may remain constant or increase gradually with milling time, and when the combustion reaction occurs there is a sudden increase in temperature. After the combustion event, the temperature drops gradually to the ambient. It may be noted that this temperature–time plot is very similar to the one observed during the SHS reaction (Fig. 14.2b) [115].

The time at which the sudden increase in temperature (see Fig. 14.2) takes place in the powder is referred to as the ignition time. Measurements of ignition time provide a useful means of characterizing the structural and chemical evolution during MA [116]. It has been noted that there is only particle refinement during milling until the ignition starts. The reduction reaction occurs only after the combustion reaction. Accordingly, it was observed that no reaction occurs between PbO_2 and TiO until 28 min and 21 s at which time combustion takes place. The reaction resulting in the formation of $PbTiO_3$ is complete at 28 min and 23 s [117]. Similarly, it was reported that no reaction occurred between Mo and Si on mechanically alloying the powder mix for 3 h and 12 min; a reaction resulting in the formation of $MoSi_2$ took place at 3 h and 13 min [118]. Similar observations have been made in several other systems also [36,51,81,100].

Intimate contact between the reactant powder particles is an essential requirement for combustion to occur. This condition is easily met during milling of ductile-brittle systems. In these mixtures the microstructure consists of brittle oxide particles dispersed in the ductile matrix; therefore, intimate contact exists between the two types

Table 14.1 Chemical Reactions Investigated by Mechanical Alloying Techniques

Reaction	Mill	BPR	% Excess reductant	ΔH(kJ)	T_{ad} (K)	t_{ig} (min)	Ref.
$3Ag_2O + 2Al \rightarrow 6Ag + Al_2O_3$	—	—	—	−532	4914	86	24
$Ag_3Sn + 2Ag_2O \rightarrow 7Ag + SnO_2$ (96 h)	Uni-Ball mill						25
$2AlCl_3 + 3CaO \rightarrow \gamma\text{-}Al_2O_3$[a] $+ 3CaCl_2$	SPEX 8000	—	—	—	—	—	26
$2BN + 3Al \rightarrow 2AlN + AlB_2$ (20 h)	Planetary ball mill	4:1					27
$BaO + ZrSiO_4 \rightarrow BaSiO_3 + ZrO_2$	Uni-Ball mill	50:1					28
$CaCl_2 + Na_2CO_3 + 3.5NaCl \rightarrow CaCO_3 + 5.5NaCl$	SPEX 8000	10:1		−100			29
$CaO + ZrSiO_4 \rightarrow CaSiO_3 + ZrO_2$	Uni-Ball mill	50:1					28
$CaWO_4 + 4C \rightarrow CaO + WC + 3CO$ (40 h)		50:1					30
$CaWO_4 + C + 3Mg \rightarrow CaO + WC + 3MgO$	Ball mill	43:1	50				31
$CaWO_4 + 3Mg \rightarrow CaO + W + 3MgO$	Ball mill	42:1					32
$CaWO_4 + \frac{1}{2}N_2 + 3Mg \rightarrow CaO + WN + 3\,MgO$	Ball mill	43:1	50				31
$CdCl_2 + Na_2S + 16NaCl \rightarrow CdS + 18NaCl$	SPEX 8000						33
$CdCl_2 + Na_2S \rightarrow CdS + 2NaCl$	SPEX 8000	10:1					34
$CdO + Ca \rightarrow Cd + CaO$	SPEX 8000	3:1	10	−377	3561	37	16
$4CeCl_3 + 6CaO + O_2 \rightarrow 4CeO_2$[b] $+ 6CaCl_2$	SPEX 8000			−276	1427		35
$2CeCl_3 + 3CaS \rightarrow Ce_2S_3 + 3CaCl_2$	SPEX 8000						33
$CeCl_3 + 3NaOH \rightarrow Ce(OH)_3 + 3NaCl$				−345			36
$CoCl_2 + 2Na \rightarrow Co + 2NaCl$	SPEX 8000	3:1	25	−495		20	37
$3CoS + 2Al \rightarrow 3Co + Al_2S_3$	SPEX 8000	11:1		−405	1710	—	19
$Cr_2O_3 + 2Al \rightarrow 2Cr + Al_2O_3$	SPEX 8000	33:1	10	−273	—	—	38
$Cr_2O_3 + 3Zn \rightarrow 2Cr + 3ZnO$	SPEX 8000	33:1	10	+49	—	—	38
$CuCl_2 + 2Na \rightarrow Cu + 2NaCl$	SPEX 8000						39
$3CuO + 2Al \rightarrow 3Cu + Al_2O_3$	SPEX 8000	33:1	10	−1197	5151	15	40
					5394	37	16, 41
$2CuO + C \rightarrow 2Cu + CO_2$ (24 h)	SPEX 8000	12:1	40	−135	—	—	42
$CuO + Ca \rightarrow Cu + CaO$ (100 min)	SPEX 8000	3:1	—	−473	4716	6.5	15, 16, 43
						19	40

$3CuO + xCu + 2Al \rightarrow (3+x)Cu + Al_2O_3$	SPEX 8000	4.4:1					44
$4CuO + 3Fe \rightarrow 4Cu + Fe_3O_4$ (20 h)	SPEX 8000	3:1	10	−488	1380	170	16, 40
	Planetary mill	30:1			1668	153	45–47
$4CuO + 3Fe \rightarrow 4Cu + Fe_3O_4$	SPEX 8000						48
$5CuO + Fe \rightarrow 5Cu + Fe_3O_4 + FeO$	SPEX 8000	15:1	10				49
$CuO + 2H \rightarrow Cu + H_2O$ (3 h)	Attritor			−87			50
$CuO + Mg \rightarrow Cu + MgO$	SPEX 8000	3:1	10	−440	4531	34	40
						90	16
$CuO + Mg \rightarrow Cu + MgO$	Vibration mill						51
$CuO + Mn \rightarrow Cu + MnO$	SPEX 8000	3:1	10	−231	2227	43	40
$CuO + Ni \rightarrow Cu + NiO$	SPEX 8000	3:1	10	−382	988	—	16
				−78	1288		40
$2CuO + Si \rightarrow 2Cu + SiO_2$ (24 h)	Stirring ball mill	30:1 to 80:1	10	−596	—	—	52
$2CuO + Ti \rightarrow 2Cu + TiO_2$	SPEX 8000	3:1	10	−633	4175	0.3	16
						2.8	40, 41
$Cu_2O + Mg \rightarrow 2Cu + MgO$	SPEX 8000			−434		20	53
$Cu_2O + Zn \rightarrow 2Cu + ZnO$	SPEX 8000			−183		44	53
$CuO + ZnO + Ca \rightarrow \beta'$-brass $+ CaO$ (24 h)	SPEX 8000	3:1	10	—			15
$3Cu_2S + 2Al \rightarrow 6Cu + Al_2S_3$	SPEX 8000	10:1		−485	1740		19
$Cu_2S + Fe \rightarrow 2Cu + FeS$	SPEX 8000	8:1		−71	590		19
$Cu_2S + Fe \rightarrow 2Cu + FeS$	Fritsch P6						54
$2ErCl_3 + 3Ca \rightarrow 2Er + 3CaCl_2$				−407			55
$FeCl_3 + 3Na \rightarrow Fe + 3NaCl$ (3 h)	SPEX 8000						56
$3Fe_3O_4 + 8Al \rightarrow 9Fe + 4Al_2O_3$	SPEX 8000	8:1	10	−1716	3805		57
		12:1	−75 to +100				58
$3Fe_3O_4 + 8Al \rightarrow 9Fe + 4Al_2O_3$	SPEX 8000			−1704			48
$Fe_3O_4 + 2Ti \rightarrow 3Fe + 2TiO_2$	SPEX 8000	8:1	10	−395	2737	20	38
$Fe_3O_4 + 4Zn \rightarrow 3Fe + 4ZnO$	SPEX 8000	16:1	10	−143	—	—	38, 59
$Fe_3O_4 + 4Zn \rightarrow 3Fe + 4ZnO$	SPEX 8000			−142			48
$Fe_2O_3 + 2Al \rightarrow 2Fe + Al_2O_3$ (0.5 h)	Fritsch P7						60
$Fe_2O_3 + 3Ca \rightarrow 2Fe + 3CaO$	SPEX 8000	3:1		−1080	3498	24	16

Table 14.1 Continued

Reaction	Mill	BPR	% Excess reductant	ΔH(kJ)	T_{ad} (K)	t_{ig} (min)	Ref.
$Fe_2O_3 + 3Mg \rightarrow 2Fe + 3MgO$ (0.5 h)	Fritsch P7						60
$2Fe_2O_3 + 3Ti \rightarrow 4Fe + 3TiO_2$	Ball mill						61
α-$Fe_2O_3 + C \rightarrow 2FeO + CO$	SPEX 8000	11:1					62
α-$Fe_2O_3 + 3H_2 \rightarrow 2Fe + 3H_2O$ (8 h)	Super Misuni NEV-MA8	100:1					63
$3FeS + 2Al \rightarrow 3Fe(Al) + Al_2S_3$	SPEX 8000	9:1		−271	1146		19
$FeS + Mn \rightarrow Fe + MnS$	SPEX 8000	10:1		−117	1300		19
$2FeS + Si \rightarrow 2Fe(Si) + SiS_2$	SPEX 8000	12:1		+88	—		19
$2FeS_2 + 6CaO \rightarrow Ca_2Fe_2O_5 + \frac{15}{4}CaS + \frac{1}{4}CaSO_4$	SPEX 8000	40:1					64
$FeTiO_3 + 7Al \rightarrow FeAl_2 + TiAl_3 + 2Al_2O_3$	Ball mill	47:1	10				65
$FeTiO_3 + 2Al + C \rightarrow Fe + TiC + Al_2O_3$	Uni-Ball mill			−605			66
$2FeTiO_3 + 4Al + N_2 + C \rightarrow 2Fe + 2Ti(C, N) + 2Al_2O_3$	Ball mill						67, 68
$FeTiO_3 + 4C \rightarrow Fe + TiC + 3CO$	Ball mill						69, 70
$FeTiO_3 + C \rightarrow Fe + TiO_2 + CO$	Uni-Ball mill						71
$FeTiO_3 + 3C + \frac{1}{2}N_2 \rightarrow Fe + TiN + 3CO$	Ball mill						69
$FeTiO_3 + BaO + \frac{1}{2}O_2 \rightarrow \frac{1}{2}Fe_2O_3 + BaTiO_3$	Uni-Ball mill						72
$FeTiO_3 + CaO + \frac{1}{2}O_2 \rightarrow \frac{1}{2}Fe_2O_3 + CaTiO_3$	Uni-Ball mill						72
$FeTiO_3 + 3Mg \rightarrow Fe + Ti + 3MgO$	Uni-Ball mill						73
$FeTiO_3 + 3Mg + C \rightarrow Fe + TiC + 3MgO$	Uni-Ball mill	43:1					74
$2FeTiO_3 + 6Mg + N_2 \rightarrow 2Fe + 2TiN + 6MgO$	Uni-Ball mill	43:1					74
$FeTiO_3 + MgO + \frac{1}{2}O_2 \rightarrow \frac{1}{2}Fe_2O_3 + MgTiO_3$	Uni-Ball mill						72
$4FeTiO_3 + 7S \rightarrow FeSO_4 + 3FeS_2 + 4TiO_2$	Ball mill			−231			75
$2FeTiO_3 + 5Si \rightarrow 2FeTiSi + 3MgO$	Uni-Ball mill						76
$FeTiO_3 + SrO + \frac{1}{2}O_2 \rightarrow \frac{1}{2}Fe_2O_3 + SrTiO_3$	Uni-Ball mill						72
$2GdCl_3 + 3Ca \rightarrow 2Gd + 3CaCl_2$				−378			55
$2GdCl_3 + 3CaO \rightarrow Gd_2O_3 + 3CaCl_2$	SPEX 8000	40:1					77

$GdCl_3 + 3NaOH \rightarrow Gd(OH)_3 + 3NaCl$	SPEX 8000	10:1					78
$3Mg + CuCl + Zr(SO_4)_2 \rightarrow$ Cu-Zr (amorphous) $+ \frac{1}{2} MgCl_2 + 2MgSO_4$ (50 h)	QM-1SP planetary ball mill	30:1					79
$0.1MgCl_2 + 0.9ZrCl_4 + 1.9Ca(OH)_2 \rightarrow (MgO)_{0.1}(ZrO_2)_{0.9} + 1.9CaCl_2.H_2O$							80
$MgO + ZrSiO_4 \rightarrow MgSiO_3 + ZrO_2$	Uni-Ball mill	50:1					28
$Na_2Cr_2O_7 + S \rightarrow Cr_2O_3 + Na_2SO_4$	SPEX 8000	10:1		−562			29
$2NbCl_5 + 5Na_2CO_3 \rightarrow Nb_2O_5 + 10NaCl + 5CO_2$ (g)	SPEX 8000	10:1		−1029			29
$3Nb_2O_5 + 10Al \rightarrow 5Al_2O_3 + 6Nb$ (3–5 h)	SPEX 8000	>7:1			2740		81
$Nb_2O_5 + 2Al + Zr \rightarrow 2Nb + Al_2O_3 + ZrO_2$ (2 h)	SPEX 8000	>7:1			2929		81
$Nb_2O_5 + 2Al + Zr \rightarrow 2Nb + Al_2O_3 + ZrO_2$	SPEX 8000	2:1					82
$Nb_2O_5 + Al_2Zr \rightarrow Al_2O_3 + ZrO_2 + 2Nb$ (2 h)	SPEX 8000	>7:1			2337		81
$Nd_2O_3 + Fe + Ca \rightarrow Nd_2Fe_{14}B + CaO$							83
$NiCl_2 + Mg \rightarrow Ni + MgCl_2$	SPEX 8000	10:1				42	84
$NiCl_2 + 2Na \rightarrow Ni + 2NaCl$	SPEX 8000	3:1	50	−507			37, 85
$3NiO + 5Al \rightarrow 3NiAl + Al_2O_3$	Fritsch P5	10:1					86
$3NiO + 3Al \rightarrow Ni_3Al + Al_2O_3$ (<1 h)	SPEX 8000	>7:1			3744		81
$2NiO + C \rightarrow 2Ni + CO_2$	SPEX 8000	12:1					87
$3PbS + 2Al \rightarrow 3Pb + Al_2S_3$	SPEX 8000	13:1		−267	1247		19
$SiC + Fe \rightarrow Fe_3C + Fe(Si)$ (100 h)	Planetary ball mill	40:1					88
$SiC + Fe \rightarrow Fe_3Si + Fe_2Si + C$ (100 h)	Planetary ball mill	40:1					88
$SmCl_3 + 3Na \rightarrow Sm + 3NaCl$	SPEX 8000	10:1	15	−202			89
$Sm_2O_3 + 17CoO + 20Ca \rightarrow Sm_2Co_{17}$[c] $+ 20CaO$							90
$Sm_2O_3 + Co + Ca \rightarrow SmCo_5 + CaO$							91
$SnCl_2 + Ca(OH)_2 + 0.5O_2 \rightarrow SnO_2 + CaCl_2 + H_2O$ (g)	SPEX 8000	10:1					92
$SnCl_2 + Ca(OH)_2 \rightarrow SnO + CaCl_2 + H_2O$	SPEX 8000	10:1		−59			36
$SnO + \frac{1}{2}O_2 \rightarrow SnO_2$[d]				−229			
$SnCl_2 + Ca(OH)_2 \rightarrow SnO + CaCl_2 + H_2O$ (g)	Centrifugal mill	9:1					93
$SnO + \frac{1}{2}O_2 \rightarrow SnO_2$[e]							
$SrO + ZrSiO_4 \rightarrow SrSiO_3 + ZrO_2$	Uni-Ball mill	50:1					28

Table 14.1 Continued

Reaction	Mill	BPR	% Excess reductant	ΔH(kJ)	T_{ad} (K)	t_{ig} (min)	Ref.
$2TaCl_5 + 5Mg \rightarrow 2Ta + 5MgCl_2$	SPEX 8000	7:1	10	−745	1616		94, 95
$TiCl_4 + 2Mg \rightarrow Ti + 2MgCl_2$	SPEX 8000	3:1 to 12:1	15	−479			96
$3TiO_2 + 13Al \rightarrow 3TiAl_3 + 2Al_2O_3$	Ball mill	47:1					97, 98
$3TiO_2 + 4Al \rightarrow 3Ti + 2Al_2O_3$	SPEX 8000	4:1	5				99
$3TiO_2 + 4Al + 3C \rightarrow 3TiC + 2Al_2O_3$ (7 h)	SPEX 8000	>7:1		−1038	2338		66, 81
$3TiO_2 + 4Al + 3C \rightarrow 3TiC + 2Al_2O_3$							100
$3TiO_2 + 4Al + 3/2N_2 \rightarrow 3TiN + 2Al_2O_3$	Ball mill						68
$3TiO_2 + 3B_2O_3 + 10\ Al \rightarrow 3TiB_2 + 5Al_2O_3$							101
$TiO_2 + B_2O_3 + {}^{10}/_{3}Al \rightarrow TiB_2 + {}^{5}/_{3}Al_2O_3$ (3 h)	SPEX 8000	>7:1			2655		81
$TiO_2 + B_2O_3 + 5Mg \rightarrow TiB_2 + 5MgO$	Ball mill	43:1					102
$TiO_2 + 3C \rightarrow TiC + 2CO$ (200 h)	Ball mill						70
$TiO_2 + 2Ca + 2B \rightarrow TiB_2 + 2CaO$	Fritsch P5	7.5:1	20	−642			103
$TiO_2 + 2Ca + C \rightarrow TiC + 2CaO$	Fritsch P5	7.5:1	20	−510			103
$TiO_2 + 2Mg \rightarrow Ti + 2MgO$	Uni-Ball mill						73
$TiO_2 + 2Mg + C \rightarrow TiC + 2MgO$	Uni-Ball mill	43:1					74
$2TiO_2 + 4Mg + N_2 \rightarrow 2TiN + 4MgO$	Uni-Ball mill	43:1					74
$3V_2O_5 + 10Al \rightarrow 6V + 5Al_2O_3$	SPEX 8000	3:1	5				99
$3V_2O_5 + 10Al \rightarrow 6V + 5Al_2O_3$	SPEX 8000	7:1	10	−3727	3468		104
$2V_2O_5 + 5C \rightarrow 4V + 5CO_2$ (48 h)	SPEX 8000	9.3:1	10				105

$2V_2O_5 + 5C \rightarrow 4V + 5CO_2$ (4 h)	SPEX 8000	10:1	10				106
$2V_2O_5 + 9C \rightarrow 4VC + 5CO_2$	SPEX 8000	4:1	50 and 100				106
$V_2O_5 + 5Mg \rightarrow 2V + 5MgO$	SPEX 8000	7:1	10	−1457	3354	—	104
$2V_2O_5 + 5Ti \rightarrow 4V + 5TiO_2$	SPEX 8000	7:1	10	−1623	2658	—	104
$WO_3 + 3Mg \rightarrow W + 3MgO$	SPEX 8000	10:1	10				107
$WO_3 + 3Mg + C \rightarrow WC + 3MgO$		10:1					108, 109
$ZnCl_2 + CaS \rightarrow ZnS + CaCl_2$	SPEX 8000	10:1					33, 110
$ZnCl_2 + Na_2CO_3 + 8.6NaCl \rightarrow ZnO^f + 10.6NaCl + CO_2$ (g)	SPEX 8000	10:1		−80			36
$ZnO + Ca \rightarrow Zn + CaO$ (24 h)	SPEX 8000	3:1	10	−285	2242	87	15, 16
$ZnO + Mg \rightarrow Zn + MgO$	SPEX 8000	7:1	10	−253	2006	45	111
$2ZnO + Ti \rightarrow 2Zn + TiO_2$	SPEX 8000	3:1	10	−249	1982		24
$3ZnS + 2Al \rightarrow 3Zn + Al_2S_3$	SPEX 8000	13:1		−107	493		19
$ZnS + 8CuO \rightarrow ZnSO_4 + 4Cu_2O$							112
$ZrCl_4 + 2CaO \rightarrow ZrO_2^g + 2CaCl_2$ (20 h)	SPEX 8000	5:1	—	−440	—	—	113
$ZrCl_4 + 2Mg \rightarrow Zr + 2MgCl_2$	SPEX 8000	10:1	10	−303			95

[a] γ-Al_2O_3 formed only after heating the milled powder to more than 300°C.
[b] CeO_2 formed on annealing the milled powder for 6 h at 400°C.
[c] Sm_2Co_{17} formed on annealing the milled powder at 800°C.
[d] SnO_2 formed on annealing the milled powder at 400°C.
[e] SnO_2 formed on annealing the milled powder for 1 h at 400°C.
[f] ZnO formed on annealing the milled powder for 0.5 h at 400°C.
[g] ZrO_2 formed only after heating the milled powder to more than 400°C.

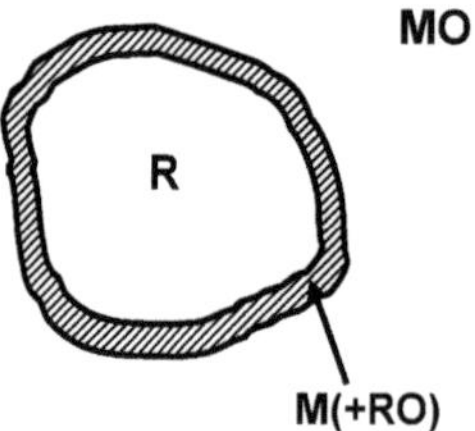

Figure 14.1 Schematic of the chemical reaction occurring between two components—metal oxide (MO) and reductant R in the mechanochemical reactions. The product phases (M and RO) form at the interface between the reactants and prevents further reaction from taking place since the reactants are not in contact with each other.

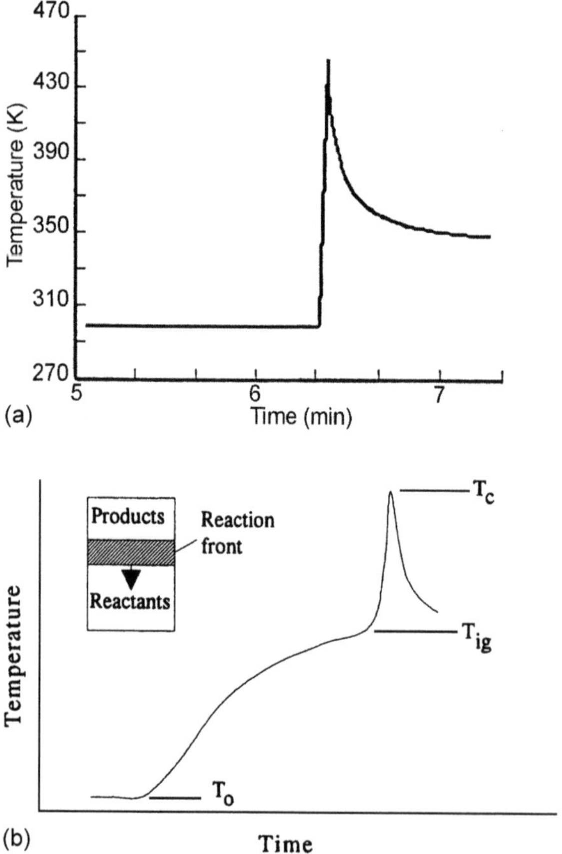

Figure 14.2 (a) Variation of milling vial temperature during mechanical alloying when a combustion reaction takes place. Note that the temperature increases gradually with time until the combustion reaction takes place. There is a sudden increase in temperature when the combustion reaction occurs. (b) Schematic representation of the temperature–time plot during an SHS reaction. Note the similarity of this curve to that shown in (a).

of particles. On the other hand, if both the oxide (which is brittle) and the reductant are hard and brittle, agglomerates of reactant particles cannot develop and combustion is not initiated. This may be why combustion was not achieved in the CuO/Cr couple, and even the gradual reaction was extremely slow between CuO and Si. This point highlights the necessity of an intimate contact for achieving faster kinetics during the reduction process.

The SHS (or combustion reaction) is known to take place in a number of powder compacts of different alloy systems [115,119]. Conventional SHS reactions are initiated when the powder compacts are heated to the ignition temperature, with relatively little loss of heat to the environment. However, since the powder during MA is in close contact with the milling tools, the heat generated is partly lost to the vial and the balls. Thus, it should be realized that SHS reactions that take place in powder compacts may not occur during MA in all the alloy systems. Consequently, some moderately exothermic SHS systems do not show the combustion reaction during MA (due to loss of heat to the surroundings). Even though preheating of the components could help in such systems, this is not generally done in the MA process.

The product of the displacement reactions normally consists of two phases: the metal (or a solid solution or a compound), and the oxide or chloride associated with the reductant. Removal of the unwanted reaction by-product can be difficult due to the high reactivity of the metal phase associated with the nanocrystalline grain sizes and intermixing of the phases induced by the MA process. The by-product phase can be removed by leaching in a dilute acid or hot water, or by vacuum distillation. It was reported that the salt (by-product $MgCl_2$) formed during the reduction of $TiCl_4$ with Mg was removed either by vacuum distillation at 900°C for 24 h, or by washing the powder in 10% nitric acid [96]. An important requirement for easy removal of the salt from the mixture is that it should be contiguous so that the leachant can easily reach it. The use of carbon or hydrogen as a reductant produces either gaseous CO_2 or water vapor as the by-product and obviates the need for leaching/distillation [50,87].

14.3 PROCESS PARAMETERS

Process parameters such as milling temperature, grinding ball diameter, ball-to-powder weight ratio (BPR), use of a process control agent, and relative proportion of the reactants seem to play an important role on the nature and kinetics of the product phase obtained by the displacement reactions. For example, a combustion reaction could be initiated during the reduction of copper oxide by iron; but the same reaction progresses gradually under slightly different milling conditions. Consequently, results from different laboratories can be effectively compared only if the exact conditions under which the reaction takes place are reported. These conditions should be optimized for the best yield.

14.3.1 Milling Temperature

McCormick et al. [96] investigated the reduction of $TiCl_4$ with Mg both at room temperature (20°C) when $TiCl_4$ is in the liquid state and at −55°C when it is in the solid state. They observed that the milling time required to synthesize Ti was reduced by a factor of 6 when milling was conducted at −55 °C (Fig. 14.3). This was explained on the basis of the greater efficiency of solid-solid collision events occurring during

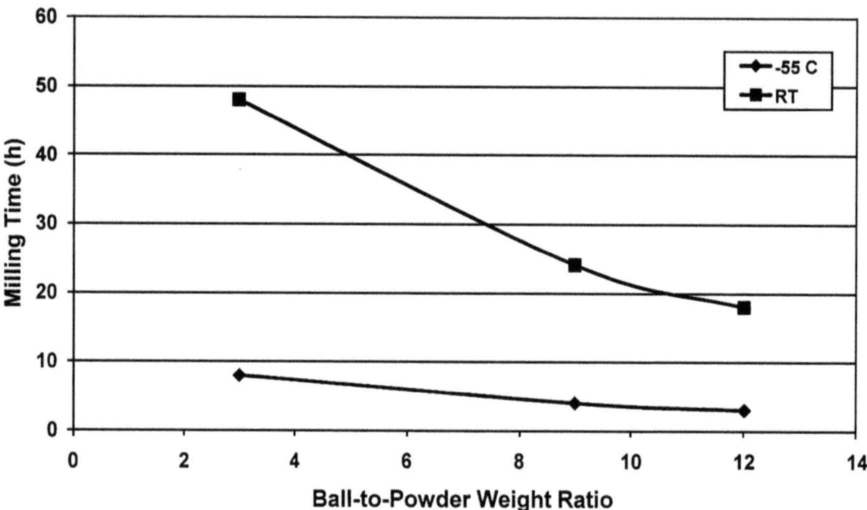

Figure 14.3 Effect of ball-to-powder weight ratio (BPR) and milling temperature on the time required to synthesize titanium during reduction of $TiCl_4$ with Mg. Note that the reduction time decreases with increasing BPR and also with decreasing temperature.

MA. However, if both the reactants were to be in the solid state at both temperatures, the enhanced diffusivity at the higher temperatures increases the reaction kinetics and consequently the times required for reduction are shorter. No such measurements have been made to date.

14.3.2 Ball-to-Powder Weight Ratio

The time required for the reduction reaction to be completed decreased with an increase in the BPR. For example, the time required to form Ti during the reduction of $TiCl_4$ with Mg at 20°C was 48 h at a BPR of 3:1, while it was only 18 h at a BPR of 12:1 (Fig. 14.3). A similar threefold reduction in time with increasing BPR was also noted at −55°C; the minimal milling time required being 3 h at a BPR of 12:1 at this temperature. Similar results were also reported for other displacement reactions, e.g., between CuO and Fe [116]. The authors also noted that the magnitude of reduction in the ignition time with increasing BPR was most dramatic at low BPR values and almost nonexistent when the BPR was increased to values greater than 20.

Xi et al. [52] studied the effect of BPR on the reduction of CuO by Si during MA in a stirring ball mill. They noted that CuO was completely reduced to the pure metal Cu only at a BPR of 80:1. At lower BPR values, CuO was reduced only partially to Cu and Cu_2O (BPR 60:1), and at a BPR of 30:1, CuO was reduced only to Cu_2O; no Cu was obtained. This result was explained on the basis that at a high BPR the collision energy is high and therefore complete reduction to the pure metal could be obtained.

Schaffer and McCormick [116] conducted a systematic investigation on the effect of BPR on the ignition temperature and time. They observed that the ignition temperature T_{ig} decreased with increasing milling time, suggesting that the reaction kinetics become faster during milling; the reaction rates decrease with time during conventional solid-state reactions. Furthermore, the rate of decrease of T_{ig} increased with increasing BPR, and this is related to the decrease in particle size. The values of the ignition time, t_{ig} also decreased with increasing BPR. Both of these effects are related to the average frequency of collisions that increases with increasing BPR. They also noted that the combustion reaction did not occur when small grinding media (1- and 3-g balls) were used, but it occurred only when large (8-g) balls were used [56].

Higher milling intensity (achieved by increasing the BPR) resulted in a combustive reaction during milling of a Mo + Si powder mixture in a SPEX mill [118,120,121]. On the other hand, formation of $MoSi_2$ was gradual when the BPR used was low [122].

14.3.3 Process Control Agent

In many of the investigations on reduction reactions, the components involved are generally brittle and therefore a process control agent (PCA) was not used. However, it was used in a few specific instances with interesting results. When CuO was ball milled with reactive metals in the absence of any PCA, a self-propagating combustion reaction occurred after an incubation period [40,41]. On the other hand, when toluene was used as a PCA, the reduction of CuO proceeded in a controlled manner and reached completion in about 24 h [40]. The use of a PCA acts likes an additive (diluent) and either delays or completely suppresses the combustion event. The PCA also inhibits interparticle welding during collisions, slowing down the reaction rate as well as decreasing the particle size. It may be noted in passing that a combustion reaction should be avoided if one is interested in producing the product material in a nanocrystalline state. This is because combustion may result in partial melting and subsequent solidification leads to the formation of a coarse-grained structure. Another requirement for formation of nanometer-sized particles is that the volume fraction of the by-product phase must be sufficient to prevent particle agglomeration.

14.3.4 Relative Proportion of the Reactants

Normally about 10–15% stoichiometric excess of the reductant has been used in most of the investigations conducted so far. This excess reductant is used partly to compensate for the surface oxidation of the reactive reductant powder particles, e.g., Al [57]. There have been only a few investigations that discussed the effect of the ratio of the constituent reactants (off-stoichiometry).

Takacs [58] investigated the displacement reaction:

$$3Fe_3O_4 + 8Al \rightarrow 9Fe + 4Al_2O_3 \tag{14.2}$$

without the use of a PCA and studied the effect of off-stoichiometry by milling mixtures corresponding to $3Fe_3O_4 + xAl$, where $2 \leq x \leq 16$. A combustion reaction occurred in all the cases except when $x = 2$. However, the nature of the product phases was quite different depending on the value of x, and the results are summarized in Table 14.2. It has been noted that a slight decrease in the amount of Al results in the formation of $FeAl_2O_4$ instead of Al_2O_3; excess Al suppresses the formation of metastable γ-Al_2O_3 and $FeAl_2O_4$ and results in the formation of random bcc α-Fe(Al) solid solution and α-Al_2O_3.

Table 14.2 Incubation Time and Phase Constitution After Milling a Mixture of $3Fe_3O_4 + xAl$ for 2 h[a]

Sample, x	Incubation time (min)	Major (minor) phases present
2	No combustion	Fe_3O_4, (Al)
6	53	α-Fe, $FeAl_2O_4$
8a	10	α-Fe, α-Al_2O_3, (γ-Al_2O_3, $FeAl_2O_4$)
11	14	α-$FeAl_{0.333}$, α-Al_2O_3
16	56	α-$FeAl_{0.890}$, α-Al_2O_3
8b	8	α-Fe, α-Al_2O_3, γ-Al_2O_3, $FeAl_2O_4$, (Fe_3O_4, Al)
8c	—	Fe_3O_4, Al
8d	23	α-Fe, α-Al_2O_3, γ-Al_2O_3, $FeAl_2O_4$

[a] Milling of samples 8b, 8c, and 8d was interrupted immediately after, somewhat before, and 5 min after combustion [58].

14.3.5 Grinding Ball Diameter

Yang and McCormick [94] investigated the effect of grinding ball diameter on the combustion reaction during milling of a mixture of $TaCl_5$ and Mg. They observed that the ignition time, t_{ig}, for the combustion reaction decreased with an increase in ball diameter.

Combustion requires that the powder mixture reaches the ignition temperature T_{ig}. Since T_{ig} is dependent on microstructural features, it decreases with increasing refinement in the microstructure. During MA the collision between the powder particles and the balls raises the powder temperature, T_c, and therefore T_c increases with increasing milling time. It has been postulated [24] that the ignition time, t_{ig} is equal to the milling time required for T_{ig} to decrease to T_c. Since increasing ball size increases the collision energy and therefore T_c, it is expected that t_{ig} decreases with an increase in the ball diameter.

As mentioned earlier, "soft" milling conditions produce metastable phases farther from equilibrium than "hard" milling conditions. Accordingly, it was reported that lower energy milling with smaller balls produced the hexagonal high-temperature form of $MoSi_2$ rather than the low-temperature tetragonal form [123].

14.4 PHASE FORMATION

It is normally expected that the displacement reactions listed in Table 14.1 follow the route indicated. In fact, the heats of formation for these reactions have been calculated assuming that these reactions take place according to that sequence. However, it has been noted that many of these reactions do not go to completion directly; instead, intermediate phases are formed. There are also instances when the phases formed are quite different from the expected ones. For example, it was noted that in some of the displacement reactions one of the product phases became amorphous and so could not be easily detected after milling. This was reported to be true for $AlCl_3$ during milling with CaO when diffraction peaks due to CaO only were observed in the X-ray diffraction patterns [26]. It was inferred that $AlCl_3$ became amorphous during milling. Similar results were reported for $TaCl_5$ [94], SiO_2 [95], $ZrCl_4$ [113], and others [36,79].

Formation of an intermediate (transition) product phase has been observed during the reduction reactions of Cr_2O_3 with Al and Zn [38], Fe_3O_4 with Al [57], Ti and Zn [38], and CuO with Fe [45–47] or graphite [42]. For example, during reduction of CuO with Fe, it was observed that the reaction proceeds in various stages that can be represented as:

$$8CuO + 3Fe \rightarrow 4Cu_2O + Fe_3O_4 \tag{14.3}$$

or $$5CuO + 3Fe \rightarrow Cu_2O + 3Cu + Fe_3O_4 \tag{14.4}$$

$$4Cu_2O + 3Fe \rightarrow 8Cu + Fe_3O_4 \tag{14.5}$$

Formation of Cu_2O as an intermediate product was also reported during the reduction of CuO with graphite [42]. Similarly, it was reported that Ti_2O_3 formed as an intermediate product (instead of TiO_2) during the reduction of V_2O_5 with titanium [104]. The amount of CuO was reported to decrease monotonically during milling and the proportion of Cu_2O increased reaching a maximum of 45 wt% after milling for 9 h. Subsequently, it decreased to zero after milling of the powder mixture for 24 h, when 100% Cu had formed (Fig. 14.4).

Reduction of Fe_2O_3 to pure Fe generally takes place in the sequence of $Fe_2O_3 \rightarrow Fe_3O_4 \rightarrow FeO \rightarrow Fe$, i.e., with intermediate phases containing lower and lower amounts of oxygen. However, in the case of mechanochemical reduction, it has been reported [61] that the pure metal Fe formed directly from Fe_2O_3 upon milling with titanium. This suggests that mechanical activation could alter the stages through which reduction reactions take place.

Investigation of the nature and role of such intermediate phases is important to understand the reaction mechanism(s). It should also be mentioned that the reactions stop with the formation of an intermediate phase when the reactions are endothermic. Accordingly, the metal Cr does not form when Cr_2O_3 is reduced with Zn [38].

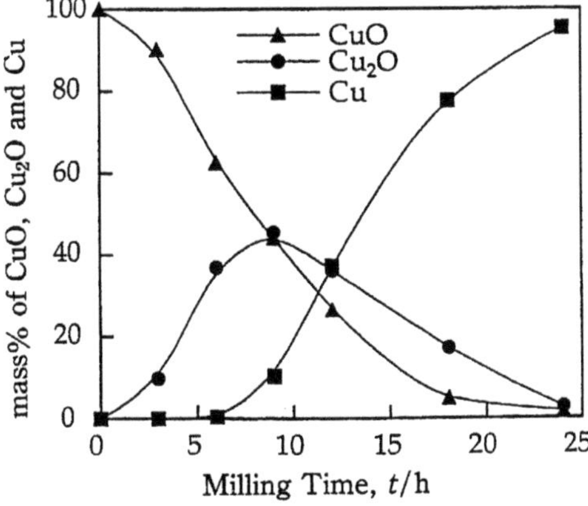

Figure 14.4 Effect of milling time on reaction kinetics during reduction of CuO with graphite. Note that the amount of CuO decreases continuously and that the transition product Cu_2O increases reaching a value of 45 wt% with milling time. Formation of Cu takes place eventually, and all of the CuO and Cu_2O gets reduced to pure Cu.

In a few instances it was also noted that instead of a pure metal a solid solution was formed. For example, an Fe(Zn) solid solution was formed during the reduction of Fe_3O_4 with Zn [38], and an Fe(Al) solid solution containing 1.7 at% Al was formed when Fe_2O_3 was reduced with Al [124] or 7 at% Al when FeS was reduced with Al [19].

It was also observed that in some cases the final phase formed is different from the expected phase. During the displacement reactions of FeS, WS_2, and MoS_2 with CaO, it was noted that the corresponding simple oxides did not form; instead, complex oxides, e.g., $CaFeO_3$, $CaWO_4$, and $CaMoO_4$, were formed [19]. The product phase in some cases has also been found to be amorphous in nature [36,52,79,94,95,113].

14.5 COMBUSTION REACTION

Milling-induced combustion was first observed by Tschakarov et al. [125] during the mechanochemical synthesis of metal chalcogenides from a mixture of elemental powders. Subsequently, a number of similar reactions were reported to occur during the chemical reduction of oxides with reactive metals under MA conditions.

The large enthalpy changes associated with the chemical reactions are responsible for the occurrence of the combustion reaction. If the temperature generated during the milling process (due to ball-to-ball and ball-to-powder collisions), T_c, is higher than the ignition temperature, T_{ig}, then the combustion reaction can occur. The ignition temperature is a function of the enthalpy change and microstructural parameters such as particle size and crystal size. Since the particle size and crystal size get refined with the progress in the MA processing, T_{ig} is found to decrease with milling time. But with increasing milling time T_c increases and reaches a steady-state value. The time at which the T_{ig} and T_c intersect is the critical milling time, t_{ig}, at which ignition would occur (Fig. 14.5).

It may also be interpreted that up to the milling time t_{ig}, intermixing of the particles, particle refinement, and accumulation of lattice defects occurs and these appear to favor the combustion reaction. Furthermore, room temperature aging was found to significantly accelerate the onset of combustion thereby substantially

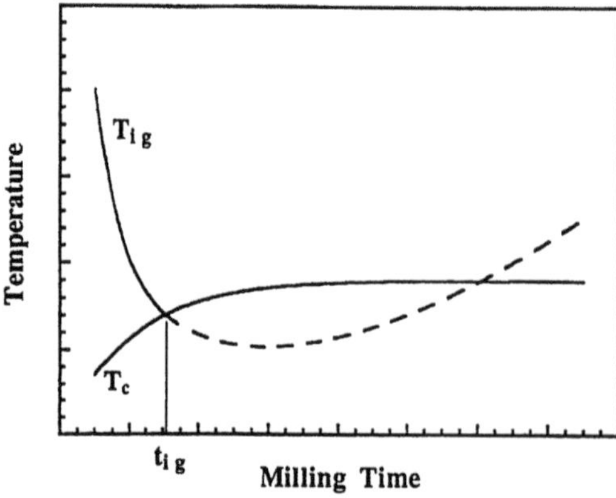

Figure 14.5 Schematic representation of the variation of T_{ig} and T_c with milling time. The time corresponding to the point of intersection at t_{ig} represents the minimal ignition time.

reducing the overall milling time [24]. A detailed review on this topic was recently published [18].

Similar to what has been noted during conventional SHS reactions, the effect of process variables on the reaction during MA has also been investigated. For example, it was reported that the ignition of the combustion reactions strongly depended on the milling intensity. High milling energy and frequency led to a shorter ignition time (activation period) [81]. The ignition temperature also decreased with mechanical activation. The ignition temperature of 1100–1300 K for the CuO + Mg reaction decreased to 400–450 K with mechanical activation [51]. The addition of diluents, e.g., Al_2O_3, to Ni-Al mixtures during milling changed the spontaneous reactions to gradual reactions [81].

14.6 REACTION MECHANISMS

The complexity of deformation and comminution processes accompanying MCP has been recognized right from the beginning. It has also been noted that mechanochemical reactions occur only when (1) the individual components are in intimate contact with each other, (2) there is a fine dispersion of the two phases involved (at least one of the phases must be uniformly dispersed), (3) the components involved are preferentially on a nanometer scale, and (4) sufficient heat is generated so that the temperature is high enough to allow ignition to occur. However, the question of how the mechanical energy is transformed to chemical energy is the most fascinating and least understood problems in MCP research. Several different approaches have been employed to explain the initiation and the mechanism of the mechanochemical reactions. These include:

- Thermal theories of initiation of mechanochemical reactions
- Mechanisms related to plastic deformation and creation of defects such as dislocations and disclinations
- Theory of surface-active states and electron excitations during fracturing of crystals

14.6.1 Thermal Theories

The simplest way to explain the initiation of mechanochemical reactions is to assume that the kinetic energy of the grinding medium transferred to the powder is converted to thermal energy. The development of surface temperature peaks between two sliding solids was investigated by Bowden and Thomas [126], and these temperature peaks have been used to explain detonation of explosives. Heinicke [5] proposed the simple "tribo plasma" model to explain most of the mechanochemical effects. According to this theory, the high temperatures developed during collisions can cause the material to evaporate and form a plasma state, which promotes chemical reactions. Some evidence of electron and photon emissions was used to support this theory.

Occurrence of chemical reactions during crack propagation through crystals was first studied by Fox and Soria-Ruiz [127] and later by Urakaev et al. [128]. It was shown that when a crack propagates through a crystal, thermal excitation at its tip may exceed 2000 K [129] for a short time (about 10^{-9} s), and this may result in the decomposition of the solid. Analysis of the decomposition products emitted from the crack propagating through an $NaNO_3$ crystal using a time-of-flight mass spectrometer revealed the presence of NO_2 [128]. NO_2 can typically form only at temperatures higher than 1000 K, whereas oxygen usually forms during low-temperature decomposition. Thus, it was concluded that thermal spikes are responsible for mechanochemical reactions. The early mechanisms of amorphization during MM were also related to possible rise in temperature and formation of liquid metal pools under ball impacts and subsequent rapid solidification of the melt by the surrounding cold solid [130].

14.6.2 Reactions Induced by Shear

Starting from the early work of Bridgeman [131] it was shown that application of shear stress to a solid or to a mixture of solids may result in phase transformations or acceleration of chemical interactions. Reactions such as synthesis of Cu_2S and other sulfides from elements, decomposition of oxides, reduction of SnO_2 into SnO, reduction of PbO and AgO to Pb and Ag, respectively, were demonstrated to take place in Bridgeman anvils. The importance of shear was demonstrated when $K_3Fe(CN)_6$ has transformed into $K_4Fe(CN)_6$ under the combined action of shear and pressure in Bridgeman anvils; pressure alone did not produce any chemical changes [132]. It was later demonstrated that the Ni-Al intermetallic compounds could be formed in Bridgeman anvils [133] as well as by MA processing [134]. It is well established in the literature on MA that plastic deformation plays a major role in further accelerating chemical reactions. Development of the "bulk" mechanical alloying technique by Aizawa et al. [135] to synthesize Cu-Ag solid solutions and intermetallics such as Mg_2Ni by a solid-state reaction between the constituent elemental powder mixtures involves multiple extrusion and forging. Since homogeneous refining and alloying take place efficiently by repeated forging, the process time can be reduced to one-fiftieth of the time necessary for conventional MM and attrition. Mg_2Ni by this process can be obtained as a highly dense powder compact with a relative mass density in the order of 85% of theoretical. The equal channel angular extrusion (ECAE) process invented in early 1970s by Segal [136] is now becoming very popular not only for refining the microstructure but also as a synthesis method [137,138]. The large volume fraction of atoms residing in the grain boundaries of these ultrafine-grained materials is expected to enhance diffusion and consequently increase the rate of mechanochemical reactions. Studies of the mechanochemical reaction mechanisms at the atomic level are very limited. Transmission electron microscopy techniques were used to demonstrate early formation of rotational disclination-type defects in Frank-Kasper cubic crystals transformed into a quasi-crystalline phase [139]. It was shown that mass transfer at the first stage of the transformation was due to rotation of crystal blocks rather than conventional diffusion process. In addition to the shear and deformation mechanisms mentioned above, enhancement of diffusion due to defect formation also plays an important role. The diffusion process is accelerated due to the increased density of defects in plastically deformed powder particles.

14.6.3 Theory of Surface Active States

Fresh surfaces formed during the comminution process are usually electrically charged, and this charge is localized reflecting the defect structure of the solid. The relationship between mechanochemical reactions and electrochemical processes in solids was discussed by Gutman [7]. Formation of free radicals and detailed studies of their mechanisms during mechanical activation of quartz were studied by Butyagin [140]. Effects of reduced coordination number for Ca on the electron distribution during Ca-O-Si bridge bonding from CaO or $Ca(OH)_2$ and SiO_2 was studied [141]. The discrete variational X alpha method was used to elucidate the formation of a heterobridging bond, Ca-O-Si, when a Si-O pendant bond of SiO_2 comes close to the Ca atom on the surface of calcium oxide or hydroxide. A decrease in the coordination number for Ca in CaO or $Ca(OH)_2$ increases the electron population density between Ca and O, and decreases the electron population density between Si and O. These simultaneous changes in electron population density favor the formation and stabilization of a Ca-O-Si heterobridging bond. The calculated results agree with the experimental observation of precursor formation towards calcium silicate during milling of a mixture of $Ca(OH)_2$ and SiO_2. Other examples may be mechanical degradation and mechanochemical reaction in heterogeneous and homogeneous systems of polyvinyl chloride and polyethylene-co-propylene polymer caused by ultrasonic irradiation at 303 K [142]. Mechanochemical reaction occurred due to the free radicals produced from the chain scissions of both polymers by ultrasonic waves. Detailed investigations of the reaction mechanisms of mechanochemical reactions present challenges due to the complexity of the processes involving chemical reactions of deformed solids; consequently, the number of detailed mechanism studies is small.

14.7 MECHANOSYNTHESIS OF CERAMICS AND NANOCOMPOSITES

Apart from the displacement reactions described above to reduce oxides, chlorides, and sulfides to pure metals, the MA technique was also used to synthesize ceramic compounds and nanocomposites using the mechanochemical reactions. In this category, a variety of borides, carbides [143,144], fluorides, nitrides, oxides, silicides [143], stannides, and composites [114,124,145,146] have been synthesized by ball milling of the respective powders for a few hours at room temperature. An important characteristic of such phases is that they are nanocrystalline in nature. Consequently, like other nanocrystalline materials [147], they exhibit properties and performance much improved over their conventional coarse-grained counterparts. These nanocrystalline oxides can be used for dispersion strengthening the metals to improve their high-temperature mechanical properties. Table 14.3 summarizes the results available so far. The main advantage of synthesizing these ceramic compounds and composites through mechanochemical displacement reactions is that they can be produced at relatively low temperatures and also in an economical way through fewer steps than in the conventional way.

Composites consisting of a metal oxide as a major phase (ranging from 48 to 73 vol%) and a metal have been synthesized according to the reaction:

$$3/y M_xO_y + 2Al \rightarrow Al_2O_3 + 3x/yM \qquad (14.6)$$

In some cases, instead of the pure metal, alloys have also been synthesized as in the case of Fe-50 at% Cr or Fe-18 at% Cr-8 at% Ni. The crystallite size of the α-Al_2O_3 phase in all the cases is in the range of 10–20 nm. The great advantage of producing composites this way is that pure metal oxide or aluminum is not required; one could use recycled aluminum for largescale production. Metal ore can also be used instead of

Table 14.3 Synthesis of Nanocomposites by Mechanochemical Synthesis Reactions

Reaction	Mill	BPR	Time for reaction (h)	Ref.
$2BN + 3Al \rightarrow 2AlN + AlB_2$	Planetary ball mill	4:1	20	27
$3CoO + 2Al \rightarrow 3Co + Al_2O_3$	Fritsch P7	42:1	1.5	124
$Cr_2O_3 + 2Al \rightarrow 2Cr + Al_2O_3$	Fritsch P7	42:1	1.5	124
$CuO + Al \rightarrow Cu + Al_2O_3$	Fritsch P7	40:1	1.5	124
$Dy_2O_3 + 4Fe + 3Ca \rightarrow 2DyFe_2$[a] $+ 3CaO$	SPEX 8000		24	148
$Fe_3C + Cr \rightarrow Cr_{23}C_6 + (Cr,Fe)_7C_3$	SPEX 8000	8:1	24	145
$FeF_3 + C \rightarrow FeF_2 + FeF_3 + (CF)_n$	SPEX 8000	14:1	24	145
$2Fe_{2.5}N + 2Al \rightarrow 2AlN + 5Fe(Al)$	SPEX 8000	10:1	24	145
$Fe_{2.5}N + Si \rightarrow Si_3N_4 + Fe(Si) + FeSi_2$	SPEX 8000	10:1	24	145
$Fe_2O_3 + 2Al \rightarrow Fe(Al) + Al_2O_3$	Fritsch P7	40:1	3	124
	SPEX 8000	11:1	24	
$Fe_2O_3 + Cr_2O_3 + 2Al \rightarrow Fe\text{-}Cr + 2Al_2O_3$	Fritsch P7	31:1	3	124
$Fe_2O_3 + Cr_2O_3 + 3NiO + 3Al \rightarrow Fe\text{-}Cr\text{-}Ni + 3Al_2O_3$	Fritsch P7	32:1	3	124
$2Fe_2O_3 + 3Si \rightarrow 4Fe + 3SiO_2$	Fritsch P5	10:1	43	149
$3MnO_2 + 2Al \rightarrow 3Mn + 2Al_2O_3$	Fritsch P7	31:1	3	124
$MoO_3 + 2Al \rightarrow Mo + Al_2O_3$	SPEX 8000	10:1	24	124
$6Nb_2O_5 + 10Al \rightarrow 12Nb + 5Al_2O_3$	Fritsch P7	42:1	1.5	124
$3NiO + 2Al \rightarrow 3Ni + Al_2O_3$	Fritsch P7	42:1	1.5	124
$2NiO + Si \rightarrow 2Ni + SiO_2$	Fritsch P5	10:1	43	149
$3SiO_2 + 2Al \rightarrow 3Si + 2Al_2O_3$	SPEX 8000	23:1	24	124
$3SiO_2 + 4N_2 \rightarrow 2\alpha\text{-}Si_3N_4$[b] $+ 3O_2\uparrow$	Planetary ball mill		623	150
$6V_2O_5 + 10Al \rightarrow 12V + 5Al_2O_3$	Fritsch P7	31:1	3	124
$WO_3 + 2Al \rightarrow W + Al_2O_3$	SPEX 8000	7:1	24	124
$WO_3 + 3Mg + C \rightarrow WC + 3MgO$	Fritsch P6	10:1	48	109
$2WO_3 + 3Ti \rightarrow 2W + 3TiO_2$	SPEX 8000	9:1	24	124
$3ZnO + 2Al \rightarrow 3Zn + Al_2O_3$	Fritsch P7	42:1	1.5	124

[a] Formed after annealing the milled powder for 2 h at 500°C.
[b] Formed after annealing the milled powder for 4 h at 900°C.

pure aluminum. This is because, as described above, oxide ores can be reduced to pure metals, as has been shown in the case of Al_2O_3-Fe composite.

The mechanically driven chemical reduction process (mechanosynthesis) has a number of advantages over the conventional metal processing techniques. It enables the reduction of metal oxides and halides directly to pure metals or alloys without first having to convert the oxides to pure metals and then to the desired alloy. For powder metallurgy applications it allows the direct formation of a powder product without first having to manufacture the bulk alloy and then convert it to powder form. In addition to potential cost savings that may arise from the reduced number of processing steps required, additional benefits accrue because the reactions occur at room temperature. In fact, the trend in recent times has been to use the mechanosynthesis route to produce commercially useful materials at reduced costs. For example, rutile and metallic iron have been produced by ball milling an ilmenite and carbon mixture at room temperature [151,152]. Furthermore, production of nanocomposites consisting of finely dispersed ferromagnetic Fe particles in a nonmagnetic matrix such as Al_2O_3 can provide interesting combination of magnetic properties, including the giant magnetoresistance effect.

REFERENCES

1. Carry-Lea, M. (1892), (1894). *Phil. Mag.* 34:46; 37:470–475.
2. Ostwald, W. (1919). *Handbuch der allgemeine Chemie*. Vol. 1. Leipzig, p. 70.
3. McCormick, P. G. (1997). In: Gschneidner, K. A. Jr., Eyring, L., eds. *Handbook on the Physics and Chemistry of Rare Earths*. Vol. 24. Amsterdam: Elsevier, pp. 47–81.
4. Tiessen, P. A., Meyer, K., Heinicke, G. (1966). *Grundlagen der Tribochemie*. Berlin: Akad Verlag.
5. Heinicke, G. (1984). *Tribochemistry*. Munchen: Hanser.
6. Boldyrev, V. V. (1991). *Siber. J. Chem.* 5:4–21.
7. Gutman, E. M. (1998). *Mechanochemistry of Materials*. Cambridge, UK: Cambridge International Science Publishing.
8. Gutman, E. M. (1994). *Mechanochemistry of Solid Surfaces*. Singapore: World Scientific.
9. Tkacova, K. (1989). *Mechanical Activation of Minerals*. Amsterdam: Elsevier.
10. Juhász, A. Z., Opoczky, L. (1990). *Mechanical Activation of Minerals by Grinding, Pulverizing, and Morphology of Particles*. Budapest: Akademia Kiadó.
11. Ivanov, E. (1993). *J. Mater. Synth. Proc.* 1:405–413.
12. Ivanov, E., Suryanarayana, C. (2000). *J. Mater. Synth. Proc.* 8:235–244.
13. Avvakumov, E. G. (1986). *Mechanical Methods of Activation of Chemical Processes*. Novosibirsk, Russia: Nauka Publishing House.
14. McCormick, P. G. (1995). *Mater. Trans. Jpn. Inst. Metals* 36:161–169.
15. Schaffer, G. B., McCormick, P. G. (1989). *Appl. Phys. Lett.* 55:45–46.
16. McCormick, P. G., Wharton, V. N., Schaffer, G. B. (1989). In: Small, W. M., ed. *Physical Chemistry of Powder Metals Production and Processing*. Warrendale, PA: TMS, pp. 19–34.
17. Gaffet, E., Bernard, F., Niepce, J.-C., Charlot, F., Gras, C., Le Caër, G., Guichard, J.-L., Delcroix, P., Mocellin, A., Tillement, O. (1999). *J. Mater. Chem.* 9:305–314.
18. Takacs, L. (2002). *Prog. Mater. Sci.* 47:355–414.
19. Matteazzi, P., Le Caër, G. (1992). *Mater. Sci. Eng.* A156:229–237.
20. Avvakumov, E. G., Senna, M., Kosova, N. V. (1998). *Soft Mechanical Synthesis: A Basis for New Chemical Technologies*. Boston, MA: Kluwer.
21. Special issues of Journal of Materials Synthesis and Processing 8: Issues 3–6, 2000.
22. Xi, S., Qu, X., Ma, M., Zhou, J., Zheng, X., Wang, X. (1998). *J. Alloys Compounds* 268:211–214.
23. Osso, D., Tillement, O., Le Caër, G., Mocellin, A. (1998). *J. Mater. Sci.* 33:3109–3119.
24. Schaffer, G. B., McCormick, P. G. (1991). *Metall. Trans.* A22:3019–3024.
25. Lorrain, N., Chaffron, L., Carry, C. (1999). *Mater. Sci. For.* 312–314:153–158.
26. Ding, J., Tsuzuki, T., McCormick, P. G. (1996). *J. Am. Ceram. Soc.* 79:2956–2958.
27. Du, Y., Li, S., Zhang, K., Lu, K. (1997). *Scripta Mater.* 36:7–14.
28. Welham, N. J. (1998). *Metall. Mater. Trans.* 29B:603–610.
29. McCormick, P. G., Tsuzuki, T. (2002). *Mater. Sci. For.* 386–388:377–386.
30. Welham, N. J. (1998). *Mater. Sci. Eng.* A248:230–237.
31. Welham, N. J. (1999). *J. Mater. Res.* 14:619–627.
32. Welham, N. J. (1999). *Mater. Sci. Technol.* 15:456–458.
33. Tsuzuki, T., McCormick, P. G. (1999). *Nanostructured Mater.* 12:75–78.
34. Tsuzuki, T., McCormick, P. G. (1997). *Appl. Phys.* A65:607–609.

35. Gopalan, S., Singhal, S. C. (2000). *Scripta Mater.* 42:993–996.
36. Tsuzuki, T., McCormick, P. G. (2000). *Mater. Sci. For.* 343–346:383–388.
37. Ding, J., Tsuzuki, T., McCormick, P. G., Street, R. (1996). *J. Phys. D: Appl. Phys.* 29:2365–2369.
38. Takacs, L. (1993). *Nanostructured Mater.* 2:241–249.
39. Ding, J., Tsuzuki, T., McCormick, P. G., Street, R. (1996). *J. Alloys Compounds* 234:L1–L3.
40. Schaffer, G. B., McCormick, P. G. (1990). *Metall. Trans.* 21A:2789–2794.
41. Schaffer, G. B., McCormick, P. G. (1990). *J. Mater. Sci. Lett.* 9:1014–1016.
42. Yang, H., Nguyen, G., McCormick, P. G. (1995). *Scripta Metall. Mater.* 32:681–684.
43. Schaffer, G. B., McCormick, P. G. (1989). *Scripta Metall.* 23:835–838.
44. Ying, D. Y., Zhang, D. L. (2000). *Mater. Sci. Eng.* A286:152–156.
45. Takacs, L., Pardavi-Horvath, M. (1994). In: Shull, R. D., Sanchez, J. M., ed. *Nanophases and Nanocrystalline Structures.* Warrendale, PA: TMS, pp. 135–144.
46. Shen, T. D., Wang, K. Y., Quan, M. X., Wang, J. T. (1991). *Scripta Metall. Mater.* 25:2143–2146.
47. Shen, T. D., Wang, K. Y., Wang, J. T., Quan, M. X. (1992). *Mater. Sci. Eng.* A151:189–195.
48. Pardavi-Horvath, M., Takacs, L. (1995). *Scripta Metall. Mater.* 33:1731–1740.
49. Forrester, J. S., Schaffer, G. B. (1995). *Metall Mater. Trans.* 26A:725–730.
50. Ďurišin, J., Orolínová, M., Ďurišinová, K., Katana, V. (1994). *J. Mater. Sci. Lett.* 13:688–689.
51. Torosyan, A. R., Baghdasaryan, V. S., Martirosyan, V. G., Aloyan, S. G., Balayan, H. G., Navasardyan, H. V., Barseghyan, S. A. (2002). *Mater. Sci. For.* 386–388:229–234.
52. Xi, S., Zhou, J., Wang, X., Zhang, D. (1996). *J. Mater. Sci. Lett.* 15:634–635.
53. Takacs, L. (1996). *Mater. Sci. For.* 225–227:553–558.
54. Baláz, P., Takacs, L., Jiang, J. Z., Soika, V., Luxová, M. (2002). *Mater. Sci. For.* 386–388:257–262.
55. McCormick, P. G., Alonso, T., Liu, Y., Lincoln, F. J., Parker, T. C., Schaffer, G. B. (1992). In: Bautista, R. G., Jackson, N., eds. *Rare Earths—Resources, Science, Technology, and Applications.* Warrendale, PA: TMS, p. 247.
56. Ding, J., Miao, W. F., McCormick, P. G., Street, R. (1995). *Appl. Phys. Lett.* 67:3804–3806.
57. Pardavi-Horvath, M., Takacs, L. (1992). *IEEE Trans. Mag.* 28:3186–3188.
58. Takacs, L. (1992). *Mater. Lett.* 13:119–124.
59. Takacs, L., Pardavi-Horvath, M. (1994). *J. Appl. Phys.* 75:5864–5866.
60. Nasu, T., Tokumitsu, K., Miyazawa, K., Greer, A. L., Suzuki, K. (1999). *Mater. Sci. For.* 312–314:185–190.
61. Tokumitsu, K., Nasu, T., Suzuki, K., Greer, A. L. (1998). *Mater. Sci. For.* 269–272:181–186.
62. Matteazzi, P., Le Caër, G. (1991). *Mater. Sci. Eng.* A149:135–142.
63. Nasu, T., Tokumitsu, K., Konno, T., Suzuki, K. (2000). *Mater. Sci. For.* 343–346:435–440.
64. Warris, C. J., McCormick, P. G. (1997). *Minerals Eng.* 10:1119–1125.
65. Welham, N. J. (1998). *J. Alloys Compounds* 270:228–236.
66. Willis, P. E., Welham, N. J., Kerr, A. (1998). *J. Eur. Ceram. Soc.* 18:701–708.
67. Kerr, A., Welham, N. J., Willis, P. E. (1999). *Nanostructured Mater.* 11:233–239.
68. Welham, N. J., Kerr, A., Willis, P. E. (1999). *J. Am. Ceram. Soc.* 82:2332–2336.
69. Welham, N. J., Willis, P. E. (1998). *Metall. Mater. Trans.* 29B:1077–1083.
70. Welham, N. J., Williams, J. S. (1999). *Metall. Mater. Trans.* 30B:1075–1081.
71. Chen, Y., Hwang, T., Marsh, M., Williams, J. S. (1997). *Metall. Mater. Trans.* 28A:1115–1121.

72. Welham, N. J. (1998). *J. Mater. Sci.* 33:1795–1799.
73. Welham, N. J. (1998). *J. Alloys Compounds* 274:260–265.
74. Welham, N. J., Llewellyn, D. J. (1999). *J. Eur. Ceram. Soc.* 19:2833–2841.
75. Welham, N. J. (1998). *Aust. J. Chem.* 51:947–953.
76. Welham, N. J. (1998). *J. Alloys Compounds* 274:303–307.
77. Tsuzuki, T., Harrison, W. T. A., McCormick, P. G. (1998). *J. Alloys Compounds* 281:146–151.
78. Tsuzuki, T., Pirault, E., McCormick, P. G. (1999). *Nanostructured Mater.* 11:125–131.
79. Wu, N. Q., Su, L. Z., Yuan, M. Y., Wu, J. M., Li, Z. Z. (1998). *Mater. Sci. Eng.* A 257: 357–360.
80. Dodd, A. C., McCormick, P. G. (1999). *Mater. Sci. For.* 312–314:221–226.
81. Tomasi, R., Pallone, E. M. J. A., Botta F., W. J. (1999). *Mater. Sci. For.* 312–314:333–338.
82. Pallone, E. M. J. A., Hanai, D. E., Tomasi, R., Botta F., W. J. (1998). *Mater. Sci. For.* 269–272:289–294.
83. Liu, Y., Alonso, T., Dallimore, M. P., McCormick, P. G. (1994). *Trans. Mater. Res. Soc. Jpn.* 14:961.
84. Baburaj, E. G., Hubert, K. T., Froes, F. H. (1997). *J. Alloys Compounds* 257:146–149.
85. Ding, J., Tsuzuki, T., McCormick, P. G. (1999). *J. Mater. Sci.* 34:5293–5298.
86. Oleszak, D. (2002). *Mater. Sci. For.* 386–388:223–228.
87. Yang, H., McCormick, P. G. (1998). *Metall. Mater. Trans.* 29B:449–455.
88. Shen, T. D., Koch, C. C., Wang, K. Y., Quan, M. X., Wang, J. T. (1997). *J. Mater. Sci.* 32:3835–3839.
89. Alonso, T., Liu, Y., Dallimore, M. P., McCormick, P. G. (1993). *Scripta Metall. Mater.* 29:55–58.
90. Liu, W., McCormick, P. G. (1999). *Nanostructured Mater.* 12:187–190.
91. Liu, Y., Dallimore, M. P., McCormick, P. G., Alonso, T. (1992). *Appl. Phys. Lett.* 60:3186–3187.
92. Tsuzuki, T., Mc Cormick, P. G. (1999). *Mater. Sci. For.* 315–317:586–591.
93. Kersen, Ü. (2002). *Mater. Sci. For.* 386–388:633–638.
94. Yang, H., McCormick, P. G. (1993). *J. Mater. Sci. Lett.* 12:1088–1091.
95. McCormick, P. G., Liu, Y., Yang, H., Nguyen, G., Alonso, T. (1993). In: deBarbadillo, J. J., et al., eds., *Mechanical Alloying for Structural Applications.* Materials Park, OH: ASM International, pp. 165–169.
96. McCormick, P. G., Wharton, V. N., Reyhani, M. M., Schaffer, G. B. (1991). In: Van Aken, D. C., Was, G. S., Ghosh, A. K., eds. *Microcomposites and Nanophase Materials.* Warrendale, PA: TMS, pp. 65–79.
97. Welham, N. J. (1998). *Mater. Sci. Eng.* A255:81–89.
98. Welham, N. J. (1998). *Intermetallics* 6:363–368.
99. Zhang, D. L., Ying, D. Y., Adam, G. (2002). *Mater. Sci. For.* 386–388:287–292.
100. Pallone, E. M. J. A., Tomasi, R., Botta F., W. J. (2000). *Mater. Sci. For.* 343–346:393–398.
101. Lü, L., Lai, M. O., Su, Y., Teo, H. L., Feng, C. F. (2001). *Scipta Mater.* 45:1017–1023.
102. Welham, N. J. (1999). *Minerals Eng.* 12:1213–1224.
103. Kudaka, K., Iizumi, K., Izumi, H., Sasaki, T. (2001). *J. Mater. Sci. Lett.* 20:1619–1622.
104. Yang, H., McCormick, P. G. (1994). *J. Solid State Chem.* 110:136–141.
105. Zhang, D. L., Zhang, Y. J. (1998). *J. Mater. Sci. Lett.* 17:1113–1115.
106. Zhang, D. L., Adam, G. (2002). *Mater. Sci. For.* 386–388:293–298.
107. Mukhopadhyay, D. K., Prisbrey, K. A., Suryanarayana, C., Froes, F. H. (1996). In: Bose, A., Dowding, R. J., eds. *Tungsten and Refractory Metals 3.* Princeton, NJ: Metal Powder Industries Federation, pp. 239–246.
108. Sherif El-Eskandarany, M., Konno, T. J., Omori, M., Ishikuro, M., Takada, K.,

Sumiyama, K., Hirai, T., Suzuki, K. (1996). *J. Japan Soc. Powder Powder Metall.* 43:1368–1373.
109. Sherif El-Eskandarany, M., Omori, M., Ishikuro, M., Konno, T. J., Takada, K., Sumiyama, K., Hirai, T., Suzuki, K. (1996). *Metall. Mater. Trans.* 27A:4210–4213.
110. Tsuzuki, T., Ding, J., McCormick, P. G. (1997). *Physica* B239:378–387.
111. Yang, H., McCormick, P. G. (1993). *J. Solid State. Chem.* 107:258–263.
112. Ryan, Z. Honors thesis, University of Western Australia, Nedlands, 1995; quoted in Ref 64.
113. Ding, J., Tsuzuki, T., McCormick, P. G. (1997). *Nanostructured Mater.* 8:75–81.
114. Takacs, L. (1993). *Mater. Res. Soc. Symp. Proc.* 286:413–418.
115. Moore, J. J., Feng, H. J. *Prog. Mater. Sci.* 39: 243-316.
116. Schaffer, G. B., McCormick, P. G. (1992). *Metall. Trans.* 23A:1285–1290.
117. Aning, A. O., Hong, C., Desu, S. B. (1995). *Mater. Sci. For.* 179–181:207–214.
118. Patankar, S. N., Xiao, S.-Q., Lewandowski, J. J., Heuer, A. H. (1993). *J. Mater. Res.* 8:1311–1316.
119. Bhaduri, S. B., Bhaduri, S. (1999). In: Suryanarayana, C., ed. *Non-equilibrium Processing of Materials.* Oxford, UK: Pergamon, pp. 287–309.
120. Yen, B. K., Aizawa, T., Kihara, J. (1996). *Mater. Sci. Eng.* A220:8–14.
121. Ma, E., Atzmon, M. (1993). *Mater. Chem. Phys.* 39:249–256.
122. Lee, P.-Y., Chen, T.-R., Yang, J.-L., Chin, T. S. (1995). *Mater. Sci. Eng.* A192/193:556–562.
123. Liu, L., Magini, M. (1997). *J. Mater. Res.* 12:2281–2287.
124. Matteazzi, P., Le Caër, G. (1992). *J. Am. Ceram. Soc.* 75:2749–2755.
125. Tschakarov, Chr. G., Gospodinov, G. G., Bontschev, A. (1982). *J. Solid State Chem.* 41:244.
126. Bowden, F. P., Thomas, P. H. (1954). *Proc. R. Soc. London* A223:29–40.
127. Fox, P. G., Soria-Ruiz, J. (1970). *Proc. R. Soc. London* A317:79–91.
128. Urakaev, F., Boldyrev, V. V., Regel, V. R., Pozdnyakov, O. E. (1977). *Kinet. Cat.* 18:350–358.
129. Weichert, R., Schonert, K. (1994). *J. Mech. Phys. Solids* 22:127–133.
130. Ermakov, A. E., Yurchikov, E. E., Barinov, V. V. (1981). *Phys. Metals Metallogr.* 52(6):50–58.
131. Bridgeman, P. (1931). *The Physics of High Pressure.* London: Bell.
132. Larsen, H. A., Drickamer, H. G. (1957). *J. Phys. Chem.* 61:1249–1254.
133. Ivanov, E., Neverov, V., Jitnikov, N., Suppes, V. (1998). *Mater. Lett.* 7:57–60.
134. Ivanov, E., Golubkova, G. V., Grigorieva, T. F. (1990). *Reacti. Solids* 8:73–76.
135. Aizawa, T., Kuji, T., Nakano, H. (1999). *J. Alloys Compounds* 291:248–253.
136. Segal, V. M. (1995). *Mater. Sci. Eng.* A197:157–164.
137. Lowe, T. C., Valiev, R. Z., eds. (2000). *Investigations and Applications of Severe Plastic Deformation. NATO Science series, 3. High Technology.* Vol. 80. Dordrecht, The Netherlands: Kluwer.
138. Valiev, R. Z., Islamgaliev, R. K., Alexandrov, I. V. (2000). *Prog. Mater. Sci.* 45:103–189.
139. Bokhonov, B., Konstanchuk, I., Ivanov, E., Boldyrev, V. V. (1992). *J. Alloys Compounds* 187:207–214.
140. Butyagin, P. Y. (1999). *Colloid J.* 61:537–544.
141. Fujiwara, Y., Isobe, T., Senna, M., Tanaka, J. (1999). *J. Phys. Chem.* 103:9842–9846.
142. Fujiwara, H., Minamoto, Y. (2000). *Polym. Bull.* 45(2):137–144.
143. Matteazzi, P., Basset, D., Miani, F., Le Caër, G. (1993). *Nanostructured Mater.* 2:217–229.
144. Matteazzi, P., Le Caër, G. (1991). *J. Am. Ceram. Soc.* 74:1382–1390.
145. Matteazzi, P., Le Caër, G. (1992). *J. Alloys. Compounds* 187:305–315.

146. Pardavi-Horvath, M., Takacs, L. (1995). *Scripta Metall. Mater.* 33:1731–1740.
147. Suryanarayana, C. (1995). *Int. Mater. Rev.* 40:41–64.
148. Milham, C. D. (1993). In: Henein, H., Oki, T., eds. *Processing Materials for Properties.* Warrendale, PA: TMS, pp. 449–452.
149. Corrias, A., Ennas, G., Musinu, A., Paschina, G., Zedda, D. (1997). *J. Mater. Res.* 12: 2767–2772.
150. Chen, Y., Ninham, B. W., Ogarev, V. (1995). *Scripta Metall. Mater.* 32:19–22.
151. Millet, P., Calka, A., Ninham, B. W. (1994). *J. Mater. Sci. Lett.* 13:1428–1429.
152. Chen, Y., Hwang, T., Marsh, M., Williams, J. S. (1997). *Metall. Mater. Trans.* 28A:1115–1121.

15

Powder Contamination

15.1 INTRODUCTION

A major concern in the processing of metal powders by mechanical alloying/milling (MA/MM) is the nature and amount of impurities that get incorporated into the powder and contaminate it. The small size of the powder particles and consequent availability of large surface area, formation of fresh surfaces during milling, and wear and tear of the milling tools all contribute to contamination of the powder. These factors are in addition to the purity of the starting raw materials and the milling conditions employed (especially the milling atmosphere). Thus, it appears that powder contamination is an inherent drawback of the technique and that the milled powder will be always contaminated with impurities unless special precautions are taken to avoid or minimize it.

As mentioned earlier, MA involves loading the constituent powder particles, grinding medium, and process control agent (PCA, at least in some instances) in a milling container and subjecting this mix to intense mechanical action. In the early stages of MA, the metal powder generally coats the surface of the grinding medium and the inner walls of the container. This was expected to prevent contamination of the powder from the milling tools and the container. Therefore, no attention was paid, especially in the early years of MA, to the problem of powder contamination. However, when different results were reported by various groups of researchers on the same alloy system, and apparently under identical milling conditions, it was recognized that powder contamination to different levels could be the reason for the differences in the obtained results. This problem is now recognized to be ubiquitous and is frequently reported in many investigations, especially when reactive metals such as titanium and zirconium are milled. The magnitude of powder contamination appears to depend on the time of milling, intensity of milling, atmosphere in which the powder is milled, nature and size of the milling medium, and difference in the strength/hardness of the powder and the milling medium.

About 1–4 wt% Fe has been found to be normally present in most of the powders milled with steel grinding medium, but amounts as large as 20 at% Fe in a W-C mixture milled for 310 h and 33 at% Fe in pure W milled for 50 h in a SPEX mill were also reported [1]. In fact, amounts as large as 60 at% Fe were reported during milling of a W-5 wt% Ni alloy for 60 h in a SPEX mill [2]. These are very high levels of contamination. Similarly, large amounts of oxygen (up to 44.8 at%) [3] and nitrogen (up to 33 at%) have also been reported in Al-6 at%Ti powders milled for 1300 h in a low-energy ball mill and in a Ti-5 at%Al powder milled in a SPEX mill for 30 h, respectively [4]. Experimentally determined levels of powder contamination during milling of different types of powder are summarized in Table 15.1. The type of mill, ball-to-powder weight ratio (BPR), and milling time employed are also listed, since these parameters determine the magnitude of contamination. The major impurities listed are oxygen and nitrogen, mainly from the atmosphere and/or PCAs used, and iron from the steel grinding medium and the steel container. The levels of carbon, chromium, and other metallic impurities are also listed in some cases, whenever reported.

The presence of contamination in the powders is suspected/inferred when (1) an unexpected phase appears to be present in the milled powder (of course, this could be a genuine metastable phase) or (2) when the lattice parameter(s) of the solid solution or intermetallic phases are substantially different from the values predicted or reported in the literature. The difference in the lattice parameters could be due to formation of supersaturated solid solutions or increased homogeneity ranges of intermetallic phases also. However, since the presence of impurities can also cause similar effects, one should be aware of this possibility.

The presence of impurities in the powder can be easily detected if the amount of the contaminant or contaminated product is reasonably large. In that case, shifts in peak positions and/or appearance of new diffraction peaks in the X-ray diffraction (XRD) patterns occur indicating either a change in lattice spacing or formation of a new phase. On the other hand, if the amount of impurities is small, specialized techniques may be necessary for their detection. Furthermore, techniques like X-ray photoelectron spectroscopy (XPS) can give additional information about the bonding state of the impurity atoms [50].

15.2 SOURCES OF CONTAMINATION

Contamination of milled metal powders can be traced to (1) chemical purity of the starting powders, (2) milling atmosphere, (3) milling equipment (milling container and grinding medium), and (4) process control agents added to the powders during milling. Contamination from source (1) can be either substitutional or interstitial in nature, whereas contamination from source (2) is essentially interstitial in nature and that from (3) is mainly substitutional, even though carbon from the steel milling equipment can be an interstitial impurity. Contamination from the PCA essentially leads to interstitial contamination, since the PCAs used are mostly organic compounds containing carbon, oxygen, and nitrogen (see Chapter 5). Irrespective of whether the impurities are substitutional or interstitial in nature, their amount increases with milling time and with increasing BPR, and reaches a saturation value. Since the powder particles become finer with milling time, their surface area increases and consequently the amount of contamination increases. Similarly, with increasing BPR, the wear and tear of the grinding medium increases and consequently contamination also is higher. However, when particle refinement reaches the limiting value and a steady-state condition is achieved, the level of contamination does not increase any further.

The main difficulty in the MA of reactive metals has been the absorption of oxygen and nitrogen from the continuous exposure of fresh metal surfaces to the atmosphere in the milling container. Even though presence of interstitial impurities such as oxygen and nitrogen is generally deleterious to most metals, their presence beyond certain limits is responsible for the formation of undesirable phases and consequent degradation of their mechanical properties. Therefore, maximum impurity levels are generally specified for acceptable microstructural and mechanical properties. For example, the suggested maximal level of combined nitrogen and oxygen in Ti-6 wt% Al-4 wt% V is 2500 ppm and that in TiAl-based alloys is only 1000 ppm. Substantial amounts of nitrogen and oxygen (the amount of nitrogen is much more than oxygen) are picked up during milling of titanium and zirconium alloys, and the presence of these impurities has been shown to lead to a change in the constitution of the alloys [51–53].

Formation of an amorphous phase in Ti-Al alloys and a crystalline phase with a face-centered cubic (fcc) structure in powders milled beyond the formation of the amorphous phase have been attributed to the presence of large quantities of nitrogen in these alloys [54]. Again, milling of a Cu-50 at% Cr powder blend in a pure argon atmosphere was shown to result in the complete formation of a bcc solid solution. On the other hand, if argon (containing air) was used, then the bcc solid solution formation did not go to completion [55]. Similarly, it was shown that milling of the Nb_3Al intermetallic in a SPEX mill produced the expected bcc solid solution phase after milling the powder for 3 h. But when milling was conducted in an improperly sealed vial, an amorphous phase was produced after 10 h, which on subsequent heat treatment transformed to the crystalline niobium nitride phase. This suggests that the amorphous phase in this system had formed because of nitrogen contamination of the powder during milling [56].

15.2.1 Starting Powders

The starting powders for MA are mostly purchased from vendors, who normally store them in appropriate containers and sometimes under protective atmospheres. However, since the powders have a large surface area, some amount of nitrogen, oxygen, and water vapor gets adsorbed onto their surfaces. This problem becomes more serious with the more reactive powders and also with finer powders, since the surface area is now much larger and therefore higher amounts of contaminants are adsorbed. Great care must be taken to choose the proper type of powder particles; coarser powder particles are to be preferred if the milling conditions permit it. It is useful to remember that the impurity content of prealloyed powders is usually less than that in blended elemental powders. The interstitial impurity levels increase further when these powders are subjected to MA/MM. Thus, the total impurity content in the milled

Table 15.1 Powder Contamination During Mechanical Alloying[a]

System	Mill	Milling medium	BPR	Milling time (h)	Contamination level (at%)			Ref.
					Oxygen	Nitrogen	Iron	
Al	SPEX 8000	Hardened steel	4:1	32	—	—	5	5
Al-20Cu-15Fe	Planetary ball mill QM-1SP	Hardened steel	10:1	300	—	—	1.2	6
Al-5Fe-5Nd	High-speed shaker mill	Steel	—	240	—	—	9	7
Al-6Ti	Attritor	Steel	—	1300	44.8	8.25	0.52	3
Al-12Ti	Attritor	Steel	—	400	27.4	22.6	0.07	3
Al-12Ti	Attritor	Steel	—	1300	36.5	16.3	0.11	3
Al-Ti	Planetary ball mill	Hardened steel	60:1	120	—	—	1.5 and 0.3 Cr	8
B	Fritsch P5	SS	10:1	15	—	—	6	9
C	Retsch PM 4000 planetary ball mill	WC	—	—	—	—	3 (W,Co)	10
Co-B	Fritsch P5	SS	10:1	—	—	—	2	9
Cu	SPEX 8000				—	—	≤1	11
Cu	Planetary ball mill G5	Tempered steel	—	240	—	—	14.2 Fe, 1.0 Cr	12
Cu-1 vol% Al_2O_3	HE mill						1.2	13
Cu-43.1Fe	Planetary ball mill G5	Tempered steel	—	240	—	—	21.9 Fe, ≤2.8 Cr	12
$Cu_{44}Nb_{42}Sn_{14}$	Lab ball mill	WC-Co	50:1	24	13.5	2.4		14
Cu-Ta	Fritsch P5	SS	4:1	30	—	—	4-5 and 1-2 Cr	15

Cu-50Zr	SPEX 8000	440C SS	10:1	13	3.3	—	0.25	16
CuZr	SPEX 8000	WC or 440C SS	10:1	13 (WC vial) (steel)	2.3	—	— 4.6	16
Fe-50Co	SPEX 8000	Steel		49	—	—	2 Cr	17
Fe-6Nb-12B	—	Hardened steel	5:1	220 (cyclohexane as PCA)	—	—	11 wt% C	18
Ir	SPEX 8000	Hardened steel	4:1	8	—	—	5	5
Mo	Netzsch 1S attritor	Steel	100:1		33 (180 h)	28 (320 h)	15 (180 h)	19
Mo-66.7Si	Planetary ball mill G5	Tempered steel	—	48	—	—	10.5	20
Nb-Al	Planetary ball mill	—	—	60	—	—	3	21
Nb-92Be	SPEX 8004	WC	7:1	6	5-6			22
Nb-25Ge	SPEX 8000	—	—	13	10	—	1.5	23
Ni	SPEX 8000				—	—	13	11
Ni-25Al	SPEX 8000	—	—	40	—	—	4	24
Ni-Al-X (X = Fe, Cr)	Fritsch P5	WC	10:1		—	—	—	25
		Hardened steel			—	—	18	
Ni-40Nb	Fritsch P6		10:1	100			2–3	26
Ni-40Nb	SPEX 8000	52100 steel	3:1	14 (air)	13.4	2.1	2.3	27
Ni-40Nb	SPEX 8000	52100 steel	3:1	11 (helium)	2.0	1.9	5.1	27
Ni-80.5W				24	—	—	6.4	28
Ni-30Zr	Vibratory ball mill	440C SS	7–81:1	32	—	—	3.7	29
Ni-Zr	Fritsch P7	Hardened steel	—	20	—	—	2	30
Am-$Ni_{68}Zr_{32}$ + Zr	Fritsch P7	Hardened steel	—	20	—	—	5	30

Table 15.1 Continued

System	Mill	Milling medium	BPR	Milling time (h)	Contamination level (at%)			Ref.
					Oxygen	Nitrogen	Iron	
NiZr	Fritsch P5	SS	30:1	10 (nitrogen) 100 (argon)	—	—	12 6	31
Pd-16Zr	SPEX 8000	—	—	24	1.1	0.5	—	32
Pd-23Zr	SPEX 8000	—	—	24	2.4	0.2	—	32
Ru	SPEX 8000	Hardened steel	—	8 (argon) 16 (argon) 32 (argon)	—	—	3 5 10	33
Ru	SPEX 8000	Hardened steel	4:1	8 (methane) 16 (methane) 32 (methane)	—	—	3 8 22	34
Ru-50 Al	SPEX 8000	Hardened steel	10:1	35 50			15 35	35
Ru-20 C	SPEX 8000	Hardened steel	—	8 (argon) 16 (argon) 32 (argon)	—	—	0.3 1.5 18	34
Si	Uni-Ball mill	—	—	300 (nitrogen)	16.63[b]	—	0.95[b]	36
Si_3N_4	Planetary ball mill Retsch PM4	WC-Co	—	195	18.8	—	13.4 (W,Co); 11.2 C	37
Ta-30Al	Rod mill Ball mill	SS	36:1	400 400	—	—	5 16	38
Ti	Vibrational ball mill	SS	45–90:1	(nitrogen)	—	—	8.8 and 3.0 Cr	39
Ti-5Al	SPEX 8000	Hardened steel	10:1	30	21	33	—	4
Ti-50Al	SPEX 8000	Hardened steel	10:1	70	10	26	—	4
Ti-60Al	Fritsch P5	—	10:1	70	6.7	7.4	—	40

Ti-20Cu-24Ni-4Si-2B	SPEX 8000	Hardened steel	5:1	24	2.9	—	0.35	41
Ti-37.5Si	SPEX 8000	Hardened steel	5:1	24	4.4	1.8	—	42
V-C	Planetary ball mill	Hardened steel	—	70	—	—	8.6 (high energy) 0.6 (low energy)	43
W	SPEX mill	440C SS	4:1	50	—	—	33	44
W-C	Ball mill	Hardened SS	30:1	310	—	—	20	1
W-24.3Cu	Fritsch P5	Steel	—	140–180	—	—	50.5	45
W-49.1Cu	Fritsch P5	Steel	—	140–180	—	—	47.2	45
W-5Ni	SPEX 8000	Hardened steel	11:1	60	—	—	60	2
W-30Ni	SPEX 8000	Hardened steel	11:1	60	—	—	50	2
W-40Ni	SPEX 8000	Hardened steel	11:1	60	—	—	40	2
Zr	Uni-Ball mill	—	—	50 (ammonia) 250 (nitrogen)	—	—	10 1.5	46
Zr-50Cu	Horizontal ball mill	SS	20:1	400	3	—	—	47
Zr-30Cu-10Al-5Ni	Retsch PM 4000 planetary ball mill	Hardened steel	14:1	100–200	1.5–2	—	0.5	48
Zr-30Ni	Fritsch P5	SS	—	35	—	—	9.9	49
Zr-70Ni	Fritsch P5	SS	—	120	—	—	13.8	49

[a]All alloy compositions are expressed in atomic percent, unless indicated otherwise.
[b]After MA in nitrogen atmosphere plus annealing for 10 h at 800°C.

powder will be the sum of the initial impurity content and that which was acquired during the milling operation. If the total impurity content is below the solid solubility level in the metal powder that is being milled, then one would notice a change in the lattice parameter of the metal as evidenced by a change in the position of the metal diffraction peaks in the XRD pattern. If the impurity content is sufficiently large, then peaks corresponding to a new interstitial phase may appear. For example, the presence of a cubic Mo_2N phase was reported in molybdenum powder mechanically alloyed in a nitrogen atmosphere [19,57] and the cubic TiN phase in Ti-Al alloys mechanically alloyed in nominally pure argon atmosphere [52,54].

15.2.2 Milling Atmosphere

The milling atmosphere is perhaps the most important factor that contributes to the contamination of mechanically alloyed powders. The importance of maintaining a clean and noncontaminating environment and gas-tight seals for the container lid during milling of reactive metal powders cannot be overemphasized. (This contamination, which is unwanted, should be distinguished from intentional milling of metal powders in reactive atmospheres to synthesize nitrides, carbides, oxides, and so forth). The powders are normally loaded into and unloaded from the milling containers in a glove box that is maintained under a protective argon atmosphere. Care must be taken to ensure that the glove box is purged with argon gas a few times prior to loading the powder into the container. It is also essential that the powder not be subsequently exposed to the ambient atmosphere.

A general practice is to mill the powders in an inert atmosphere, such as argon, to minimize powder contamination. It has been reported that even repeated flushing with argon gas does not remove oxygen and nitrogen adsorbed on the powder and mill tool surfaces, and that some trace contaminant species remain in even high-purity argon [58]. It is therefore the overall nitrogen and oxygen contents of the atmosphere sealed within the container that are relevant.

The integrity of the milling container seal preventing access of external atmosphere into the powder being milled is also very important. It has been observed that if the container is not properly sealed, the atmosphere surrounding the container, usually air (containing essentially nitrogen and oxygen), leaks into the container and contaminates the powder. Thus, when reactive metals like titanium and zirconium are milled in improperly sealed containers, the powders are contaminated with nitrogen and oxygen. In most cases, the nitrogen pickup has been found to be much more than that of oxygen. The formation of a cubic phase at long times of milling titanium alloys has been attributed to the formation of titanium nitride having a cubic structure [54]. In fact, the high levels of oxygen and nitrogen reported in powders milled under argon atmosphere could only be explained by leakage of air through the seals into the container [59]. Thus, the integrity of the milling container seals seems to be most crucial in minimizing or avoiding contamination of metal powders and in producing clean material.

Pickup of impurities during milling would reduce the pressure within the container allowing the outside atmosphere to continuously leak into the container through an ineffective seal. Thus, it has been noted experimentally that difficulty in opening the container lid, due to the vacuum present inside, is an indication that contamination of the powder is minimal.

The milling containers are sometimes opened before the milling operation is complete so that one could follow the different stages of alloying and also follow the changes in the lattice parameter(s), crystallite size, and lattice strain as a function of milling time. Even though small quantities of the powders are generally taken out inside a glove box, frequent opening of the milling container could contaminate the powder. In fact, it has been reported [19] that the uptake of iron and oxygen impurities in the powder correlated well with the number of interruptions (opening of the container lid) of the milling process. Furthermore, the amount of iron impurity present in the powder appears to be coupled with that of oxygen. That is, the amount of oxygen, and consequently iron, is higher when the powder is milled in an oxygen atmosphere than when it is milled in an argon atmosphere (Fig. 15.1) [19]. On the other hand, uptake of nitrogen during milling seems to be directly related to milling time and not dependent on the number of interruptions.

15.2.3 Milling Equipment

During MA, the powder particles get trapped between the grinding media and undergo severe plastic deformation; fresh surfaces are created due to fracture of the powder particles. In addition, collisions occur between the grinding medium and the vial, and also among the grinding balls. All these effects cause wear and tear of the grinding

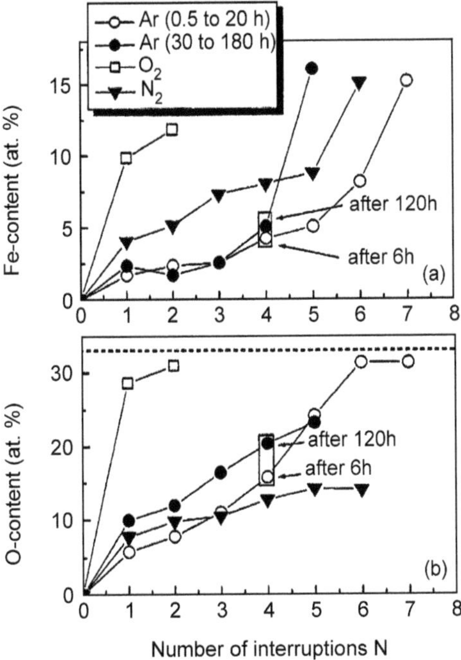

Figure 15.1 (a) Iron content of the molybdenum samples milled under different atmospheres, as indicated in the figure, vs. the number of interruptions, *N*. (b) The corresponding dependence of the oxygen content on the number of interruptions. The horizontal dashed line indicates the approximate, apparent asymptote for oxygen uptake (about 33 at%).

medium and the vial, resulting in the incorporation of these impurities into the powder. It is essentially the metallic impurities, e.g., Fe from the steel containers and W and Co from the WC containers, that get incorporated into the powder and contaminate it. The extent of contamination increases with increasing milling energy, use of higher BPR, higher speed of milling, and so forth. Even though it is well known that adsorption of elements takes place faster at higher temperatures, there has not been any report on the levels of contamination of powders milled at different temperatures.

Figure 15.2 shows the increase in iron contamination when blended elemental Al-Ta powders corresponding to the composition of Al-50 at% Ta were milled in a planetary-type ball mill. It may be noted that the iron contamination increases with milling time. Furthermore, it may be noted that the level of contamination increases with increasing BPR. Whereas the maximal iron content was only about <1 at% when milled at a BPR of 36:1, it was as much as 5 at% when milled at a BPR of 108:1 [60].

The extent of contamination could also be different depending on whether one starts with prealloyed powders (mechanically disordered) or blended elemental powders (mechanically alloyed). It was reported [61] that the amount of iron in the milled Al-50 at% Ta powder was about 1 at% when mechanically disordered (prealloyed powders), and only about 0.5 at% when mechanically alloyed (blended elemental powders). In both the cases, the rod mill was used and a BPR of 36:1 was used. It is normally expected that MA introduces more contamination due to the generation fresh surfaces and consequently large surface area of both the metal powders. But surprisingly, the rate and amount of contamination were reported to be higher for MA than for mechanical disordering (MD) (Fig. 15.3).

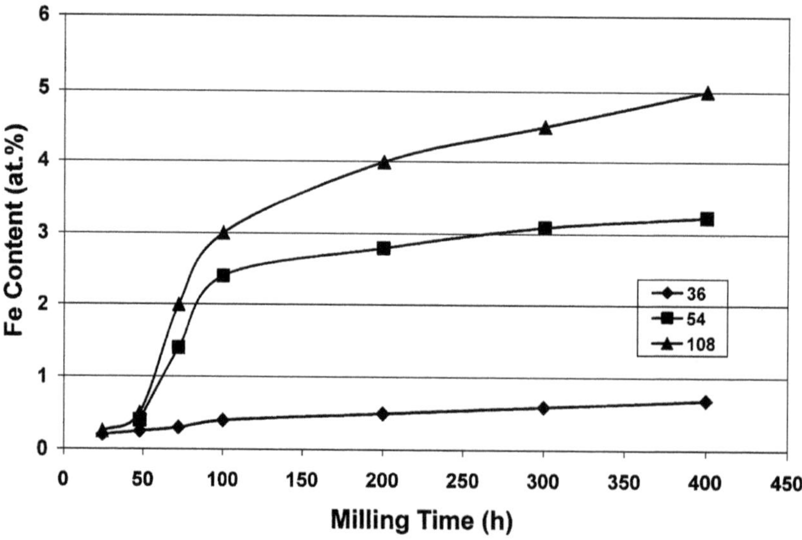

Figure 15.2 Amount of iron impurity in the mechanically alloyed Al-50 at% Ta powder as a function of milling time and BPR (36:1, 54:1, and 108:1) [60]. It may be noted that the iron content increases with both milling time and BPR.

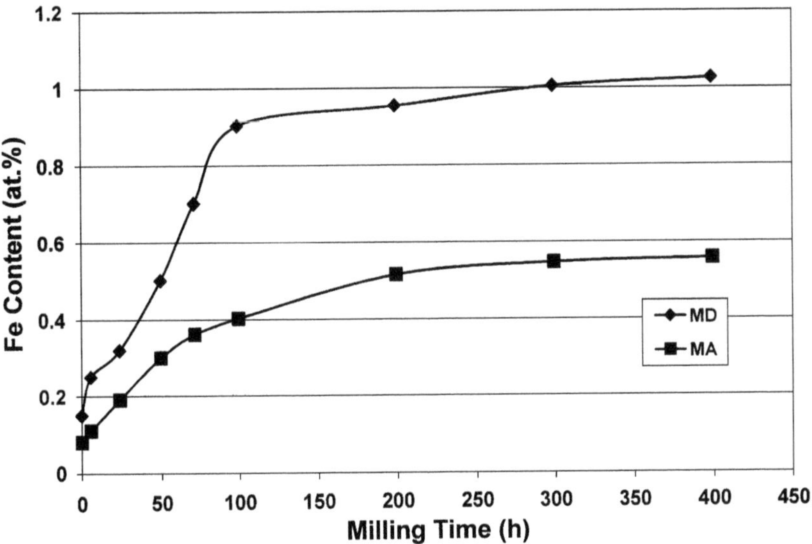

Figure 15.3 The level of iron contamination in Al-50 at% Ta powder as a function of milling time and nature of the starting powder. The contamination level is higher when mechanically disordered than when mechanically alloyed [61].

The type of mill also seems to be important in determining the magnitude of contamination of the milled powder. It has been reported [38] that during MA of Al-70 at% Ta powders, the magnitude of iron contamination is higher in the ball mill than in the rod mill (Fig. 15.4). Another point of interest is that the level of contamination decreases in the second and third runs. In this case, the powder is milled for the first time, and for the second and subsequent runs, the same grinding media and the vials are used. Since the alloy powder is coated onto the grinding medium, thus preventing wear and tear, the amount of contamination is lower in later runs. Comparing the efficiency of obtaining clean powders using a SPEX and the Zoz Simoloyer mills, it was noted that the iron content in the milled Ag-70 at% Cu powder was less in the Zoz mill (Fig. 15.5) [62].

15.3 ELIMINATION/MINIMIZATION OF CONTAMINATION

Several attempts have been made in recent years to minimize powder contamination during MA. An important point to remember is that if the starting powders are highly pure, then the final product is going to be clean and pure, with minimal contamination from other sources, provided that proper precautions were taken during milling. It is also important to remember that all applications do not require ultraclean powders and that the purity desired depends on the specific application. Nonetheless, it is a good practice to minimize powder contamination at every stage and to ensure that additional impurities are not picked up subsequently. We shall now discuss the different ways of minimizing contamination by controlling the milling tools and/or the atmosphere.

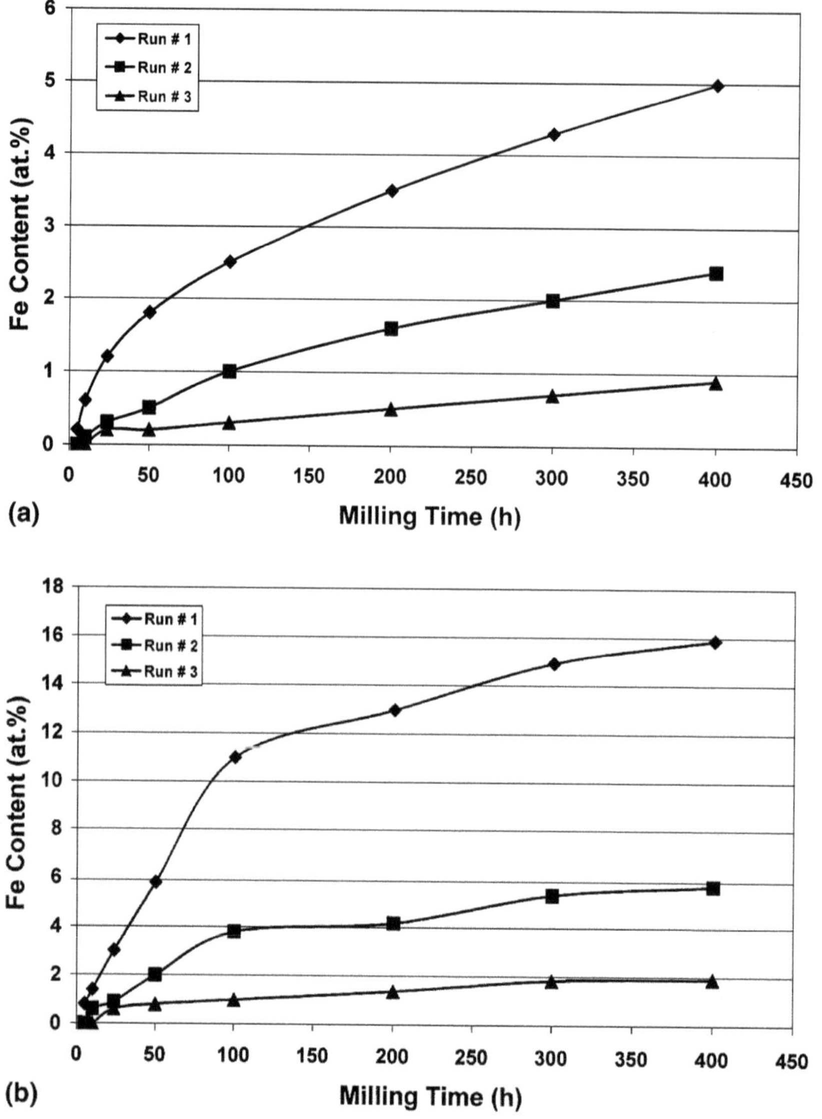

Figure 15.4 Level of iron contamination in the Al-70 at% Ta powder mechanically alloyed using (a) rod mill and (b) ball mill. Note that the contamination level is less in the rod mill than in the ball mill and that it is reduced in the second and third runs of the milling process in both the mills [38].

15.3.1 Milling Equipment

One way to minimize contamination from the grinding medium and the milling container is to use the same material for the container and the grinding medium as the powder being milled. Thus, one could use copper balls and copper container for milling copper and copper alloy powders [63]. In this case there will also be wear and tear of the grinding medium and this gets incorporated into the powder. Thus, even

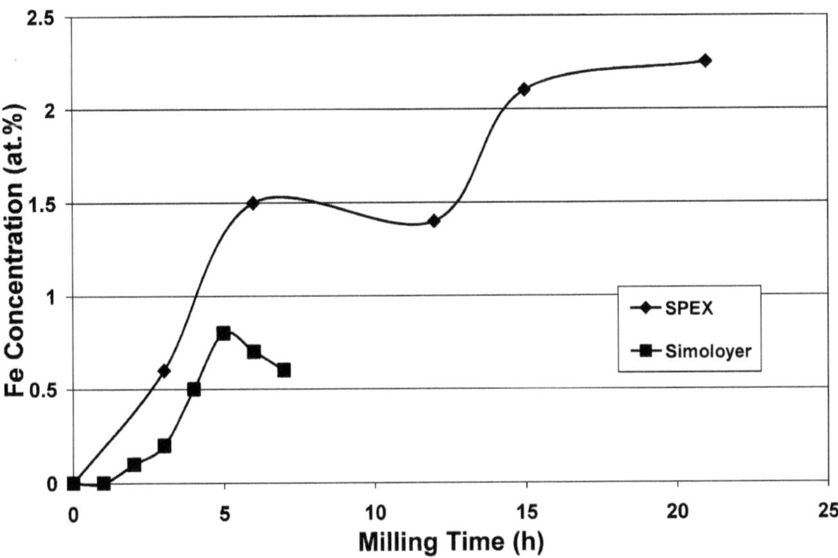

Figure 15.5 Comparison of iron contamination when the Ag-70 at% Cu powder was milled in the SPEX and Zoz Simoloyer mills. Note that the contamination is less in the Zoz Simoloyer mill [62].

though there is no "contamination," the chemistry of the final powder will be different from the starting powder; the metallic content in the final powder (from the container and balls) would be higher than in the initial powder. This change in composition may be tolerated if the desired phase has sufficient compositional homogeneity range. If it does not, the excess metal uptake must be compensated for. This can be done, if one has prior knowledge of the extent of increase of the metallic content in the final powder or by conducting some preliminary experiments.

The above solution is possible in some cases, but in many cases it is difficult due to the nonavailability of special milling tools (grinding media and containers). The problem is becoming more and more complex since the technique of MA is now being applied to a variety of materials such as metals, alloys, ceramics, composites [64], and polymers [65,66], and it is almost impossible to get containers and grinding media of all these types of materials. If a container of the same material to be milled is not available, then a thin adherent coating on the internal surface of the container (and also on the surface of the grinding medium) with the material to be milled will minimize contamination. Thus, as shown in Figure 15.4, the extent of iron contamination during milling of blended elemental Ta-Al powders is less in the second run and much less in the third run [38]. The idea here is to mill the powder once, allowing the powder to be coated onto the grinding medium and the inner walls of the container. The milled loose powder is then discarded and a fresh batch of powder is milled, but with the old grinding media and container. If this operation is repeated several times, then contamination due to the grinding medium and the vessel is likely to be minimal, provided that the powder does not cake up at the bottom of the vial and peel off.

In general, a simple rule for minimizing contamination from the milling container and the grinding medium is that the container and grinding medium should be harder and stronger than the powder being milled. Thus, in one study the reason for the high amount of iron contamination during milling of tungsten and tungsten alloy powders is that a steel container and steel milling medium, which are much softer than tungsten, were used [2]. However, a disadvantage of using a very hard grinding medium, such as tungsten carbide or Al_2O_3, during milling of very soft metal powders is that these materials are brittle and so tend to break up or chip out, getting incorporated into the powder and contaminating it. In fact, the amount incorporated into the powder can be so high that clear diffraction peaks from the contaminant could be observed in the XRD patterns of the milled powder [29].

Use of vacuum or an inert gas is desirable to prevent or minimize the powder contamination. It has been reported that a Teflon seal could be used to prevent air from leaking from outside into the container [67].

15.3.2 Milling Atmosphere

Attempts have been made in the recent past to improve the container seal integrity to prevent the outside atmosphere from leaking inside. Goodwin and Ward-Close [68], using two different types of milling atmospheres, compared the powder contamination levels during milling of both prealloyed and blended elemental powders. In the first atmosphere, referred to as AT1, a steel milling container was loaded in a glove box containing research grade argon (99.995% pure) and the container was sealed using an "O" ring of circular cross-section. In the second atmosphere, AT2, higher purity argon (guaranteed 99.998%), which is subsequently passed through moisture and oxygen filter towers to further purify it, was used. Furthermore, the gasket used to obtain a more leak-tight seal was a flat neoprene one. The contamination levels obtained using these two types of atmosphere are listed in Table 15.2. It has been shown that by careful control of the purity of the atmosphere and improvements of the seal quality, the interstitial contaminants in the titanium alloy could be reduced to as low as 100 ppm of oxygen and 35 ppm of nitrogen [53]. With additional control of the atmosphere this could be reduced even further. This process, which was demonstrated on a laboratory scale, was subsequently scaled up. It was reported [58] that a large-scale clean milling facility was developed at the Defence Evaluation and Research Agency, Farnborough, United Kingdom, which utilizes a proprietary design of a mill capable

Table 15.2 Oxygen and Nitrogen Contents of Ti-24 Al-11 Nb (at%) Powders Milled for 24 h in a SPEX Mill with a BPR of 4:1 and Using Two Different Types of Atmosphere[a]

	Oxygen content (wt%)		Nitrogen content (wt%)	
Powder (at%)	AT1	AT2	AT1	AT2
Prealloyed Ti-24Al-11Nb	3.6	0.10	6.8	0.015
Blended elemental Ti-24Al-11Nb	4.8	0.48	7.6	0.035

[a] No PCA was used during milling [68].

of producing kilogram quantities of clean mechanically alloyed material. The oxygen contamination levels were reported to be equal to, or even less than, those demonstrated on the laboratory scale. For example, the combined oxygen and nitrogen level in clean processed Ti-48Al-2Mn-2Nb (at%) alloy was about 950 ppm and exhibited a tensile strength of 639 MPa. A similarly processed Ti-6 wt% Al-4 wt% V alloy had a tensile strength of 1070 MPa and 10.2% ductility, suggesting that MA processing could produce a clean material without adversely affecting the mechanical properties of the alloy. In addition, iron pickup during the MA process was eliminated by careful substitution of compatible materials in the mill interior.

Contamination levels obtained by milling prealloyed γ-Ti-48Al-2W (at%) powders in different atmospheres were also evaluated [59]. This was done by loading the powder in the particular atmosphere, milling it for 4 h, opening the lid, and resealing the container lid in the same atmosphere. The milling and resealing operations were done repeatedly to achieve a total milling time of 20 h. From a knowledge of the milling container volume, the maximal expected amount of pickup of nitrogen and oxygen from the pure oxygen, nitrogen, or air atmospheres were calculated to be 2.99 wt% oxygen, 2.69 wt% nitrogen, and 0.69 wt% oxygen/1.97 wt% nitrogen, respectively. These values should be compared with the actual values observed, which are listed in Table 15.3. It is to be noted that the amount of nitrogen is very high when the powder was milled in a pure nitrogen or air atmosphere. This result is not surprising. As expected, the nitrogen content in the powder was very low when it was milled in an oxygen atmosphere. However, it is most surprising that the oxygen content in the powder was higher when it was milled both in air or nitrogen atmosphere than when milled in a "pure" oxygen atmosphere. This result can only be rationalized based on the assumption that air was continuously entering the milling container during milling due to an improper seal. During milling the powder would have reacted with nitrogen and oxygen inside the container and created a vacuum. The vacuum in the milling container was higher the more the oxygen and nitrogen in the initial atmosphere were consumed during milling. Air is then drawn from outside at a rate dictated by the magnitude of the pressure difference across the seal as well as the integrity of the vial seal.

One of the most efficient methods to minimize or avoid contamination from the milling atmosphere is to place the mill inside a chamber that has been evacuated and then filled with high-purity argon gas. Since the whole chamber is maintained under

Table 15.3 Chemical Analysis of the Prealloyed γ-Ti-48Al-2W (at%) Powder Milled in Different Atmospheres for a Total of 20 h in a SPEX Mill with a BPR of 10:1 [59]

	Maximal expected value (wt%)		Measured values (wt%)	
Atmosphere	Oxygen	Nitrogen	Oxygen	Nitrogen
Starting powder	—	—	0.093	<0.001
Air	0.69	1.97	2.08	4.66
Oxygen	2.99	—	1.49	0.63
Nitrogen	—	2.61	4.72	9.87
Argon	—	—	0.80	1.52

argon gas atmosphere (continuously purified to keep oxygen and water vapor below 1 ppm each) and the container is inside the mill, which is inside the chamber, contamination from the atmosphere is minimal. Thus, Klassen et al. [69] reported that the oxygen content in their milled $Ti_{51}Al_{49}$ powder was only 0.30 wt% after milling for 100 h in a Fritsch P5 mill that was enclosed inside a glove box maintained under argon atmosphere. Iron contamination from the wear and tear of the milling tools, of course, could not be avoided. They reported about 1.4 at% Fe, but only 0.3 at% oxygen and 0.12 at% nitrogen, in a Ti-45Al-2.4Si (at%) powder processed in this way [70].

15.3.3 Process Control Agent

Contamination from PCAs is perhaps the most ubiquitous. Since most of the PCAs used are organic compounds, which have low melting and boiling points, they decompose during milling due to the heat generated. The decomposition products consisting of carbon, oxygen, nitrogen, and hydrogen react with the metal atoms and form undesirable carbides, oxides, nitrides, and so forth. For example, use of Nopcowax 22-DS during milling of aluminum powder in a SPEX mill has resulted in the formation of oxide and carbide dispersoids [71]. Formation of carbide and hydride phases was reported during the milling of Ti-Al powders using hexane as a PCA [72]. Formation of Fe_3O_4 and Fe_3C contaminants was reported during milling of Cu-Fe powders with ethanol as a PCA [12]. Similarly, other PCAs can introduce different types and amounts of impurities into the milled material. Mishurda et al. [73] studied the effect of several PCA's on Fe and interstitial contamination during milling of prealloyed Ti-48Al-2W (at%) powder in a SPEX mill. The results of contamination, summarized in Table 15.4, show that boric acid and borax were quite effective in reducing the iron contamination from 1.04 wt% (when no PCA was used) to 0.44 wt% and 0.26 wt%, respectively. Use of a vial that was seasoned twice (meaning that the powder was milled in the container using a particular grinding medium, during which

Table 15.4 Chemical Analysis (wt%) of Prealloyed Ti-48Al-2W (at%) Powder Milled for 24 h in a SPEX Mill Using Different PCA's [73]

		Chemical composition (wt%)			
Condition	PCA	Iron	Carbon	Oxygen	Nitrogen
Gas atomized (as received)	—	0.08	0.02	0.084	0.006
Milled	None	1.04	0.08	1.81	5.09
	Seasoned[a]	0.46	0.03	1.60	4.82
Milled	Stearic acid	0.78	0.78	1.79	4.82
	Seasoned	0.43	0.85	1.74	4.06
Milled	Boric acid	0.44	0.08	1.28	0.41
	Seasoned	0.06	0.03	1.32	0.71
Milled	Borax	0.26	0.02	1.35	1.20
	Seasoned	0.09	0.03	1.46	1.47

[a] "Seasoned" means that the powder coated the grinding medium during the first run.

time the powder coats the grinding medium; this powder was subsequently discarded and the same set of milling container and grinding medium was again used) resulted in even lower values of iron contamination—0.06 wt% with boric acid and 0.09 wt% with borax. Similarly, it was shown [12] that milling of Cu-Fe alloy powders without any PCA introduced considerable contamination from the steel milling tools. This reached as much as 21.9 at% Fe and 2.8 at% Cr for a Cu-43.1 at% Fe composition. However, when powder of the same composition was milled under identical conditions, but with ethanol as a PCA, the contamination was found to be much less, i.e., 2.9 wt% Fe and under 0.1 wt% Cr. This has been explained on the basis of decreased cold welding between the milled powder and the balls and milling container walls when the PCA was used. Thus, a proper choice of PCA can substantially reduce the powder contamination. Of course, the best situation would be not to use PCA at all, if it could be avoided.

It is also important to remember that cross-contamination could occur if a container that was used earlier to mill some powder is used again to mill another powder without having undergone proper cleaning.

Another way of minimizing powder contamination is to control the milling time, ensuring of course that steady-state conditions were reached. If one knows a priori what phase is desired in the milled powder, then milling can be stopped once that phase has been obtained in the powder. Since powder contamination increases with prolonged milling time, it is advisable not to continue milling the powder beyond the time required. A case in point is that formation of the nitride contaminant phase in Ti-Al alloys could be avoided if milling were stopped after the formation of the amorphous phase [54].

As pointed out earlier [19], the number of interruptions also increases the level of contamination. Accordingly, it was suggested that it is desirable to conduct the milling operation without opening the vial in between; this minimizes the powder contamination.

In conclusion, it may be mentioned that to minimize powder contamination the following steps may be followed:

1. Use milling tools that are harder than the powder being milled.
2. Use grinding media and vials that have the same chemical composition, at least the major component, as the powder being milled.
3. Avoid using a PCA so that cold welding may increase. This will minimize fracturing of the powder particles; consequently, the surface area of the powder particles is smaller.
4. Minimize the number of interruptions. Continuous milling without opening the vial will also minimize powder contamination.

The level of contamination may be different under different processing conditions and is dependent on type of mill, intensity of milling, nature of the powder, nature of the grinding medium and container, ambient atmosphere, BPR, temperature of milling, seal integrity, and others. Although not reported by anyone so far, milling at low temperatures (e.g., cryomilling) could certainly reduce the levels of powder contamination. Claims have been made in the literature about the superiority of certain mills and practices over others [62]. However, systematic investigations on milling the same powder and under identical conditions in different mills and evaluating the contamination levels have not been undertaken; and this appears to be an important aspect of the production of clean powders.

REFERENCES

1. Wang, G. M., Campbell, S. J., Calka, A., Kaczmarek, W. A. (1997). *J. Mater. Sci.* 32:1461–1467.
2. Courtney, T. H., Wang, Z. (1992). *Scripta Metall. Mater.* 27:777–782.
3. Saji, S., Abe, S., Matsumoto, K. (1992). *Mater. Sci. For.* 88–90:367–374.
4. Guo, W., Martelli, S., Padella, F., Magini, M., Burgio, N., Paradiso, E., Franzoni, U. (1992). *Mater. Sci. For.* 88–90:139–146.
5. Eckert, J., Holzer, J. C., Krill, C. E. III, Johnson, W. L. (1992). *Mater. Sci. For.* 88–90:505–512.
6. Zhang, F. X., Wang, W. K. (1996). *J. Alloys Compounds* 240:256–260.
7. Cardoso, K. R., Garcia-Escorial, A., Botta, W. J., Lieblich, M. (2001). *Mater. Sci. For.* 360–362:161–166.
8. Fan, G. J., Gao, W. N., Quan, M. X., Hu, Z. Q. (1995). *Mater. Lett.* 23:33–37.
9. Corrias, A., Ennas, G., Marongiu, G., Musinu, A., Paschina, G. (1993). *J. Mater. Res.* 8:1327–1333.
10. Hermann, H., Gruner, W., Mattern, N., Bauer, H.-D., Fugaciu, F., Schubert, Th. (1998). *Mater. Sci. For.* 269–272:193–198.
11. Shen, T. D., Koch, C. C. (1996). *Acta Mater.* 44:753–761.
12. Gaffet, E., Harmelin, M., Faudot, F. (1993). *J. Alloys Compounds* 194:23–30.
13. Correia, J. B., Marques, M. T., Oliveira, M. M., Matteazzi, P. (2001). *Mater. Sci. For.* 360–362:241–246.
14. Inoue, A., Kimura, H. M., Matsuki, K., Masumoto, T. (1987). *J. Mater. Sci. Lett.* 6: 979–981.
15. Fukunaga, T., Nakamura, K., Suzuki, K., Mizutani, U. (1990). *J. Non-Cryst. Solids* 117–118:700–703.
16. Jang, J. S. C., Koch, C. C. (1989). *Scripta Metall.* 23:1805–1810.
17. Gonzalez, G., Sagarzazu, A., Villalba, R., Ochoa, J., D'Onofrio, L. (2001). *Mater. Sci. For.* 360–362:349–354.
18. Caamaño, Z., Pérez, G., Zamora, L. E., Suriñach, S., Muñoz, J. S., Baró, M. D. (2001). *J. Non-Cryst. Solids* 287:15–19.
19. Lucks, I., Lamparter, P., Mittemeijer, E. J. (2001). *Acta Mater.* 49:2419–2428.
20. Gaffet, E., Malhouroux-Gaffet, N. (1994). *J. Alloys Compounds* 205:27–34.
21. Hellstern, E., Schultz, L., Bormann, R., Lee, D. (1988). *Appl. Phys. Lett.* 53:1399–1401.
22. Di, L. M., Bakker, H. (1991). *J. Phys. C: Condens. Matter.* 3:9319–9326.
23. Kenik, E. A., Bayuzick, R. J., Kim, M. S., Koch, C. C. (1987). *Scripta Metall.* 21: 1137–1142.
24. Cardellini, F., Contini, V., Mazzone, G. (1995). *Scripta Metall. Mater.* 32:641–646.
25. Murty, B. S., Joardar, J., Pabi, S. K. (1996). *Nanostructured Mater.* 7:691–697.
26. Schumacher, P., Enayati, M. H., Cantor, B. (1999). *Mater. Sci. For.* 312–314:351–356.
27. Koch, C. C., Cavin, O. B., McKamey, C. G., Scarbrough, J. O. (1983). *Appl. Phys. Lett.* 43:1017–1019.
28. Mi, S., Courtney, T. H. (1998). *Scripta Mater.* 38:637–644.
29. Haruyama, O., Asahi, N. (1992). *Mater. Sci. For.* 88–90:333–338.
30. Weeber, A. W., Bakker, H. (1988). *J. Phys. F: Metal Phys.* 18:1359–1369.
31. Aoki, K., Memezawa, A., Masumoto, T. (1993). *J. Mater Res.* 8:307–313.
32. Krill, C. E., Klein, R., Janes, S., Birringer, R. (1995). *Mater. Sci. For.* 179–181:443–448.
33. Hellstern, E., Fecht, H. J., Fu, Z., Johnson, W. L. (1989). *J. Appl. Phys.* 65:305–310.
34. Eckert, J., Holzer, J. C., Li, M., Johnson, W. L. (1993). *Nanostructured Mater.* 2: 433–439.
35. Liu, K. W., Müklich, F., Birringer, R. (2001). *Intermetallics* 9:81–88.
36. Calka, A., Williams, J. S., Millet, P. (1992). *Scripta Metall. Mater.* 27:1853–1857.

37. Lönnberg, B. (1994). *J. Mater. Sci.* 29:3224–3230.
38. Sherif El-Eskandarany, M., Aoki, K., Suzuki, K. (1990). *J. Less Common Metals* 167:113–118.
39. Ogino, Y., Miki, M., Yamasaki, T., Inuma, T. (1992). *Mater. Sci. For.* 88–90:795–800.
40. Guo, W., Iasonna, A., Magini, M., Martelli, S., Padella, F. (1994). *J. Mater. Sci.* 29: 2436–2444.
41. Zhang, L. C., Xu, J. (2002). *Mater. Sci. For.* 386–388:47–52.
42. Counihan, P. J., Crawford, A., Thadhani, N. N. (1999). *Mater. Sci. Eng.* A267, 26–35.
43. Calka, A., Kaczmarek, W. A. (1992). *Scripta Metall. Mater.* 26:249–253.
44. Yang, E., Wagner, C. N. J., Boldrick, M. S. (1993). *Key Eng. Mater.* 81–83:663–668.
45. Gaffet, E., Louison, C., Harmelin, M., Faudot, F. (1991). *Mater. Sci. Eng.* A134, 1380–1384.
46. Calka, A., Nikolov, J. I., Ninham, B. W. (1993). In: deBarbadillo, J. J., et al., eds. *Mechanical Alloying for Structural Applications.* Materials Park, OH: ASM. International, pp. 189–195.
47. Ahn, J.-H., Zhu, M. (1998). *Mater. Sci. For.* 269–272:201–206.
48. Daledda, S., Eckert, J., Schultz, L. (2001). *Mater. Sci. For.* 360–362:85–90.
49. Mizutani, U., Lee, C. H. (1990). *J. Mater. Sci.* 25:399–406.
50. Sato, K., Ishizaki, K., Chen, G.-H., Frefer, A., Suryanarayana, C., Froes, F. H. (1993). In: Moore, J. J., et al., eds. *Advanced Synthesis of Engineered Structural Materials.* Materials Park, OH: ASM International, pp. 221–225.
51. Klassen, T., Oehring, M., Bormann, R. (1994). *J. Mater. Res.* 9:47–52.
52. Chen, G.-H., Suryanarayana, C., Froes, F. H. (1995). *Metall. Mater. Trans.* 26A:1379–1387.
53. Goodwin, P. S., Ward-Close, C. M. (1995). *Mater. Sci. For.* 179–181:411–418.
54. Suryanarayana, C. (1995). *Intermetallics* 3:153–160.
55. Ogino, Y., Yamasaki, T., Murayama, S., Sakai, R. (1990). *J. Non-Cryst. Solids* 117/118:737–740.
56. Mukhopadhyay, D. K., Suryanarayana, C., Froes, F. H. (1995). In: Phillips, M., Porter, J., eds. *Advances in Powder Metallurgy and Particulate Materials—1995.* Vol. 1. Princeton, NJ: Metal Powder Industries Federation, pp. 123–133.
57. Calka, A., Nikolov, J. I., Li, Z. L., Williams, J. (1995). *Mater. Sci. For.* 179–181:295–300.
58. Goodwin, P. S., Hinder, T. M. T., Wisbey, A., Ward-Close, C. M. (1998). *Mater. Sci. For.* 269–272:53–62.
59. Goodwin, P. S., Mukhopadhyay, D. K., Suryanarayana, C., Froes, F. H., Ward-Close, C. M. (1996). In: Blenkinsop, P. A., Evans, W. J., Flower, H. M., eds. *Titanium '95: Science and Technology.* Vol. 3. London: Institute of Materials, pp. 2626–2633.
60. Sherif El-Eskandarany, M., Aoki, K., Itoh, H., Suzuki, K. (1991). *J. Less Common Metals* 169:235–244.
61. Sherif El-Eskandarany, M., Aoki, K., Suzuki, K. (1991). *J. Less Common Metals* 177:229–244.
62. Zoz, H., Vernet, I., Jaramillo, D. (2003). V. Unpublished results.
63. Suryanarayana, C., Ivanov, E., Noufi, R., Contreras, M. A., Moore, J. J. (1999). *J. Mater. Res.* 14:377–383.
64. Suryanarayana, C. (2001). *Prog. Mater. Sci.* 46:1–184.
65. Shaw, W. J. D. (1998). *Mater. Sci. For.* 269–272:19–30.
66. Koch, C. C., Smith, A. P., Bai, C., Spontak, R. J., Balik, C. M. (2000). *Mater. Sci. For.* 343–346:49–56.
67. Chen, Y., Hwang, T., Marsh, M., Williams, J. S. (1997). *Metall. Mater. Trans.* 28A:1115–1121.
68. Goodwin, P. S., Ward-Close, C. M. (1993). In: deBarbadillo, J. J., et al., eds. *Mechanical Alloying for Structural Applications.* Materials Park, OH: ASM International, pp. 139–148.

69. Klassen, T., Oehring, M., Bormann, R. (1997). *Acta Mater.* 45:3935–3948.
70. Bohn, R., Klassen, T., Bormann, R. (2001). *Acta Mater.* 49:299–311.
71. Singer, R. F., Oliver, W. C., Nix, W. D. (1980). *Metall. Trans.* 11A:1895–1901.
72. Keskinen, J., Pogany, A., Rubin, J., Ruuskanen, P. (1995). *Mater. Sci. Eng.* A196: 205–211.
73. Mishurda, J. C., Suryanarayana, C., Froes, F. H. (1994). Moscow, ID: University of Idaho. unpublished research.

16

Modeling Studies and Milling Maps

16.1 INTRODUCTION

From the description of the process (Chapter 5) and mechanism (Chapter 6) of mechanical alloying (MA), it is easy to realize that MA is a process involving a number of both independent and interdependent variables. As for any other process, modeling of MA has been carried out for the purpose of identifying the salient factors affecting the process and establish process control optimization and instrumentation. By modeling the process effectively, it is possible to bring down the number of actual experiments needed to optimize the process and achieve a particular application. A highly successful model would also predict the outcome of the process. In MA, for example, if one can predict the nature of the phase produced under a given set of milling conditions, it would be considered a very successful model.

16.2 PROCESS VARIABLES

Mechanical alloying is a stochastic process, and the number of variables involved in the process is very large. For a particular alloy system the variables include the type of mill, intensity of milling (velocity of the grinding medium, frequency of impacts, energy transfer efficiency, and so on), size, shape and type of milling media, ball-to-powder weight ratio (BPR), the atmosphere under which the powder is milled, purity of the powders, milling time, milling temperature, and nature and amount of the process control agent (PCA) used. All these have a significant effect on the constitution of the powder. Even on a local scale, the nature of impacts between two balls, the frequency of impacts, and the amount of powder trapped between two balls (grinding media) during a collision could vary from point to point. Furthermore, the motion of balls in different types of mills is quite complex. Thus, modeling the MA process is a difficult task. In spite of this, some attempts have been made in recent times and moderate success has been achieved in modeling the physics and mechanics of the process. From the actual experiments conducted, attempts have also been made to correlate the phases formed with the process parameters during milling, and milling maps have been constructed. However, we are far from a situation where we can predict the final chemical constitution of the powder (type and description of phases formed).

16.3 EARLY MODELING ATTEMPTS

Modeling of the MA process has been conducted by several groups in recent years. The very first attempt at estimating the time required for alloying to occur was modeled by Benjamin and Volin [1]. Assuming that the rate at which material is trapped between two colliding balls is independent of time and that the energy required per unit strain for a constant volume of material is a linear function of the instantaneous hardness of the powder, they calculated the strain, ε, as a function of time, t, as:

$$\varepsilon = \frac{K}{4.57}\ln(1 + 0.365t) \tag{16.1}$$

where K is a constant. They suggested that the rate of structural refinement was dependent on the rate of mechanical energy input to the process and the work-hardening rate of the material being processed. They also concluded that the specific times to develop a given structure in any system would be a function of the initial particle size and characteristics of the ingredients as well as the particular apparatus and operating parameters. Subsequently more rigorous modeling was done.

Another early modeling attempt was to estimate the temperature rise during collision between two balls and/or the powder [2, 3; see also Chapter 8]. Subsequently, Maurice and Courtney [4] presented a first attempt to define the basic geometry, mechanics, and physics of MA. The model defined the volume of material affected per collision, and from this they approximated information about impact times, powder strain rates and strains, powder temperature increase, powder cooling times, and milling times. Efforts to improve this model were undertaken later. The most important groups who conducted very detailed modeling studies are those led by Courtney [4–6], Gaffet [7], Magini [8], Koch [9], and Watanabe [10]. A few others have also attempted the modeling [11–15].

Modeling of MA/MM is done to identify and define the important process parameters controlling the structure and properties of the resulting powder. For example, it is very difficult to accurately predict the nature of the phase formed under the given conditions of MA/MM. This is due to the fact that phase transformations that occur during MA/MM are mostly dependent on the energy transferred to the powder, deformation behavior of these powders, and the chemical reactions that take place in them. Sufficient information is not readily available on the deformation behavior of metal powders at high strain rates, impact velocities, and impact frequencies, as encountered during MA. Due to these limitations, it is not possible to make accurate predictions but rather only to obtain the general trends. Therefore, it is important to realize that these models cannot provide information on all aspects of the process. As Courtney and Maurice [16] succinctly put it:

> It is important to appreciate what can be expected of even the most successful efforts of this kind. "Absolute" predictions are unlikely; this is exacerbated in the case of MA by its inherently stochastic nature. In addition, models require input "data"; material properties (often not known to precision, particularly at the high strain levels to which powder is often subjected during MA) and process characteristics (collision frequency and velocity, also seldom known *a priori*). Thus, realistic goals of process modeling are to correctly predict general trends, and perhaps even to predict resulting properties/dimensions within an order of magnitude. The benefits of successful models lie not in their abilities to predict outcomes, but in that they help identify critical process and material variables, and reduce the amount of testing needed for process optimization.

16.4 TYPES OF MODELING

Modeling of the MA process has been broadly classified into two categories: local and global. Local modeling considers a "typical" collision occurring in a specific milling device with a stipulated impact velocity and frequency. The size of the grinding medium, ball-to-powder weight ratio (BPR), and the mechanical characteristics of the powder (hardness, ductility, fracture toughness, etc.) are also specified beforehand. During the collision involving powder trapped between two grinding media, the powder hardness, size, and shape are altered. Thus, "local" modeling attempts to calculate (1) the deformation the particle experiences, (2) the change in powder particle shape, (3) the probability of a particle coalescing, and (4) the probability of a particle fracturing.

Milling parameters such as impact frequency and angle, and media velocity are taken into account during global modeling. Since many media impacts are ineffective in bringing about deformation, fracture, and coalescence of powder particles, one seeks to specify the frequency of *effective* media impacts by considering the collision velocities, frequencies, and powder coating thickness. Thus, "global" modeling determines distribution of (1) impact angles, (2) velocities, (3) rolling forces, and so forth over the entire mill. It is important to appreciate that both local and global modeling are interrelated and complementary to each other. While local modeling describes the events that take place in the powder entrapped between two grinding media during a single collision, global modeling deals with macroscopic factors and mill-specific features. A combination of these two models can provide a complete picture of the total milling process. Figure 16.1 shows the difference between these two types of modeling.

16.4.1 Local Modeling

Both analytical and phenomenological approaches have been taken for local modeling. The collision between two grinding media was considered and the powder coating was ignored. It was reasoned that the powder only mildly perturbs the collision, i.e., the plastic deformation (and other) work performed on the powder is assumed to be only a small fraction of the precollision media kinetic energy. Thus, the collision between grinding media involving entrapped powder between them is similar to the Hertzian collision between balls without the powder. The collision volume is consid-

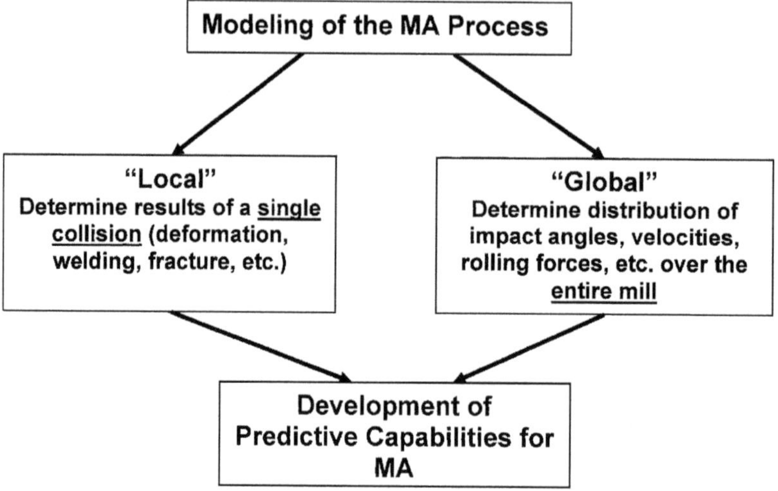

Figure 16.1 Difference between "local" and "global" modeling of MA/MM.

ered to be a cylinder of radius equal to the Hertz radius, r_h, and height h_0. Estimating the media collision velocity, v, as ranging from 0.5 m/s (e.g., for an attritor) to 5 m/s (e.g., for a SPEX mill) and to decrease linearly from the maximal value at the time of impact to zero after an elapsed time, τ, Maurice and Courtney [4] calculated the strain rates to be of the order of 10^3–$10^4 s^{-1}$. Thus, the deformation strain in a single collision was calculated as

$$\varepsilon = -\ln\left(1 - \frac{v\tau}{2h_o}\right) \tag{16.2}$$

The estimated powder strains ranged from several tenths of a percent (for an attritor) to unity (for a SPEX mill with a high BPR). Courtney and Maurice [17] subsequently estimated the processing times by assuming that a critical amount of powder deformation (Σ) is needed to achieve alloying. The alloying time, t_p, was shown to vary as

$$t_p \approx \frac{\Sigma h_o}{v\tau r_h^2 h_o} \tag{16.3}$$

Using the equations for the velocity dependence of the collision time and radius, it was shown that the alloying time could be related to the precollision velocity as

$$t_p \sim v^{-2.6} \tag{16.4}$$

The above model did not include the criteria for fracture and welding of powder particles, nor the frequency of such events. Furthermore, the above treatment provided only a "snapshot" of a single collision, and no attempt was made to obtain information about the powder particle size, shape, hardness, and microstructural refinement scale fully. These aspects were subsequently undertaken.

A rigorous treatment of the single collision between grinding media coated with a uniform thickness of the powder helped in developing criteria for welding and fracture events [18]. Here it was considered that the powder particles are generally not spherical but more like oblate spheroids. At the initial impact of the grinding medium, the powders deform elastically; with continued milling the stress reaches the yield strength of the powder and consequently the powders get deformed plastically. With further deformation, the radius of the plastically deformed zone spreads radially outward. Thus, the powder-affected volume consists of a plastic cylindrical core surrounded by an elastic annulus. The strain the powder experiences depends on the value of the center-to-center approach of the balls [$\alpha(r)$] which is a function of the distance (r) from the contact center. It is given by

$$\alpha(r) = Rv\left(\frac{\rho}{H_v}\right)^{1/2} - \frac{r^2}{R} \tag{16.5}$$

where R is the ball radius, v the precollision velocity, H_v the powder hardness, and ρ the density of the grinding media. The strain experienced by the powder is then expressed as

$$\varepsilon(r) = -\ln\left\{1 - \frac{\alpha(r)}{h_o}\right\} \tag{16.6}$$

and the maximal strain (at $r = 0$) is

$$\varepsilon(0) = -\ln\left\{1 - \frac{Rv}{h_0}\left(\frac{\rho}{H_v}\right)^{1/2}\right\} \tag{16.7}$$

Thus, if h_o, the powder cylinder height, increases linearly with R, the radius of the grinding medium (which is to be expected), the strain is independent of the size of the grinding medium.

Powder coalescence takes place by cold welding of the plastically deformed particles. If the powder particles contain a surface oxide layer, this brittle oxide layer is broken down during deformation and the fresh metal surface establishes contact with other metal surfaces. For alloying of different materials, it is necessary that both components be deformed before welding occurs. Thus, if we have a situation where one metal is softer than the other, then the softer metal has to get work hardened to a level comparable to that of the harder metal before cold welding can take place.

The next step is fracture of the cold-welded particles, which was considered to take place through "forging" fracture. Based on macroscopic models of forging and invoking sticking friction and work piece barreling, strain is greatest at the particle circumference. Thus, edge fracture initiates there. If one disregards sticking friction, strain is greatest at the contact center and crack initiation is predicted to take place at the center of the particle, as might be expected for a brittle material. A crack begins when the critical tensile strain is attained over a critical length and the crack propagation occurs when the plastic energy release exceeds a value characteristic of the material. Thus, if the particle is sufficiently small so that the crack length is larger than the particle size, i.e., the crack length is not reached, then the particle will not fracture. Thus, there appears to be a minimal particle size below which fracturing does not take place.

Maurice and Courtney [19] developed two computer programs. MAP1, the simpler of the two, considers milling of a single species with the option of adding dispersoids, and MAP2, the more sophisticated, can be used to model the kinetics of MA of two species (A and B) to form a third (i.e., alloy) species (C). These two programs describe the changes in powder size, shape, hardness, and microstructural scale observed during MA. The predicted results were compared with the experimental observations of Benjamin and Volin [1] on SPEX milling of Fe-Cr powders. Even though experimental results and predictions display the same trends (they are typically within a factor of two or three of each other), some discrepancies were found [20]. For example, the predicted welding frequencies exceed predicted fracture frequencies for the first 12 min of processing; the average particle size (about 100 μm) is less than that predicted by the model (about 130 μm), and the time at which the maximum is observed (30–40 min) is threefold longer than the corresponding model predicted time. The hardness predicted by the model was less than the observed value.

Maurice and Courtney [21] showed that the processing time dependency on impact velocity for different combinations of metals of different hardness and fracture toughness is in the range of v^{-2} to v^{-3}. The model also successfully predicted the influence of atmosphere and premilling of Ni on the formation of Ni_3Si_2 during MA [22].

Aikin and Courtney [23,24] subsequently developed a phenomenological description of the process based on kinetic principles. Using these models, the authors estimated the plastic strain and strain rate, the local temperature rise due to a collision, the particle hardness due to work hardening, the lamellar thickness, and the impact frequency and velocity. They showed that the fracturing and welding probabilities scale directly with the BPR and vary almost linearly with mill power. In fact, in a recent publication Maurice and Courtney [21] compared the alloying behavior in the SPEX mill and the attritor and identified some of the commonalities and differences between the two devices. While the lamellar thickness of the component metals and the hardness are almost the same in both milling devices, the time required for completion of the process was estimated as 15 min in the SPEX mill and 11,000 min (183 h) in the attritor—nearly three orders of magnitude longer. The composite particle diameter was also calculated to be 52 μm in the SPEX mill and 910 μm in the attritor. Maurice and Courtney [21] showed that these differences arise because of the differences in the average impact velocity (0.5 m/s in the attritor and 4 m/s in the SPEX mill) and the average impact frequency (0.88/s in the attritor and 7/s in the SPEX mill). Although it has been long known that the SPEX mill is much more energetic than the attritor, this provides a more quantitative comparison. Similarly, Abdellaoui and Gaffet [25] calculated the velocity of the ball, kinetic energy per impact, shock frequency, and power for different types of mills. Their results are presented in Table 16.1.

The kinetics of alloying has also been studied. It was shown that for the elements having the greatest difference in initial hardness, alloying has hardly begun after eight impacts at a velocity of 2 m/s. This was reflected in a fine particle size (essentially the initial size). However, for the same number of impacts and at the same velocity, alloying was essentially complete for elemental combination having the most similar properties, inferred from a large particle size, reflecting the cold welding accompanying the formation of the composite particles. The maximum in particle size for different material combinations is different and is found approximately at the velocity needed to "alloy" in the given time.

The precision of the material properties and process parameters used in the computational applications varies, sometimes substantially. A partial listing of these properties and parameters, along with the comments regarding the precision up to which they are known, is presented in Table 16.2. Some properties are known precisely (e.g., modulus and density); others are likely known only to a factor of 2 at best (e.g.,

Table 16.1 Kinetic Energy, Shock Frequency, and Shock Power for Different Types of Mills Used for Mechanical Alloying

	Parameter			
Mill	Velocity of the ball (m/s)	Kinetic energy (10^{-3} J/hit)	Shock frequency (Hz)	Power (W/g/ball or rod)
Attritor	0–0.8	<10	>1000	<0.001
Vibratory mills:				
Pulverisette "O"	0.14–0.24	3–30	15–50	0.005–0.14
SPEX Mill	<3.9	<120	200	<0.24
Planetary ball mills				
Pulverisette P5	2.5–4.0	10–400	~100	0.01–0.8
Pulverisette G7	0.24–6.58	0.4–303.2	5.0–92.4 (5 balls)	0–0.56
Pulverisette G5	0.28–11.24	0.53–884	4.5–90.7 (5 balls)	0–1.604
Horizontal rod mill	0–1.25	0–190	0–2.4 (1 rod)	0–0.1

hardening laws at high strain, fracture initiation strain, etc.). Processing variables (collision frequencies and velocities) are known even less accurately. Global modeling and "intuitive" extrapolations from such modeling provide "ballpark" figures for process parameters. Global modeling can improve the accuracy of local modeling by providing refined values of the process parameters used in local modeling.

16.4.2 Global Modeling

Global modeling is useful in two respects. First, knowledge of device-specific media dynamics provides input information for local modeling. The availability of more accurate information could improve process modeling and lead to process optimization. Second, the device efficiency can be improved from a better understanding of device-specific media motion.

Table 16.2 Some Input Parameters Required for Numerical Application of the Model [6]

Parameter		Precision known to
Material property	Modulus	High
	Density	High
	Initial particle size	High
	Initial hardness	High
	Tensile strength	Moderate
	Fracture strain	Moderate to low
	Fracture toughness	Moderate to low
	Plastic flow law parameters	Moderate to low
Process variable	Ball-to-powder ratio	High
	Powder coating thickness	Moderate
	Impact velocity	Moderate to low
	Impact frequency	Moderate to low

Davis et al. [3] modeled the kinetics of ball milling by studying the actual collision processes inside a SPEX mill. They obtained information on (1) the velocities acquired by the milling media, (2) the amount of energy transferred to the powders during an impact, and (3) the amount of energy transferred to the powders as heat. The vial motion was recorded on a videotape after slowing its apparent velocity by use of a high precision stroboscope. This videotape was then analyzed by a "motion analysis" computer translation system, which converted the analog motion of the vial into digital coordinate displacements and velocities, based on appropriate time constants and length calibrations. The ball motion was also studied by observation through a transparent lucite vial. If any two balls were within one diameter of their loci, or if any one ball was within one radius of the vial interior surface, then an impact was recorded, the ball impact dissipated kinetic energy, and mean free paths were recorded. The number of impacts occurring for 0.5 s and 1.0 s of mill operation at several different ball loads is shown in Table 16.3. A majority of the impacts occurred in the 10^{-3}- to 10^{-2}-J range of energy dissipated during the collision. The impact velocity was predicted to be 6 m/s or less. The model also predicted that a majority of the impacts were glancing in nature and not head-on; this was subsequently confirmed by others [4].

Basset et al. [12] estimated impact velocities from the indentation on a copper plate fixed at one end of the vial in a SPEX mill. The estimated velocities were in the range of 1.8–3.3 m/s and were found to be strongly dependent on the ball size. Hashimoto and Watanabe [10,26–28] have also analyzed the motion of the balls, using discrete element method, in vibratory, tumbler, and planetary ball mills. Le Brun et al. [11] analyzed the ball motion in a planetary ball mill and observed three modes of ball motion—chaotic, impact + friction, and friction, with an increasing ratio of the angular velocities of the vial and the disk.

Rydin et al. [29] used high-speed cinematography to study the global dynamics of a transparent attritor. They noted that substantial segregation of powder occurred at the edge of the attritor; this region was termed "dead zone." It was reported that an attritor is a very inefficient device because direct impacts were restricted to its core (where little powder was present), but the region containing proportionally the most powder was characterized by a low frequency of ball sliding. Sliding is caused

Table 16.3 Number of Impacts for Various Ball and Kinetic Energy Values in a SPEX 8000 Mill[a]

	Kinetic energy of impact (J)				
Number of balls	10^{-7} to 10^{-4}	10^{-4} to 10^{-3}	10^{-3} to 10^{-2}	10^{-2} to 10^{-1}	10^{-1} to 1.0
Mill operation for 0.5 s					
5	0	43	297	3	0
10	0	78	505	13	1
15	4	124	928	24	0
Mill operation for 1.0 s					
5	0	78	612	3	0
10	0	148	1201	13	2
15	4	229	1873	24	0

[a] The weight of each ball used is 2 g [3].

by the close-packed array of balls. Accordingly, Rydin et al. [29] suggested that the efficiency of the attritor could be improved by using a mixture of differently sized balls in the mill. It was shown [30] that when differently sized balls were used, the smaller balls segregated to the tank bottom at low device rotational velocities. Increasing this velocity reduced ball segregation, and at a critical rotational velocity (which is dependent on ball size, ball size ratio, and tank diameter), complete ball mixing occurred. It was also reported that differently sized balls double *both* the welding and fracture rates.

16.5 MILLING MAPS

The above modeling studies have been successful in predicting the trends of impact velocities and frequencies, but have not taken into account the type of phase formed under different milling conditions. The ability to predict the nature of the phase formed is a great advantage in designing the microstructure and constitution of the alloy. Some investigations have been carried out with this objective in view. In all these investigations, the energy (or other parameter) calculations are always compared with the experimentally obtained phase constitution and conclusions are drawn.

Abdellaoui and Gaffet [25,31] studied the amorphization behavior of $Ni_{10}Zr_7$ in different types of mills and evaluated the shock power that is transmitted to the powder. Based on a mathematical treatment of the process taking place, they calculated the shock frequency and the shock power. Irrespective of the device used to accomplish the milling process, it was shown that the injected shock power is the unique physical parameter governing the phase transformation. For the prealloyed $Ni_{10}Zr_7$

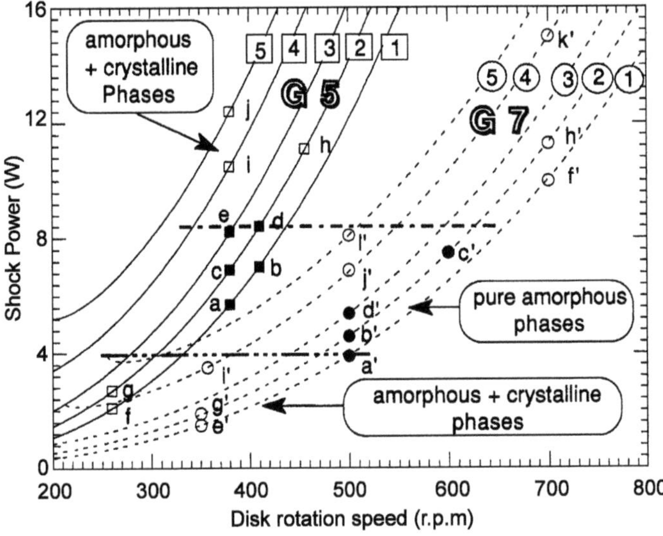

Figure 16.2 Injected shock power as a function of the disk and vial rotation speeds for the Fritsch Pulverisette G5 (—) and G7 (- - - -) planetary ball mills. The vial rotation speed values (in rpm) are referred by the numbers 1–5 written near each corresponding curve: 1 = 150, 2 = 250, 3 = 350, 4 = 500, and 5 = 600 rpm.

intermetallic compound, a homogeneous amorphous phase was obtained only when the injected shock power was in the range 0.4–0.8 W/g. At power values less than 0.4 or greater than 0.8 W/g, a mixture of amorphous and crystalline phases was obtained (Fig. 16.2).

Since phase transformations in mechanically alloyed powders occur due to the energy transferred from the milling media to the powder, it would be useful to calculate the energy transferred. This, of course, depends on the type of mill and the operating parameters. It was shown that collision is predominant when the grinding medium fills only a small part of the mill; collision between the balls and the opposite wall is the dominant energy transfer event here. However, when the mill begins to be filled with the grinding medium, shear stress between neighboring balls is the dominant energy transfer event. In this case, a "filling factor" should be taken into consideration [32]. By considering that collision is the basic event by which energy gets transferred from the mill to the powder, one could calculate (1) kinetic energy of the

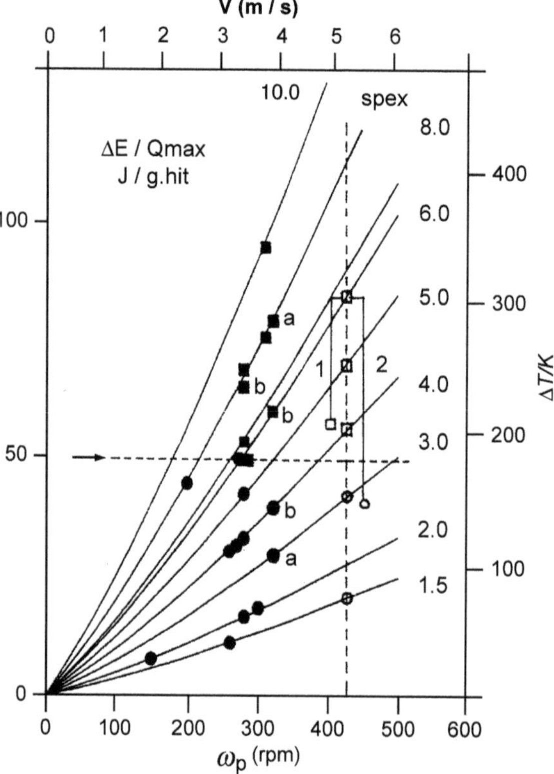

Figure 16.3 Energy transferred per hit and per unit mass for the Pd-Si system as a function of the rotation speed of the planetary ball mill for different ball diameters (given in mm near each curve). The filled symbols represent experiments carried out in the planetary ball mill and the open symbols for the SPEX mill. The horizontal broken line refers to the critical energy above which an intermetallic forms and below which an amorphous phase forms. The experiments have been carried out with a Pd-20 at% Si composition except in (a) = Pd-13.5 at% Si and (b) = Pd-17 at% Si.

ball, (2) fraction of the kinetic energy transferred to the powder, and (3) quantity of material entrapped in the collision event. It was shown that the energy dissipated per hit during milling increases with increasing rotation speed and also with increasing size of the grinding ball [8]. These calculations assumed that the collisions are perfectly inelastic. However, it was shown through video recordings that movements of the balls are somewhat different from the expected ones and that one should consider a "slip factor" that takes into account the sliding phenomena between balls and the container wall [33]. The energy released by one ball as a function of the specific power (power per unit weight of the powder) was calculated. By comparing these calculations with the experimentally determined phase constitution, a critical energy was identified below which the phase formed is amorphous and above which it is intermetallic. This has been shown to be true in the Fe-Zr system [32] as well as in the Pd-Si system [8]. A similar calculation for the case of a SPEX mill is difficult because the mill is subjected to impulses in three directions. Since the vibration frequency is fixed for the SPEX mill, the situation can be compared with that in a planetary ball mill at fixed rotation speed. An "energy map" featuring the phase constitution in the Pd-Si system is shown in Figure 16.3. It may be noted that these maps define a threshold value for the milling conditions above which an intermetallic phase forms. Below this critical value, an amorphous phase forms. These results confirm the earlier observations that soft MA conditions favor the formation of a nonequilibrium phase.

A milling map describing the phases present in a Ti-33 at% Al powder mix at different milling times and for different BPR values is shown in Figure 16.4 [34]. It was shown that the time required for the formation of a particular phase was shorter at higher BPR since the higher BPR value translates to higher mechanical energy input

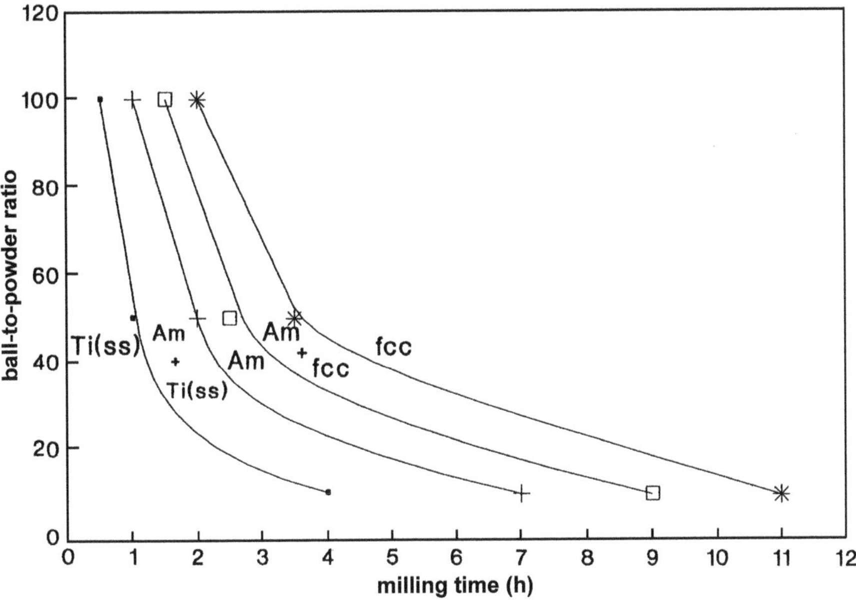

Figure 16.4 Milling map for blended elemental Ti-33 at% Al powder relating the BPR and milling time to the phases present.

per unit mass of the powder. At a constant BPR, the sequence of phase formation can be determined as a function of milling time. Similarly, at a constant milling time, the phases occurring with increasing BPR can be identified. Thus, this map is somewhat similar to a phase diagram, except that the milling map is for a particular alloy composition while the phase diagram is for the whole alloy system. It has been reported earlier that increased speed of milling [32,35] or very high BPR [36] generates more heat and leads to the formation of crystalline phases either by crystallization of the amorphous phase or by other mechanisms. Thus, the *y* axis on the milling map may be equated to "temperature" and so the milling map can be considered as a constant composition section of a ternary phase diagram where the three axes are milling time, composition, and BPR. These milling maps should be very useful in quickly identifying the phases present at any milling time for a specific BPR and in defining the final phases produced by the MA process.

REFERENCES

1. Benjamin, J. S., Volin, T. E. (1974). *Metall. Trans.* 5:1929–1934.
2. Schwarz, R. B., Koch, C. C. (1986). *Appl. Phys. Lett.* 49:146–148.
3. Davis, R. M., McDermott, B., Koch, C. C. (1988). *Metall. Trans.* 19A:2867–2874.
4. Maurice, D. R., Courtney, T. H. (1990). *Metall. Trans.* 21A:289–303.
5. Courtney, T. H. (1994). *Rev. Particulate Mater.* 2:63–116.
6. Courtney, T. H. (1995). *Mater. Trans. Jpn. Inst. Metals* 36:110–122.
7. Gaffet, E., Abdellaoui, M., Gaffet, N. M. (1995). *Mater. Trans. Jpn. Inst. Metals* 36:198–209.
8. Magini, M., Iasonna, A. (1995). *Mater. Trans. Jpn. Inst. Metals* 36:123–133.
9. Koch, C. C. (1995). *Mater. Trans. Jpn. Inst. Metals* 36:85–95.
10. Watanabe, R., Hashimoto, H., Lee, G. G. (1995). *Mater. Trans. Jpn. Inst. Metals* 36:102–109.
11. Le Brun, P., Froyen, L., Delaey, L. (1993). *Mater. Sci. Eng.* A161:75–82.
12. Bassett, D., Matteazzi, P., Miani, F. (1994). *Mater. Sci. Eng.* A174:71–74.
13. Dallimore, M. P., McCormick, P. G. (1996). *Mater. Trans. Jpn. Inst. Metals* 37:1091–1098.
14. Bhattacharya, A. K., Arzt, E. (1992). *Scripta Metall. Mater.* 27:635–639.
15. Bhattacharya, A. K., Arzt, E. (1993). *Scripta Metall. Mater.* 28:395–400.
16. Courtney, T. H., Maurice, D. R. (1996). *Scripta Mater.* 34:5–11.
17. Courtney, T. H., Maurice, D. R. (1989). In: Clauer, A. H., deBarbadillo, J. J., eds. *Solid State Powder Processing*. Warrendale, PA: TMS, pp. 3–19.
18. Maurice, D. R., Courtney, T. H. (1994). *Metall. Mater. Trans.* 25A:147–158.
19. Maurice, D. R., Courtney, T. H. (1995). *Metall. Mater. Trans.* 26A:2431–2435.
20. Maurice, D. R., Courtney, T. H. (1995). *Metall. Mater. Trans.* 26A:2437–2444.
21. Maurice, D. R., Courtney, T. H. (1996). *Metall. Mater. Trans.* 27A:1981–1986.
22. Kosmac, T., Maurice, D. R., Courtney, T. H. (1992). *J. Am. Ceram. Soc.* 76:2345–2352.
23. Aikin, B. J. M., Courtney, T. H. (1993). *Metall. Trans.* 24A:647–657.
24. Aikin, B. J. M., Courtney, T. H. (1993). *Metall. Trans.* 24A:2465–2471.
25. Abdellaoui, M., Gaffet, E. (1996). *Acta Mater.* 44:725–734.
26. Hashimoto, H., Watanabe, R. (1990). *Mater. Trans. Jpn. Inst. Metals* 31:219–224.
27. Hashimoto, H., Watanabe, R. (1992). *Mater. Sci. For.* 88-90:89–96.
28. Hashimoto, H., Watanabe, R. (1994). *Mater. Trans. Jpn. Inst. Metals* 35:40–45.
29. Rydin, R. W., Maurice, D. R., Courtney, T. H. (1993). *Metall. Trans.* 24A:175–185.
30. Cook, T. M., Courtney, T. H. (1995). *Metall. Mater. Trans.* A26:2389–2397.

31. Abdellaoui, M., Gaffet, E. (1994). *J. Alloys Compounds* 209:351–361.
32. Burgio, N., Iasonna, A., Magini, M., Martelli, S., Padella, F. (1991). *Il Nuovo Cimento* 13D:459–476.
33. McCormick, P. G., Huang, H., Dallimore, M. P., Ding, J., Pan, J. (1993). In: deBarbadillo, J. J., et al., eds. *Mechanical Alloying for Structural Applications*. Materials Park, OH: ASM International, pp. 45–50.
34. Suryanarayana, C., Chen, G.-H., Froes, F. H. (1992). *Scripta Metall. Mater.* 26:1727–1732.
35. Eckert, J., Schultz, L., Urban, K. (1990). *Z. Metallkde.* 81:862–868.
36. Sherif El-Eskandarany, M., Aoki, K., Itoh, H., Suzuki, K. (1991). *J. Less Common Metals* 169:235–244.

17

Applications

17.1 INTRODUCTION

From the previous chapters it becomes amply clear that the technique of mechanical alloying (MA) has been used extensively in synthesizing a variety of materials including stable and metastable phases. The types of material include nanocrystalline, crystalline, quasi-crystalline, and amorphous alloys. The incentive and the most important reason for the invention and development of the MA process was the production of oxide dispersion strengthened (ODS) materials in which fine particles of Y_2O_3 or ThO_2 were uniformly dispersed in nickel- or iron-based superalloys. In the mid-1980s it was realized that MA was also capable of producing true alloys from elements that are not easily formed by conventional means or sometimes are even impossible to prepare, e.g., elements that are immiscible under equilibrium conditions. Investigations have revealed that metastable phases such as supersaturated solid solutions, nonequilibrium crystalline or quasi-crystalline intermediate phases, and amorphous alloys can be synthesized by MA. In addition, nanostructures with a grain size of a few nanometers, typically less than 100 nm, were also produced by MA. These metastable phases have interesting combinations of physical, chemical, mechanical, and magnetic properties and are being widely explored for potential applications [1–11]. It will be impossible to go into full details of each and every one of these developments, and so only a brief survey of the present and potential applications of the mechanically alloyed/milled materials is presented here. The two production facilities of Inco Alloys International in the United States have a combined annual capacity approaching 300,000 kg. The yield of the final product varies greatly with the product form and size, but the final wrought capacity is greater than 200,000 kg.

Mechanically alloyed materials (including those synthesized by mechanochemistry and mechanical activation of solids) find applications in a variety of industries. The applications include synthesis and processing of advanced materials (magnetic materials, superconductors, and functional ceramics), intermetallics, nanocomposites, catalysts, hydrogen storage materials, food heaters, gas absorbers, and also in the modification of solubility of organic compounds, waste management, and production of fertilizers. However, the major industrial applications of mechanically alloyed materials have been in the areas of thermal processing, glass processing, energy production, aerospace, and other industries. These applications are based on the ODS effect achieved in mechanically alloyed nickel-, iron-, or aluminum-based alloys. Applications in other areas are based on chemical homogeneity and fine dispersion of one phase in the other, enhanced chemical activity of the mechanically alloyed products, and interesting properties of materials with nanometer-sized grains. These will be briefly described now.

17.2 OXIDE DISPERSION STRENGTHENED MATERIALS

ODS alloys contain a solid-solution-strengthened matrix in which a fine dispersion of an oxide is achieved. Usually about 1–2 wt% of an oxide is added to the alloy. The oxide particles are very fine, about 5–50 nm, and have a spacing of the order of 100 nm. The commonly used oxides are Y_2O_3 (yttria), ThO_2 (thoria), and La_2O_3 (lanthana).

The mechanically alloyed ODS alloy powders are degassed in vacuum and sealed in mild steel cans for extrusion. Extrusion is carried out in the temperature range of 1000–1100°C and at an extrusion ratio of 13:1. The microstructure of the alloy at this stage consists of very fine equiaxed grains of about 0.5 μm resulting from dynamic recrystallization during hot deformation. The extruded alloy is then subjected to hot rolling at 1025°C with 20% thickness reduction per pass and a total reduction in thickness reaching up to 90%. Optimal high-temperature strength is developed by optimizing the thermomechanical processing steps to produce a stable, recrystallized grain structure that is coarse and highly elongated in the direction of hot working. Grain aspect ratios may be as high as 10:1 to 15:1. The highly directional structure leads to varying degrees of anisotropy in mechanical and physical properties. A high-temperature recrystallization treatment is required to develop the desired grain structure; the usual treatment is 1315°C for 1 h followed by air cooling. The details of temperature, extrusion ratios, and degree of deformation in thermomechanical processing will be different for different alloys, and the reader is advised to refer to the manufacturer's brochures for full details.

Mechanically alloyed ODS materials are strong both at room and elevated temperatures (Table 17.1). The elevated temperature strength of these materials is derived from more than one mechanism. First, the uniform dispersion of very fine oxide particles, which are stable at high temperatures, inhibit dislocation motion in the metal matrix and increase the resistance of the alloy to creep deformation. Another function of the dispersoid particles is to inhibit the recovery and recrystallization processes because of which a very stable large grain size is obtained; these large grains resist grain rotation during high-temperature deformation. A stable large grain size can also be obtained by secondary recrystallization mechanisms. Second, the very homogeneous distribution of alloying elements during MA gives both the solid solution–strengthened and precipitation-hardened alloys more stability at elevated

Table 17.1 Room Temperature and Elevated Temperature Mechanical Properties of Commercial Oxide Dispersion Strengthened Nickel-and Iron-Based Superalloys[a]

Alloy	Test temp. °C (F)	0.2% YS (MPa)	UTS (MPa)	% El	% RA
MA 6000	Room temperature	1220	1253	7.2	6.5
	871 (1600)	675	701	2.2	4.6
	982 (1800)	445	460	2.8	1.9
MA 754	Room temperature	586	965	21	33
	871 (1600)	214	248	31	58
	982 (1800)	169	190	18	34
MA 956	Room temperature	517	655	20	35
	1000 (1832)	97	100	—	—
	1100 (2192)	69	72	12	30

YS, yield strength; UTS, ultimate tensile strength; El, elongation; RA, reduction in area.
[a] See Tables 17.2 and 17.3 for chemical compositions of alloys.

temperatures and overall improvement in properties. Mechanically alloyed materials also have excellent oxidation and hot corrosion resistance. The increased resistance to oxidation-sulfidation attack is due to the homogeneous distribution of the alloying elements and the improved scale adherence due to the dispersoid itself [12,13]. Elliott and Hack [14] have presented an overview of the use of mechanically alloyed products for aerospace applications, and deBarbadillo and Smith [15] have described the unique challenges to manufacture and use posed by ODS alloys. A recent survey of the applications of mechanically alloyed products can be found in Refs. 16–18. The different brochures produced by INCO should also be consulted for recent developments and applications of mechanically alloyed products.

17.2.1 ODS Nickel-Based Alloys

Today there are two manufacturers of ODS Ni-based superalloys: Inco Alloys International and Plansee GmbH Lechbruck. While the Inco alloys have the prefix MA, those from Plansee have the prefix PM. Typical compositions of the commercially available mechanically alloyed nickel-based superalloys from Inco Alloys International are presented in Table 17.2. These alloys typically have a solid solution strengthened Ni-Cr matrix with small amounts of Al, Ti, W, and Mo. About 1 wt% of

Table 17.2 Nominal Compositions (wt%) of Mechanically Alloyed Nickel-Based Superalloys

Alloy	Ni	Cr	Al	Ti	Mo	W	Y_2O_3	Ta
Inconel alloy MA754	Bal.	20	0.3	0.5	—	—	0.6	—
Inconel alloy MA757	Bal.	16	4.0	0.5	—	—	0.6	—
Inconel alloy MA758	Bal.	30	0.3	0.5	—	—	0.6	—
Inconel alloy MA760	Bal.	20	6.0	—	2.0	3.5	0.95	—
Inconel alloy MA6000	Bal.	15	4.5	2.5	2.0	4.0	1.1	2.0
TMO-2[a]	Bal.	6	4.2	0.8	2.0	12.4	1.1	4.7

[a] This alloy additionally contains 9.7 wt% cobalt.

Y_2O_3 is also added. Tungsten and molybdenum provide additional solid solution strengthening, while chromium along with titanium and tungsten improve oxidation and sulfidation resistance.

The most significant advantage of oxide dispersion–strengthened superalloys is the increased stress rupture properties. Figure 17.1 compares the specific rupture strength (strength/density) for a 1000-h life as a function of temperature for several nickel-based superalloys used for turbine blade applications. Mar-M200 is a nickel-based alloy containing by weight percent 9.0 Cr, 5.0 Al, 2.0 Ti, 12.0 W, 10.0 Co, 1.0 Nb, and 1.8 Hf, while PWA454 is a nickel-based alloy containing 10.0 Cr, 5.0 Al, 1.5 Ti, 12.0 Ta, 4.0 W, and 5.0 Co, and TD (thoria-dispersed) Ni is nickel containing 2.0 wt% ThO_2. It is clear from this figure that the MA 6000 alloy can maintain a given stress for a much longer time than a conventional alloy for similar vane applications. This is mainly due to the benefits of the combined strengthening modes in the mechanically alloyed material. Due to the increased Al and Ti contents in MA 6000 (and PM 3030 from Plansee), formation of cuboidal, coherent Ni_3Al (γ') particles leads to the precipitation hardening effect in addition to the dispersion hardening effect.

Mechanically alloyed nickel-based superalloys are considered mainly for three groups of applications—gas turbine vanes, turbine blades, and sheets for use in oxidizing/corrosive atmospheres. The largest use of MA 754 is as vanes and bands

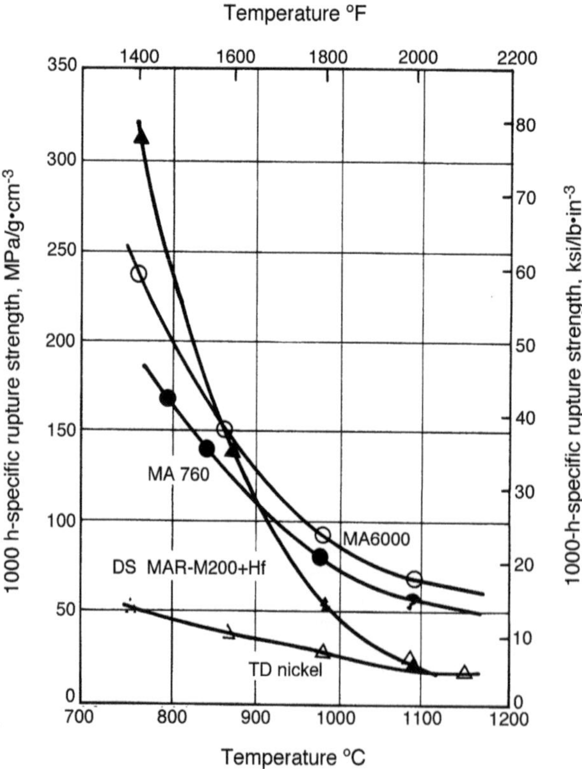

Figure 17.1 Comparison of 1000-h specific rupture strength of Inconel MA 6000 with dispersion strengthened Mar-M200 + Hf, Thoria-dispersed (TD)-Ni, and single-crystal PWA 454.

for aircraft gas turbine engines (Fig. 17.2). For applications requiring good resistance to thermal fatigue, such as gas turbine vanes, Inconel MA 754 is given a strong texture. Majority of the grains are aligned so that their <100> axes are parallel to the principal working direction and along the length of the bar. Such texture results in low modulus of elasticity (149 GPa) in the longitudinal direction. The low modulus improves resistance to thermal fatigue by lowering stresses for given thermal strains. Components are fabricated from bar stock using state-of-the-art machining and brazing processes typical for conventional wrought superalloys. Alloy MA 758 is used in a number of industrial applications where its high chromium content makes it resistant to extremes of temperature and environment. The alloy is used in glass industry for high-temperature components requiring both elevated temperature strength and resistance to extremely corrosive molten glass. Alloy MA 758 is also used for internal combustion engine components, mainly in critical fuel injection parts. One novel industrial application of alloy MA 754 is for a high-temperature atmosphere circulation fan in a "floating" furnace design being commercialized in Japan. Large rounds are used for the hub and plate material is used for blades of the fan, which operates at temperatures over 1100°C.

New product forms of the commercial alloys continue to be developed. Large-diameter, thin-wall tubing of alloy MA 754 has been produced and evaluated for radiant tube applications, and alloy MA 758 has been used as tubing in heat exchangers and process equipment operating at very high temperatures.

The alloy MA 754 is used for brazed nozzle guide vane and band assemblies in U.S. military aero engines. The principal advantages of the alloy for these applications are thermal fatigue resistance, long-term creep strength, and high melting point.

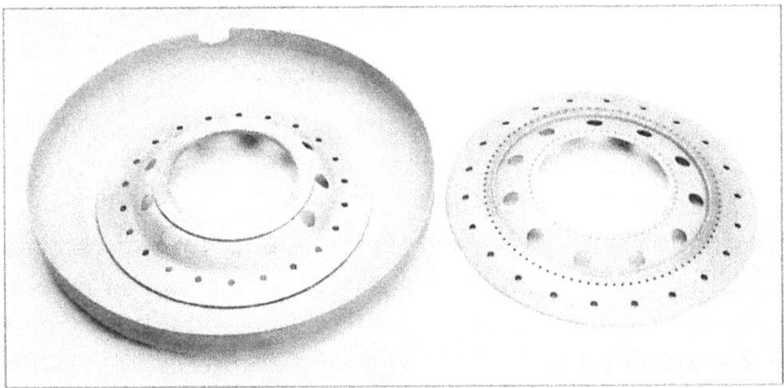

Figure 17.2 Some typical gas turbine engine components made of Inconel alloy MA 754.

The MA 6000 alloy is a more complex alloy developed as a blade material for advanced gas turbines. It is used for first- and second-stage turbine vanes and blades machined from solid bar. Unlike cast alloys, MA 6000 exhibits nearly flat rupture-life curves at high temperatures due to the combination of ODS and high grain-to-width ratios (typically more than 10 to 1). Because of its composition, MA 6000 has excellent resistance to oxidation and sulfidation. The characteristics of this alloy allow blade cooling to be reduced or eliminated as the metal temperature can be increased by 100°C or more in engines where the stresses are medium or low.

17.2.2 ODS Iron-Based Alloys

Table 17.3 lists the chemical compositions of mechanically alloyed iron-based superalloys. These alloys are chromium-rich ferritic stainless steels containing aluminum for oxidation resistance and Y_2O_3 for creep resistance. They combine the high-temperature strength and stability of ODS with excellent resistance to oxidation, carburization, and hot corrosion. These alloys are suitable for use in gas turbine combustion chambers. Incoloy alloy MA 956 is particularly well suited for use in heat processing applications. For example, vacuum furnace fixtures made of MA 956 have shown excellent durability and can compete with wrought molybdenum, which is also used in these applications. In comparison to molybdenum, MA 956 is about 30% lower in density, providing weight savings and cost advantages. Furthermore, since MA 956 has a lower vapor pressure than molybdenum, it does not coat the inside of the vacuum chamber or the parts being heat treated. Thus, MA 956 rods, flats, and sheets are used in numerous atmosphere and vacuum furnace applications including muffles, baskets, trays, and thermowells. MA 956 in tubing form has also been used for high temperature, severe service applications such as coke injection lance pipes in steel making. MA 956 is also being used in glass processing industry because of its resistance to attack by molten glass. Because of this corrosion resistance, the alloy is being evaluated for applications such as firing-kiln rollers, muffle tubes, and furnace racks. Other applications include molten-glass resistance heaters, thermocouple protection tubes, glass processing components used in nuclear waste disposal, and the bushings used to make single C-strand and multistrand fibers.

The MA 957 alloy, containing Mo and Ti in addition to aluminum, has received international consideration for fuel cladding applications in liquid-metal fast breeder reactors [19]. The alloy shows excellent long term microstructural stability in irradiation environments and is expected to retain superb high-temperature strength. The microstructure consists of a metal matrix with uniformly distributed Y_2O_3 dispersoids on the order of 5 nm in diameter. It also contains a highly elongated subgrain structure that is introduced by thermomechanical processing. Since such a material could be considered for first-wall applications of a fusion reactor, the composition of the alloy

Table 17.3 Nominal Compositions (wt%) of Mechanically Alloyed Oxide Dispersion Strengthened Iron-Based Superalloys

Alloy	Fe	Cr	Al	Ti	Mo	Y_2O_3
Incoloy alloy MA 956	Bal.	20	4.5	0.5	—	0.5
Incoloy alloy MA 957	Bal.	14	—	1.0	0.3	0.25

has been altered to be in line with low activation data. Accordingly, Mo is substituted by W. The optimized composition for this purpose without containing any austenite phase in the alloy was found to be Fe-13.5Cr-2W-0.5Ti-0.25Y_2O_3 (wt%) [20].

More recently, MA 957 has been evaluated for use as the fuel cladding in fast neutron, breeder reactors. Conventional austenitic alloys are unsuitable for this application due to the dimensional swelling phenomenon caused by the high neutron fluxes. Mechanically alloyed materials are also being evaluated for heat exchanger components in high-temperature gas-cooled reactors. Figure 17.3 shows some typical high-temperature applications of the alloy MA 956.

Dour Metal developed a series of dispersion-strengthened steels through MA methods. These steels, designated as ODM (oxide dispersion microforged) alloys, contain Cr and Al as major solute elements and 1.5 wt% Mg, 0.6 wt% Ti, and 0.5 wt% Y_2O_3. Both Cr and Al impart oxidation resistance to the steels. Table 17.4 lists the compositions and mechanical properties in the hot isostatically pressed condition. These alloys have satisfactory mechanical properties at temperatures up to 1100°C, which is the upper limit for use of superalloys from the viewpoint of both oxidation resistance and mechanical properties. Ductility of the ODM alloys has been improved by giving a suitable heat treatment to favor transgranular rupture. Compared to MA 956, the time to rupture for the ODM alloys is two orders of magnitude higher. Because of the combination of these excellent mechanical properties, ODM alloys are expected to find applications in energy conversion systems (particularly heat exchanger tubes), and also in systems requiring protection from corrosive atmospheres [21].

Some attempts have also been made to produce high-wear-resistant steels via MA by incorporating hard phases such as NbC, TiC, TiN, and Al_2O_3 into iron. The steel contains about 0.9 wt% C and 0.6 wt% P. Since P causes grain boundary embrittlement, its amount must be carefully controlled. The wear behavior of these alloys was found to be the best when the amount of the refractory compound is about 10 vol% [22,23]. Use of coarse hard-phase particles reduced dry wear, while fine hard phase particles reduced wear for oil lubrication at higher surface pressures and longer times.

17.2.3 ODS Aluminum-Based Alloys

The success of mechanically alloyed superalloys led to the development of dispersion-strengthened aluminum alloys. Table 17.5 lists the compositions of the mechanically alloyed dispersion-strengthened aluminum alloys. Since an aluminum oxide layer is always present either on the surface of the powder particles at the start of processing or during milling, its incorporation into the alloy during consolidation contributes to significant improvements in the mechanical properties of the alloy. Furthermore, since aluminum is a ductile metal, process control agents (PCAs) are added to assist in minimizing cold welding during processing. Consequently, oxide and carbide (Al_2O_3 and Al_4C_3) particles are formed due to reaction of Al with oxygen and carbon during MA by the decomposition of the PCA. Both the oxide- or carbide-type dispersions are about 30–50 nm in size and stabilize the ultrafine grain size. This results in 50% increase in strength, higher fracture toughness, and improved resistance to stress corrosion cracking and fatigue crack growth of the mechanically alloyed materials. Additional strengthening is also achieved due to the dislocation substructure developed during thermomechanical processing. These alloys also exhibit excellent thermal

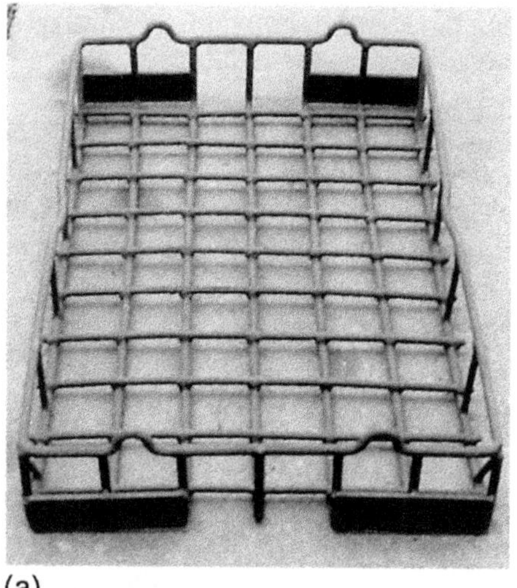

(a)

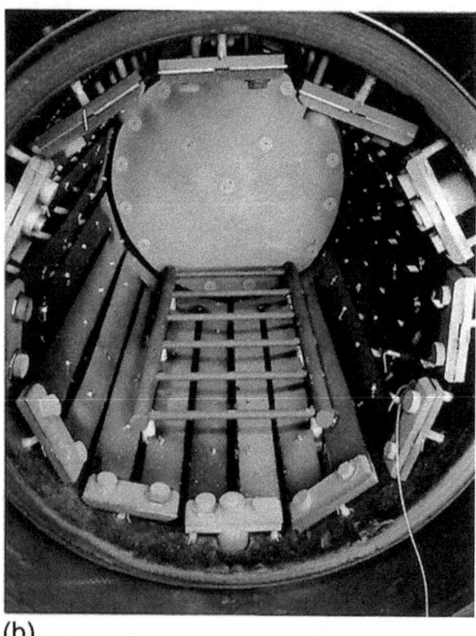

(b)

Figure 17.3 (a) Furnace baskets made of Incoloy alloy MA 956 operating in air at temperatures above 1200 °C. (b) Removable hearth of vacuum heat treating furnace fabricated from 1.0- and 0.5-in. diam. rods of Incoloy alloy MA 956. The hearth legs are encased in alumina sleeves to prevent direct contact with the graphite heating elements.

Table 17.4 Compositions and Mechanical Properties of the Oxide Dispersion Microforged Alloys (Fe-*x*Cr-*y*Al-1.5Mg-0.6Ti-0.5Y_2O_3, all in wt%) in the Hot Isostatically Pressed Condition

			Microcrystal 900°C			Macrograin 900°C		
Alloy	% Cr	% Al	0.2% YS (MPa)	UTS (MPa)	% El	0.2% YS (MPa)	UTS (MPa)	% El
ODM 331	13	3	28	41	157	169	176	10
ODM 361	13	6	42	61	86	166	177	12
ODM 061	20	3	49	71	88	159	173	16
ODM 751	16.5	4.5	32	47	134	—	—	—
MA 956	20	4.5				108	115	8

shock resistance and high sound damping capacity. Additional benefits can be accrued by adding carbide-forming elements such as titanium. ODS aluminum alloys, with the trade name DISPAL (DISPersion strengthened ALuminum), have been jointly developed by Krebsöge, Erbslöh-Aluminium, and Eckart-Werke. These materials can be used in combustion engines (e.g., motor cycle pistons). The major obstacle to their widespread use is their cost; thus, they are used only in some specific applications.

As in the case of rapid solidification processed (RSP) aluminum alloy development, progress in mechanically alloyed aluminum alloys also has occurred in three different groups. These are (1) high-strength alloys, (2) ultra-light-weight alloys, and (3) high-temperature alloys.

High-Strength Alloys

IncoMAP alloy AL-9052 is a solid solution alloy that provides a good combination of tensile strength, fatigue strength, and toughness, normally observed in precipitation-hardened 7XXX series alloys. The high strength of this alloy arises from solid solution strengthening due to Mg; ultrafine grain size of the matrix; fine dispersion of MgO, Al_2O_3, and Al_4C_3 particles; and substructure strengthening. Since this is a solid solution–type alloy, it has also good resistance to corrosion and stress corrosion cracking. This alloy has a density 5% less than that of conventional age-hardenable aluminum alloys of comparable strength such as 2024. With its combination of light weight, high strength, and corrosion resistance, IncoMAP alloy AL-9052 is appropriate for aerospace applications where marine corrosion is also a factor.

Recently, high-strength Al-Ti alloys have been developed using MA by dispersing nanometer or submicrometer-sized Al_3Ti intermetallic particles (in addition to the

Table 17.5 Nominal Compositions (wt%) of Mechanically Alloyed Dispersion Strengthened Aluminum-Based Alloys

Alloy	Al	Mg	Li	C	O
IncoMAP alloy AL-9021	Bal.	1.5	—	1.1	0.8
IncoMAP alloy AL-9052	Bal.	4.0	—	1.1	0.6
IncoMAP alloy AL-905XL	Bal.	4.0	1.3	1.1	0.6

Al_2O_3 and Al_4C_3 dispersoids from addition of PCAs) in an Al matrix [24]. Similar approaches could be used to develop high-strength alloys in other systems. In fact, it has been reported that Nb additions at the level of about 10 wt% produced a fine-grained microstructure with Al_3Nb, Al_2O_3, and Al_4C_3 and NbC dispersoids, which possessed a high strength of 400 MPa at room temperature [25]. Addition of 1 wt% Y_2O_3 increased the strength further to 450 MPa, without any deleterious effect on the ductility [25].

Ultra-Lightweight Alloys

Addition of lithium to mechanically alloyed aluminum alloys has produced an ultra-light-weight alloy IncoMAP alloy AL-905XL. Its density is 8% lower and stiffness 10% greater than the age-hardenable conventional alloy 7075-T73 of comparable strength. The excellent combination of the properties makes this alloy very attractive for airframe applications. In particular, the freedom from age-hardening treatments makes it possible to produce forgings and heavy sections with homogeneous metallurgical structures.

High-Temperature Alloys

Reactive milling of aluminum alloys leads to incorporation of dispersoids, such as Al_2O_3 and Al_4C_3, into the aluminum matrix, which increases the strength of the alloy. This strength is retained to relatively high temperatures. In addition to these alloys, which are DISPAL alloys, high-strength Al-Fe-Ce alloys synthesized through rapid solidification processing have also been developed by MA methods. Addition of transition metals, such as Fe, and rare-earth elements, such as Ce, to aluminum has resulted in alloys that can be used up to 590 K, competing with the performance of titanium alloys [26]. Strengthening in these alloys is achieved by the transition metal intermetallic compounds. The low solubility and diffusivity of the transition and rare-earth elements result in slow coarsening of these precipitates [27]. Many investigations on rapidly solidified Al-Fe-Ce alloys have revealed that the strength decreases rapidly at elevated temperatures [28]. MA of the rapidly solidified alloys resulted in a uniform dispersion of the thermodynamically stable carbides and oxides in the aluminum matrix, which helped in supplementing the strengthening achieved by the intermetallics. This led to increased stiffness and strength at temperatures, much higher than those obtained for the nonmechanically alloyed rapidly solidified alloys [29]. The strength of the mechanically alloyed alloy was found to be six times stronger, even at

Table 17.6 Steady-State Creep Rates of Mechanically Alloyed (MA) and Non–Mechanically Alloyed (Non-MA) Al-8 wt% Fe-4 wt% Ce Alloys at Different Temperatures and Stresses

		Creep rate (s^{-1})	
Temp. (°C)	Stress (MPa)	MA	Non-MA
350	103	8.3×10^{-10}	4.1×10^{-7}
380	83	1.6×10^{-9}	2.6×10^{-7}
380	103	1.4×10^{-9}	4.9×10^{-5}

773 K, than the rapidly solidified alloy. Comparison of the creep response of the nonmechanically alloyed and mechanically alloyed Al-Fe-Ce alloys at the same conditions of stress and temperature shows that the mechanically alloyed material shows significant enhancement in creep resistance [30]. Table 17.6 summarizes the steady-state (minimum) creep rates at different temperatures and stresses.

17.3 MAGNESIUM-BASED ALLOYS

Several mechanically alloyed magnesium alloys have also found useful applications. We will, however, concentrate on two specific applications here: supercorroding alloys and hydrogen storage materials.

17.3.1 Supercorroding Alloys

A useful application of the MA technique has been in the production of supercorroding magnesium alloys that operate as short-circuited galvanic cells to corrode (react) rapidly and predictably with an electrolyte such as seawater to produce heat and hydrogen gas [31,32]. Such an alloy system is suitable as a heat source for warming deep-sea divers, as a gas generator to provide gas for buoyancy, or as a fuel in hydrogen engines or fuel cells. The corrosion rate of alloys can be maximized by providing (1) a short electrolyte path, (2) a large amount of exposed surface area, and (3) a strong bond (weld) between the cathode and the anode. It is also useful to provide a very low resistance path for external currents to flow through the corroding pairs. All these requirements can be met with MA processing. Consequently, magnesium-based alloys containing Fe, Cu, C, Cr, or Ti have been evaluated for such applications. The Mg-5 to 20 at% Fe alloy is ideal because of its extremely fast reaction rate, high power output, and high percentage of theoretical completion of the actual reaction. For corrodable release links an alloy with a slower reaction rate, such as Mg-5 at% Ti is useful. The corrosion rates can be precisely controlled by adjusting the alloy composition.

One of the recent applications of this phenomenon of increased activity for the mechanically alloyed Mg alloys is the development of meals, ready to eat (MRE) used to provide hot food for the U.S. soldiers during the Gulf War (Desert Storm Operation). This is a very good example of commercialization of a simple idea with tremendous potential.

In this food heater (Fig. 17.4), the starting materials are finely ground, mechanically alloyed powders of Mg and Fe. A mixture of this powder is pressed into the pocket. When water is added to the finely ground mix, the following reaction takes place generating heat:

$$Mg/Fe + 2H_2O \rightarrow Mg(OH)_2 + H_2 \uparrow + \text{heat} \qquad (17.1)$$

A galvanic couple exists between the two dissimilar metals (Mg and Fe) and this reacts with water to produce heat. This heat raises the temperature of the food stored inside the packet making hot meals available, ready to eat.

17.3.2 Hydrogen Storage Materials

In recent years, there has been intense activity on the synthesis of metal hydrides through MA [33–35]. Although not yet commercialized, metal hydrides have been attracting very serious attention because metal hydrides are materials for safe storage of hydrogen and they can store hydrogen with a higher volume density than liquid hydrogen. For a material to be seriously considered for hydrogen storage, it should have (1) high hydrogen storage capacity, (2) fast kinetics of storage and removal (reversible hydrogenation and dehydrogenation), (3) low hysteresis, (4) possibility of hydriding/dehydriding at relatively low temperatures and pressures, preferably at room temperature and atmospheric pressure, (5) long lifetime, (6) high chemical stability and thermally easily decomposable, (7) no or minimal need to activate the material, (8) easy availability, (9) relative immunity to impurities, and (10) low cost (including that of fabrication). Among the different alloy systems investigated so far (based on Ti, La, Zr, Hf, V, Pd, and Mg), magnesium alloys have the pronounced advantage because of their large hydrogen capacity (7.6 wt%), low cost, and easy availability [36]. However, Mg is hard to activate, its temperature of operation is high, and the sorption kinetics are slow. For example, Mg desorbs hydrogen at a minimal temperature of 300°C, whereas a temperature of 180°C is desired for applications. Furthermore, due to its low melting point and high vapor pressure, preparation of Mg-based alloys by conventional metallurgical processes is difficult. It is easier to produce

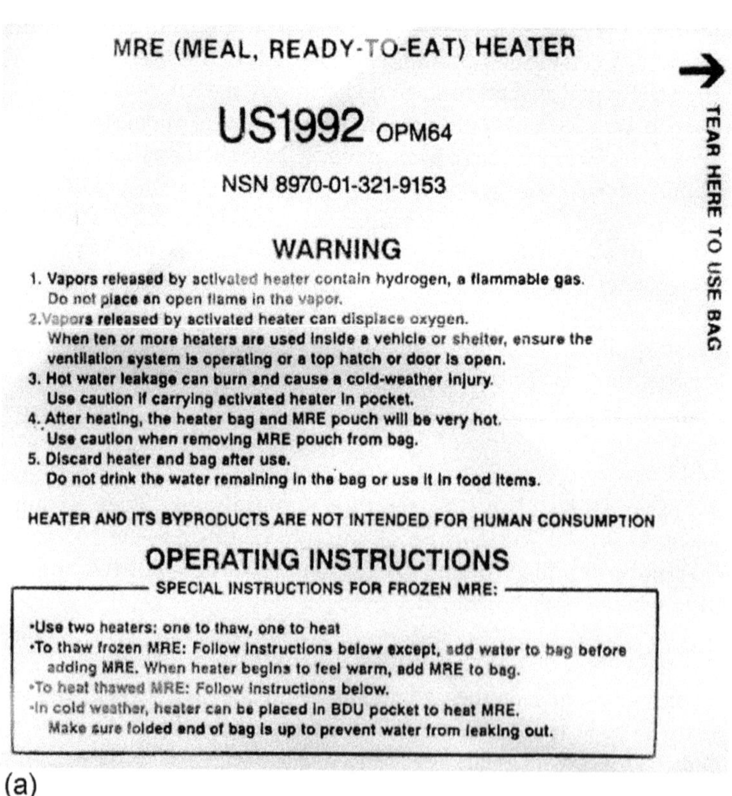

(a)

Figure 17.4 Cover of the MRE (Meal, Ready to Eat) heater packet used to provide hot meals to U.S. soldiers during the Desert Storm operation. The food heater contains mechanically alloyed powders of Mg and Fe, which on contact with water produce heat.

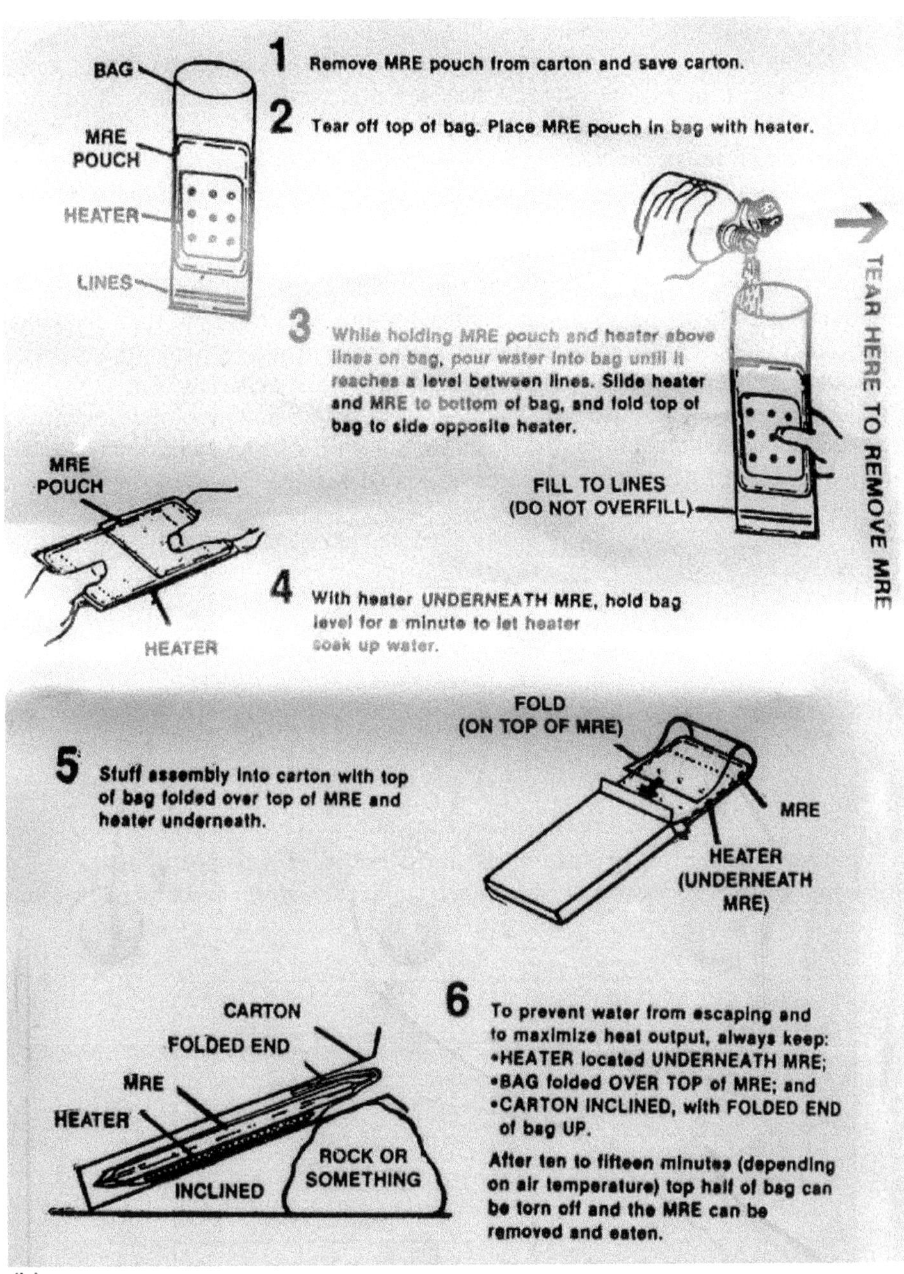

(b)

Figure 17.4 Continued.

Mg-Ni alloys by MA than by ingot metallurgy methods because of the large difference in the melting points of the two metals and the high vapor pressure of magnesium. These are some of the reasons why MA has been utilized to produce Mg-based alloys for use as hydrogen storage materials [37].

Figure 17.5 shows the hydrogen absorption behavior at 300°C of the intermetallic Mg_2Ni as a function of grain size. It may be noted that while the coarse-grained polycrystalline intermetallic powder absorbs hydrogen slowly, the nanocrystalline intermetallic obtained by MA absorbs hydrogen faster. Furthermore, the maximal amount of hydrogen absorbed is also higher (3.5 wt%) for the nanocrystalline compound in comparison to about 3 wt% for the coarse-grained polycrystalline compound. In fact, it has been noted that the rate of absorption is higher the smaller is the grain size. A similar trend was also noted for other materials. The desorption behavior of hydrogen has also been found to be faster for nanocrystalline materials.

It was reported that even faster hydrogen sorption kinetics and easier activation could be achieved by preparing a multicomponent Mg-based system. The additives could be classified into six groups [36]:

1. Ni which forms Mg_2Ni
2. Elements such as Ce, Nb, and Ti, which form hydrides and which can serve as "hydrogen pumps"
3. Low-temperature metal hydrides such as $LaNi_5$ and FeTi, which give rise to multiphase systems
4. Metal catalysts such as Fe, Co, or Cr, which do not form hydrides themselves
5. Elements that form covalent bonds with Mg, such as Si and C, and
6. Metal oxides

Figure 17.6 shows the absorption and desorption behavior of hydrogen in magnesium with microcrystalline and nanocrystalline structures and with and without

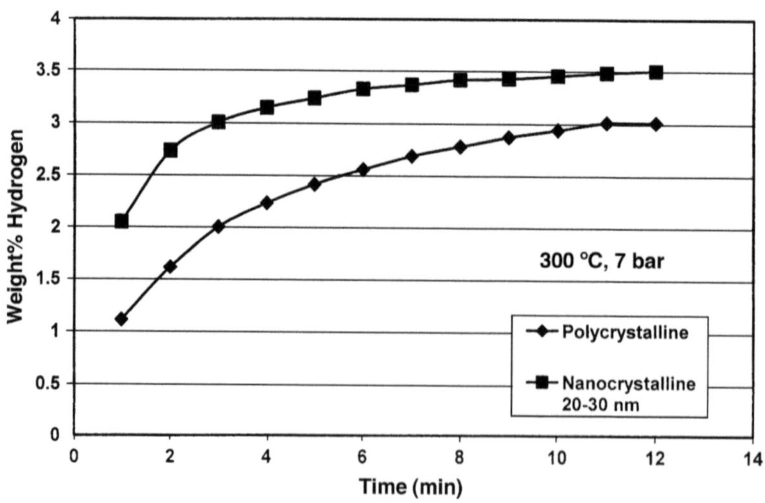

Figure 17.5 Effect of grain size on hydrogen absorption in mechanically alloyed Mg_2Ni powder at 300°C and 7 bar pressure.

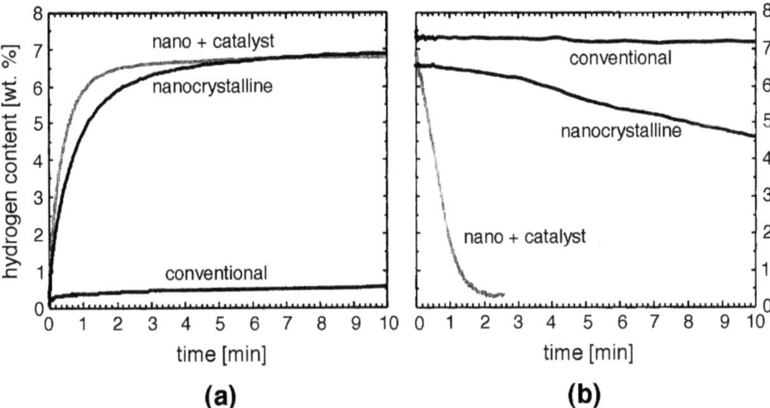

Figure 17.6 (a) Hydrogen absorption of nanocrystalline Mg with and without the catalyst in comparison to coarse-grained Mg. (b) Hydrogen desorption behavior of nanocrystalline MgH_2 with and without the oxide catalyst in comparison to coarse-grained MgH_2.

an oxide catalyst in the nanocrystalline powder. It may be noted again from Figure 17.6a that the hydride formation reaction in the microcrystalline material is slow. Less than half of the full hydrogen storage capacity was realized in less than 30 min. In contrast, nanocrystalline Mg could be completely converted to the hydride phase in a few minutes. This has been related to the larger grain boundary area in the nanocrystalline material and consequent higher diffusion rate. The presence of a high density of heterogeneous nucleation sites aids the situation further. It has also been known that diffusion of hydrogen through MgH_2 is more than a factor of 5000 times slower than diffusion through Mg. Formation of the hydride along the grain boundaries, may result in slow filling of the interior of the micrometer-sized Mg crystallites, i.e., growth of the hydride slows down since the hydride shell of the crystallites acts as a diffusion barrier for hydrogen. In the nanocrystalline material, the small dimensions of the crystallites eliminate this problem, and consequently hydride formation is faster.

The desorption behavior is also accelerated in the case of nanocrystalline materials (Fig. 17.6b). While the unmilled microcrystalline material does not desorb any hydrogen within 1 h, the hydrided milled powder (with a nanocrystalline structure) can desorb completely in about 30 min. The reaction kinetics of hydrogen absorption and desorption are further enhanced by the addition of different metal oxides, e.g., Cr_2O_3, which act as catalysts. Figure 17.6a shows that 95% of the full capacity (of hydrogen absorption) is reached within 30 s, in the presence of a catalyst. The effect of a catalyst is even more dramatic for the desorption reaction (Fig. 17.6b). While desorption from pure nanocrystalline Mg may require about 30 min (the microcrystalline material does not show any desorption in this time range), complete desorption is possible in less than 5 min in the presence of a catalyst. Another advantage of the addition of a catalyst is that significant reaction rates can be achieved even at reduced temperatures [38]. Although at a lower rate, absorption was found to be possible at 40°C. For technical applications, the heat release upon absorption increases the temperature of the hydride bed, and this continues to accelerate the process. Thus, the fast kinetics, in combination with one of the highest reversible storage capacities, qualifies these light-weight hydride materials for application in zero emission vehicles [39].

While investigating the application of MA to hydrogen storage materials, Schulz and coworkers at Hydro Quebec, Canada [40] demonstrated results that are useful for pilot plant scale studies. Intensive milling of magnesium hydride with vanadium produced a nanocomposite of β-MgH_2 + γ-MgH_2 + $VH_{0.81}$. The MgH_2 + 5 at% V composite produced by mechanical milling desorbed hydrogen at 473 K under vacuum and reabsorbed hydrogen rapidly even at room temperature. The activation energy for hydrogen desorption was 62 kJ mol^{-1}. The fast kinetics of the ball-milled MgH_2-V composite primarily comes from the catalytic effect of vanadium and the small powder particle size. The formation enthalpy and entropy of the composite were identical to that of pure magnesium.

Several magnesium-based and titanium-based intermetallics are being evaluated for this application. It was reported that MA of MgH_2 and vanadium had the best ever reported hydrogen storage (about 5.5 wt%) properties for an Mg-based system [34].

With increasing interest in "green" energy sources and relatively low-purity requirements, hydrogen storage materials hold the most promise for the next MA application to be commercialized.

17.4 OTHER APPLICATIONS

17.4.1 Spray Coatings

Nanostructured materials show high potential for various engineering applications due to their improved mechanical properties [41–43], and these could be conveniently produced by MA/MM methods (see Chapter 13). For example, it has been shown that mechanically alloyed and hot isostatically pressed or sintered WC-Co composites retain a nanocrystalline microstructure and show higher hardness and wear resistance than coarse-grained material of equivalent ceramic content [44]. Since for most applications it is not necessary to produce costly massive components, coating techniques can be used to protect local areas of materials against wear or corrosion. Several hard coatings based on WC, TiC, or Cr_3C_2 have been developed in recent years. The wide variety of thermal spray techniques and especially the opportunity to use powder as a feedstock has opened a wide field of applications for single-phase or composite materials.

The major advantages of mechanically alloyed/milled powders include achievement of a homogeneous and fine distribution of metallic and ceramic components in an economical and cost-effective manner. Due to high particle velocities, high-velocity oxy fuel (HVOF) spraying exposes the coating material to high flame temperatures only for less than a millisecond, and consequently only slight changes in microstructure occur. The large mechanical impact due to the high velocity guarantees the formation of dense, solid coatings without requiring complete melting of the powder stock.

Gärtner et al. [45] studied the microstructure and mechanical and tribological properties of thermally sprayed WC- or TiC-based coatings and noted that, in contrast to conventional powder, nanocrystalline cermet powders generally showed more homogeneous coating microstructure. The hardness of the nanocrystalline composite coatings was reported to be higher than that of the coarse-grained powder. Stated differently, hardness similar to that of the coarse-grained powder coating can be obtained in the nanocrystalline coating, even with significantly higher metallic contents. The resistance of nanocrystalline composites against abrasive or adhesive wear was comparable or even superior to that of coatings of conventionally processed powders. Furthermore, since the microstructure of nanostructured coatings was homogeneous and less porous than that from conventional powder, corrosion resistance is also expected to be better.

Mechanical alloying can be used to produce powders of MCrAlY type, where M is Fe, Co, Ni or a combination of these elements. These powders can be plasma-sprayed in a low-pressure chamber to provide oxidation- and corrosion-resistant coatings on gas turbine blades and vanes [2]. These coatings provide corrosion and wear resistance. Uniform dispersion of oxide, nitride, and carbide phases formed in situ during MA can be beneficial in increasing the hardness of the coatings. It has also been noted that these coatings are useful for environmental protection [46]. Additions of yttrium can form Y_2O_3 dispersoids and such composites can be used as diffusion barrier coatings [47]. This type of barrier reduces the adverse effects of concentration gradients between the substrate and an overlay coating.

Mechanically alloyed powders can also be used as feedstock for powder injection, since MA can easily achieve a very intimate mixture of the constituents. It has been recently reported that mechanically alloyed Fe-SiC composite powders can be coated with a detonation gun to improve the wear resistance of medium carbon steel [48]. There was very little SiC phase in the coating layer when the powder was not mechanically alloyed. MA helped in uniformly distributing up to about 60 vol% of SiC powder in the iron matrix.

17.4.2 Thermoelectric Power Generator Materials

Thermoelectric power generator has assumed lot of importance in recent years as a reliable source of electric power. A number of different materials have been investigated for use in the thermoelectric power generation. β-$FeSi_2$ is one of the most promising materials for energy conversion in the temperature range of 200–900°C. This material exhibits high electrical conductivity and large thermoelectric power. $FeSi_2$ is a traditional thermoelectric material, which is nonpoisonous, has good resistance to oxidation, and can be operated in air without any protection. However, the biggest disadvantage of this material is that it is porous and brittle in the as-cast condition. Furthermore, the presence of the eutectic and peritectoid reactions during cooling from the liquid to the β phase causes difficulties in obtaining a homogeneous bulk material due to problems of segregation. Thus, conventional processing includes several steps: melting and casting, grinding the ingot to fine size, screening the powder, sintering the powder of uniform size, homogenization heat treatment, and annealing to produce the material in the β form to be used as thermoelectric material.

MA-processed powders will not have the segregation problem because no liquid phase is involved during processing. In addition, the number of processing steps is also minimized. For example, Fe, Si, and the dopant powders (to make the material either p-type or n-type semiconductor) can be mechanically alloyed and then subjected to sintering followed by β annealing to produce the desired material [49,50]. Figure 17.7 shows two different ways of making the β-$FeSi_2$ thermoelectric generator material.

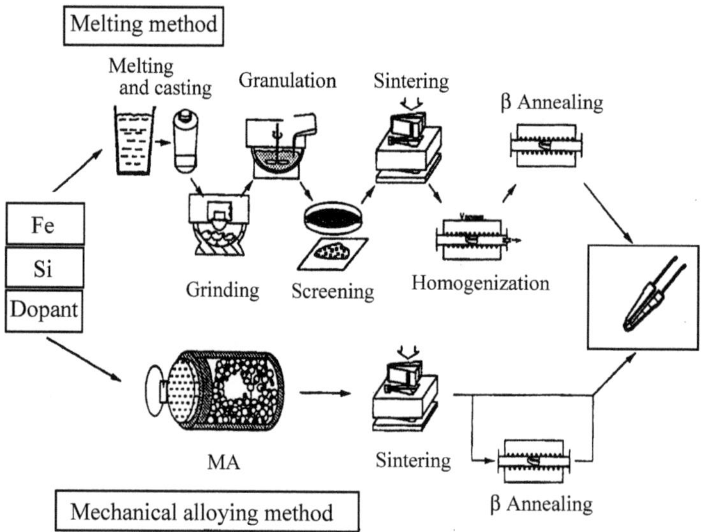

Figure 17.7 Schematic illustration showing the traditional (ingot solidification method) and modified (MA) methods of forming the β-$FeSi_2$ thermoelectric generator material.

The thermoelectric performance of materials is represented by the effective maximal power (P_{eff}) and maximal conversion efficiency (η_{max}). Both factors should be large for the material used in a thermoelectric power generator. Improvement in both these factors can be achieved by reducing the thermal conductivity of the material. Since MA is known to increase solid solubility limits, reduce grain size, and increase defect density, mechanically alloyed materials should exhibit lower thermal conductivity than the ingot-processed material.

Table 17.7 shows the thermoelectric properties of the specimens prepared from ingot metallurgy and MA processes. It may be noted that the mechanically alloyed powder shows a higher conversion efficiency and exhibits higher values for almost all the properties; the power produced is slightly less.

17.4.3 Waste Utilization

Approximately 100 million tons of mining waste is produced annually in the United States as a result of the crushing and processing of rock minerals to produce construction and road products. An estimated one-fourth of this is sold as an inert filler while the remaining material must be stored or disposed. The stockpiling and disposal of these wastes produced as a result of aggregate crushing and production operations is considered one of the major problems facing the aggregate industry today. Methods have been developed to add this material as nonreactive filler to concrete and asphalt products, with little added value. Other limited studies have evaluated use as soil admixtures or amendments, and use in flowable fill materials. However, until recently little attention was paid to chemically changing these materials to alter their reactive characteristics.

Lessard and Havens-Cook [51] have recently developed a process to convert by-product granite dust ("fines") into reactive "pozzolan" using an attritor mill.

Table 17.7 Thermoelectrical Properties of Fe-Mn-Al-Si Alloy Powders Processed by Ingot Metallurgy (IM) and Mechanical Alloying (MA) Methods[a]

Specimen	IM ($Fe_{30.7}Mn_{2.7}Si_{64.6}Al_2$)	MA ($Fe_{28}Mn_2Si_{67}Al_3$)
Seebeck potential, E_o (mV)	1.5×10^2	1.9×10^2
Seebeck coefficient, α (mV/K)	0.21	0.27
Average specific resistance, ρ_m	5.7	11
Thermal conductivity, K (W/m.K)	6.9	≈5
Figure of merit, Z (1/K)	1.1×10^{-4}	1.3×10^{-4}
Effective maximal power, P (W.cm/cm^2)	0.97	0.86
Maximal conversion efficiency, η (%)	1.9	2.3

[a] ΔT, Temperature difference one which operation takes place at 720 K.

According to the American Society for Testing and Materials, (ASTM), pozzolan materials are "siliceous or siliceous and aluminous materials which in themselves possess little or no cementitious value but will, in finely divided form and in the presence of moisture, chemically react with calcium hydroxide at ordinary temperatures to form compounds possessing cementitious properties." When the granite rock dust milled for 1 h was mixed with lime, water, and sand, the compressive strength of the samples exceeded the minimal required value of 600 psi, specified by ASTM. It was also noted that the compressive strength increased with increasing milling time. In fact, one of the samples prepared with mechanically activated granite fines had a compressive strength almost twice that made from commercially available fly ash.

Similar to the above-mentioned granite dust, about 300 million tires are discarded every year in Europe alone. Most of these are landfilled at present, but this situation cannot continue for long due to the new proposed regulations. Therefore, reuse of the scrap tire rubber is desirable. Although energy recovery from tire scraps is a practical option, it is preferable to recycle rubber in new articles rather than to simply use it as fuel; this could also mitigate environmental pollution. Accordingly, Magini et al. [52] used the process of MA to convert the scrap tire rubber into value-added product. The major problem in this process is that the vulcanized rubber must be devulcanized. Vulcanization is the chemical process by which polymer molecules are joined together by cross-linking into larger molecules to restrict molecular movement, i.e., they become rigid. This happens when the carbon-carbon double bonds of polyisoprene rubber open up and cross-link with sulfur atoms. These bonds need to be broken down for utilizing this scrap rubber into useful forms; this was achieved by MA [52]. Progress of de-cross-linking at different milling times was monitored by measuring the extracted sol fractions and cross-linking densities of the milled samples. The cross-linking density decreased and the sol fraction increased when milling was conducted in air in the presence of an antioxidant agent. When this milled rubber was blended with virgin rubber, great homogeneity was observed with the surrounding matrix. Optimization of the process parameters must be done to achieve the best possible degree of devulcanization by milling.

17.4.4 Metal Extraction

It was described in Chapter 14 that mechanochemical reactions could be utilized to reduce oxides, chlorides, and sulfides to pure metals by reacting the ores with a more reactive metal in a high-energy ball mill. This process is now utilized on a commercial scale to produce pure metals.

Mechanical milling has been used in mineral processing basically to increase the surface area and improve the chemical reactivity of the milled materials. It is now clear that mechanical activation increases defect concentration in the milled material and thus changes the reactivity of the solids. Although mineral processing requires very high productivity and very low cost, Balaz [53] exploited the principles of mechanical activation for extracting metals from minerals. The sulfide concentrates from different sources (e.g., from Chile, Peru, Slovakia) were successfully used for extracting antimony and arsenic from them. The mechanochemical treatment improved the degree of recovery and the rate of leaching of both metals. The process of mechanochemical leaching of tetrahedrite was patented and tested on a pilot plant scale in eastern Slovakia. Tests have shown that with alkaline Na_2S solution applied at 357–361 K and atmospheric pressure, 20–40 min was sufficient for almost full recovery of antimony from tetrahedrite (Fig. 17.8). The other valuable metals (Cu, Ag, and Au) form the main economic components of the solid residue. The liquid/solid ratio (4.8:2.5) and power inputs (150–250 kWh per ton of a product applied in optimized experiments) were acceptable from the viewpoint of plant operation. The process could be conducted at a lower temperature and for much shorter leaching times in comparison with the Sunshine process currently used in the United States. This is probably the only known example of detailed understanding of mechanochemical effects and their application on pilot plant scale technology.

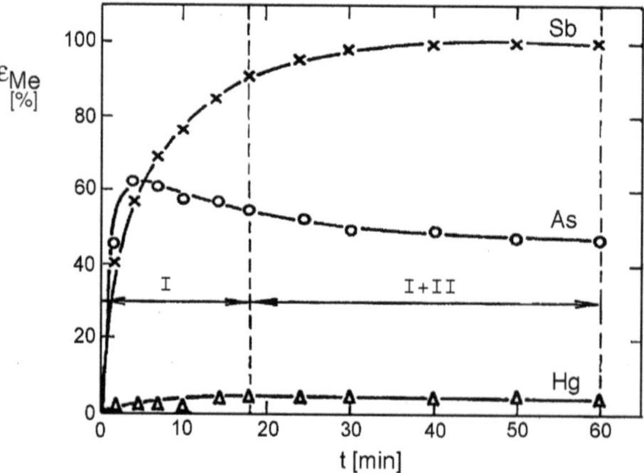

Figure 17.8 Recovery of Sb, As, and Hg into leach (ε_{Me}) vs. time of leaching, t. (I) Mechanochemical leaching and (II) combined mechanochemical and chemical leaching. The leaching conditions are: leaching agent 300 g/L Na_2S + 50 g/L NaOH, temperature 361 K, liquid/solid ratio 3.3, and power input 245 kWh/ton.

Table 17.8 Comparison of Traditional and Mechanochemical Methods for Production of Phosphate Fertilizers

Process characteristics	Traditional method	Mechanochemical method
Pittrman's probe in % citric acid S	80–85	55–60
Ammonium citrate	60	40–50
Increase of harvest related to superphosphate	1.00	0.95
Duration of production process	20 d (periodical)	20–24 h (continuous)
Consumption of energy per ton	450 kWh/t	500 kWh/t
Consumption of reagents per ton	0.7 t H_2SO_4 (60%)	—
Duration of agrochemical action of the fertilizer	1–2 years	5–7 years
Release of fluorine into atmosphere	Yes	No
Increase of acidity of soils	Yes	No
Possibility to use in small deposits (<1 million tons) of phosphorus rocks	No	Yes

Mechanochemical methods were also used to produce phosphorus fertilizer on a commercial scale using a vibrocentrifuge mill. Traditionally, phosphate fertilizers are produced utilizing the reactions:

$$Ca_3F(PO_4)_3 + 5H_2SO_4 + 2.5H_2O \rightarrow 5(CaSO_4.0.5H_2O) + 3HPO_4 + HF \qquad (17.2)$$

$$Ca_3F(PO_4)_3 + 7H_3PO_4 + 5H_2O \rightarrow 5[Ca(H_2PO_4).H_2O] + HF \qquad (17.3)$$

The same product could also be obtained by mechanical activation of phosphates and apatites. However, mechanochemical methods have advantages over the traditional methods as summarized in Table 17.8. Different machines consume different amounts of power for the production of the fertilizers, and the capacity of the mills is also different. It was reported that the vibrocentrifuge mill was the best suited for this purpose because it could produce about 5000 kg of fertilizer per hour and could also be used continuously. The first factory to produce the phosphate fertilizers was built in Mongolia to use the Berenghoi phosphorus ore. Since then about 1.5 million tons of the fertilizer has been produced.

17.4.5 Processing of Polymers

Application of MA to polymeric materials is a recent development, first reported in 1992 by Shaw [54]. Encouraged by the success of their experiments on MM of polyamide (PA), Shaw and coworkers investigated polyethylene (PE), acrylonitrile-butadiene-styrene (ABS), polypropylene (PP), polystyrene (PS), and combinations of these [55–61]. Shaw recently summarized the results of his group [62], and other investigators have also started similar investigations [63–69].

It was originally thought that MA of polymers could lead to (1) degradation due to continual breaking of the long polymer chains, resulting in a large number of short chains with no useful physical properties; (2) inability to recombine the resultant particles to form a solid component due to their low diffusivity; and (3) difficulty in combining different polymers (which may be incompatible) to produce polymeric alloys. However, actual experiments on different polymeric materials have proved these suspicions to be unwarranted.

Polymeric materials are usually processed at low temperatures so that processing takes place below the glass transition temperature, T_g. It was reported that mechanically alloyed polymers have lower amounts of crystallinity, and that both weight-averaged and number-averaged molecular weights decreased with increasing milling time. Furthermore, ambient temperature milling produced much more pronounced reduction in molecular weight than cryomilling. It was noted that MA of polymers did not result in chain degradation but rather in new bond formations. Some polymers showed chain grafting resulting in increase of molecular weight; others showed chain breakage and decrease in molecular weight. Cross-linking and interpenetrating networks, and some attachment to the carbon rings were also noted. In most of the cases, the strength of the milled polymer was higher than the unmilled one. The strength could be increased further by heat treatment and increasing the crystallinity. The mechanically alloyed polymers also showed increased impact resistance, although the reasons for this are not known. Koch et al. [70] reported that milling initially disrupted polyethylene terephthalate (PET) crystals by shearing parallel to the (010), planes which left the intermolecular spacings within the (010) planes intact. After about 1 h of milling the PET chains underwent rotational disordering about the chain axis and the intermolecular spacing within the (010) planes increased. Extended milling produced an oriented amorphous morphology, regardless of the crystallinity of the PET sample before milling. This oriented amorphous morphology crystallized upon annealing into extended-chain crystals, as evidenced by an increase in their melting temperature.

Normally incompatible polymers could be made to react on an atomic level by MA. Thus, PA/PE and PA/ABS combinations could be successfully processed. Alloying was inferred from the shifting of the T_g peaks of the two individual components toward each other, as determined from the differential scanning calorimetric traces. This is somewhat similar to alloy formation in normally immiscible metals as described in Chapter 9. Mechanical properties of these polymeric alloys are some combination of the individual constituent properties but are not necessarily based on a rule of mixtures as found in composite materials [62].

There seem to be many similarities in the milling behavior of polymeric and metallic materials. In both cases, grain refinement (or chain breakage) occurs. The mechanical properties of the milled materials are improved over those in the unmilled condition. The temperature for full consolidation of the mechanically processed polymers and metals seems to be lower than for the unmilled powder. This may be related to the formation of ultrafine grains or small chains and enhancement of diffusivity. Lastly, traditionally immiscible metals could be alloyed; a similar phenomenon seems to happen in polymers also.

Polymeric materials were found to form bonds with metals, ceramics, and other polymers, thus opening up the possibility of synthesizing many different types of composites.

17.4.6 Room Temperature Solders

Reactions between Ga or liquid Ga–based eutectics with metal powders such as Cu, Ag, Au, Ni, and Sn have been investigated for possible use as low-temperature solders or as dental filling materials. However, it was reported that neither pure metal powders nor equilibrium commercial Cu-Sn alloys mixed with Ga-based liquids are suitable for practical use due to their long solidification times and poor mechanical properties [71]. These solders, required to join sputtering targets to the copper backing plates, should remain workable at room temperature for at least 15 min after initial mixing and then solidify at room temperature in a reasonable time (preferably less than 24 h) to yield good mechanical properties. Since reaction of the metallic powder with a liquid alloy is a diffusion-controlled process, the process of MA could be utilized to change the reactivity of the solid.

Ivanov et al.[71] mechanically alloyed Cu-20 wt% Sn powders in three different types of mills and noted that the solid solubility of Sn in Cu could be increased to 16.9 wt% (or 9.8 at%) Sn. The minimum crystallite size obtained was 7 nm. The mechanically alloyed Cu-Sn alloy powder was mixed with a Ga-based eutectic alloy to react in the following manner:

$$\begin{aligned} Cu_{88}Sn_{12} + Ga_{74}In_{16}Sn_{10} &\rightarrow CuGa_2 + \beta\text{-InSn} \\ &+ Cu + \text{unknown phase} \end{aligned} \tag{17.4}$$

A strong correlation was observed between the reaction rate (measured as the intensity of the $CuGa_2$ peak in the XRD pattern) and the crystallite size of the milled powder; the reaction rate increased significantly with milling time, due to a decrease in the crystallite size (Fig. 17.9).

Different sputtering targets (Si, W, W-10Ti, indium tin oxide, and $MoSi_{2.7}$) were bonded to copper backing plates using this room-temperature solder. In all the cases, the solder was found to be good to achieve room temperature bonding, in spite of the fact that these materials showed large differences in coefficients of thermal expansion. The compositions and shapes of these targets do not allow for joining using conventional methods.

17.4.7 Biomaterials

The search for biomaterials has been directed to those that exhibit a good physiological response and offer the additional advantage of excellent biocompatibility, endurability, and osteoconductive properties. The ideal material should be susceptible to total resorption by the ossean cells and be biocompatible with the natural bone. Among the possible materials, hydroxyapatite (HAp), tricalcium phosphate, and bioactive glasses have been successfully exploited. But new, simple, and economic methods for synthesizing these materials are desirable, and in this respect mechanochemical processing appears to be very promising. HAp has been successfully synthesized by MA methods [72–75]. Kim et al. [72] synthesized HAp using mixtures of $Ca(OH)_2$-P_2O_5 and CaO-$Ca(OH)_2$-P_2O_5 in a planetary ball mill. Yeong et al.[73]

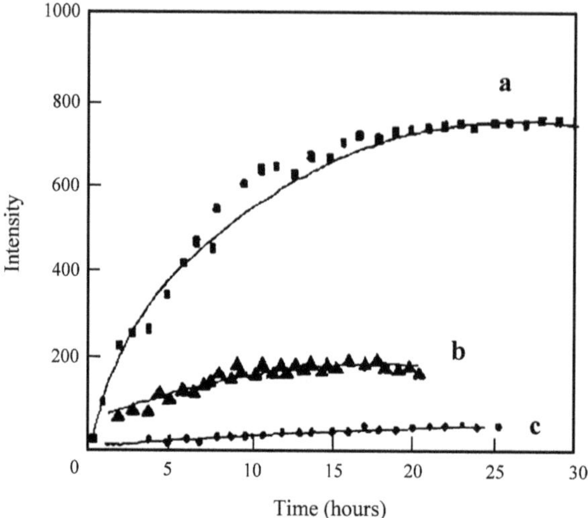

Figure 17.9 Reaction rate of mechanically alloyed Cu-20 wt% Sn powders. (a) Mechanically alloyed for 90 min, (b) mechanically alloyed for 30 min, and (c) commercial alloy. The reaction rates have been normalized to the surface area of the powders. Note that the reaction rate is higher the longer the MA time due to a decrease in the crystallite size.

milled CaO and brushite together and calcined it at 800 °C to obtain HAp. Gonzalez et al.[75], employing a Fisher mortar dry grinder, used stoichiometric mixtures of $Ca(OH)_2$ and either P_2O_5 or $(NH_4)_2HPO_4$ to obtain HAp after calcination according to the following equations:

$$10Ca(OH)_2 + 3P_2O_5 \rightarrow Ca_{10}(PO_4)_6.2(OH) + 9H_2O \tag{17.5}$$

$$10Ca(OH)_2 + 6(NH_4)_2HPO_4 \rightarrow Ca_{10}(PO_4)_6.2(OH) + 12NH_4OH + 6H_2O \tag{17.6}$$

The average particle size of the product obtained after calcination at 800 °C was 200 nm. Short needle-like HAp crystals were obtained for short milling periods (less than 5 h) even without further calcination from the $Ca(OH)_2$ + $(NH_4)_2HPO_4$ mixture. The presence of water contained in the starting mixtures played a key role in promoting the synthesis reaction.

17.4.8 Bearings

Today the state of the art is the application of slide bearings based on Stellite against Stellite, which is, in general, a 50–60 wt% Co-matrix alloy containing Cr and W carbides and other composites. (Stellite is a trademark registered in the United States and owned by Deloro Stellite Company). Stellite is used as the bearing material although its chemical properties and physical properties—in particular, the poor lubricating properties—are not satisfactory. To increase the sliding behavior in the bearing system, about 0.15–0.2 wt% of lead was added to the hot bath. Due to environmental regulations, use of lead had to be reduced dramatically. To improve the situation, the technique of MA has been used [76] to produce advanced Stellite-based bearing materials. A lubricating phase is introduced into Stellite powder material by MA and the composite powder particles are coated by high-energy milling in order to produce bearing bushes of approximately 12 kg by sintering, liquid phase sintering, and hot isostatic pressing.

17.4.9 Miscellaneous Applications

Some miscellaneous applications have also been proposed for mechanically alloyed materials. One such application involves use of MA to decrease the extent of damage caused by aspirin to the stomach. When aspirin is consumed, it will dissolve in the fluids of the stomach and cause damage to it. This is essentially due to the fast rate of dissolution, reported to be $40 \times 10^{-2}\, s^{-1}$. However, when aspirin is encapsulated into cellulose, a chemical bond is created between cellulose and aspirin. The dissolution rate of this mixture was then found to be an order of magnitude slower, $3.9 \times 10^{-2}\, s^{-1}$. Thus, damage to the stomach can be minimized.

Yet another application where MA has been found to be very useful is the economical production of large quantities of powders for industrial applications, e.g., in paints and soldering materials. A production facility of 600 kg/day was established to supply powder to Fukuda Metal Foil and Powder Co. Ltd. in Japan. Zoz GmbH in Wenden, Germany has collaborated with the Fukuda to produce this powder on a semicontinuous basis [77]. Such a process appears to be beneficial in countries where labor costs are high. For example, it was calculated that, per ton of powder, the total processing cost in Germany would be DM 759 (about 380 euros) for continuous processing and DM 1399 (about 700 euros) for the batch operation. On the other hand, the cost in China would be DM 663 (about 330 euros) for the continuous process and DM 641 (about 320 euros) for the batch process, a clear indication of the low labor costs in China.

Copper-based [78] and nickel-based [79] shape memory alloys have also been synthesized by MA methods.

Mechanochemical processing of nanopowders using exchange reactions is under development, by Advanced Powder Technology Ltd. in Australia., with the use of conventional ball mill as a low-temperature chemical reactor.

17.5 CONCLUDING REMARKS

The technique of MA/MM has been very useful in the synthesis of advanced materials for different applications. The major application for this technique still is in the production of ODS superalloys. Other applications are emerging and new applications are also being proposed. In this context, it should be remembered that mechanically alloyed materials have a very small grain/crystallite size and that it is possible to achieve a uniform dispersion of the different constituent phases. However, contamination of the powder during milling and/or consolidation is still a matter of concern. Therefore, potential applications of mechanically alloyed materials should take these factors into consideration. As an example, one could search for new applications where homogenization of the product phase and formation of new active interfaces are important. These could include hydrogen storage alloys, composite materials, and modification of drugs. Consolidation of the powder is involved in these cases, and care should be taken not to introduce additional contamination in this process. Alternatively, one could use this technique to produce advanced materials that show high tolerance to gas content. One could also consider using the MA technique to synthesize materials that could be used in the powder form without the need for consolidation. Such applications include catalysts, paints, pigments, fertilizers, solder components, and carbides for inclusion as hardening phases and the like. New applications are continuously being explored.

REFERENCES

1. Benjamin, J. S. (1976). *Sci. Am.* 234(5):40–48.
2. Gilman, P. S., Benjamin, J. S. (1983). *Annu. Rev. Mater. Sci.* 13:279–300.
3. Koch, C. C. (1991). In: Cahn, R. W., ed. *Processing of Metals and Alloys, Vol. 15 of Materials Science and Technology—A Comprehensive Treatment.* Weinheim: VCH, pp. 193–245.
4. Bakker, H., Zhou, G. F., Yang, H. (1995). *Prog. Mater. Sci.* 39:159–241.
5. Suryanarayana, C. (1995). *Bibliography on Mechanical Alloying and Milling.* Cambridge, UK: Cambridge International Science Publishing.
6. Lu, L., Lai, M. O. (1998). *Mechanical Alloying.* Boston: Kluwer.
7. Murty, B. S., Ranganathan, S. (1998). *Int. Mater. Rev.* 43:101–141.
8. Soni, P. R. (2000). *Mechanical Alloying.* Cambridge, UK: Cambridge International Science Publishing.
9. Sherif El-Eskandarany, M. (2001). *Mechanical Alloying for Fabrication of Advanced Engineering Materials.* New York: William Andrew.
10. Suryanarayana, C. (2001). *Prog. Mater. Sci.* 46:1–184.
11. Takacs, L. (2002). *Prog. Mater. Sci.* 47:355–414.
12. Hack, G. A. J. (1987). *Metals Mater.* 3:457–462.
13. Fischer, J. J., deBarbadillo, J. J., Shaw, M. J. (1991). *Heat Treating* 23(5):15–16.
14. Elliott, I. C., Hack, G. A. J. (1990). In: Froes, F. H., deBarbadillo, J. J., eds. *Structural Applications of Mechanical Alloying.* Materials Park, OH: ASM International, pp. 15–24.
15. deBarbadillo, J. J., Smith, G. D. (1992). *Mater. Sci. For.* 88–90:167–174.
16. Suryanarayana, C. (1998). In: *Powder Metal Technologies and Applications. ASM Handbook.* Vol. 7, Materials Park, OH: ASM International, pp. 80–90.
17. Ivanov, E., Suryanarayana, C. (2000). *J. Mater. Synth. Proc.* 8:235–244.
18. Suryanarayana, C., Ivanov, E., Boldyrev, V. V. (2001). *Mater. Sci. Eng.* A304–306:151–158.
19. Gelles, D. S. (1990). *ISIJ Int.* 30:905.
20. Mukhopadhyay, D. K., Suryanarayana, C., Froes, F. H., Hebeisen, J., Gelles, D. S. (1996). In: Froes, F. H., Hebeisen, J., Widmer, R., eds. *Hot Isostatic Pressing.* Materials Park, OH: ASM International, pp. 175–179.
21. Kazimierzak, B., Prigon, M., Lecomte-Mertens, Ch., Coutsouradis, D. (1990). In: Bachelet, E., Brunetaud, R., Coutsouradis, D., et al., eds. *High-Temperature Materials for the Power Engineering 1990–II.* Dordrecht: Kluwer, pp. 131–142.
22. Kohler, E., Gutsfeld, C., Thummler, F. (1990). *Powder Metall. Int.* 22(3):11–14.
23. Gutsfeld, C., Thummler, F. (1990). *Metal Powder Rep.* 45:769–771.
24. Lee, K. M., Moon, I. H. (1994). *Mater. Sci. Eng.* A185:165–170.
25. Hawk, J. A., Lawless, K. R., Wilsdorf, H. G. F. (1988). *Scripta Metall.* 22:1317–1322.
26. Griffith, W. M., Sanders, R. E. Jr, Hildeman, G. J. (1982). In: Koczak, M. J., Hildeman, G. J., eds. *Elevated Temperature Aluminum Alloys for Aerospace Applications.* Warrendale, PA: AIME, pp. 209–221.
27. Angers., L. 1985. Ph.D. dissertation, Northwestern University, Evanston, IL.
28. Yaney, D. L., Nix, W. D. (1987). *Metall. Trans.* 18A:893–902.
29. Öveç oglu, M. L., Nix, W. D. (1989). In: Arzt, E., Schultz, L., eds. *New Materials by*

Mechanical Alloying Techniques. Oberursel: Deutsche Gesellschaft für Metallkunde. pp. 287–295.
30. Ezz, S. S., Lawley, A., Koczak, M. J. (1988). In: Kim, Y. -W., Griffith, W. M., eds. *Dispersion Strengthened Aluminum Alloys*. Warrendale, PA: TMS, pp. 243–263.
31. Black, S. A. (1979). *Development of Supercorroding Alloys for Use as Timed Releases for Ocean Engineering Applications*. Port Hueneme, CA: Civil Engineering Laboratory (Navy), p. 40.
32. Sergev, S. S., Black, S. A., Jenkins, J. F. US Patent 4,264,362, August 13, 1979.
33. Suryanarayana, C., Ivanov, E., Konstanchuk, I. G. (2001). In: Kim, N. J., Lee, C. S., Eylon, D., eds. *Proc LIMAT 2001*. Vol. 1. Pohang, Korea: Pohang University of Science and Technology, pp. 261–268.
34. Schulz, R., Liang, G., Huot, J. (2001). In: Dinesen, A. R., Eldrup, M., Juul Jensen, D., Linderoth, S., Pedersen, T. B., Pryds, N. H., Schrøder Pedersen, A., Wert, J. A., eds. *Science of Metastable and Nanocrystalline Alloys: Structure, Properties, and Modelling*. Roskilde, Denmark: Risø National Laboratory, pp. 141–153.
35. Klassen, T., Oelerich, W., Bormann, R. (2001). *Mater. Sci. For.* 360–362:603–608.
36. Konstanchuk, I. G., Ivanov, E., Boldyrev, V. V. (1998). *Russian Chem. Rev.* 67:69–79.
37. Ivanov, E., Konstanchuk, I. G., Stepanov, A., Boldyrev, V. V. (1987). *J. Less Common Metals* 131:25–29.
38. Oelerich, W., Klassen, T., Bormann, R. (2001). *Adv. Eng. Mater.* 3:487–490.
39. Klassen, T., Bohn, R., Fanta, G., Oelerich, W., Eigen, N., Gärtner, F., Aust, E., Bormann, R., Kreye, H. (2003). *Z. Metallkde.* 94:610–614.
40. Liang, G., Huot, J., Boily, S., Van Neste, A., Schulz, R. (1999). *J. Alloys Compounds* 291:295–299.
41. Suryanarayana, C. (1995). *Int. Mater. Rev.* 40:41–64.
42. Gleiter, H. (2000). *Acta Mater.* 48:1–29.
43. Suryanarayana, C., Koch, C. C. (2000). *Hyperfine Interactions* 130:5–44.
44. Schlump, W., Willbrand, J., Grewe, H. (1994). *Metall.* 48:34.
45. Gärtner, F., Bormann, R., Klassen, T., Kreye, H., Mitra, N. (2000). *Mater. Sci. For.* 343–346:933–940.
46. Rairden, J. R., Habesch, E. M. (1981). *Thin Solid Films* 83:353–360.
47. Gedwill, M. A., Glasgow, T. K., Levine, S. R. (1982). *Metallurgical Coatings*. Vol. 1. New York: Elsevier.
48. Jia, C., Li, Z., Xie, Z. (1999). *Mater. Sci. Eng.* A263: 96–100.
49. Umemoto, M., Shiga, S., Raviprasad, K., Okane, I. (1995). *Mater. Sci. For.* 179–181: 165–170.
50. Umemoto, M. (1995). *Mater. Trans. Jpn. Inst. Metals* 36:373–383.
51. Lessard, P. C., Havens-Cook, M. (2002). *Mater. Sci. For.* 386–388:589–596.
52. Magini, M., Cavalieri, F., Padella, F. (2002). *Mater. Sci. For.* 386–388:263–268.
53. Balaz, P. (2000). *Extractive Metallurgy of Activated Minerals*. Amsterdam: Elsevier.
54. Pan, J., Shaw, W. J. D. (1992). *Microstruct. Sci.* 19:659–668.
55. Pan, J., Shaw, W. J. D. (1992). In: Reinhart, T. S., Rosenow, M., Cull, R. A., Struckholt, E., eds. *Advanced Materials: Meeting the Economics Challenge*. Vol. 24. Covina, CA: SAMPE, pp. T762–T775.
56. Pan, J., Shaw, W. J. D. (1993). *Microstruct. Sci.* 20:351–365.
57. Shaw, W. J. D., Pan, J., Gowler, M. A. (1993). In: deBarbadillo, J. J., et al., eds. *Mechanical Alloying for Structural Applications*. Materials Park, OH: ASM International, pp. 431–437.
58. Shaw, W. J. D., Gowler, M. A. (1993). In: Henein, H. Oki, T., eds. *Processing Materials for Properties*. Warrendale, PA: TMS, pp. 687–690.
59. Pan, J., Shaw, W. J. D. (1994). *Microstruct. Sci.* 21:95–106.
60. Pan, J., Shaw, W. J. D. (1994). *J. Appl. Polym. Sci.* 52:507–514.

61. Pan, J., Shaw, W. J. D. (1995). *J. Appl. Polym. Sci.* 56:557–566.
62. Shaw, W. J. D. (1998). *Mater. Sci. For.* 269–272:19–30.
63. Ishida, T., Tamaru, S. (1993). *J. Mater. Sci. Lett.* 12:1851–1853.
64. Ishida, T. (1994). *J. Mater. Sci. Lett.* 13:623–628.
65. Farrel, M. P., Kander, R. G., Aning, A. O. (1996). *J. Mater. Synth. Process* 4:151–161.
66. Namboodri, S. L., Zhou, H., Aning, A. O., Kander, R. G. (1996). *Polymer* 35:4088–4091.
67. Bai, C., Spontak, R. J., Koch, C. C., Saw, C. K., Balik, C. M. (2000). *Polymer*. 41:7147–7157.
68. Smith, A. P., Shay, J. S., Spontak, R. J., Balik, C. M., Ade, H., Smith, S. D., Koch, C. C. (2000). *Polymer*. 41:6271–6283.
69. Smith, A. P., Ade, H., Balik, C. M., Koch, C. C., Smith, S. D., Spontak, R. J. (2000). *Macromolecules* 33:1163.
70. Koch, C. C., Smith, A. P., Bai, C., Spontak, R. J., Balik, C. M. (2000). *Mater. Sci. For.* 343–346:49–56.
71. Ivanov, E., Patton, V., Grogorieva, T. (1996). In: Suryanarayana, C., et al., eds. *Processing and Properties of Nanocrystalline Materials.* Warrendale, PA: TMS, pp. 189–197.
72. Kim, W., Zhang, Q., Saito, F. (2000). *J. Mater. Sci.* 35:5401–5405.
73. Yeong, B., Junmin, X., Wang, J. (2001). *J. Am. Ceram. Soc.* 84:465–467.
74. Shuk, P., Suchanek, W. L., Hao, T., Gulliver, E., Riman, R. E., Senna, M., TenHuisen, K. S., Janas, V. F. (2001). *J. Mater. Res.* 16:1231–1234.
75. Gonzalez, G., Villalba, R., Sagarzazu, A. (2002). *Mater. Sci. For.* 386–388:645–650.
76. Zoz, H., Ren, H. (2000). *Mater. Sci. For.* 343–346:955–963.
77. Zoz, H., Reichardt, R., Ernst, D. (1998). *Metall.* 51:521–527.
78. Zhang, S., Lu, L., Lai, M. O. (1993). *Mater. Sci. Eng.* A171:257–262.
79. Crone, W. C., Yahya, A. N., Perepezko, J. H. (2002). *Mater. Sci. For.* 386–388:597–602.

18

Safety Hazards

18.1 INTRODUCTION

The safety hazards related to processing and handling of production quantities of metal or metallic powders have been the subject of extensive study and concern for many years. Injury to personnel and production facilities due to fire and/or explosion are often reported and could be catastrophic. Frequently, research personnel assume that such major accidents occur only in production facilities and not in laboratories. This is not right. Many accidents have been reported in situations where only gram quantities of powders are involved. Weber et al. [1] have summarized some of the safety hazards associated with milling of metal powders in laboratories and production facilities.

Powder handling requires safety precautions and cleanliness since certain powders can have harmful effects on personnel exposed to them. Respirable powders are a health concern and can cause disease or lung dysfunction. The particle size and specific gravity of the powder largely determine the deposition site for an inhaled particle. Coarse particles (larger than 10 μm) are trapped on the mucous membranes and do not reach the lungs; however, particles in the size range of 0.01–10 μm can reach the lungs and may be dissolved in the body. The specific consequences of such dissolution depend on powder chemistry. However, it has been reported that powder particles in the size range of 0.1–1.0 μm (100–1000 nm) require the greatest care in handling.

Powders of some metals are toxic, especially when small; these include As, Be, Cd, Co, Cr, Pb, Ni, and Te. Hence, it is desirable to avoid contact with those metals in the powdered form. The maximal occupational air exposure for many of the toxic metals is less than 10^{-4} g/m^3. Therefore, using protective equipment, safe handling of powders (such as in glove boxes) or robotic handling may be practiced. On a laboratory scale, however, handling of powders inside an evacuated (or inert gas-filled) glove box is recommended. It may also be emphasized that not all powders are hazardous; in fact, some powders are intentionally added to foods as vitamin and mineral supplements.

Another hazard with metal powders, especially when they are fine, is that they are pyrophoric (burn in air) in the presence of oxygen. Therefore, a dust-free environment is recommended. Metals such as Al, Zr, Ta, Th, Ti, and Mg can ignite in air at concentrations on the order of 40 g/m^3, and this can occur at relatively low temperatures. The general precautions that must be taken in handling of powders in general have been described [2].

Mechanically alloyed powders are produced by severe plastic deformation of coarse powder particles. As mentioned in earlier chapters, the characteristics of mechanically alloyed/milled powder particles are small crystallite (grain) size, lattice strain, and presence of different metastable phases. All these effects increase the free energy of the system and make the material highly active. Due to the small particle size, these powders have a large surface area and therefore chemical reactions could occur much faster than in coarse-grained materials. Furthermore, chemical reactions could occur in these powders resulting in the evolution of heat (due to occurrence of exothermic reactions). Therefore, in comparison to conventional powders, safety hazards are accentuated in the mechanically alloyed powders due to the increased stress and strain, temperature, and chemical reactivity.

18.2 HAZARDS RELATED TO MECHANICAL ALLOYING PROCESSES

In order to avoid damage to equipment and personnel during mechanical alloying (MA), one should be aware of the different processes that occur during MA/mechanical milling of powders. These have been described in earlier chapters and include powder deformation; fragmentation and cold welding of powder particles; solid-state diffusion to promote formation of solid solutions, intermetallics, and amorphous phases; occurrence of mechanochemical reactions; and formation of ultrafine grains, reaching down to nanometer dimensions. Thus, the safety hazards associated with MA are related to one or more of these processes.

During the-above mentioned processes, some phenomena occur irrespective of the nature of the final product phase, e.g., evolution of heat. Other phenomena that occur are gas evolution, especially during mechanochemical reactions, and increased reaction rates during processing. All of these either alone or in combination could cause explosions.

18.2.1 Heat Evolution

As mentioned in Chapter 8, the temperature of the powder increases during MA. This temperature increase could be due to the energy added via the mechanical system used to run the mill (e.g., kinetic energy transferred to the milling powder and also the frictional forces of the machine). In addition, chemical reactions occurring in the powders could also raise the temperature of the powder. Temperature rise due to the first part could be minimized by a good basic design of the mill. Furthermore, in practice, a fan is used to cool the system during the milling of powders. However, the more important aspect to be kept in mind is the temperature rise due to the chemical reactions that occurs during milling. If the temperature reached by the powder is much higher than some critical "reaction temperature" for one or more of the components being milled, then unexpected and possibly uncontrolled reactions could occur. The extent to which the heat generated could be hazardous depends on the amount of powder being processed, the rate at which the reaction occurs, and the ability of the equipment to remove the heat.

It was also mentioned earlier that the instantaneous temperature experienced by the powder particles at the time of impact of the grinding media could be a few hundred degrees Celsius. This could lead to increased reaction rates and even melting of the powder constituents.

18.2.2 Gas Evolution

Gases may be produced during milling due to decomposition of process control agents, vaporization and/or sublimation of a deliberately added component, or chemical reactions among the components. All these effects could increase the pressure in the milling container. It is also possible in some instances when the milling container lid is not tight that air or other ambient gases may leak into the container and change the partial pressure of the gases inside the container.

In addition to the nature and volume of the gases inside the container, it is important to realize that some of the gases produced may be flammable, and so the potential for fire or explosion exists. The partial pressures of the gases should not exceed the lower limit of inflammability for the atmosphere–temperature–pressure combination.

18.2.3 Explosions

The presence of fuel and oxidant clouds, combined with an ignition source, is the necessary condition for explosions. It is very easy to realize these conditions during MA. Presence of extremely fine particles with a large surface area is a potential source of fuel. The oxidant is a gas such as oxygen. The ignition source could be the high temperatures reached by the powder or the static electricity generated when solids move one over the other, i.e., during flow of powders. Explosion due to static electricity occurs whenever charge buildup takes place. This can be minimized by using only electrically conducting containers, vessels, scoops, and tools, and bonding and grounding such items. Other sources of ignition are (1) electric devices that produce a spark such as electric switches; (2) impact sparks resulting from metal contact with metal concrete; (3) friction heating as might be experienced with sleeve bearings; and (4) failure of equipment, such as breakage of an attritor arm.

18.3 HANDLING OF MECHANICALLY ALLOYED POWDERS

Special precautions should also be taken during handling of powders subsequent to MA. This depends on powder characteristics such as particle size and shape, composition, surface condition, and gas content. Knowledge of these parameters facilitates safe handling of powders as well as their conversion to a consolidated form.

Pyrophorocity may be defined as the ability of a powder or dust to ignite when exposed to a gaseous environment. Most commonly the gas is oxygen, but other gases, such as chlorine, nitrogen, carbon monoxide, and carbon dioxide, may also be vigorous oxidants. Mechanically alloyed powders are very fine and consequently have a large surface area. Such powders have an increased tendency to pyrophorocity. Thus, fine and/or irregularly shaped powder particles are more sensitive. The cleaner and fresh powder surfaces produced during MA further accentuate their sensitivity to pyrophorocity. For example, many metal powders in the 1-μm size range, (a size commonly seen during MA), such as iron and nickel, may be pyrophoric.

Mechanically alloyed powders tend to contain very large volumes of gases. The gases can arise from occlusion of the milling atmosphere, volatile content or process control agents, and products of reactions between the milled constituents. The hazard potential for these high gas levels is significant, as many milled powders are subjected to elevated temperatures during subsequent processing. Containment of these evolved gases can lead to high pressures and can cause explosions.

Immediately after MA the powders are hot (or at least warm). Therefore, the lid should not be immediately opened and the powder exposed to atmosphere. In such an event, the powder may ignite because of interaction with the oxygen. This is much more important when dealing with reactive powders such as magnesium.

Safety hazards such as explosions have been reported during milling of powders or subsequent handling of milled powders. Some of these have been cataloged by Weber et al. [1].

18.4 ACCIDENT AVOIDANCE

Knowledge about causes of accidents, education, and specialized test techniques will help the people involved in conducting MA to avoid injury to equipment and personnel, and even to prevent loss of life. Many sources of education and information are available to assist in the development and design of safe equipment and powder production practices. When using commercially available equipment, it is necessary to consult the manufacturers' brochures and data sheets.

An important source of information is the National Fire Protection Association (NFPA) at 1 Batterymarch Park, Quincy, MA 02169-3000, USA (Tel: 617-770-3000 or 1-800-344-3555; FAX: 617-770-0700; www.nfpa.org). The mission of this association is "to reduce the worldwide burden of fire and other hazards on the quality of life by providing and advocating scientifically-based consensus codes and standards, research, training, and education." Two publications that are of primary relevance to the current discussion are the *Fire Protection Handbook* [3] and the *National Fire Codes* [4]. The handbook is a general source guide and the codes are annual compilation of codes, standards, and guides. Other professional associations, such as the Aluminum Association, (Washington, DC, USA), American Powder Metallurgy Institute and the Metal Powder Industries Federation (Princeton, NJ, USA) are also valuable sources of information related to powder safety.

REFERENCES

1. Weber, J. H., deBarbadillo, J. J., Mehltretter, J. C. Paper presented at the Second International Conference on Mechanical Alloying for Structural Applications, Vancouver, BC, Canada, September 20–22, 1993 (not included in the Proceedings).
2. German, R. M. (1994). *Powder Metallurgy Science*. 2nd ed. Princeton, NJ: Metal Powder Industries Federation, pp. 159–160.
3. Fire Protection Handbook. (2003). Quincy, MA: National Fire Protection Association (in two volumes and about 3100 pages long).
4. National Fire Codes, Quincy, MA: National Fire Protection Association (available on CD-ROM and with automatic updates).

19

Concluding Remarks

From the previous chapters it becomes clear that mechanical alloying (MA) and the related processes are simple, elegant, and useful and that they continue to attract the serious attention of researchers. Even though the technique was originally developed to produce oxide dispersion strengthened (ODS) nickel-based superalloys for aerospace applications, the technique has now matured to be applied to the synthesis of a variety of equilibrium and metastable alloy phases including solid solutions, quasicrystalline and crystalline intermetallic phases, and amorphous alloys. New and varied applications are being explored. This has resulted in significant research in recent years. It is estimated that so far more than 7000 research/review papers have been published in this area, with nearly 500 papers being published annually during the past 4–5 years.

One of the greatest advantages of MA is in the synthesis of novel alloys that are either impossible or difficult to prepare by any other technique, such as alloying of normally immiscible elements. This is because MA is a completely solid-state processing technique and limitations imposed by phase diagrams do not apply here. The synthesis of nanostructured materials—both monolithic and composites—by MA has become an accepted practice in recent years due to the ease with which they are obtained. However, the number of investigations that deal with the decomposition behavior of metastable phases obtained by MA is still small.

On an industrial scale, MA is an accepted process. The ODS alloys processed by MA have a higher temperature operating capability at the same stress or increased load bearing capability at the same temperature than alloys without an oxide dispersion. The ODS alloys continue to find applications in a wide variety of industries.

In spite of a lot of research effort, the mechanism of phase formation during MA is not well understood. It is most often proposed that the process of MA introduces a variety of defects (vacancies, dislocations, grain boundaries, stacking faults, and so forth) which raise the free energy of the system making it possible to produce metastable phases. But there are very few investigations that deal with the characterization and quantification of the defects produced in mechanically alloyed powders. Furthermore, properties of the mechanically alloyed products (after consolidation of the powders) are reported only in a few cases, except for the ODS alloys. This is understandable because researchers are still struggling to find out the mechanism of formation of the different phases.

MA is a complex process that involves many variables, and many of them are interdependent. Therefore, modeling of the MA process is difficult. In spite of some success, one has to go much farther in developing models that can reach the final goal of predicting the nature of phases produced under a given set of milling conditions.

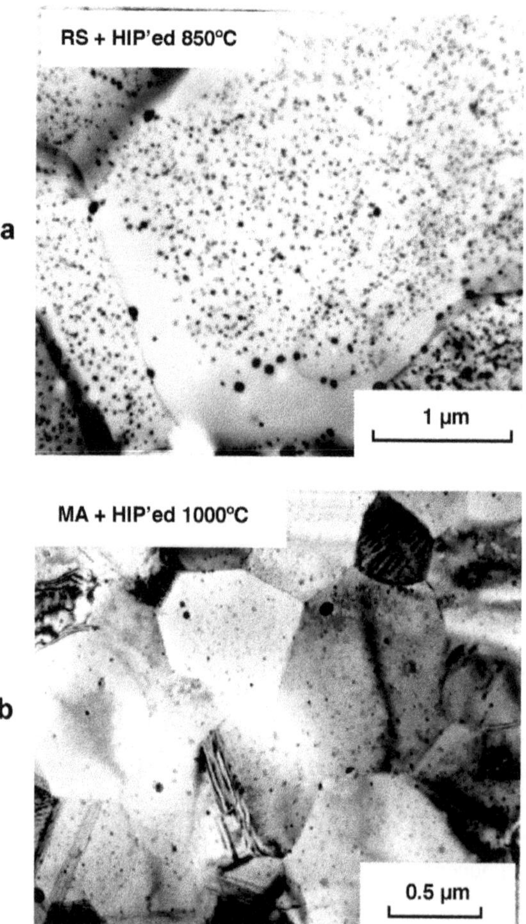

Figure 19.1 Transmission electron micrographs showing the difference in the matrix grain size, and size and distribution of dispersoids in rapidly solidified (RS) and mechanically alloyed (MA) materials.

There have not been enough investigations to predict the formation of different types of phases based on thermodynamic or other models.

As mentioned in the text in different places, mechanically alloyed powders contain a variety of metastable phases. Since there are many other nonequilibrium processing techniques that produce similar phases, comparisons are frequently made. The major comparison has been between MA and rapid solidification processing (RSP) techniques. Each of these two techniques has its own advantages and limitations. An important difference between mechanically alloyed and rapidly solidified alloys containing dispersoids appears to be in the size and distribution of dispersoids. Figure 19.1 shows a pair of transmission electron micrographs comparing the matrix grain sizes, as well as size and distribution of dispersoids in hot isostatically pressed compacts of Ti_3Al-based alloys containing Er_2O_3 dispersoids. The rapidly solidified alloy was Ti_3Al to which 2 wt% Er was added and this was hot isostatically pressed at 850 °C after rapid solidification. The mechanically alloyed material was Ti-25Al-10Nb-3V-1Mo (at%) to which 2 wt% Er was added and the alloy powder was hot isostatically pressed at 1000 °C after MA. Even after hot isostatic pressing at a higher temperature, in comparison to the RSP alloy, the mechanically alloyed material showed a finer matrix grain size, more uniform distribution of the dispersoids, absence of large dispersoids at the grain boundaries, and absence of dispersoid-free zones near the grain boundaries.

Many other differences are also reported in the constitution of the alloys processed by RSP and MA methods. And these have been discussed in the respective chapters. For example, mechanically alloyed materials show the formation of amorphous phases in much wider composition ranges than in rapidly solidified alloys. Furthermore, it also appears that it is easier to produce an amorphous phase (including the bulk amorphous alloy compositions) by MA than by RSP. Section thickness limitations in bulk amorphous alloys obtained by solidification methods do not apply in the case of MA. This is because amorphous alloys produced by MA can be consolidated into any thickness. This should also be possible in the amorphous alloys produced by solidification methods. However, detailed investigations on the mechanism of formation of metastable phases and their decomposition to equilibrium condition (including crystallization of amorphous alloys) have not been undertaken for mechanically alloyed powders.

It is clear from what has been described in the earlier chapters that research in the area of MA is continuing to flourish and that new and exciting discoveries are taking place. However, the industrial applications of mechanically alloyed materials still appear to be confined to ODS alloys, even though new application avenues are being explored. One such area with immediate application potential is synthesis of materials with novel compositions for hydrogen storage. There is so much more to learn about the "science" of MA that the future of MA is assured for several years to come.

Index

Milton Keynes UK
Ingram Content Group UK Ltd.
UKHW052023071024
449327UK00027B/2404